# PRINCIPLES OF ROCK FRAGMENTATION

GEORGE B. CLARK

A WILEY-INTERSCIENCE PUBLICATION
**JOHN WILEY & SONS**
New York   Chichester   Brisbane   Toronto   Singapore

Copyright © 1987 by John Wiley & Sons, Inc.

All rights reserved. Published simultaneously in Canada.

Reproduction or translation of any part of this work
beyond that permitted by Section 107 or 108 of the
1976 United States Copyright Act without the permission
of the copyright owner is unlawful. Requests for
permission of further information should be addressed to
the Permissions Department, John Wiley & Sons, Inc.

*Library of Congress Cataloging in Publication Data:*

Clark, George Bromley, 1912–
  Principles of rock fragmentation.

  "A Wiley-Interscience publication."
  Bibliography: p.
  Includes index.
  1. Boring.  2. Rock-drills.  3. Rock excavation.
4. Rocks—Fracture. I.  Title.
TN279.C58  1987      624.1′52      86-33973
ISBN 0-471-88854-0

Printed in the United States of America

10  9  8  7  6  5  4  3  2  1

*To the engineers, scientists, and practioners of all countries
who are earnestly seeking to better our understanding and practical use
of the principles and processes of rock fragmentation*

# PREFACE

The mechanics of rock fragmentation involve dynamic applications of energy. This principle applies to the very old processes of fire-setting, picking by hand, as well as mechanical drilling, rock boring, rock splitting, and blasting with conventional and nuclear explosives. While these methods are relatively inefficient, some are much more useful than others. Basic understanding of the energy release mechanisms of high explosives is better developed now than before the discovery and use of the thermohydrodynamic theory of detonation. In the future, significant advances in the efficiency of application of energy transfer in rock fragmentation will probably be made in increments, but these small advances may be critically important. A better understanding of the transfer of explosive energy—particularly to both soil and rock—the resulting wave propagation, and the effects of these waves on rock structures is critical to the solutions of problems of current interest to both civil and military engineers. Because of the nature of experimentation with explosives, the use of the principles of similitude in future rock fracture and fragmentation studies will be predictably greater.

GEORGE B. CLARK

*Wheat Ridge, Colorado*
*July 1987*

# CONTENTS

**ix**

**X    CONTENTS**

# PRINCIPLES OF ROCK FRAGMENTATION

# 1

# ROCK FRAGMENTATION PROCESSES

## 1.1   FRAGMENTATION

From the inception of productive applications of modern processes of rock fragmentation, their use has been essential to the progress of civilization in terms of the technical, material, and economic progress of the nations of the earth. Much of the excavation in modern mining and civil engineering could not have been accomplished without the related development of methods of drilling and blasting rock. Recent and current research have laid the background for further technical and operational improvement of these processes and have revealed the critical need for further research. Mechanical methods of tunnel excavation, which have been applied at a usable level only within the past 30 yr, have likewise been developed to a remarkably effective productivity. Although these machines have a specialized application and produce only a limited relative volume of excavated rock materials, they have a critical place in the excavation industry. Meaningful laboratory and field studies have been made of TBM cutters, drills, and other types of rock cutters. While such studies have revealed many of the basic principles of the mechanisms of excavation and machine operation, which has resulted in significant improvements, much further research and evaluation are needed.

One of the purposes of this book is to begin to meet the urgent need for a one-source treatment of the elementary phenomena and the mechanisms of rock fragmentation in relation to commercial mining and related excavation processes, civil and military, in earth materials other than soils, sand, and gravels. As part of the background information for engineering and operation purposes, in addition to mechanisms of rock fracture by various processes, the behavior of fragmented earth material under the force of gravity, a body force rather than an applied force—all of this requires more study, research, and scientific understanding. To date the latter has been studied only in a preliminary, semi-quantitative way. Experimental model studies of the behavior of fragmented, particulate material, with the exception of sand and soil, have not utilized body force loading. This may have markedly different effects from externally applied loading.

## 1.2 DRILLING

The information in the technical literature of the drilling of rock for excavation purposes is widely scattered in books, journals, symposia proceedings, and other literature. Many of the research results of drilling for petroleum remain locked up in proprietary company files. Most of the drilling for excavation is done by percussive drills, pneumatic or hydraulic. The operation of pneumatic drills may be analyzed on a simplified basis, and the energy transfer from compressed air to piston to drill rod can be evaluated by simple computer programs. The energy transfer in hydraulic drills is similar, but a usable analysis of the same does not appear to be available. Further, the energy transfer at the rock–bit interface requires additional research and analysis and is a function of drill characteristics and rock properties in terms of force–displacement curves, but rock property effects are extremely variable.

Other processes of fragmentation are of equal importance in their place, including methods of mechanical breakage; fracture by heat, crushing, and grinding; water jet cutting; and other novel methods.

There are two basic engineering factors that have led to the development of more effective rock drills, primarily of the pneumatic type. The first, where the greatest progress has been made, is the improvement of the mechanisms of drill operation and metallurgical properties of drill components, drill steel, and drill bits. The second is the increased knowledge of the basic processes of rock fragmentation and energy utilization in achieving effective drilling and blasting.

Rock is drilled for several purposes, and a number of methods have been devised and successfully used for making holes in rock. Such purposes include geologic exploration for mineral deposits, drilling for petroleum, making holes for placement of explosives for blasting, making holes for rock bolts, drilling for research purposes, and other similar objectives. Though many novel methods of cutting holes have been subjected to experimentation, the most successful for practical appli-

**TABLE 1.1. Classification of Drilling Methods**[a]

| Method | Machine |
|---|---|
| Percussive | Churn drill, cable tool, |
| Drop tool | pneumatic or hydraulic |
| Hammer | rock drill |
| Rotary, drag | |
| Blade | Auger, high-pressure rotary |
| Stone | Diamond |
| Shot | Calyx |
| Sawing | Wire rope, chain rotary saw |
| Rotary, roller | Rolling cutter |
| Rotary–percussion | Drag, roller |
| Thermal | |
| Flame | Jet piercer, channeler |

[a]See Hartman (1968).

cations are the mechanical and, to a lesser extent, the thermal methods. High-pressure water jets are effective mostly for specialized cutting in softer rocks.

The types of drills that have been used for drilling operations in rock include (1) percussive or impact, (2) abrasion, (3) cutting, (4) crushing, (5) combinations of the above, and (6) thermal spallation (Table 1.1).

The churn drill, which raises and drops a large, heavy bit in the bottom of the hole with cuttings being removed by a bailer, is seldom used at the present time. Pneumatic percussive drills, until the recent development of hydraulically powered impact tools, were restricted to the use of compressed air to energize the drill. A combination of rotation and percussion is employed by a rotary-percussion tool with asymmetrical cutting edges on the bit, the drill being powered pneumatically or hydraulically. Rotary drills may use drag bits, auger bits, diamond bits, or roller cone cutters, the latter using either disc cutters or button inserts to crush the rock. The flame jet drill is usually limited to hard siliceous rocks which spall readily.

## 1.3  EXPERIMENTAL METHODS

Experimental (novel) methods have been used for melting rock by means of lasers and electron beams, for cutting by sonic vibration, for drilling with shaped (cavity) explosive charges, and for cutting with very high-pressure water jets. None of these appear to have potential applications except water jets at about 15,000 psi. The promising areas for the use of water jets are for cutting softer materials, such as coal; drilling of holes in softer rocks, such as shale and sandstone; or assisting certain types of mechanical cutters, such as those on tunnel-boring machines.

## 1.4 EXPLOSIVES

It has been over three decades since the discovery and application of the hydro-dynamic theory of plane shock waves to solid explosives for the calculation of their detonation and explosion parameters, such as detonation velocity, pressure, and temperature and the explosion state pressure, energy, and temperature. Many computer codes have been developed utilizing different equations of state for gases at high temperatures and pressures and various thermodynamic and thermochemical approaches for the analysis of explosions. As described in this book, Cook's method using a modified Abel equation of state has been found to be useful for engineering calculations for near oxygen-balanced explosives and those containing only small amounts of metallic elements such as aluminum or sodium, as well as for water-based explosives. This modified equation uses experimentally determined values of co-volume as a function of pressure, together with chemical equilibria as a function of temperature with related thermochemical data. The application of Cook's method thus requires a knowledge of advanced chemistry and computer programming. Its presentation is intended primarily for advanced undergraduate engineers, graduate engineers, and field engineers with a similar educational background.

In the development of blasting agents the fortuitous adoption of a manufacturing process for fertilizer grade ammonium nitrate (AN), which produced porous prills, was a major factor in its sensitization and consequent use as a blasting agent ingredient. ANFO, an oxygen-balanced mixture of AN and fuel oil, is a simple do-it-yourself explosive whose advantages outweigh its disadvantages, primarily in material costs and safety. It is relatively inexpensive, strong enough in specific energy content for many blasting processes, and usually not cap-sensitive. This makes it safe to handle, and it has good fume properties when properly designed. It contains a single fuel and oxidizer and illustrates many of the important performance parameters of commercial explosives and blasting agents.

The development of slurries came about as one of the technological events associated with the use of ANFO. A principal thermodynamic question arose as to whether water added to granular explosives or blasting agents in the oxygen-balanced range could increase their density, and consequently their detonation velocity and pressure, and could be heated by the energy of combustion sufficiently to maintain critical hydrodynamic conditions. This was found to be possible, first by computation and then by experimentation, within usable limits and was followed by the addition of gelling agents, other nitrates, metallic elements, sensitizers, and other substances. Slurries have also been called water gels, but the term ''slurries'' appears to be the most appropriate technologically. The water-explosive compositions are easily amenable to computer analysis.

## 1.5 ROCK BREAKAGE BY EXPLOSIVES

Most rock breakage, except coal, is accomplished by mechanical drilling and rock cutting and by blasting with explosives, with minor amounts accomplished by water jets and such means as rock splitters. The results of experimentation and analysis are available on most of the industrial methods of rock breakage. Most theory involves idealized or simplified assumptions and consequently may be of limited value for application to nonidealized rock. Theoretical models have been devised for drill bit action at the rock–bit interface, as well as for roller cones and for rotary bits and cutters on TBMs. Each has its values and limitations. Some of the basic ideas involved in these theories are given below in appropriate chapters, especially where they support the results of experimentation.

The basic mechanics of rock breakage by explosives have been the subject of a large amount of experimentation. However, much of this is qualitative or semi-quantitative in nature, and, though the results are useful, they do not represent definitive scientific results. Some hypotheses have been put forward as to the details of rock breakage in benches but have not been supported by the results of experimentation or field observation.

Perhaps some of the most definite studies done on rock breakage are the instrumented cratering shots in six types of rock by the U.S. Bureau of Mines personnel. This work has been the subject of considerable analysis and discussion in the literature.

## 1.6  TUNNEL BORING

Mechanical tunnel boring was first meaningfully initiated in the 1950s and has been used to perform a relatively small but very significant amount of specialized excavation in the United States and abroad. Its particular advantages are the continuity of operation and that the rock excavation, mucking, and support operations can be carried on simultaneously. This method of excavation and rock removal results in more even, stable walls than does drill-and-blast, and the relatively quiet method of rock breakage permits its use much closer to dwellings and similar structures. The large capital investment for TBMs, however, allows them to be used on only relatively large jobs, and the machine wear and consequent costs may be excessively high for excavation in rock masses that are difficult to cut mechanically.

## 1.7  AUTOMATED DRILL AND BLAST

The operational features of a TBM are so advantageous in construction that research efforts have been designed to adapt some of these principles to automated drill-and-blast systems. Several concepts have been investigated, and some have been patented. These utilize the basic concept of near-simultaneous integral operations, with similar types of equipment for drilling, explosive loading, and mucking. Some provide for remote control, and some for shielded operations. They have shown promising developments, but research has been impeded largely because of lack of funding. A major problem to be solved is automated-explosives handling, but the problems encountered appear to be subject to solution by further well-conceived research.

## 1.8  ROCK SPLITTING

Various methods of splitting rock have been used since early history. The operational principle of mechanical rock splitters is based on the same concept as manual wedge and feathers, which are employed in quarrying of dimension stone. The most commonly known type of mechanical splitter utilizes a hydraulically driven wedge forced between two feathers placed in a hole in rock, the rear end of the feathers being anchored to the hydraulic cylinder frame. The wedge must be well lubricated, because it exerts side pressures up to 150,000 psi with a cylinder pressure of 2000 psi.

The principle of free faces of rock breakage caused by internal stresses applies to the breaking capabilities of rock splitters in a manner analogous to breakage by explosives. Rock is split most effectively if it is in the form of an isolated boulder, but splitters have been used to break benches and the faces in trenches and small tunnels, where explosives could not be employed because of environmental reasons. Here also the rate of advance was not a critical factor. Of all methods of fragmentation, splitters are the most energy-efficient. Although the splitter method of excavation does not have the disadvantages of drill and blast, it requires too much time for some of the integral operations to be competitive with conventional methods.

An analysis made of the stresses induced by a splitter (Chollette et al., 1976) for two distributions showed that the stress concentration factors varied from 1.0 to 1.65. An extended experimental program with a commercial splitter on operational parameters included effects of lubricants, autolubrication, and longitudinal impact superposed on the thrust of the wedge. Such impact increases the upper limit of stalling of the wedge and may increase its speed if hydraulic flow can be increased (Clark and Maleki, 1978).

## 1.9  GRAVITY AND MUCK FLOW

The effects of gravity on the behavior of fragmented geologic materials from fine to coarse particulate sizes, including crushed and blasted rock and rock fractured by geologic processes, may be critical design factors in excavation, support of underground openings, and materials handling. Gravity-induced phenomena may be critical in the behavior of fragmented material in bins and chutes, in the flow of materials in caving methods of mining, in excavated areas in fault zones in tunnels, in open-pit mines, and in zones cratered by nuclear explosives. With lateral confinement, fragmented materials are subject to arching. Experimentation on this phenomenon to the present time appears to have been limited to tests on sand.

Some effective experimentation on gravity effects on spoil piles in surface mining has been done in centrifuges, and this relatively novel method appears to be the most viable for meaningful research on gravitational effects on flow and related behavior of particulate material. These effects are critical in true cratering by large explosive charges in both solid rock and soil. Centrifugal testing has also been found to be necessary to simulate stress fields caused by gravity around underground openings.

## 1.10 PRACTICE AND RESEARCH

There have been many books and technical articles published on the practices of drilling, blasting, and mucking, so its treatment here would be duplication of information available in existing sources. This consists of information from detailed minutae of blasting arithmetic to the valuable results of technical analysis of research and practice, as well as effective engineering and scientific research.

In the whole field of rock excavation there is a need for both applied and basic research. One of the purposes of compiling the technological information on rock excavation in this book is to stimulate more creative thinking into current and future needs of research. There are many gaps that remain to be filled in the national and international technological libraries and in the data banks for sources of background material for industrial development.

# 2

# WAVE EQUATIONS FOR ROCK FRAGMENTATION

In percussive drilling, energy is transferred to the rock by the mechanics of elastic waves (pulses) traveling in steel rods of relatively small diameter, or rod waves. For most practical purposes, the waves in rock are also considered to be elastic, their geometry being either plane, cylindrical, or spherical, and simple waves or pulses are assumed to be mostly longitudinal in character. In these types of waves, the particle velocity is in the same direction as the direction of wave propagation. In shear waves, which are of relatively minor importance in blasting, the direction of the particle velocity is perpendicular to the direction of propagation. Strong waves in explosives, where the pressures are high enough to cause the material to behave hydrodynamically, are properly designated as shock waves, and hydrodynamic equations apply. Pressures generated by explosives are strong enough to crush rock and to cause it to behave plastically in the immediate vicinity of boreholes, but true shocks are not induced by chemical explosives in rock.

The equations for plane, cylindrical, and spherical waves are easily derived and can be readily evaluated for rod waves without resorting to the solution of a differential wave equation. However, for dilatation waves for transient conditions, the differential equations can only be solved by use of transform calculus or other specialized methods. The wave equations for plane shock waves are derived in the chapter on explosives.

## 2.1  STRESS WAVES IN ROCK

There are several types of waves that may be generated in rock, such as large-scale disturbances resulting from earthquakes, nuclear explosions, or more local waves generated by explosives for seismic prospecting. The waves that are generated by confined explosives immediately around blast holes are very intense. Geometrically, they are approximately plane, cylindrical, or spherical in shape. Usually, only the symmetrical forms are considered in theoretical analyses for the sake of simplicity. These types serve to illustrate the basic mechanics of wave propagation and consequent fracturing of the rock near an explosive because of its confinement and nearness to free faces. Also, the propagation of waves in the rock and air at some distance from the blast may cause damage to structures or cause environmental disturbances that must be controlled.

Most studies of waves in rock assume that they are elastic in character and that the corresponding relationships between stress, strain, Poisson's ratio, dilatation, and so on, can be applied with reasonable accuracy to both transients and waves of long duration. Usually transient models are assumed to have the form of rectangular or triangular pulses, and important parameters are pulse length, rise time, fall time, and peak values of stress, strain, particle velocity, or acceleration. The rise and fall times for a rectangular or square pulse are approximately zero.

### 2.1.1  Transient Waves

Transient parameters (except shock waves in explosives) are governed by the same basic wave equations that govern other types of waves, all of which travel with a characteristic velocity, the value of which depends upon the density and elastic contents of the medium. For a longitudinal wave, the stress $\sigma$ is the tensile or compressive stress in the direction of velocity and propagation of a longitudinal wave. The strain is likewise the unit deformation at a point in the wave in the direction of propagation. In a shear wave, the stress $\tau$ and other parameters are normal to the direction of propagation. The particle velocity $v$ is the velocity of movement of a particle as the wave moves it in the direction of the wave propagation for a longitudinal wave and normal to the direction of propagation in a shear wave.

The direction of particle acceleration is the same as that for velocity but is never in phase with the velocity. This latter condition is obvious from the mathematical relationship $a = (dv)/(dt)$. Thus, if the velocity is represented by a sin $\omega t$ curve, the acceleration is $\omega$ cos $\omega t$.

For a longitudinal rectangular plane wave for velocity, the relations between the

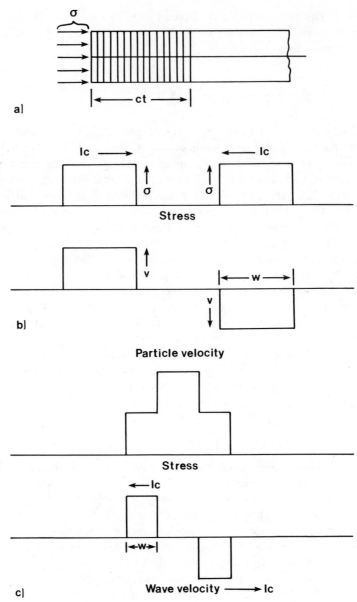

**Figure 2.1** (*a*) Square wave induced by impact; (*b*) colliding compressive waves; (*c*) tensile and compressive waves colliding.

basic wave parameters are shown in Figure 2.1. Thus, the curves for particle velocity, stress, and strain are the same shape.

### 2.1.2 Transients in Rock

All of the waves generated in rock by explosives are transient in character; that is, they are of relatively short duration and have different shapes depending on such items as the properties of the explosive, the (elastic) properties of the rock, the stemming, the rock structure near the detonating explosive, the loading density of the explosive, and similar factors. Most analyses of transient waves also assume elastic behavior.

Such waves are also studied assuming that the blast geometry creates (1) plane, (2) cylindrical, or (3) spherical waves, but most investigations assume that a wave is plane after it has traveled a short distance from the explosion cavity.

### 2.2 PLANE WAVES

The wave equation for a plane wave may be developed by considering an element of the elastic material (rock) through which the wave is passing. Two types of plane waves must be considered. The first is that in a rod or bar of small diameter in which the lateral effects of a longitudinal wave may be neglected; that is, the Poisson's ratio effects due to lateral extension caused by longitudinal stress are negligible.

### 2.2.1 Rod Wave

A small longitudinal section of a solid rod subject to a stress wave moves in accordance with Newton's law of motion $F = ma$ (Figure 2.2). The mass of a unit cross section is equal to the volume times the density:

$$m = \rho dx \tag{2.1}$$

The summation of the forces in the $x$ direction is

$$F = \left( \sigma + \frac{\partial \sigma}{\partial x} dx \right) - \sigma = \frac{\partial \sigma}{\partial x} dx \tag{2.2}$$

Application of Newton's law gives

$$\frac{\partial \sigma}{\partial x} dx = \rho dx \frac{\partial^2 u}{\partial t^2} \tag{2.3}$$

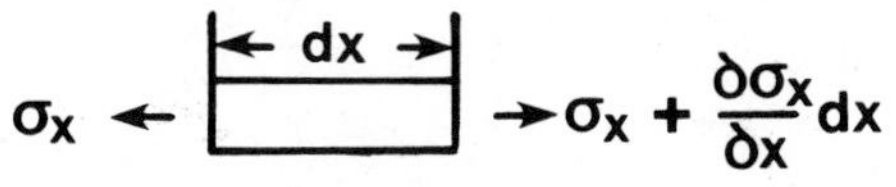

Figure 2.2  Stresses on an infinitesimal section, rod wave.

where $u$ is the displacement in the $x$ direction. Equation 2.3 in the form

$$\frac{\partial \sigma_x}{\partial x} = \rho \frac{\partial^2 u}{\partial t^2} \qquad (2.4)$$

holds for any solid material. If the material is assumed to be elastic, Hooke's law relating stress and strain may be applied, that is,

$$E = \frac{\sigma}{\epsilon} \qquad (2.5)$$

where $E$ = Young's modulus and $\epsilon$ = strain.

In terms of infinitesimals, the strain in the $x$ direction is defined as

$$\epsilon_x = \frac{\partial u}{\partial x} \qquad (2.6)$$

and

$$\sigma = E \frac{\partial u}{\partial x} \qquad (2.7)$$

Equation 2.7 combined with equation 2.4 yields

$$\frac{\partial^2 u}{\partial x^2} = \frac{\rho}{E} \frac{\partial^2 u}{\partial t^2} \qquad (2.8)$$

Equation 2.8 is the well-known plane wave equation for a wave in a bar of small diameter, and it may be utilized for the solution of the parameters of such a wave if the boundary conditions are known. It may also be written as shown with the stress $\sigma$, strain $\epsilon$, or particle velocity $v$ substituted for the displacement $u$. The equation can also be written in the form

$$\frac{\partial^2 u}{\partial x^2} = \frac{1}{c^2} \frac{\partial^2 u}{\partial t^2} \qquad (2.9)$$

where

$$c = \sqrt{\frac{E}{\rho}} \qquad (2.10)$$

and $c$ is the (bar) velocity of a longitudinal wave.

It can be demonstrated by substitution that any function $f(x + ct)$ or $f_1(x - ct)$ is a solution to the equation, or

$$u = f(x + ct) + f_1(x - ct) \qquad (2.11)$$

is also a solution. The first term represents a wave traveling in the $x$ direction toward the origin, and the second, a wave traveling away from the origin.

The velocity of propagation $c$ of a rectangular pulse can be obtained as follows (Timoshenko, 1934): It is assumed that a uniformly distributed compressive stress is suddenly applied to the end of a bar (Figure 2.1), producing at the first instant a uniform compression of a thin layer at the end of the bar. The stress, maintained at the surface of the bar, is then transmitted to adjacent layers, causing a compressive stress to travel along the bar with a velocity $c$. After a time $t$, a length of the bar equal to $ct$ will be compressed.

When the stress is first applied at the end of the bar, the face of the bar will move forward with a velocity $v$, which is the *particle velocity*. For a constant force, the particles will move forward until the material is compressed an amount determined by Young's modulus; then the face will stop moving if the force is discontinued. Meanwhile, however, the disturbance or wave front will move down the bar with a velocity $c$. The particle velocity can thus be determined from the fact that the compressed zone shortens an amount $(\sigma/E)ct$—that is, the unit strain multiplied by the length $ct$. The particle velocity, or the distance moved by the end of the bar divided by the time $t$, is

$$v = \frac{c\sigma}{E} \qquad (2.12)$$

The wave velocity $c$ is determined from considerations of momentum—that is, mass $\times$ velocity. The shaded portion of the bar (Figure 2.1$a$) is accelerated to a velocity $v$ in a time $t$, and the momentum is, for a unit cross section, equal to $\rho c \times v$. This is equated to the impulsive force, force multiplied by time. Hence,

$$\sigma t = \rho c v t \qquad (2.13)$$

or

$$v = \frac{\sigma}{\rho c} \qquad (2.14)$$

and, from equation 2.10, also

$$v = \frac{\sigma}{\sqrt{E\rho}} = \epsilon c \qquad (2.15)$$

Thus, the stress in a wave in an elastic bar is related by (1) the impact of the Young's modulus and the bar, and (2) the ratio of the particle and wave velocities. The particle velocity is determined by the product of the strain and the wave velocity. Likewise, the stress is equal to the product of the density, wave velocity, and particle velocity. Although these relationships were derived for a suddenly applied constant stress on the end of a bar, they are also found to apply for nontransient types of stress waves.

In both cylindrical and spherical waves, the stresses ($\sigma_r$) on an element are compressive in the radial direction and usually tensile ($\sigma_\theta$) in the tangential direction as long as the wave front has appreciable curvature. Hence, in a plane longitudinal wave in a mass of rock or other elastic medium, lateral effects must be taken into account.

### 2.2.2  Drill Steel

When a piston in a percussion drill strikes the end of a drill rod, the impact generates a wave in both the piston and the rod. The amount of energy transmitted and that reflected depend upon several factors, such as the relative cross-sectional areas, lengths, densities, elastic properties, and geometry of the piston and the rod.

To illustrate the analysis, we choose a piston and rod of the same diameter (Figure 2.3), the rod being of infinite length so that no reflected waves occur in it. This also serves as a basis for analyzing more complex geometries for pistons whose shape is designed to determine the shape of the wave form in the steel rod. The effects of several piston shapes have been analyzed by Dutta (1968), two of which are presented here.

For a piston of length $l$ with a cross-sectional area $A_1$ striking an infinitely long rod of the same cross section and elastic properties, the impact generates compressive waves in each of equal magnitude and shape, their parameters being governed by equation 2.8 and the initial conditions at impact.

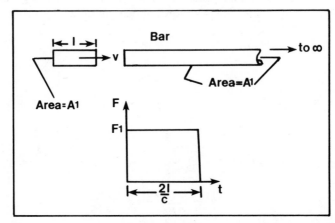

**Figure 2.3**  Force wave form for piston and bar of same cross-sectional area (Dutta, 1968).

The reflected (square) compressive wave in the piston reaches the free end of the piston in a time $l/c$ where $l$ is the piston length and $c$ is the wave velocity in the steel piston. The wave is then reflected as a tensile wave, which doubles the particle velocity at the free end, and the tensile wave is reflected to the contact face between the piston and the rod, arriving there at a time equal to $2l/c$. Since the wave is tensile, the contact between the piston and the rod is broken, resulting in a square wave of length $2l/c$ in the rod (Figure 2.3).

The force $F_1$ during the contact is constant, and if the particle velocity in the piston is $v_p$ and that in the bar is $v_b$ and the impact velocity of the piston is $v_o$, then

$$v_o - v_p = v_b \tag{2.16}$$

It has been shown that the stress in a plane wave is related to the particle velocity by

$$\sigma = \rho c v \tag{2.17}$$

Consequently, whether the cross section areas are equal or not (Figure 2.4), the initial force at the interface between the piston and rod is

$$F_p = \rho_p c_p v_p A_p = \rho_b c_b v_b A_b = F_b \tag{2.18}$$

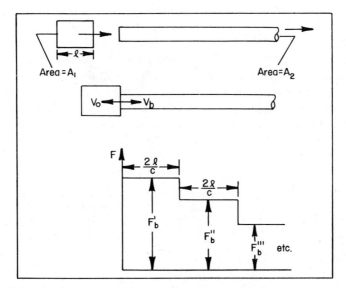

**Figure 2.4**  Change of cross-sectional areas in a force transmission system (Dutta, 1968).

If the densities and wave velocities of the steel piston and bar are assumed to be equal, then

$$v_p A_p = v_b A_b \tag{2.19}$$

or

$$v_p = \frac{A_b v_b}{A_p} = \gamma v_b \tag{2.20}$$

where $\gamma$ = the ratio of the two areas. Hence, from equation 2.16

$$v_b = v_o - \gamma v_b \tag{2.21}$$

or

$$v_b = v_o \frac{1}{1 + \gamma} \tag{2.22}$$

and, from equation 2.22,

$$v_p = v_o \frac{\gamma}{1 + \gamma} \tag{2.23}$$

For pistons and rods of equal areas

$$v_p = v_b = \tfrac{1}{2} v_o \tag{2.24}$$

The reflected wave in the piston will have a particle velocity of $-2v_p$ and will arrive back at the interface at a time $2l/c$, which will result in a change in force being generated in both the piston and the bar. That is, the particle velocity at the interface, which was equal to $v_o$, will now be equal to $v_o - 2v_p$ and, for equal cross-sectional areas, will be equal to zero (Figure 2.4).

For the more general case of unequal areas, the new particle velocities in the piston, $v_p'$, and in the bar, $v_b'$, are (Figure 2.4)

$$v_p' = (v_o - 2v_p)\frac{\gamma}{1 + \gamma} = v_o \frac{\gamma}{1 + \gamma}\frac{1 - \gamma}{1 + \gamma} \tag{2.25}$$

and

$$v_b' = (v_o - 2v_b)\frac{1}{1 + \gamma} = v_o \frac{1}{1 + \gamma}\frac{1 - \gamma}{1 + \gamma} \tag{2.26}$$

The new force $F_p'$ in the piston is $A_p \rho_p c_p$ multiplied by new reduced velocity or

$$F_p' = A_p \rho_p c_p v_o \frac{\gamma}{1 + \gamma} \frac{1 - \gamma}{1 + \gamma} = F_p \frac{1 - \gamma}{1 + \gamma} \qquad (2.27)$$

which can be written

$$F_p' = (-F_p) \frac{\gamma - 1}{\gamma + 1} = (-F_p) R_c \qquad (2.28)$$

where

$$R_c = \frac{\gamma - 1}{\gamma + 1} \qquad (2.29)$$

with $R_c$ being defined as the reflection coefficient. After reflection from the free end of the piston $F_p$ becomes $(-F_p)$, and when it reaches the interface a portion of it is reflected back into the piston as defined by $(-F_p R_c)$ while the remainder of the force is transmitted to the bar.

Likewise the change in force in the bar is, from equation 2.26,

$$F_b' = A_b \rho_b c_b v_b \frac{1}{1 + \gamma} \frac{1 - \gamma}{1 + \gamma} \qquad (2.30)$$

or, from equations 2.22 and 2.18,

$$F_b' = F_b \frac{1 - \gamma}{1 + \gamma} \qquad (2.31)$$

Equation 2.31 can be rewritten in the following form:

$$F_b' = F_p + (-F_p) \frac{2\gamma}{1 + \gamma} \qquad (2.32)$$

where $(2\gamma)/(1 + \gamma)$ is defined as the transmission coefficient $T_c$.

Thus, for each successive pulse reflection at the interface, the transmitted force in the (infinite) bar is reduced by an amount equal to $(2\gamma)/(\gamma + 1) \times$ the last reflected wave in the piston, resulting in a decreasing step wave form (Figure 2.4):

$$F_b'' = F_p' + (-F_p') T_c \qquad (2.33)$$

The above development may be applied to a wave or pulse traveling in a rod when it meets a change in cross section where part of the wave will likewise be

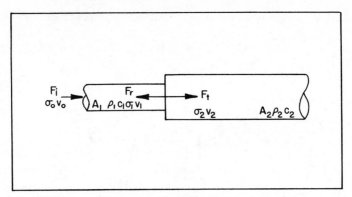

**Figure 2.5**   Nomenclature for area change (Dutta, 1968).

reflected and part transmitted (Figure 2.5). Continuity of forces at the change in cross section requires that

$$F_i + F_r = F_t \tag{2.34}$$

where $F_i$ = incident force = $\sigma_o A_p = A_p \rho_p c_p v_o$, $F_r$ = reflected force = $\sigma_p A_p = A_p \rho_p c_p v_o$, and $F_t$ = transmitted force = $\sigma_b A_b = A_b \rho_b c_b v_o$. Continuity of motion also requires that

$$v_o = v_p + v_b \tag{2.35}$$

and, as above, assuming that the rod is of the same material on both sides of the area change, again one obtains $\gamma = A_b/A_p$, and from equation 2.34

$$A_b v_b = A_p(v_o + v_p) \tag{2.36}$$

or from equation 2.35

$$\gamma(v_o - v_p) = v_o + v_p \tag{2.37}$$

which results in

$$v_p' = \left(\frac{\gamma - 1}{\gamma + 1}\right) v_o \tag{2.38}$$

and

$$v_b' = \frac{2}{\gamma + 1} v_o \tag{2.39}$$

where $v_p'$ is the particle velocity in the piston and $v_b'$ is that in the bar.

Therefore, from equations 2.38 and 2.34, the force reflected at the first impact is

$$F_r = A_p \rho_p c_p v_o \left(\frac{\gamma - 1}{\gamma + 1}\right) = F_i \left(\frac{\gamma - 1}{\gamma + 1}\right) \qquad (2.40)$$

and

$$F_r = F_i R_c \qquad (2.41)$$

Likewise, the transmitted force for the first impact is

$$F_t = A_b \rho_b c_b v_o \frac{2}{\gamma + 1} \qquad (2.42)$$

or

$$F_t = A_p \rho_p c_p v_o \frac{2\gamma}{\gamma + 1} \qquad (2.43)$$

which gives

$$F_t = F_i T_c \qquad (2.44)$$

Thus, for the configuration shown in Figure 2.5, where the bar is semi-infinitely long to the right, the particle velocity for each successive reflection can be determined, as well as the resulting particle velocity and force transmitted to the bar. The particle velocity of the pulse impinging on the surface at the change of area ($A_c$) is altered by the multiplying factor $(2\gamma)/(\gamma + 1)$, and the reflection of the pulse from the free end diminishes the particle velocity by an amount of twice the velocity of the pulse reflected from the interface for each pulse. The calculated force or stress wave from one impact of a conventional piston by two similar methods (Figure 2.6) compares favorably with an experimentally measured wave for a selected piston, even though several simplifying assumptions were made.

The foregoing analysis illustrates the basic processes of one-dimensional propagation of waves for simple systems. In a percussion drill, it would also be necessary to consider the effects of waves reflected from the end of the drill rod, the effects of the drill chuck, and other factors that were not taken into account in the above. This and more complex analyses lend themselves readily to computer calculations that can be checked experimentally (Simon, 1963).

If the medium is of infinite extent normal to the wave direction, or its lateral extent is very large, an element will be subject to a lateral stress equal to $\sigma v/(1 - v)$ but will have no lateral motion, and the lateral strain will be zero. If the mass is of limited lateral extent but not a ''thin rod,'' the element will expand in the lateral direction, and the lateral strain will have a finite value. In either of these

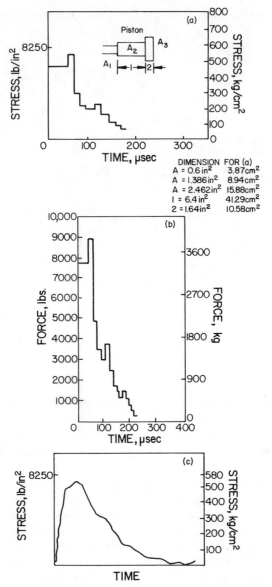

**Figure 2.6**   Comparison of wave forms for a Holman SL200 rock drill (Dutta, 1968). (*a*) calculated, (*b*) theoretical computer, (*c*) experimental.

two cases, as shown in the following equation, the longitudinal velocity of the wave is given by

$$c = \sqrt{\frac{\lambda + 2G}{\rho}} \qquad (2.45)$$

where $\lambda$ = Lame's constant and $G$ = shear modulus, as compared to $c = E/\rho$ for a thin rod. When the curvature of the wave front is small (radius of curvature large) and the wave front approaches a plane geometry, the lateral stress is compressive, and the lateral strain is zero.

In a spherical wave the tensile tangential stress is quite large near an explosive cavity, but it diminishes rapidly, and, as the wave front curvature becomes more nearly plane, the tangential stress changes from tension to compression. That is, it assumes the properties of a plane wave.

The relationship between the wave energy, the wave parameters, and the physical properties of the material may be obtained as follows: The energy in a wave of constant stress throughout a length $ct$ (a rectangular wave) consists of two parts. The first is the strain (potential) energy of deformation which is equal to the volume of the material multiplied by the strain energy, the latter being equal to the area under the stress curve, or

$$PE = \frac{ct\epsilon\sigma}{2} = \frac{ct\sigma^2}{2E} = \frac{Ect\epsilon^2}{2} \tag{2.46}$$

and the kinetic energy ($\frac{1}{2}mv^2$) which is equal to

$$KE = \frac{ct\rho v^2}{2} \tag{2.47}$$

It can be easily shown that the two types of energy for a simple stress wave are equal and that the total energy of the wave is equal to the work done by the force per unit cross-sectional area, $\sigma$, multiplied by the distance over which it acts, $\sigma\epsilon ct = \sigma^2 ct/E$, and is half potential and half kinetic.

The plane wave equation 2.9 is linear, and, hence, if solutions for two interacting waves are obtained, the principle of superposition may be applied, and their sum is also a solution. That is, stress, strain, and particle velocity are additive, and the principle of superposition may be applied in the analysis of waves traveling along a bar. So, if two waves traveling in opposite directions meet and pass through each other, the resulting stress, particle velocity, strain, and energy distribution can be obtained by addition or subtraction, as appropriate.

For two equal square compressive waves passing in opposite directions through each other, as the two waves overlap, the stresses and strains are additive (Figure 2.1), but the particle velocities are in opposite directions and cancel each other. Thus, while the two waves overlap, the kinetic energy is all changed to potential energy (energy must be conserved). After the waves have passed completely through each other, they resume their initial character. As the waves begin to penetrate each other, the changes indicated take place in the portions of the waves that overlap until the waves occupy the same section of the bar. As the waves begin to exit from the common space, the changes take place in the opposite manner of the changes at initial penetration, and the waves assume their original character. It should be emphasized that equations 2.12–2.16 must be modified for waves in the area of interference.

**Problem 1.** If two equal waves, one in compression and one in tension, travel in opposite directions, what are the values of stress, strain, particle velocity, and energy before they meet and when they completely overlap.

**Problem 2.** The same for two equal right triangular waves: (1) both in compression, and (2) one in compression and one in tension.

**Problem 3.** The total energy, kinetic plus potential, is to be conserved. Show why the potential energy for compressive intersecting square stress waves is equal to $(ct\sigma^2)/E$ instead of $(ct(2\sigma)^2)/(2E)$.

### 2.2.3   Plane Dilatation Waves

The equation for a plane dilatation wave is derived in much the same manner as that for a wave in a rod. Thus, for a wave traveling in the $x$ direction in an infinite medium the particle velocity is in the $x$ direction. The material is constrained in the $y$ and $z$ directions so that the displacement and strain in these directions is zero. As with a wave in a bar, the equation of equilibrium leads to

$$\frac{\partial \sigma_x}{\partial x} = \rho \frac{\partial^2 u}{\partial t^2} \tag{2.48}$$

but, for an infinite medium, the relationship between stress and strain is given by

$$\sigma_x = \lambda e + 2\mu\epsilon_x \tag{2.49}$$

The strains and displacements in the $y$ and $z$ directions are zero, although the stresses are $\sigma_y = \lambda\epsilon_x$, $\sigma_z = \lambda\epsilon_x$, and hence

$$\sigma_x = (\lambda + 2\mu)\epsilon_x = (\lambda + 2\mu)\frac{\partial u}{\partial x} \tag{2.50}$$

Substitution of this value for $\sigma_x$ in the equation gives

$$(\lambda + 2\mu)\frac{\partial^2 u}{\partial x^2} = \rho \frac{\partial^2 u}{\partial t^2} \tag{2.51}$$

or

$$\frac{\partial^2 u}{\partial x^2} = \frac{1}{c^2}\frac{\partial^2 u}{\partial t^2} \tag{2.52}$$

and

$$c^2 = \frac{\lambda + 2\mu}{\rho} \tag{2.53}$$

whereas for a wave in a rod the velocity is $c^2 = E/\rho$. Thus, the dilatation velocity is greater than the bar velocity. For a Poisson's ratio of $\nu = 0.25$, the velocity ratio is 1.095, and, for $\nu = 0.30$, the ratio of velocities is 1.16.

The equation of a shear wave is of the same form as for bar and dilatation waves, but the shear wave velocity is found to be $c^2 = \mu/\rho$, where $\mu$ is the shear modulus.

If any type of impulse function, such as a Dirac delta function, a unit step, or the like, is applied to the plane wave equations, the solutions obtained simply indicate that a pulse of the same form is induced in the elastic medium. However, this is not the case for spherical and cylindrical waves, as is shown in the next section.

### 2.2.4 Wave Reflection from a Free Surface

When a plane wave or pulse meets a free surface at normal incidence, the stress and strain are reflected 180° out of phase; that is, a tensile wave is reflected in

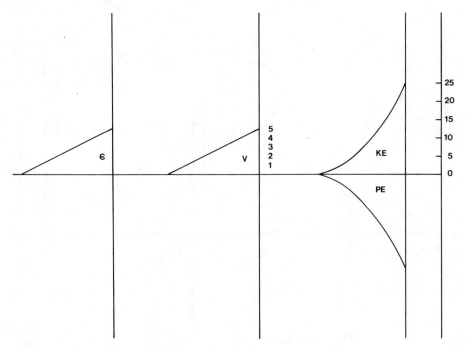

**Figure 2.7** Stress, particle velocity, and energy density of sawtooth wave as it meets free boundary.

compression, and a compressive wave is reflected in tension. The wave resulting from reflection may be treated just as though it had met its mirror image of opposite sign at the free boundary (Figure 2.1). For example, the stress and strains cancel each other where waves of tension and compression moving in opposite directions overlap, and the particle velocity and kinetic energy are additive. Thus, where two opposite traveling square waves overlap, the strain and the strain energy are zero, and the strain energy is converted to kinetic energy in the area of interference. The reverse process takes place after the reflected wave passes through the incident wave, and each assumes its original form.

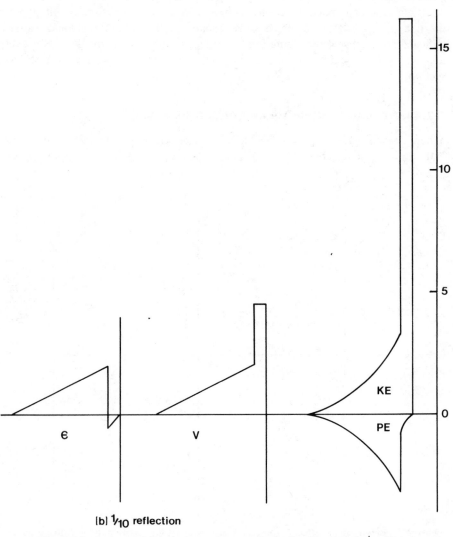

(b) 1/10 reflection

**Figure 2.8**   Strain, particle velocity, and energy distribution for sawtooth $\frac{1}{10}$ reflected.

For a sawtooth wave, the reflection geometry is more complex. Before the wave meets the free surface, the strain and particle velocity are proportional to each other at their respective positions in the wave. The potential energy is proportional to the square of the strain, and the kinetic energy is proportional to the square of the particle velocity. The total energy is proportional to the sum of the areas under the PE and KE energy curves (Figures 2.7–2.13).

The strain wave (and the stress) is reflected as shown, the value at the front of the reflected wave being equal to the difference between the incident and reflected strains, and the value of the normal strain (stress) at the free surface being zero. However, in the same area of overlap, the particle velocity is additive, yielding a flat top wave as the reflected portion of the pulse travels back into the material. The corresponding energy distribution is determined from the strain and velocity

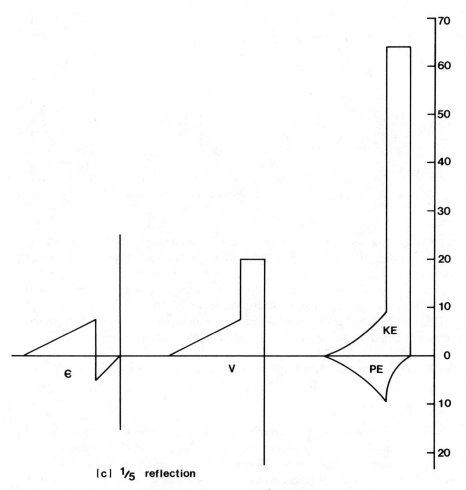

[c] $\frac{1}{5}$ reflection

**Figure 2.9** Parameters for $\frac{1}{5}$ reflection.

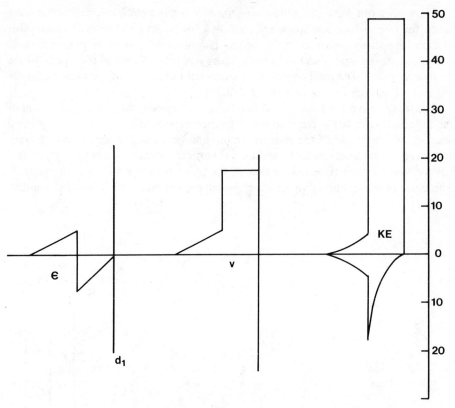

**Figure 2.10**  Parameters for continued reflection.

distribution, the total potential (strain) and kinetic energies being proportional to the areas under the respective energy distribution curves. It should be noted that the relationships $\sigma = \rho c v$ and $v = \epsilon c$ do not hold for the interfering portion of waves. That is, the velocity may increase for interfering stress waves of opposite sense, but, for this case, the stress and strain will decrease.

The total energy at any position in the incident wave before reflection for an elemental length $ds$ is the sum of the kinetic and potential energies:

$$dE_k + dE_p = \frac{E}{2} \epsilon^2 \, ds + \frac{\rho v^2}{2} \, ds \tag{2.54}$$

For equal overlapping or interfering waves of opposite stress traveling in opposite directions, particle velocities are additive at any position in the wave, and the kinetic energy at that position is determined by the equation $KE = 1/2 \, mv^2$. For reflected waves, a segment of the mass is affected by two waves. The total

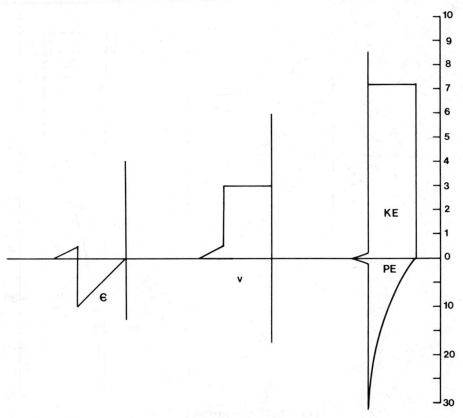

**Figure 2.11** Parameters for $\frac{2}{3}$ reflection.

energy at a given plane can then be determined by means of equation 2.46. Where an incident compressive sawtooth wave is reflected from a free surface in tension, the energy distribution for both the kinetic and potential energy is parabolic just before reflection. There is no normal strain at the surface; hence, at the surface all of the strain energy is converted to kinetic energy, which is twice the energy per unit mass in the incident wave. The occupied mass in the interfering wave is half that occupied by the same portion of the noninterfering wave. The velocity is also doubled. The strain energy distribution changes rapidly as the wave is reflected (Figures 2.7–2.12) back into the solid from the free face with corresponding changes in particle velocity, stress, and strain.

The total wave energy partitioning also changes as the wave moves back into the rock. This may be found by utilizing the equation

$$E_t = \frac{E}{2c^2}\left[\int (c\epsilon)^2\, ds + \int v^2\, ds\right] \tag{2.55}$$

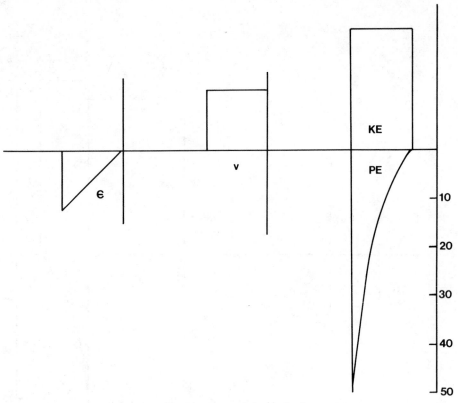

**Figure 2.12**   Parameters for $\frac{1}{2}$ reflection.

and integrating over the segments of the wave forms. The total kinetic energy of the pulse increases rapidly as the wave front moves back into the solid (Figure 2.13), representing more than 90% of the wave energy where the pulse has traveled about one third of its length back into the solid.

**Problem:** Assume a strain pulse has the form $\epsilon = Ke^{-\alpha t}$. For a value of $K = 1$ and $\alpha = 1.5$, determine the strain, particle velocity, and energy distribution at 5 points for this shape of pulse.

The total energy may be obtained by integrating to determine the sum of the areas under the sections of the energy curves. For a sawtooth curve, the energy partitioning shows that the potential energy decreases rapidly as the waves overlap, increasing again to the normal value for a free pulse after the waves have completely passed through each other.

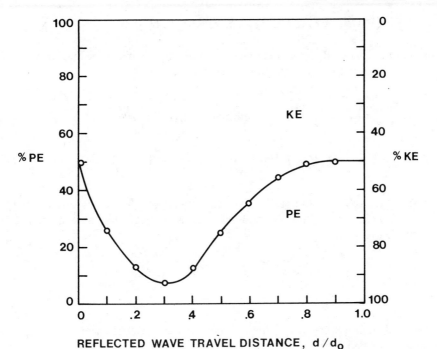

**Figure 2.13** Sawtooth pulse energy distribution, reflected wave.

**Problem 1.** Perform the necessary integrations to determine the energy partitioning for a sawtooth pulse.

**Problem 2.** Determine the energy partitioning during reflection of a decaying compressive exponential pulse $\epsilon = Ke^{-\alpha t}$.

Thus, when half of a sawtooth pulse has been reflected (that is, the wave completely overlaps), a maximum peak strain results, and approximately 75% of the *total* pulse energy is in the form of kinetic energy. For a reflected square pulse that completely overlaps, all of the strain energy is converted to kinetic energy. Also, if fracture takes place at the front of the reflection wave in a square wave overlap, the total energy trapped in the slab is all kinetic and equals twice the initial kinetic energy contained in the slabbed section. The momentum is also doubled, so that the slab velocity is twice the initial particle velocity.

The relationship of the reflection slabbing mechanism to the energy partitioning—that is, between potential and kinetic energy—is not clear. For a triangular pulse, if slabbing occurs before the front of the reflected wave reaches the tail of the incident wave, more than 50% of the total energy is trapped in the slab. If slabbing occurs when the wave just overlaps, all of the wave energy is trapped in the slab, about 75% of it being kinetic.

## 2.3 SPHERICAL ELASTIC WAVE EQUATION

Consider the spherical element (Figure 2.14) in which the sum of the forces in the $r$ direction is equal to the mass times the acceleration, $A$:

$$\left( \sigma_r + \frac{\partial \sigma_r}{\partial r} dr \right) (r + dr)\, d\psi\, (r + dr)\, d\theta - \alpha_r (r\, d\psi)\, (r\, d\theta)$$

$$- 2\sigma_\psi \sin \frac{d\psi}{2} (r + dr)\, d\theta\, dr - d\sigma_\theta \sin \frac{d\psi}{2} (r + dr)\, d\psi\, dr = mA \quad (2.56)$$

The angles $(d\psi/2)$ and $(d\theta/2)$ are substituted for the values of $\sin(d\psi)/2$ and $\sin(d\theta)/2$, respectively. Second-order terms are neglected, and equation 2.56 is simplified to yield

$$\frac{\partial \sigma_r}{\partial_r} + \frac{2\sigma_r}{r} - \frac{\sigma_\psi}{r} - \frac{\sigma_\theta}{r} = \frac{\rho \partial^2 u}{\partial t^2} \quad (2.57)$$

For a dilatation wave where there is no rotation of the element, $\sigma_\theta = \sigma_\psi$, and equation 2.57 becomes

$$\frac{\partial \sigma_r}{\partial r} + \frac{2(\sigma_r - \sigma_\theta)}{r} = \frac{\rho \partial^2 u}{\partial t^2} \quad (2.58)$$

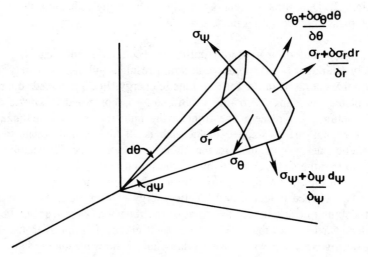

**Figure 2.14**   Stresses on a spherical element.

which is the equation of motion for a symmetrical spherical wave in terms of stress and displacement, but for an elastic material

$$\sigma_r = \lambda e + 2\mu\epsilon_r,$$

$$\sigma_\theta = \lambda e + 2\mu\epsilon_\theta,$$

$$\epsilon_r = \frac{\partial u}{\partial r},$$

$$\epsilon_\theta = \frac{u}{r},$$

$$\vdots$$

$$e = \epsilon_r + \epsilon_\theta + \epsilon_\psi = \epsilon_r + 2\epsilon_\theta = \frac{\partial u}{\partial r} + \frac{2u}{r} \qquad (2.59)$$

Substitution of equations 2.59 into equation 2.58 gives

$$(\lambda + 2)\frac{\partial^2 u}{\partial r^2} - \frac{2\partial u}{r\partial r} - \frac{2u}{r^2} = \frac{\rho\partial^2 u}{\partial t^2} \qquad (2.60)$$

or

$$\frac{\partial^2 u}{\partial r^2} + \frac{2\partial u}{r\partial r} - \frac{2u}{r^2} = \frac{1}{c^2}\frac{\partial^2 u}{\partial t^2} \qquad (2.61)$$

The displacement $u$ in terms of the displacement potential $\phi$ is

$$u = \frac{\partial \phi}{\partial r} \qquad (2.62)$$

Equation 2.61 becomes

$$\frac{\partial^2}{\partial r^2}\left(\frac{\partial \phi}{\partial r}\right) + \frac{2}{r\partial r}\left(\frac{\partial \phi}{\partial r}\right) - \frac{2}{r^2}\left(\frac{\partial \phi}{\partial r}\right) = \frac{1}{c^2}\frac{\partial^2}{\partial t^2}\left(\frac{\partial \phi}{\partial r}\right) \qquad (2.63)$$

or

$$\frac{\partial}{\partial r}\left[\frac{\partial^2 \phi}{\partial r^2} + \frac{2}{r}\frac{\partial \phi}{\partial r}\right] = \frac{1}{c^2}\frac{\partial}{\partial r}\frac{\partial^2 \phi}{\partial t^2} \qquad (2.64)$$

Thus the equation of motion of a spherical elastic wave is

$$\frac{\partial^2(r\phi)}{\partial r^2} = \frac{1}{c^2}\frac{\partial^2(r\phi)}{\partial t^2} \tag{2.65}$$

The boundary conditions for a unit impulse (Dirac) pressure at an internal spherical surface $r = r_o$ are

$$(\lambda + 2\mu)\frac{\partial^2\phi}{\partial r^2} + \frac{2\lambda\partial\phi}{\partial r} = P_o\sigma_r = -P_o\delta(t), \qquad \text{for } r = r_o, \quad t = 0$$

$$\frac{\partial(r\phi)}{\partial t} = 0, \qquad \text{for } r < r_o, \quad t = 0$$

$$(r\phi) = 0, \qquad \text{for } r < r_o, \quad t = 0 \tag{2.66}$$

and

$$\lim_{r \to \infty} (r\phi) = 0$$

The transformed equations and boundary conditions are

$$\frac{d^2(r\phi)}{dr^2} - \frac{s^2(r\phi)}{c^2} = 0 \tag{2.67}$$

and

$$(\lambda + 2\mu)\frac{d^2\overline{\phi}}{dr^2} + \frac{2\lambda d\overline{\phi}}{dr} = -P_o \tag{2.68}$$

and, again,

$$\lim_{r \to \infty} (\overline{r}\phi) = 0 \tag{2.69}$$

the total derivatives indicating the operation is in the $s$ plane.

The solution of equation 2.67 is

$$\overline{r}\phi = Ae^{-s\overline{r}/c} + Be^{s\overline{r}/c} \tag{2.70}$$

The application of transformed conditions to equation 2.70 yields $B = 0$ for $\overline{r}\phi \to 0$ as $r \to \infty$ and

$$A = \frac{-aP_oe^{sr_o/c}}{\dfrac{(\lambda + 2\mu)s^2}{c^2} + \dfrac{4\mu s}{r_oc} + \dfrac{4\mu}{r_o^2}} \tag{2.71}$$

Equation 2.70 becomes

$$\bar{r}\phi = \frac{-r_o P_o e^{-s(\bar{r}-r_o)/c}}{\dfrac{(\lambda + 2\mu)s^2}{c^2} + \dfrac{4\mu s}{r_0 c} + \dfrac{4\mu}{r_o^2}} \tag{2.72}$$

The inverse may be obtained by finding the residues at the poles. Equation 2.71 is rewritten as

$$\bar{r}\phi =$$

$$\frac{-r_o P_o e^{-sr_o(\bar{r}-r_o)/c}}{\left[ s + \dfrac{2c}{r_o(\lambda + 2\mu)} \left( \mu + i \sqrt{(\lambda + \mu)\mu} \right) \right]\left[ s + \dfrac{2c}{r_o(\lambda + 2\mu)} \left( \mu - i \sqrt{(\lambda + \mu)\mu} \right) \right]} $$

$$\tag{2.73}$$

The inverse is

$$r\phi = \frac{-P_o r_o^2(\lambda + 2\mu)}{2c \sqrt{(\lambda + \mu)\mu}} \exp\left[ -\frac{2c}{r_o(\lambda + 2)} \left( t - \frac{r - r_o}{c} \right) \right] x$$

$$\cdot \left\{ \frac{1}{2i} \exp\left[ \frac{2ic}{r_o} \frac{\sqrt{(\lambda + \mu)\mu}}{(\lambda + 2\mu)} \left( t - \frac{r - r_o}{c} \right) \right] \right\}$$

$$\cdot \left\{ -\frac{1}{2i} \exp\left[ \frac{-2ic}{r_o} \frac{\sqrt{(\lambda + \mu)\mu}}{(\lambda + 2\mu)} \left( t - \frac{r - r_o}{c} \right) \right) \right\} \tag{2.74}$$

The complex exponents are combined and rewritten in trigonometric terms to give

$$\sin\left( \frac{2c}{a} \frac{\sqrt{(\lambda + \mu)\mu}}{(\ + 2\mu)} \left[ t - (r - r_o)/c \right] \right)$$

substitution of

$$\omega, \left[ t - \frac{(r - r_o)}{c} \right] = \tau, \quad \frac{2c\mu}{r_o(\lambda + 2\mu)} = \frac{2c}{r_o} \frac{\sqrt{(\lambda + \mu)\mu}}{\gamma + 2} = k \tag{2.75}$$

and equations 2.75 with equation 2.74 yields the final form of the inverse

$$r\phi = \frac{-r_o^2 P_o}{2c} \frac{(\lambda + 2\mu)}{\sqrt{(\lambda + \mu)\mu}} e^{-k\tau} \sin \omega\tau \tag{2.76}$$

The displacement, strain, and particle velocity are, then, respectively,

$$u = \frac{\partial\phi}{\partial r} = \frac{r_o^2 P_o}{2c} \frac{(\lambda + 2\mu)}{\sqrt{(\lambda + \mu)\mu}} e^{-k\tau} \left[\left(\frac{1}{r^2} - \frac{k}{rc}\right)\right.$$
$$\left.\cdot \sin \omega\tau + \frac{\omega}{rc} \cos \omega\tau\right] \tag{2.77}$$

$$\epsilon = \frac{\partial u}{\partial r} = \frac{r_o^2 P_o}{2c} \frac{(\lambda + 2)}{\sqrt{(\lambda + \mu)\mu}} e^{-k\tau} \left[\left(\frac{2}{r^2 c} \frac{-2\omega k}{rc^2}\right) \cos \omega\tau\right.$$
$$\left.+ \left(\frac{2}{r^3} - \frac{2k}{r^2 c} + \frac{k^2 - \omega^2}{rc^2}\right) \sin \omega\tau\right] \tag{2.78}$$

and

$$v = \frac{\partial u}{\partial t} = \frac{r_o^2 P_o}{2c} \frac{(\lambda + 2\mu)}{\sqrt{(\lambda + \mu)\mu}} e^{-k\tau} \left[\left(\frac{2k}{rc} - \frac{\omega}{r^2}\right) \cos \omega\tau\right.$$
$$\left.+ \left(\frac{\omega^2 - k^2}{rc} + \frac{k}{r^2}\right) \sin \omega\tau\right] \tag{2.79}$$

The above equations are for a unit impulse $P_o \delta(t)$. For a unit step function, the right-hand side of equation 2.75 may be multiplied by $1/s$, or equation 2.76 integrated with respect to $t$. For an exponential pressure function $P_o e^{-\alpha t}$, whose transform is $P_o/(s + \alpha)$, or, for a pressure function of the form $P_o(e^{-\alpha} - e^{-\beta t})$, the multiplier for equation 2.73 is $P_o[1/(s + \alpha) - 1/(s + \beta)]$. The wave parameters in the time plane are then obtained by inversion of the transforms.

A convenient form of expression for a pressure function is the simple exponential, and a combination of the sum of two exponentials allows the pressure function to be varied with a range of rise and fall times (Sharpe, 1942; Duvall, 1953). The pressure function

$$P = P_o e^{-\alpha t} \tag{2.80}$$

yields

$$\Phi = \frac{ap_0/pr}{\left(\dfrac{\omega}{\sqrt{2}} - \alpha\right)^2 + \omega^2} \left\{ -e^{-\alpha\tau} + e^{-\omega\tau/\sqrt{2}} \right.$$

$$\left. \left[ \left(\frac{I}{\sqrt{2}} - \frac{\alpha}{\omega}\right) \sin \omega\tau + \cos \omega\tau \right] \right\} \tag{2.81}$$

where

$$\omega = \frac{2\sqrt{2}c}{3^a},$$

$$\tau = t - \frac{r-a}{c},$$

$$c = \sqrt{\frac{3\mu}{p}} \tag{2.82}$$

For a pressure pulse of the form

$$p = p_0(e^{-\alpha\tau} - e^{-\beta t}) \tag{2.83}$$

the potential is

$$\Phi = \frac{ap_0/pr}{\left(\dfrac{\omega}{\sqrt{2}} - \alpha\right)^2 + \omega^2} \left\{ -e^{-\alpha\tau} + e^{-\omega\tau/\sqrt{2}} \right.$$

$$\left[ \left(\frac{I}{\sqrt{2}} - \frac{\alpha}{\omega}\right) \sin \omega\tau + \cos \omega\tau \right] \bigg\}$$

$$+ \frac{ap_0/pr}{\left(\dfrac{\omega}{\sqrt{2}} - \beta\right)^2 + \omega^2} \left\{ +e^{-\beta\tau} - e^{-\omega\tau/\sqrt{2}} \right.$$

$$\cdot \left[ \left(\frac{I}{\sqrt{2}} - \frac{\beta}{\omega}\right) \sin \omega\tau + \cos \omega\tau \right] \bigg\} \tag{2.84}$$

From the equation for displacement potential the displacement, $u$; velocity, $v$; acceleration, $A$; radial strain, $\epsilon$; and dilation, $\Delta$, can be derived by the following:

$$u = \frac{\partial \Phi}{\partial r},$$

$$v = \frac{\partial u}{\partial t},$$

$$A = \frac{\partial v}{\partial t},$$

$$\epsilon = \frac{\partial u}{\partial r},$$

$$\Delta = \frac{\partial u}{\partial r} + 2\frac{u}{r} \qquad (2.85)$$

To analyze the spherical wave equation, Hornsey and Clark (1968) wrote the equation for displacement potential in the following form for $P = P_o e^{-\alpha t}$:

$$\phi = \frac{K}{r}\left\{-e^{-\alpha\tau} + e^{-M\tau}\left[\cos \omega\tau + L \sin \omega\tau\right]\right\} \qquad (2.86)$$

where

$$K = \frac{aP_o}{\rho\left[\alpha^2 - \dfrac{4\alpha c\mu}{a(\lambda + 2\mu)} + \dfrac{4c^2\mu}{a^2(\lambda + 2\mu)}\right]}$$

$$M = \frac{2c\mu}{a(\lambda + 2\mu)}$$

$$\omega = \frac{2c\left[(\lambda + \mu)\mu\right]^{1/2}}{a(\lambda + 2\mu)}$$

and

$$L = \left[\frac{\mu}{(\lambda + \mu)}\right]^{1/2} - \frac{a}{\omega}$$

The displacement, particle velocity, and acceleration are then found to be

$$u(r, t) = \frac{\partial \phi}{\partial r} = \frac{K}{r^2} \left\{ e^{-\alpha \tau} - e^{-M\tau} \left[ \cos \omega \tau + L \sin \omega \tau \right] \right\}$$

$$+ \frac{K}{rc} \left\{ -\alpha e^{-\alpha \tau} + e^{-M\tau} \left[ (M - \omega L) \cos \omega \tau + (ML + \omega) \sin \omega \tau \right] \right\}$$

$$(2.87)$$

$$v(r, t) = \frac{\partial u}{\partial t} = \frac{K}{r^2} \left\{ -\alpha e^{-\alpha \tau} + e^{-M\tau} \left[ (M - \omega L) \cos \omega \tau \right.\right.$$

$$+ (ML + \omega) \sin \omega \tau \right] \Big\} + \frac{K}{rc} \left\{ \alpha^2 e^{-\alpha \tau} + e^{-M\tau} \right.$$

$$\cdot \left[ (\omega^2 + 2M\omega L - M^2) \cos \omega \tau + (\omega^2 L - 2M\omega - M^2 L) \sin \omega \tau \right] \Big\}$$

$$(2.88)$$

$$A(r, t) = \frac{\partial y}{\partial t} = \frac{K}{r^2} \left\{ \alpha^2 e^{-\alpha \tau} + e^{-M\tau} \left[ (\omega^2 + 2M\omega L - M^2) \right.\right.$$

$$\cdot \cos \omega \tau + (\omega^2 L - 2M\omega - M^2 L) \sin \omega \tau \right] \Big\}$$

$$+ \frac{K}{rc} \left\{ -\alpha^3 e^{-\alpha \tau} + e^{-M\tau} \left[ (M^3 - 3M^2 \omega L - 3M\omega^2 + \omega^3 L) \right.\right.$$

$$\cdot \cos \omega \tau + (M^3 L - 3M\omega^2 L + 3M^2 \omega - \omega^3) \sin \omega \tau \right] \Big\} \qquad (2.89)$$

$$\epsilon(r, t) = \frac{\partial u}{\partial r} = \frac{2K}{r^3} \left\{ -e^{-\alpha \tau} + e^{-M\tau} \left[ \cos \omega \tau + L \sin \omega \tau \right] \right\}$$

$$- \frac{2K}{r^2 c} \left\{ -\alpha e^{-\alpha \tau} + e^{-M\tau} \left[ (M - \omega L) \cos \omega \tau + (ML + \omega) \sin \omega \tau \right] \right\}$$

$$- \frac{K}{rc^2} \left\{ \alpha^2 e^{-\alpha \tau} + e^{-M\tau} \left[ (\omega^2 + 2M\omega L - M^2) \cos \omega \tau \right.\right.$$

$$+ (\omega^2 L - 2M\omega - M^2 L) \sin \omega \tau \right] \Big\} \qquad (2.90)$$

and

$$\sigma(r, t) = (\lambda + 2\mu) \epsilon + 2\lambda\mu / r \qquad (2.91)$$

These are identical for the second term of the double exponential forcing function.

### 2.3.1   Scaling Laws

The elastic wave parameters obey the scaling laws of dimensional analysis. If the explosive cavity radius is scaled by a factor $k$ and the pressure pulse is scaled timewise by the same factor $k$, a simple dimensional relationship exists. If the elastic constants, density, and pressure $P_o$ are constant, the scaling law can be written:

$$v(r, t, a, \alpha, \beta/\alpha) = v(kr, kt, ka, \alpha/k, \beta/\alpha)$$

$$\sigma(r, t, a, \alpha, \beta/\alpha) = \sigma(kr, kt, ka, \alpha/k, \beta/\alpha)$$

$$\epsilon(r, t, a, \alpha, \beta/\alpha) = \epsilon(kr, kt, ka, \alpha/k, \beta/\alpha)$$

$$A(r, t, a, \alpha, \beta/\alpha) = kA(kr, kt, ka, \alpha/k, \beta/\alpha)$$

$$u(r, t, a, \alpha, \beta/\alpha) = \frac{1}{k}u(kr, kt, ka, \alpha/k, \beta/\alpha) \qquad (2.92)$$

The division of $\alpha$ by $k$, for a constant $\beta/\alpha$ ratio, causes the dual exponential pressure pulse to scale timewise by the factor $k$. It was also convenient to use the product of the cavity radius $a$ and the parameter $\alpha$, since the scaling laws are applicable for a given rock medium if the product $a\alpha$ and the $\beta/\alpha$ ratios are constant.

For the analysis, four media were chosen:

| Medium | Poisson's Ratio | Dilational Wave Velocity, Fps | Density Slug/Ft$^3$ |
|---|---|---|---|
| Tuff | 0.09 | 6,262 | 4.00 |
| Sandstone | 0.00 | 9,314 | 4.98 |
| Granite | 0.15 | 12,284 | 5.24 |
| Limestone | 0.19 | 15,248 | 5.10 |

These represent measured elastic rock properties. The properties of the sandstone, granite, and limestone were based on average properties for these materials, and those of the tuff are of Rainier tuff. Most of the study was made using granite, with comparisons of peak values only for the other media.

**LIST OF SYMBOLS (EQUATIONS 2.86–2.92)**

| | |
|---|---|
| $a$ | Effective cavity radius |
| $a\alpha$ | Product of $a$ and $\alpha$ |
| $A$ | Radial acceleration |
| $b$ | Constant |
| $c^m$ | Dilational wave velocity $= [(\lambda + 2\mu)/\rho]^{1/2}$ |

| | |
|---|---|
| $e$ | Exponential |
| $E$ | Modulus of elasticity (Young's modulus) |
| $k$ | Scale factor |
| $K, L, M$ | Constants |
| $m, n$ | Summation indices |
| $p$ | Subscript indicating peak value |
| $P(t)$ | Pressure pulse function |
| $P_o$ | Constant pressure |
| $r$ | Radial distance |
| $R$ | Dimensionless radial distance $= \omega_o(r - a)/c$ |
| $s$ | Laplace transform variable |
| $t$ | Time |
| $u$ | Radial displacement |
| $v$ | Radial particle velocity |
| $\alpha, \beta$ | Constants (pressure pulse parameters) |
| $\gamma$ | Constant $= 1 - \alpha/\omega_o$ |
| $\epsilon$ | Normal radial strain |
| $\lambda$ | Lame's constant (elastic) |
| $\mu$ | Elastic shear modulus |
| $\lambda', \mu'$ | Corresponding viscoelastic moduli |
| $\nu$ | Poisson's ratio |
| $\rho$ | Density |
| $\Sigma$ | Summation |
| $\sigma$ | Normal radial stress |
| $\tau$ | Time $= t - (r - a)/c$ |
| $\phi$ | Displacement potential ($u = \partial\phi/\partial r$) |
| $\omega$ | Circular frequency |
| $\omega_o$ | Transition frequency |

## 2.4 CYLINDRICAL WAVE EQUATION

The stresses acting on an infinitesimal element of a cylinder due to an internally applied pressure may be utilized to derive the equation of equilibrium for static conditions or the wave equation for dynamic loading inside of a radial cylindrical cavity. If the stresses are assumed to be symmetrical with respect to the axis, then the shear stresses on the element surfaces are zero and the tangential stresses are independent of the angle $\theta$.

The sum of the forces in the radial direction is

$$\left(\sigma_r + \frac{\partial\sigma_r}{\partial r}\, dr\right)(r + dr)\, d\theta - \sigma_r\, rd\theta - \sigma_\theta\, drd\theta = 0 \qquad (2.93)$$

for the static case, which also gives the symmetrical wave equation

$$\frac{\partial \sigma_r}{\partial \theta} + \frac{\sigma_r - \sigma_\theta}{r} = \rho \frac{\partial^2 u}{\partial t^2} \tag{2.94}$$

This equation is similar to that for a spherical wave for which the second term is multiplied by 2, which makes the spherical equation much easier to solve.

For cylindrical polar coordinates

$$\sigma_r = \lambda(\epsilon_r + \epsilon_\theta) + 2\mu\epsilon_r$$

$$= (\lambda + 2\mu) \frac{\partial u}{\partial r} + \lambda \frac{u}{r},$$

$$\sigma_\theta = \lambda(\epsilon_r + \epsilon_\theta) + 2\mu\epsilon_\theta$$

$$= (\lambda + 2\mu) \frac{u}{r} + \lambda \frac{\partial u}{\partial r} \tag{2.95}$$

and equation 2.94 becomes, in terms of displacement,

$$\frac{\partial^2 u}{\partial t^2} + \frac{1}{r} \frac{\partial u}{\partial r} - \frac{u}{r^2} = \frac{1}{c^2} \frac{\partial^2 u}{\partial t^2} \tag{2.96}$$

If a pressure is suddenly applied to the inside of a cylindrical cavity of radius $r_o$, then the wave parameters may be determined by LaPlace transforms. The boundary condition equation is, for a Dirac delta function,

$$\sigma_r = (\lambda + 2\mu) \frac{\partial u}{\partial r} + 2\mu \frac{u}{r} = -P_o \delta(t) \tag{2.97}$$

for $t > 0$, $r = r_o$.

The transform of the wave equation is

$$\frac{\partial^2 u}{\partial r^2} + \frac{1}{r} \frac{\partial u}{\partial r} - \left(\frac{1}{r^2} + \frac{s^2}{c^2}\right) u = 0 \tag{2.98}$$

This is a Bessel equation, the solution of which is

$$u = AK_1 \left(\frac{sr}{c}\right) + BI_1 \left(\frac{sr}{c}\right) \tag{2.99}$$

where $K_1$ and $I_1$ are modified Bessel functions of the second kind of order 1. For $u$ to be finite as $r \to \infty$, $B = 0$.

Application of the boundary condition gives

$$\sigma_r = (\lambda + 2\mu)\frac{s}{c} K_1'\left(\frac{sr}{c}\right) + \frac{2\mu}{r} K_1\left(\frac{sr}{c}\right)$$

$$= \frac{s}{c}(\lambda + 2\mu) K_o\left(\frac{sr}{c}\right) + \frac{2\mu c}{rs} K_1\left(\frac{sr}{c}\right), \qquad (2.100)$$

where

$$K_1'(z) = \frac{1}{z} K_1(z) - K_o(z)$$

at $r = r_o$ and $\sigma_r = -P_o \delta(t)$.

Hence,

$$A = \frac{P_o}{s(\lambda + 2\mu) K_o\left(\dfrac{sr_o}{c}\right) + \dfrac{2c}{r_o s} K_1\left(\dfrac{sr_o}{c}\right)} \qquad (2.101)$$

and

$$u(r, s) = \frac{PcK_1\left(\dfrac{sr}{c}\right)}{s\left[(\lambda + 2\mu) K_o\left(\dfrac{sr_o}{c}\right) + \dfrac{2c}{sr_o} K_1\left(\dfrac{sr_o}{c}\right)\right]} \qquad (2.102)$$

Selberg (1951) and Azo (1966) derived a similar equation for the displacement potential $\phi$, where $u = (\partial\phi)/(\partial r)$ for a unit step function for which

$$\phi(r, s) = \frac{P_o c^2 K_o\left(\dfrac{sr}{c}\right)}{s^3\left[(\lambda + 2\mu) K_o\left(\dfrac{sr}{c}\right) + \dfrac{2c}{sr_o} K_1\left(\dfrac{sr_o}{c}\right)\right]} \qquad (2.103)$$

from which they obtained the transform for the stress $\sigma_r$. The inverse was obtained by integration in the complex plane. Selberg and Azo both used an asymptotic expansion of a Bessel function, Selberg using one complex root and Azo (1966) utilizing two complex roots. The latter results comply with the requirement that complex roots occur in conjugate pairs. However, integration of the $f(s)$, which involves this type of Bessel function, is complicated and tedious.

Numerical methods for the inversion of LaPlace transforms have been devised and used successfully for inversion of complicated functions for which no transform pairs are tabulated or are not easily derived.

One such method is by means of series expansion, somewhat similar to that of Heaviside. That is, the denominator of $f(s)$ can be expanded in a series of the type $\Sigma A_n/s^n$ which can then be inverted term by term, or the term $1/s$ is equivalent to an integration in the $t$ plane and $(1/s)^n$ is equivalent to $n$ integrations, in both cases the constants of integration being zero.

The displacement given by equation 2.102 is for the response to a unit impulse $P_o\delta(t)$. The responses to other functions can be obtained by multiplying the transform for $u(r, s)$ by the transform of the desired pressure function. The operations to determine stress, strains, and particle velocity can be carried out either in the $s$ plane or the $t$ plane.

It is more convenient to work with the Bessel function of equation 2.102 by placing it in a different form. Writing the transform in terms of its $t$ plane equivalent

$$\mathcal{L}\left\{u(r, t)\right\} = \frac{PcK_1\left(\dfrac{sr}{c}\right)}{s(\lambda + 2\mu) K_o\left(\dfrac{sr_o}{c}\right) \dfrac{2\mu c}{sr_o} K_1\left(\dfrac{sr_o}{c}\right)} \tag{2.104}$$

The scaling property of transforms may be employed to advantage. That is, if

$$\mathcal{L}\left\{F(t)\right\} = f(s) \tag{2.105}$$

then

$$\mathcal{L}\left\{F(at)\right\} = \frac{1}{a}f\left(\frac{s}{a}\right) \tag{2.106}$$

In the above equation let $a = r_o/c$.

Also, if $\lambda = \mu$ and the numerator and denominator are multiplied by $e^s$, equation 2.104 then becomes

$$\mathcal{L}\left\{u\left(r, \frac{r_o t}{c}\right)\right\} = \frac{P_o cK_1\left(\dfrac{sr}{r_o}\right)}{3\mu\left[sK_o(s) + \dfrac{2}{3} K_1(s)\right]} \tag{2.107}$$

or

$$\frac{3\mu}{P_o c} \mathcal{L} \left\{ u\left(r, \frac{r_o t}{c}\right) \right\} = \frac{e^s K_1 \left(\dfrac{sr}{r_o}\right)}{\left[ s e^s K_o(s) + \dfrac{2}{3} e^s K_1(s) \right]} \tag{2.108}$$

Multiplying $f(s)$ by $e^{-s}$ gives

$$\frac{3\mu}{P_o c} \mathcal{L} \left\{ u\left(r, \frac{r_o(t-1)}{c}\right) \right\} = \frac{K_1 \left(\dfrac{sr}{r_o}\right)}{\left[ s e^s K_o(s) + \dfrac{2}{3} e^s K_1(s) \right]} \tag{2.109}$$

The factor $1/[s e^s K_o(s) + 2/3 e^s K_1(s)]$ may be expanded in the form $f(s) = \Sigma((A_n)/(s_n))$ where the $A_n$ is determined by curve fitting and $s$ is treated as a real variable. A method for more accurate inversion was developed by Daneshy (1969).

## 2.5  STRAIN ENERGY IN SPHERICAL ELASTIC WAVES

The strain energy transferred to the rock by confined explosives is one measure of blasting effectiveness. A comparison of the measured elastic strain energy with the energy given off by the explosive indicates that only 5 to 9% of the available explosive energy is transmitted as elastic strain energy with approximately the same amount as kinetic wave energy. However, in the field experimentation by Fogelson et al. (1959), virtually all of the usable strain measurements were at scaled distances of five or greater in a granite gneiss. Most rock crater breakage with spherical charges occurs within a scaled distance of three. Typical strain gage readings at small scaled distances in granite showed permanent deformation (Duvall and Atchison, 1957).

Langefors and Kihlstrom (1978) also estimate that only about 3% of the explosive energy is converted into elastic wave energy. However, if one considers the fact that explosives must be loaded in concentrated columns, and not evenly throughout the rock mass or in an ''ideal'' manner, and that there is crushing and plastic deformation, the energy efficiency of explosive breakage is much higher than that indicated by the elastic wave energy alone.

Hence, it is assumed that for most blasting the rock immediately around the cavity is crushed, changing to a zone that is plastic–elastic, and thence to a zone that is elastic in its response. The energy utilization for the first two of the three zones has not been determined theoretically or experimentally. Both theory and experimental results are available for the zone of elastic behavior (Fogelson et al., 1959).

For a spherical wave that is symmetrical with respect to the cavity, the elastic strain energy density $E_d$ is given in terms of the three components of strain by

$$E_d(t, r) = \frac{\lambda}{2} (\epsilon_r + \epsilon_\theta + \epsilon_\phi) + \mu(\epsilon_r^2 + \epsilon_\theta^2 + \epsilon_\phi^2) \qquad (2.107)$$

Equation 2.107 can be rewritten as follows, because $\epsilon_\theta = \epsilon_\phi$, and assuming $\lambda = \mu$,

$$E_d(t, r) = \lambda \left( \frac{3}{2} \epsilon_r^2 + 2\epsilon_r\epsilon_\theta + 4\epsilon_\theta^2 \right) \qquad (2.108)$$

For a representative type of pressure pulse, the values of $\epsilon_r$ and $\epsilon_\theta$ can be readily calculated (Duvall, 1953), and the plotted values of the terms $3/2\epsilon_r^2$ and $(3/2\epsilon_r^2 + 2\epsilon_r\epsilon_\theta + 4\epsilon_\theta^2)$ show that the areas under the curves for these two functions differ by only 3% (Figure 2.15). Hence, values of the radial strain may be employed to approximate the total strain energy at distances greater than five radii. Therefore,

$$E_d = \frac{\lambda + 2\mu}{2} \epsilon_r^2 \qquad (2.109)$$

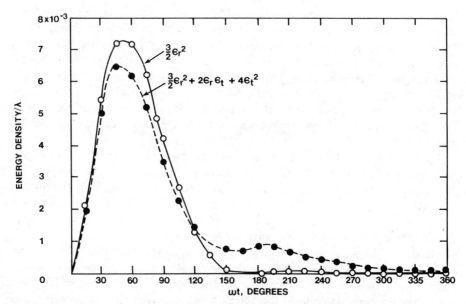

**Figure 2.15**  Comparison between radial strain energy density and total potential energy density (Fogelson et al., 1959).

where the radial strain energy travels outward at a velocity

$$c = \left( \frac{\lambda + 2\mu}{\rho} \right)^{1/2} \tag{2.110}$$

The radial strain energy passing through a unit area of spherical surface per unit time is $cE_d dt$, and the amount for a pulse is

$$E_d = \int_0^t cE_d dt = \frac{\rho c^3}{2} \int_0^t \epsilon_r^2 \, dt \tag{2.111}$$

and that through a spherical shell is

$$E_s = 4\pi R^2 E_a = 2\pi R^2 \rho c^3 \int_0^t \epsilon_r^2 \, dt \tag{2.112}$$

The total energy in the wave is the sum of the strain energy plus the kinetic energy due to particle motion. For a plane wave, the kinetic and strain energies are equal, and, if one assumes the same relation for a spherical wave, the total energy is equal to twice the strain energy.

A linear array of strain gages was installed in boreholes in granite gneiss so that the radial strain from confined charges of four types of explosives could be recorded. The strain pulses were plotted with the values of strain squared and the strain energy $E_s$ taken as a constant times the area under the curve. The available explosive energy was approximated utilizing a simplified method, and the total elastic wave energy was divided by the explosive energy. The results indicated that 5–9% of the explosive energy was transmitted through the plastic zone to be recorded as elastic strain energy, or 10–18% as total elastic energy.

It should be noted that in earlier work (Duvall and Atchison, 1957), strain-time recordings within small-scale distances in four rock types, including granite, showed that permanent deformation took place. This indicates that the rock close to the explosive acted in an elastic–plastic manner. Plastic energy is not recoverable, and, hence, the energy loss in the plastic zone, depending on the ratio of plastic to elastic strain, could be equal to or greater than the elastic energy in addition to the energy lost in crushing.

# 3

# PERCUSSIVE DRILLING— CHARACTERIZATION AND EARLY RESEARCH

The first important automated rock drill for small hole drilling for blasting was powered by a pneumatic agent such as steam or air, and, although hydraulic drills (driven by high-pressure oil) are gaining in usage, the pneumatic drill will probably continue to be widely used. Compressed air is supplied to the drill usually at 90–100 psi, which is a relatively efficient pressure level. Higher pressures are used in some drilling operations, although experience with drilling in many types of rocks and an analysis of available data indicate that higher penetration rates and lower specific drilling energies may be obtainable at pressures higher than 100 psi, but high operating pressures cause more wear of equipment and consequent repairs and drill down time. Although drill machines, drill steel, and drill bits have been markedly improved in recent decades, the most effective range of pressure has been found to be about 100 psi for overall effectiveness, including operation, maintenance, and repair.

## 3.1  OPERATING PRINCIPLES

An analysis of the ideal basic physical–mechanical principles of the operation of pneumatic percussive drills and recommended design factors for down-the-hole

**46**

drills was presented by Pfleider and Lacabanne (1961). The operational parameters of a pneumatic piston tool may be expressed mathematically and analyzed for optimum conditions using simplifying assumptions. The derived equations give good approximations to operating conditions and are useful for the evaluation of the performance characteristics of a percussive drill for both design purposes and test evaluation.

Two operational mechanical parts of a pneumatic piston drill that determine its output are the (1) cylinder and (2) piston, whose dimensions are (3) piston working face area, $A$, (4) piston stroke, $S$, and (5) piston weight, $W$ (Figure 3.1). The piston is accelerated during the power stroke by air pressure on the back side of the piston and is returned by pressure on the face, the air being regulated by automatic valves. The work output is a function of all the factors listed previously plus the inlet air pressure and the back pressure at the discharge valve.

The performance equations are developed using the assumptions that (1) the air pressure during power stroke is uniform over the piston face, (2) the pressure has a mean, effective value, $P$, for the full stroke, and (3) the acceleration of the piston during the power stroke is linear. The pertinent operating factors and equations (Table 3.1) show the interrelationships of all of the operating parameters and their

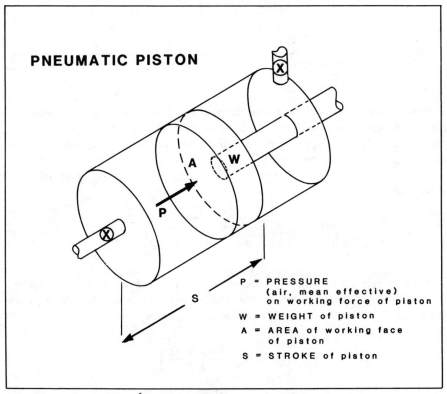

**PNEUMATIC PISTON**

P = PRESSURE
(air, mean effective)
on working force of piston

W = WEIGHT of piston

A = AREA of working face
of piston

S = STROKE of piston

**Figure 3.1** Pneumatic drill piston (Pfleider and Lacabanne, 1961.)

**TABLE 3.1. Ideal Performance Equations**

| Factor | Equation | | Eq. No. |
|---|---|---|---|
| Force on piston power stroke | $F_p = pA$ | lb | (3.1) |
| Work | $E_1 = \dfrac{pAS}{12}$ | ft-lb | (3.2) |
| Acceleration | $a = \dfrac{pAg}{W}$ | ft/s$^2$ | (3.3) |
| Time | $t_1 = \left(\dfrac{WS}{6pAg}\right)^{1/2}$ | s | (3.4) |
| Terminal velocity | $V_s = \left(\dfrac{pASg}{6W}\right)^{1/2}$ | ft/s | (3.5) |
| Time, forward and return | $t = (1 + k_1)\left(\dfrac{WS}{6pAg}\right)^{1/2}$ | s | (3.6) |
| Blows/min | $f = \dfrac{60}{1 + k_1}\left(\dfrac{6pAg}{WS}\right)^{1/2}$ | | (3.7) |
| Work/min (power) | $E_r = \dfrac{\text{Work}}{\text{Blow}} \times \dfrac{\text{Blows}}{\text{Min}}$ | | |
| | $E_r \, \alpha \, (pAS)\left(\dfrac{pA}{WS}\right)^{1/2}$ | | |
| | $E_r \, \alpha \, \dfrac{p^{3/2}A^{3/2}S^{1/2}}{W^{1/2}}$ | ft-lb min | (3.8) |

*Source:* Pfleider and Lacabanne (1961).

effect on drilling. The work output of the piston is thus a function of air pressure, piston area, stroke, and weight. For a given drill, the piston energy per stroke increases directly with the pressure, and the number of blows per minute increases as the square root of the pressure.

Hustrulid (1971b) and others have used an additional factor, $k_2$, in equation 3.7 for the return stroke:

$$f = \frac{60}{(1 + k_1 + k_2)}\left(\frac{C_{(e,m)}pAg}{WS}\right)^{1/2} \quad (C_e = 6; C_m = 1/2) \quad (3.9)$$

(constants in English and metric systems), where $k_1$ and $k_2$ represent time factors for the return stroke and piston at rest. If the time for the forward stroke, back stroke, and time at rest are equal, equation 3.9 becomes

$$(\overline{f}) \to \overline{f} = K_o\left(\frac{C_{(e,m)}pAg}{WS}\right)^{1/2},$$

$$20 \leq K_o \leq 30 \quad (3.10)$$

The velocity is usually less than ideal, so equation 3.5 may be written

$$V_s = \beta_o \left( \frac{pASg}{C_{(e,m)}W} \right), \tag{3.11}$$

where $\beta_o$ is an operating factor less than 1 for a given set of drilling conditions.

It has been shown by Hustrulid (1968) that the minimum force $F_1$ required to keep the bit in contact with the rock is

$$F_1 = \frac{\overline{f}}{30} (1 + \beta) \int_0^\tau \sigma_i \, dt \tag{3.12}$$

where $\sigma_i$ = incident stress at the rock–bit interface as a function of time (psi), $\tau$ = duration of stress pulse against the rock (s), and $\beta$ = coefficient of momentum transfer from drill steel to piston, usually $0 < \beta < 0.2$.

Equation 3.12 may be simplified to give

$$F_t = \frac{\overline{f}}{30} (1 + \beta) \frac{WV_s}{g} \tag{3.13}$$

which is combined with equations 3.10 and 3.5 to yield

$$F_t = \frac{K_o}{30} \beta_o (1 + \beta) \frac{WV_s}{g} \tag{3.14}$$

One method (Cook et al., 1968) of predicting the penetration rate for a given machine–bit–rock combination is by use of the equation

$$PR = \frac{C_{(e,m)} E_r \times \overline{f} \times T_R}{A_e \times E_v}, \tag{3.15}$$

where $C_e = 12$, $C_m = 1$, $T_R$ = coefficient of energy transfer from drill bit to rock, $A_H$ = cross-sectional area of drill, and $E_v$ = specific energy of rock breakage for a given rock–bit combination.

For calculations, the following approximate values have been employed: $T_R = 0.8$ and $E_v = C_o$ (compressive strength of rock). Substitution for $\overline{f}$ and $E_r$ gives

$$PR = \frac{K_o T_R}{A_H E_v} \left[ \frac{C_{(e,m)} Sg}{W} (PA)^2 \right]^{1/2} \beta_o^2 \tag{3.16}$$

($C_e = 6$; $C_m = \frac{1}{2}$).

Equations 3.14–3.16 can be used to calculate an approximate relationship among thrust, air pressure, rock properties, and the drilling rate.

## 3.2   INDEXING

For efficient drilling with single- or multiple-type cutting wedges, the bit must rotate as each blow is applied to the drill rod. This is illustrated by the hand-drilling method used before drilling machines were invented. That is, a single-wedge short drill steel was struck with a hand-held hammer, the steel being rotated part of a revolution for each blow. If the bit is not rotated, a groove is broken out of the rock, and chipping and penetration cease after a few blows. Rotating of the bit presents a new surface to the bit for each blow, causing chipping, crushing, and consequent penetration.

Early indexing tests were conducted using parallel alignment of single-edge bits (Simon, 1956). A bit, fastened to a heavy weight, was dropped on the smooth surface of a rock specimen. When the bit wedge was dropped at two positions close together, the material between was broken out. As the distance between impact points was increased beyond a critical value, the volume of rock broken out between the two grooves decreased rapidly (Figure 3.2). The tests with parallel indexing serve to illustrate the principle of indexing that is applicable to several types of mechanical rock breakage. That is, there are limiting geometrical spacings related to energy, force, stress, impulse, method of energy application, and other factors where effective breakage can occur. For example, a linear geometry analogous to the above is formed in the effective cutting and breakage between grooves in tunnel faces by rolling cutters on the head of a tunnel-boring machine.

For the angular indexing that occurs in percussive and rotary-percussive drilling in a borehole, it has been found (Hartman, 1966) that the angle of indexing between each blow of a percussive hammer is not critical. The geometry of parallel indexing does not simulate that of angular indexing closely enough for the results to be directly comparable.

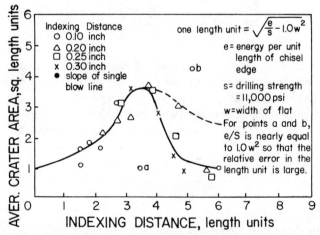

**Figure 3.2**   Linear indexing for the 0.03-in.-wide die, based on experimental data. Dashed line correction for down-hole statistics (Simon, 1956).

Most of the analyses of breakage of rock by bits and other mechanical agents have been based on static or quasi-static models of stress distribution, with certain assumptions of types or modes of failure, and these have proved to be of use in the description, understanding, and prediction or drillability of rock. But a dynamic analysis that includes the processes of fracture initiation and propagation, plus the effects of intense stress pulses, energy levels, and effects of rock defects, would appear to lend itself to a better, more fundamental scientific understanding of rock fragmentation processes in drilling.

Both pneumatic and more recently developed hydraulic drills strike a large number of blows per minute. As each impact of the bit chips out rock, this process will leave radial ridges and grooves in the bottom of the drill hole. For a single edge bit or one with edges crossed at 90°, it would appear advisable to index at an odd fraction of the 90° angle. Laboratory tests (Hartman, 1959) indicate that angles of 30–45° provide near-optimum indexing. However, no quantitative field tests appear to have been made to verify this laboratory finding. In an operating drill, it is probable that, although the rotating mechanism provides for a relatively constant revolution of the steel for each blow, there are many factors that could cause the indexing angle to vary, such as resistance to rotation, effects of reflected stress waves in the rod, inertia of rotating parts, rock defects, and the like.

When these and related factors are considered, it is logical that there should be a range of indexing angles at which the cutting edges of the bit will impact on ridges or new surfaces in the bottom of the drill hole at least as often as they

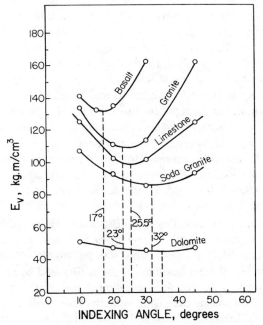

**Figure 3.3**  $E_v$ versus indexing angle (Paithankar and Misra, 1980).

impact in grooves, although the grooves are usually shallow enough that a second impact on a groove will chip out additional rock. The same holds true for a disc cutter on a tunnel-boring machine. That some cross bits are made with the wedges at angles other than 90° indicates designs that are used to reduce the probability of repeated blows in grooves in the bottom of the drill holes.

Paithankar and Misra (1980) found, as did U.S. Bureau of Mines investigators, that percussive drillability of rock from small-scale laboratory tests does not correlate well with the measured "standard" physical properties. They performed full-scale drillability tests in five different rocks—basalt, granite, soda granite, limestone, and dolomite—and compared these rates with those from laboratory tests with a microbit. The apparatus consisted of a WC microbit (110° wedge angle), 10 mm diameter, impacted by a drop weight, giving 0.14 kg-m energy. It was assumed that (approximately) all of the impact energy was transmitted to the rock. Cuttings were removed after each blow to avoid regrinding, and the volume of cuttings was measured to determine the specific energy $E_v$. The specific energy of drilling, measured at the rock–bit interface, was found to be a function of rock properties and indexing angle (Figure 3.3).

## 3.3   CHARACTERIZATION TESTS

Definitive laboratory and field tests were made by the Bureau of Mines with two sizes of mounted pneumatic–percussive drills to analyse (1) drill operating characteristics (Bruce and Paone, 1969), (2) laboratory drilling characteristics in rock (Paone et al., 1969), (3) longitudinal strain waves in drill steel (Lundquist and Anderson, 1969), (4) statistical regression analysis of penetration rates and rock properties (Selim and Bruce, 1970), (5) field drilling characteristics and correlation with Coefficient of Rock Strength (CRS) (Schmidt, 1972), and (6) further studies of the CRS and penetration rates (Tandanand and Unger, 1975). Similar field tests were made underground in quartzite by Hustrulid (1971b) in which the penetration rates of three drills were measured as a function of air pressure, thrust, and bit type. In the Bureau of Mines tests, the rate of penetration was correlated with the Coefficient of Rock Strength, whereas Hustrulid used an assumed value of coefficient of energy transfer from the steel to the rock (equation 3.16). Both gave approximate correlations, but neither gave accurate results for rocks over a wide range of properties.

The results of drill characteristics tests (Bruce and Paone, 1969) are largely in agreement with the results of other investigations where the same factors were experimentally evaluated. The experiments with pneumatic drill mechanisms were designed in part to test the applicability of the theory and equations described in Table 3.1. The laboratory tests on drill operating characteristics by Bruce and Paone (1969) consisted of measuring the piston velocity under drilling conditions, which is very difficult to do. Also, the thrust and blow frequency were measured, and the piston energy and travel time were calculated. Similar measurements were made by Hustrulid (1965), Ditson (1948), Cheetham and Inett (1953–55), Wells (1949), and Ryd and Holdo (1953). Some drill operating characteristics are not predicted by equations 3.1–3.8, such as optimum values of thrust and operating pressure.

The tests by Bruce and Paone were performed with two sizes of drills (Table 3.2) to determine piston velocity and energy, blow frequency, work rate, and effective stroke length. Operating air pressure, air consumption, speed of rotation, and blow frequency were measured without the drill rod in the chuck. The ratio of striking velocity of maximum piston velocity ranged from 89 to 94% for drill A and from 88 to 99% for drill B. The maximum piston velocity decreases with feed pressure (FP) for operating pressures ranging from 50 to 100 psi for both drills (Figure 3.4), while the blow frequency in general increases with the feed pressure (Figure 3.5). Thus, the piston energy also decreases with feed pressure but increases with operating pressure. That is, from equations 3.1–3.8, the piston energy (kinetic) is $E = WV_s^2/2g$. The relation of piston velocity to operating pressure at a given level of feed pressure is given by equation 3.5 as $V_s \propto (pAS/W)^{1/2}$. Since $A$, $W$, and $p$ are fixed, and it is assumed that the operating air pressure is proportional to a "mean effective air pressure," the equation reduces to $V_s \propto (S)^{1/2}$. The stroke length decreases as feed pressure increases (Figure 3.6). Also as the feed pressure increases, the maximum piston velocity decreases for constant operating pressure (Figure 3.4).

The blow frequency $f$ is given by equation 3.7 in terms of operating pressure and drill parameters, $f \propto (pA/WS)^{1/2}$; however, an increase of feed pressure (FP) (Figure 3.7) causes the stroke $S$ to decrease and the blow frequency to increase (Figure 3.5), and some of the curves display an unexplained anomalous low point.

The stroke data (Figure 3.7) show that an increase in feed pressure shortens the

**TABLE 3.2. Characteristics of Drills A and B**

|  | Drill A | Drill B |
|---|---|---|
| Twist on rifle bar | 1:30 | 1:38 |
| Piston weight[a] (lb) | 4.44 | 10.19 |
| Piston weight[a] (g) | 2015 | 4622 |
| Nominal stroke (in.) | 2.625 (6.67 cm) | 3500 (8.89 cm) |
| Bore diameter (in.) | 2.625 (6.67 cm) | 3500 (8.89 cm) |
| Flushing | Air | Air |
| Bit diameter (in.) | $1\frac{1}{2}$ (3.81 cm) | $1\frac{1}{2}$ (3.81 cm) |
| Bit type | Timken MCB, H-thread carbide | Timken MCB, H-thread carbide |
| Chuck size | $\frac{7}{8}$-in. hex, $4\frac{1}{4}$-in. shank thrust = 4.4 × press (2.22 cm hex; 10.80 cm shank) | $1\frac{1}{4}$-in. round lugged, thrust = 10 × press (3.18 cm round lugged) |

[a]Includes piston, rifle nut, and linkages.
*Source:* Bruce and Paone (1969).

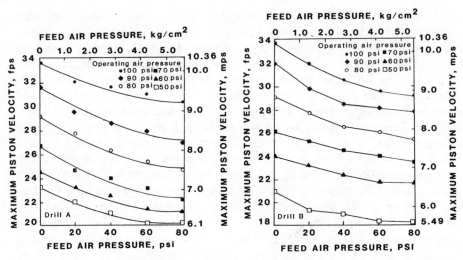

**Figure 3.4** Maximum piston velocity versus feed air pressure for operating air pressures (Bruce and Paone, 1969).

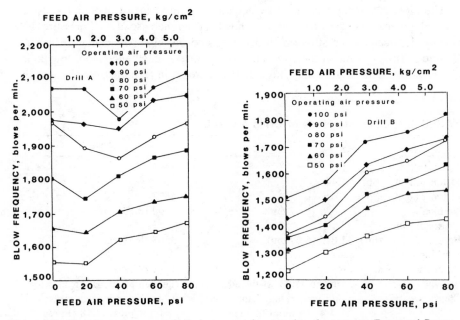

**Figure 3.5** Blow frequency versus feed air pressure for operating air pressures (Bruce and Paone, 1969).

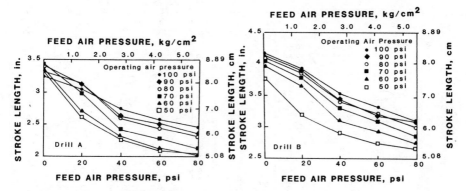

**Figure 3.6**  Stroke length versus feed air pressure for operating air pressures (Bruce and Paone, 1969).

stroke length, but the relation is not linear. The stroke length is a characteristic of a given drill and has a physical maximum because of the drill cylinder geometry, and the work rate increases somewhat with operating pressure (Figure 3.8). The relation of the maximum piston velocity to the operating pressure (Figure 3.9) for constant feed pressure is $V_s = K(pS)^{1/2}$, the constant $K$ being evaluated from measured values of $V_s$, $p$, and $S$. The relation of idling blow frequency to operating

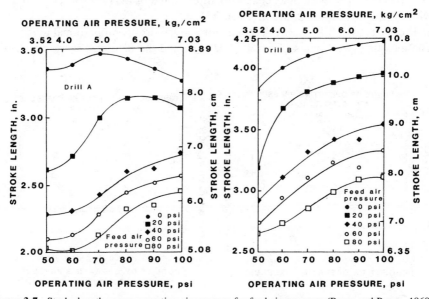

**Figure 3.7**  Stroke length versus operating air pressure for feed air pressures (Bruce and Paone, 1969).

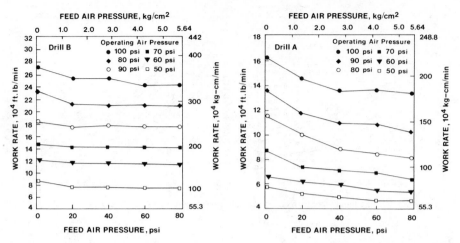

**Figure 3.8**   Work rate versus feed air pressure for operating air pressures (Bruce and Paone, 1969).

air pressure (Figure 3.10) is similar to that observed by Cheetham and Inett (1953–55) and is given by $f \propto p^{1/2}$.

The air consumption (Figure 3.11) for idling for both drills is a linear function of the operating air pressure, and the rotational speed decreases nonlinearly with increasing air pressure (Figure 3.12). Neither of the drills used in this series of tests was equipped with independent rotation. Some of the characteristics of a drill with independent rotation are given on $p$.

From equation 3.8, the work rate $E_r$ for a given drill is proportional to $p^{3/2}S^{1/2}$. The measured work rates ($E_r = 1/2\ MV_S^2 f$) and the predicted work rate $E$ were

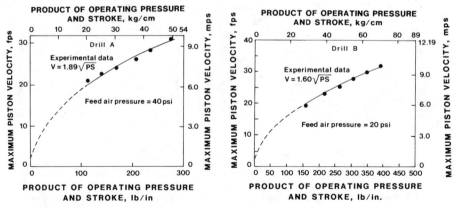

**Figure 3.9**   Maximum piston velocity versus product of operating air pressure and stroke length (Bruce and Paone, 1969).

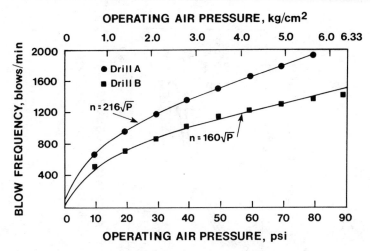

**Figure 3.10** Blow frequency versus operating air pressure for operation with no drill rod (zero thrust) (Bruce and Paone, 1969).

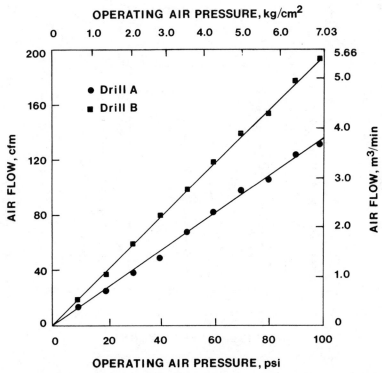

**Figure 3.11** Rotational speed versus operating air pressure for operation with no drill rod (zero thrust) (Bruce and Paone, 1969).

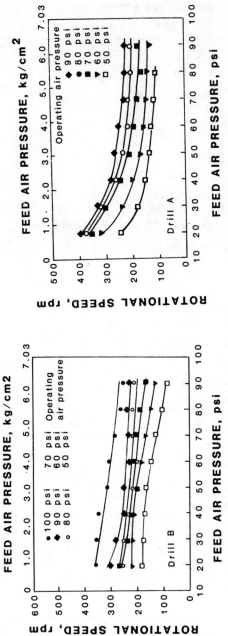

Figure 3.12 Rotational speed versus feed air pressure for operating air pressures (Bruce and Paone, 1969).

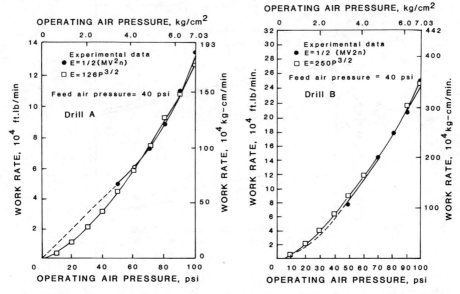

**Figure 3.13**  Relationship between work rates calculated by two methods (Bruce and Paone, 1969).

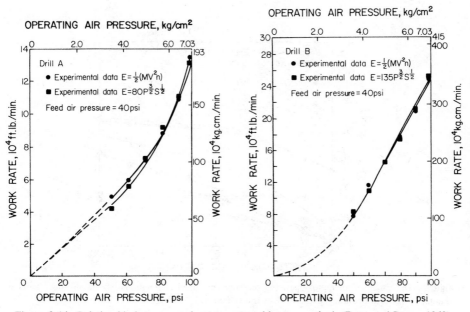

**Figure 3.14**  Relationship between work rates computed by two methods (Bruce and Paone, 1969).

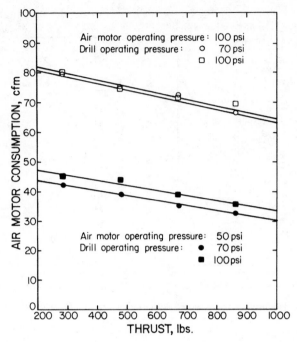

**Figure 3.15** Effect of thrust, air motor pressure, and drill operating pressure on volume of air used by air motor, drilling Charcoal granite (Unger and Fumanti, 1972).

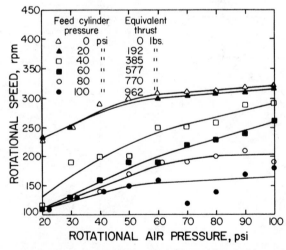

**Figure 3.16** Effect of air motor operating pressure and thrust on rotational speed, drilling Charcoal granite, 60-psi drill operating pressure (Unger and Fumanti, 1972).

also plotted against that based on $Kp^{1/2}S^{1/2}$ (Figures 3.13, 3.14), which shows a close agreement between the two.

Wells (1950) developed the following equation:

$$E_t = CW(Sf)^2 \tag{3.17}$$

where $C$ = constant. The striking energy computed from Well's equation as a function of operating pressure compares well with the observed energy at fixed feed pressures (Figure 3.15).

Ditson (1948) developed a similar equation:

$$E_r = \frac{(W \times 10^{-6})(Sf)^2}{2.21} \tag{3.18}$$

Measured values for maximum piston energy, those obtained by substituting the U.S. Bureau of Mines data into Well's and Ditson's equations (Figures 3.16, 3.17), and energy values reported by Hustrulid (1965) measured from the strain wave in the drill steel indicate that Hustrulid's energy values are lower than those obtained by the Bureau. Ditson's equation (3.6) gives somewhat more conservative energy values than those found by Bruce and Paone (1969) above 20- to 30-psi feed pressure, but higher values for lower than 20-psi feed pressure for drill A. For drill B, however, the values of Ditson's equation are higher than the Bureau's.

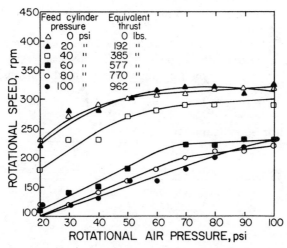

**Figure 3.17** Effect of air motor operating pressure and thrust on rotational speed, drilling Charcoal granite, 90-psi drill operating pressure (Unger and Fumanti, 1972).

**TABLE 3.3. Drill Specifications**

| Item | Units | Dimensions |
|---|---|---|
| Bore diameter | in. | $3\frac{1}{2}$ |
| Piston weight | lb | $9\frac{2}{3}$ |
| Nominal stroke | in. | $3\frac{1}{2}$ |
| Nominal blow frequency | blows/min | 1940 |
| Nominal striking energy | ft-lb/blow | 117 |
| Work rate[a] | ft-lb/min | 227,000 |
| Chuck size | ft-lb/min | 1-in. hex, 6-in. shank |

[a]Statistics noted are for operating pressure of 100 psi.
*Source:* Unger and Fumanti (1972).

## 3.4  PNEUMATIC DRILLS—INDEPENDENT ROTATION

The U.S. Bureau of Mines (Unger and Fumanti, 1972) conducted experiments in the laboratory with a percussive drill on eight rock types having a wide range of physical properties. The drill had independent rotation and was operated with selected combinations of rotational speed, thrust, and operating pressure. As with conventional percussive drills, penetration rates and energy-per-unit volume of drilling correlated well with the Coefficient of Rock Strength for the rocks tested. Size distribution analyses of some of the drill cuttings showed that the particle size varies with rotational speed and "rock hardness." The rotational speed increases the penetration rate in softer rocks. In the following section, tests are described of an independently rotated percussive drill used for drilling small-diameter blast holes.

Testing was done in the laboratory under controlled conditions with short holes, but the results can be adapted to field conditions. The physical property measurements were those of specimens taken from rock blocks after they had been drilled.

**TABLE 3.4. Air Motor Specifications**

| Rotational speed, rpm | Torque, ft-lb | Power output, $10^3$ ft-lb/min |
|---|---|---|
| 250 | 6.0 | 11 |
| 200 | 13.5 | 20 |
| 150 | 23.0 | 25 |
| 100 | 30.0 | 22 |
| 50 | 36.5 | 14 |
| Stall | 55.8 | 0 |

[a]Values shown are for 100 psi operating pressure with a 2-in. diameter root.
*Source:* Unger and Fumanti (1972)[a].

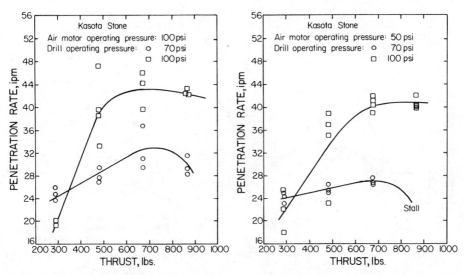

**Figure 3.18**  Penetration rate as a function of thrust and operating pressure in Kasota stone with air motor operating pressures of 50 and 100 psi (Unger and Fumanti, 1972).

The drilling was done with production-type percussive drill with an air motor for independent rotation. Manufacturer's specifications for the drill motor are given in Tables 3.3 and 3.4.

The consumption of air was 177 and 250 cfm for 70- and 100-psi operating pressure, respectively (Figure 3.18). Air motor pressures of 50 and 100 psi were

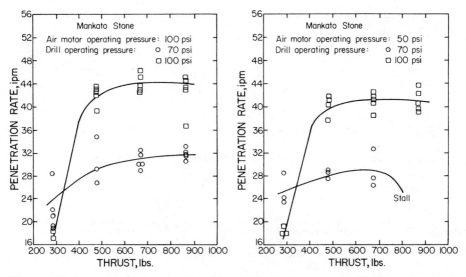

**Figure 3.19**  Penetration rate as a function of thrust and operating pressure in Mankato stone with air motor operating pressures of 50 and 100 psi (Unger and Fumanti, 1972).

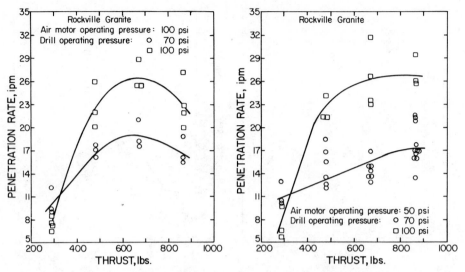

**Figure 3.20** Penetration rate as a function of thrust and operating pressure in Rockville granite with air motor operating pressures of 50 and 100 psi (Unger and Fumanti, 1972).

used for the drillability studies. There is a slight increase in air consumption, since this operating pressure is increased from 70 to 100 psi at each thrust (Figure 3.18), probably owing to a slight increase in rotational speed of the air motor and faster penetration at 100 psi. The rotation speed increases with both rotation air pressure and thrust (Figures 3.19, 3.20). For a constant rotation speed to be maintained,

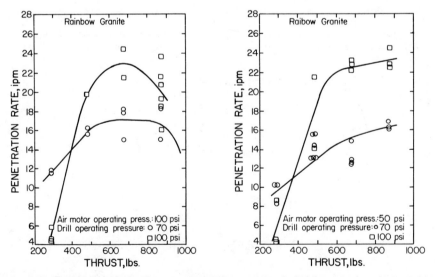

**Figure 3.21** Penetration rate for thrusts and operating pressures in Rainbow granite, air motor operating pressures of 50 and 100 psi (Unger and Fumanti, 1972).

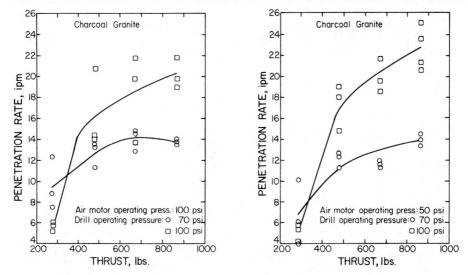

**Figure 3.22** Penetration rate for thrusts and operating pressures in Charcoal granite, air motor operating pressures of 50 and 100 psi (Unger and Fumanti, 1972).

the pressure to the rotational motor must be increased with the thrust. Rotational speed and blows per minute were measured with appropriate electronic equipment.

The drill was mounted horizontally on a steel framework to drill rock specimens (about $3\frac{1}{2}$-ft cubes). Each hole was collared, the operating air motor and feed cylinder pressures were set at predetermined values, and periodic readings were

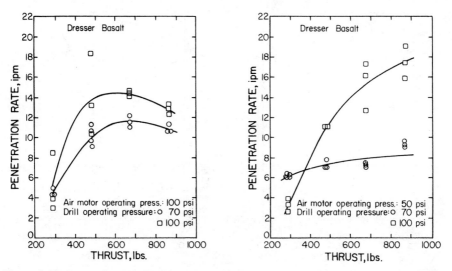

**Figure 3.23** Penetration rate for thrusts and operating pressures in Dresser basalt with air motor operating pressures of 50 and 100 psi (Unger and Fumanti, 1972).

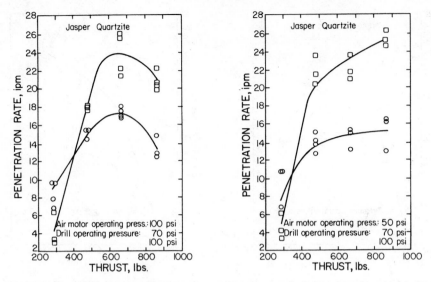

**Figure 3.24** Penetration rate for thrusts and operating pressures in Jasper quartzite with air motor operating pressures of 50 and 100 psi (Unger and Fumanti, 1972).

taken to check rotational speed and blow rate. Drilling time and the distance drilled were then recorded.

Drilling was done with a 1-in. hexagonal, 4-ft-long collared shank steel with an $\frac{11}{32}$-in. flushing hole. Drill bits were 2-in.-diameter, four-wing cross bits with tungsten carbide inserts, which were either new or reground to like-new conditions.

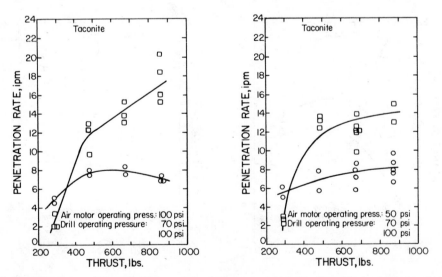

**Figure 3.25** Penetration rate for thrusts and operating pressures in taconite with air motor operating pressures of 50 and 100 psi (Unger and Fumanti, 1972).

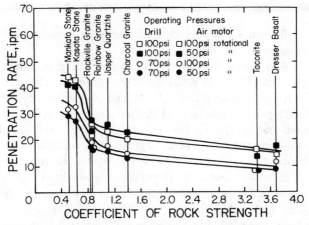

**Figure 3.26** Relation between penetration rate and coefficient of rock strength for variable rpm and drill operating pressures (Unger and Fumanti, 1972).

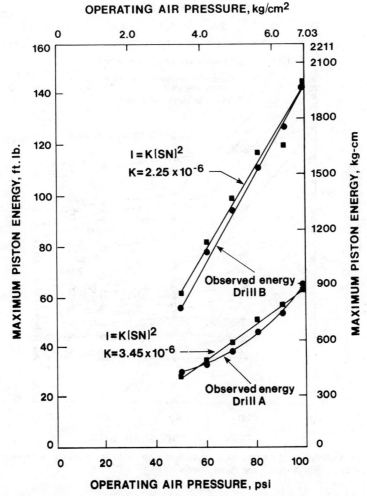

**Figure 3.27** Maximum piston energy versus operating air pressure for observed energy and energy predicted using Wells' equation (Bruce and Paone, 1969).

The holes were flushed with air, with a vacuum system being used to collect the cuttings and to suppress the dust.

A 2 × 2 × 4 factorial experiment with two levels of operating air pressures (70 and 100 psi), two levels of rotational motor pressures (50 and 100 psi), and four feed cylinder air pressures (30, 50, 70, and 90 psi) was performed for each of the eight rock types—Kasota stone, Mankato stone, Rockville granite, Rainbow granite, Charcoal granite, Dresser basalt, Jasper quartzite, and taconite. Each drilling condition was repeated at least twice to give a more accurate average penetration rate. The distance drilled in each individual test was the travel within the drill frame or about $1\frac{1}{2}$ ft. The results are presented graphically in Figures 3.21–3.29.

For some rocks at low-thrust values and 70-psi drill operating pressure, penetration rates were higher than the rates at 100 psi. This reverse trend, similar to percussive drilling, is probably due to inadequate thrust, which allowed bouncing

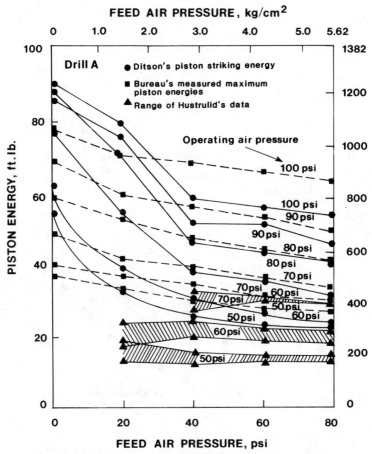

**Figure 3.28** Comparison of piston energy versus feed air pressure for operating air pressures by three different methods, drill A (Bruce and Paone, 1969).

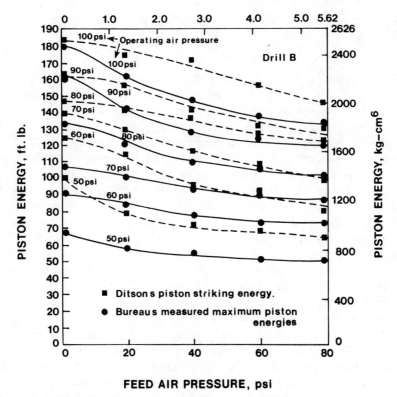

**Figure 3.29**  Comparison of piston energy versus feed air pressure for operating pressures by two different methods, drill B (Bruce and Paone, 1969).

of the bit and rod, which in turn prevented effective coupling between the drill and rock. These low-thrust values are not in the usual operating range but are given to complete the curves. A thrust of 600–800 lbs is required to obtain high penetration rates in most of the rocks tested.

## 3.5  STRAIN ENERGY—DRILL RODS

In specialized studies of percussive drilling (Lundquist and Anderson, 1969), a comparison was made of the measured longitudinal strain energy in the drill rod with the maximum piston energy. Strain gages were attached to the drill steel to measure the strain-time wave form and amplitude at seven levels of operating pressure and eight levels of thrust. The strain energy in the steel was found to be approximately 14 ft-lb (1.94 kg-m) less than the maximum piston energy for the two drills tested.

Percussive drillability or penetration rate is defined, based on the assumptions described earlier, by

$$PR = \frac{12E_r \times f \times T_r}{E_v \times A_H} \tag{3.19}$$

Thus, the evaluation of five factors is required for the calculation of penetration rate. However, only two, the blow frequency and the cross-sectional area of the hole, can be measured accurately. The others are interrelated and can be approximated or measured in different ways. For example, an assumed energy transfer ratio of unity and empirically determined blow energies have been used with measured penetration rates to calculate an apparent specific energy of drilling. Specific energy ideally should be an intrinsic property of the rock and the particle size distribution, but the specific energy also depends on the type and newness of the bit, the indexing angle, the drilling energy level, and the capabilities of the drill.

Specific energy values measured by quasi-static or impulsive methods of indenting a smooth rock surface can differ by a factor of two from those obtained for downhole geometry (Haimson, 1965). Static and dynamic values also may differ by a factor of about two (Haimson, 1968; Fairhurst, 1961). Neither of these methods accounts for the energy transmitted from the bit to the rock that is not used in crushing but is dissipated into the rock as wave energy, or that which is retained in the drill steel and dissipated as heat.

It is assumed that the specific energy value is a "true" specific energy and that the energy transfer ratio is the ratio of the energy output of the drill to the energy used in rock breakage only. The energy output of a drill has also been evaluated in terms of the kinetic energy of the piston, and Bruce and Paone (1969) found by means of empirical formulas that the striking velocity is from 88 to 99% of the calculated maximum. Ditson's expression for blow energy (equation 3.18) was deduced based on the assumption that the power stroke time is 45% of the total cycle time and that the striking velocity is 1.75 times the average velocity on a forward stroke. These constants were determined by means of a study of stroke-time diagrams. However, both factors may vary appreciably, depending on the drill design and operating conditions.

Pfleider and Lacabanne (1961) assumed a constant acceleration during the power stroke and assumed that the striking velocity and the terminal velocity of the power stroke are equal. Based on these assumptions only one empirical constant is required for calculations, and the formula for the blow energy is equivalent to that of Ditson.

From drilling data for two drills at constant thrust, Bruce and Paone (1969) used the approximate relation

$$E_r = CW(S \times t)^2 \tag{3.20}$$

The constant $C$, however, was about 1.5 times greater for drill A than for drill B, which indicates that it is a function of drill size and design.

Based on the assumption of constant acceleration, the maximum velocity is twice the average piston velocity. Ditson's assumption of a value of 1.75 gives a striking velocity of 87.5% of the maximum which agrees with the lower values found by Bruce and Paone (1969). The latter found it necessary to determine more precisely the piston energy that is transferred as longitudinal strain energy to the drill steel in order to predict the energy delivered to the bit for useful work in drilling. The remainder of the piston energy is eventually dissipated as heat in the drill rod, in the drill machine, and in wave energy in the rock.

### 3.5.1   Stress Pulse Measurement

The drill used in this investigation was drill A (Table 3.2). The longitudinal strain was measured with four pairs of strain gages, connected as a full resistance bridge mounted to detect only longitudinal strain, bonded to the drill rod 5 ft (1.52 m) from the drill end to prevent the reflected pulse from overlapping the incident pulse. A 100-ft (30.48-m) conducting cable connecting the strain gages and the electronic equipment was used in lieu of commutators and was allowed to wrap around the drill steel as it rotated, permitting several measurements before unwinding was necessary. A 7 × 8 factorial experiment with seven levels of operating pressure from 40 to 100 psi and eight levels of thrust from 132 to 439 lb (59.8 to 199 kg) (corresponding to 30 and 100 psi, 2.1 to 7 kg/cm$^2$, on the thrust cylinder) was used with five replications of each point.

The longitudinal strain pulse at first incidence was recorded (Figure 3.30) to determine the energy. The shape of the pulse remained constant, and only the amplitude varied with changing thrust and operating air pressure; hence, the pulse energy is proportional to the square of the peak amplitude (Figures 3.30, 3.31).

Measurements of strain waves in the drill rod (Cook et al., 1968) show that when the thrust is too low and the bit is not in contact with the rock, the incident and reflected waves are of nearly the same amplitude (Figure 3.32). However, when the proper thrust is used, the reflected wave is of lower magnitude, and more energy is delivered to the rock.

The traces for five piston blows in each sweep show the incident wave and a

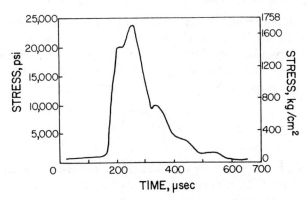

**Figure 3.30**  Longitudinal strain wave form at first incidence (Lundquist and Anderson, 1969).

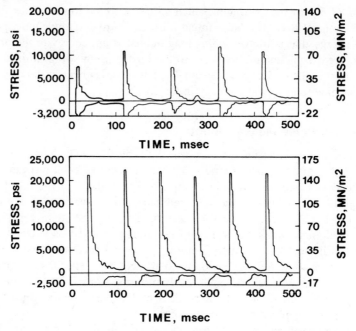

**Figure 3.31** Strain wave envelope for multiple piston blows (Lundquist and Anderson, 1969).

number of reflections. Hustrulid (1965) demonstrated that the useful energy is transferred to the rock in the first two (compressive) incidences of the pulse and that the remaining energy is dissipated as heat. The peak strain amplitudes were converted to peak stress values, and these and the number of blows were averaged for each data point.

### 3.5.2 Analysis of Strain Data

Since the impedance ($\rho c$) and the wave velocity ($c$) of steel do not vary with impact velocity or wave energy (Fairhurst, 1961; Simon, 1963), the pulse shape is also independent of impact velocity. Thus, the stress at a point in the drill rod is directly proportional to the strain at the same point at a time ($t$) after impact. The energy of the strain wave is then given by

$$E_w = \frac{Ac}{E} \int_o^{t_f} \sigma^2 \, dt \qquad (3.21)$$

where $E_w$ = pulse or wave energy, $A$ = cross-sectional area of drill steel, $c$ = wave velocity in steel, $E$ = Young's modulus of steel, $\sigma$ = stress, $t$ = time, and $t_f$ = time duration of the pulse.

Horizontal sweep speed – 200 s/cm
Vertical sensitivity –10740 psi/cm

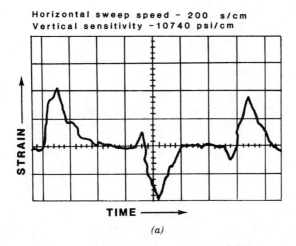

*(a)*

Horizontal sweep speed – 200 s/cm
Vertical sensitivity – 10740 psi/cm

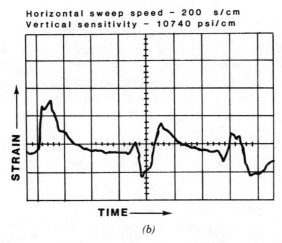

*(b)*

**Figure 3.32** Strain waves recorded for (*a*) a properly thrusted drilling machine and (*b*) an underthrusted drilling machine (Hustrulid, 1971).

Hence, the energy of a pulse is proportional to the square of the peak stress, and the constant of proportionality between the energy content and the square of the peak stress that is calculated from one pulse can be used to determine the energy of similar pulses.

A pulse (Figure 3.30) was numerically squared and integrated and the constant $K$ in

$$\int_o^{tf} \sigma^2 \, dt = K\sigma_p^2 \tag{3.22}$$

was found to be $0.9578 \times 10^{-4}$. The strain energy was then calculated from the peak stress as follows:

$$E_w = \frac{Ac}{E} \int_o^{tf} \sigma^2 \, dt = \frac{Ac}{E} K\sigma_p^2$$

$$= \frac{(0.60)(2.007 \times 10^5)(0.9578 \times 10^{-4})}{(29.5 \times 10^6)} \sigma_p^2$$

$$= (0.391 \times 10^{-6})\sigma_p^2 \tag{3.23}$$

where $\sigma_p$ = average peak stress, psi.

Pulse energies for each combination of thrust and pressure (Figure 3.33) were computed from the average peak stresses (Table 3.5), and the energies were correlated by statistical regression using the equation

$$W = A + BT + CT^2 \tag{3.24}$$

where $T$ = thrust (psi) and $A$, $B$, and $C$ = coefficients of equation (3.24) for each operating pressure (Table 3.6, Figure 3.34).

The maximum piston energies measured earlier by Bruce and Paone (1969) are consistently higher than the stress energies (Table 3.7, Figure 3.34) in the drill stress, the difference representing a loss in transmission of energy from the piston to the drill steel.

The relation between the maximum piston energy and the pulse strain energy was computed using linear regression for those points for which operating condi-

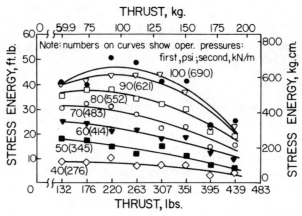

**Figure 3.33** Longitudinal strain energy for thrusts and operating pressures (Lundquist and Anderson, 1969).

**TABLE 3.5. Average Peak Stress, psi (MN/m²)**

| Operating Pressure psi | kg/cm² | Average Peak Stress[a] | 132 (60) | 176 (80) | 220 (100) | 263 (119) | 307 (139) | 351 (159) | 395 (179) | 439 (199) |
|---|---|---|---|---|---|---|---|---|---|---|
| 40 | 2.8 | $\overline{X}$ (psi) | 16,330 | 15,770 | 17,680 | 14,660 | 14,970 | 14,890 | 10,580 | 11,670 |
|  |  | $\overline{X}$ (kg/cm²) | (1,148) | (1,109) | (1,243) | (1,031) | (1,052) | (1,047) | (744) | (820) |
|  |  | s (%) | 8.09 | 15.51 | 4.96 | 21.80 | 22.63 | 14.73 | 30.40 | 27.92 |
|  |  | n | 7 | 7 | 7 | 5 | 5 | 6 | 5 | 5 |
| 50 | 3.5 | $\overline{X}$ (psi) | 23,640 | 22,850 | 21,350 | 18,910 | 21,250 | 18,650 | 15,100 | 14,150 |
|  |  | $\overline{X}$ (kg/cm²) | (1,662) | (1,606) | (1,501) | (1,329) | (1,494) | (1,311) | (1,062) | (995) |
|  |  | s (%) | 5.64 | 10.29 | 14.02 | 14.71 | 5.75 | 24.43 | 11.53 | 10.86 |
|  |  | n | 7 | 7 | 7 | 6 | 6 | 6 | 6 | 6 |
| 60 | 4.2 | $\overline{X}$ (psi) | 27,810 | 27,210 | 25,610 | 25,750 | 23,540 | 23,990 | 19,380 | 17,340 |
|  |  | $\overline{X}$ (kg/cm²) | (1,955) | (1,913) | (1,800) | (1,810) | (1,655) | (1,686) | (1,362) | (1,219) |
|  |  | s (%) | 3.18 | 4.79 | 7.60 | 6.42 | 6.11 | 6.41 | 7.77 | 29.54 |
|  |  | n | 7 | 7 | 7 | 6 | 7 | 6 | 6 | 5 |
| 70 | 4.9 | $\overline{X}$ (psi) | 30,470 | 31,380 | 30,970 | 29,500 | 26,300 | 26,550 | 21,430 | 21,730 |
|  |  | $\overline{X}$ (kg/cm²) | (2,142) | (2,206) | (2,177) | (2,074) | (1,849) | (1,866) | (1,507) | (1,528) |
|  |  | s (%) | 6.22 | 2.97 | 5.03 | 6.16 | 7.76 | 12.70 | 13.17 | 9.57 |
|  |  | n | 7 | 7 | 7 | 7 | 7 | 7 | 6 | 7 |
| 80 | 5.6 | $\overline{X}$ (psi) | 32,980 | 34,470 | 34,140 | 32,350 | 31,380 | 30,320 | 25,580 | 24,360 |
|  |  | $\overline{X}$ (kg/cm²) | (2,318) | (2,423) | (2,400) | (2,274) | (2,206) | (2,131) | (1,798) | (1,713) |
|  |  | s (%) | 5.66 | 3.77 | 5.41 | 2.74 | 8.53 | 7.89 | 5.62 | 8.98 |
|  |  | n | 7 | 7 | 7 | 7 | 7 | 7 | 7 | 7 |
| 90 | 6.3 | $\overline{X}$ (psi) | 35,050 | 35,650 | 36,400 | 36,610 | 36,650 | 33,510 | 26,080 | 26,580 |
|  |  | $\overline{X}$ (kg/cm²) | (2,464) | (2,506) | (2,559) | (2,574) | (2,576) | (2,356) | (1,833) | (1,858) |
|  |  | s (%) | 9.79 | 2.10 | 3.51 | 2.80 | 6.19 | 5.83 | 6.91 | 9.58 |
|  |  | n | 7 | 7 | 7 | 6 | 7 | 7 | 7 | 7 |
| 100 | 7.0 | $\overline{X}$ (psi) | 35,400 | 34,800 | 39,250 | 38,650 | 34,520 | 34,520 | 26,550 | 27,930 |
|  |  | $\overline{X}$ (kg/cm²) | (2,489) | (2,446) | (2,759) | (2,717) | (2,427) | (2,427) | (1,866) | (1,963) |
|  |  | s (%) | 11.14 | 7.11 | 1.18 | 2.08 | 8.56 | 9.04 | 9.38 | 6.57 |
|  |  | n | 7 | 7 | 7 | 7 | 7 | 7 | 7 | 7 |

[a]$\overline{X}$ is average peak stress; s, percent standard deviation; n, number of blows.

Source: Lindquist and Anderson (1969).

**TABLE 3.6. Regression Coefficients for Longitudinal Strain Energy Data**

| Operating Pressure | | Regression Coefficients | | |
|---|---|---|---|---|
| psi | (kN/m²) | A | B | C |
| 40 | (276) | 7.26 | 0.09 | −0.0012 |
| 50 | (345) | 20.14 | .04 | −0.0009 |
| 60 | (414) | 25.21 | .05 | −0.0020 |
| 70 | (483) | 30.57 | .14 | −0.0031 |
| 80 | (552) | 28.62 | .44 | −0.0055 |
| 90 | (621) | 17.85 | 1.03 | −0.0101 |
| 100 | (690) | 13.52 | 1.24 | −0.0116 |

*Source:* Lundquist and Anderson (1969).

tions of the two studies were similar (Figure 3.35). The pulse strain energy as a function of maximum piston energy is given by

$$E_w = 1.073 \ MPE - 14.693 \tag{3.25}$$

where $MPE$ = maximum piston energy (ft-lb).

The standard error of the regression coefficient is 0.076, or the regression coef-

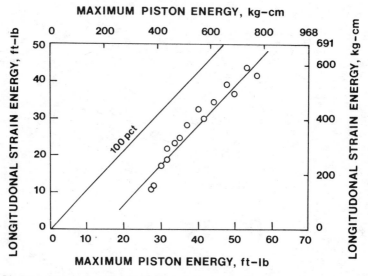

**Figure 3.34** Regression study of longitudinal strain energy on maximum piston energy (Lundquist and Anderson, 1969).

**TABLE 3.7. Maximum Piston Energy, ft-lb (J)**

| Operating Pressure | | Thrust, lb (N) | | | | |
|---|---|---|---|---|---|---|
| psi | (kN/m²) | 0 (0) | 88 (390) | 176 (781) | 263 (1171) | 351 (1562) |
| 50 | (345) | 37.40 (50.71) | 33.43 (45.32) | 30.57 (41.45) | 28.27 (38.33) | 27.93 (37.87) |
| 60 | (414) | 40.11 (54.38) | 37.56 (50.92) | 35.24 (47.78) | 31.75 (43.05) | 30.87 (41.85) |
| 70 | (483) | 49.19 (66.70) | 42.30 (57.35) | 40.10 (54.25) | 37.30 (50.57) | 34.07 (46.19) |
| 80 | (552) | 59.03 (80.03) | 53.52 (72.56) | 47.91 (64.96) | 44.52 (60.36) | 41.82 (56.70) |
| 90 | (621) | 69.43 (94.13) | 50.05 (81.42) | 56.40 (76.47) | 53.85 (73.02) | 50.00 (67.79) |
| 100 | (690) | 78.60 (106.57) | 70.70 (95.86) | 68.73 (93.19) | 66.18 (89.73) | 63.22 (85.71) |

*Source:* Lundquist and Anderson (1969).

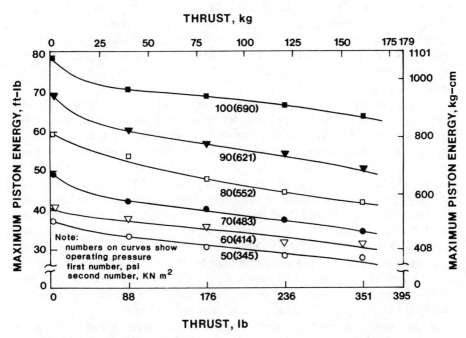

**Figure 3.35** Maximum piston energy versus thrust and operating pressure (Lundquist and Anderson, 1969).

ficient is unity at the 95% confidence level. The regression coefficient for the data points is 0.969, and the standard error of estimate for equation 3.25 is 2.670 ft-lb.

Maximum piston energies and data calculated from piston velocity measurements are useful in that they give an approximate measure of energy output. The constants in the elementary formulas for drill performance are based on simplifying assumptions and vary with drill characteristics.

Measurements of pulse strain energy in the drill rod are the most accurate means of determining the energy output of a drill. Hustrulid (1965, 1968) also found that energy transfer from the piston to the drill steel can be predicted from the shape of the longitudinal strain wave using force-displacement data for the bit–rock combination. Thus, if drill output wave forms and strain energy values for operating conditions could be provided by drill manufacturers, they could be used in predicting drilling rates. From the above data, the pulse strain energy, thought to be a constant percentage of the piston energy, was better represented for the drill employed by a constant loss in the transfer of energy from piston to the drill steel (equation 3.25).

## 3.6   EARLY RESEARCH

The percussive drilling tests reported by investigators, before the more detailed tests begun by the U.S. Bureau of Mines in 1969, were concerned with machine characteristics, drilling rates, and some of the research effort with bit wear. Later, the Bureau's investigations were concerned with details of pneumatic drill performance, a wider range of rocks was tested, and rock properties and penetration rates were analyzed, but bit wear was not considered by the Bureau, all tests being done with sharpened bits.

In early research by Cheetham and Inett (1953), tests were made on the factors that affect percussive drilling in Darley Dale sandstone and bit wear in sandstone and gray granite. The factors affecting rock disintegration by percussive drilling were given as (1) efficiency of conversion of compressed air energy for transmission in the steel rod and the frequency of application; (2) the kinetic relation of moving parts of the drill—drill body, piston, and steel; and (3) efficiency of transmission of energy to rock. These factors in turn were considered to be affected by (1) rock properties; (2) rock removal and fragment transport; (3) design of bit and steel; (4) drill machine design, which affects energy per blow, efficiency of energy delivery, blow frequency, and steel rotation; and (5) nature and magnitude of thrust.

It was stated that drill process control can be obtained by (1) standardization of drill equipment, (2) control of compressed air, (3) control of flushing, (4) control of lubrication, and (5) control of thrust. These factors were evaluated in terms of number of blows, piston, stroke, rotation, and penetration rate.

A drill frame and platform were designed to mount the drill, apply a known thrust, and hold the rock specimen in place for drilling. Compressed air was metered and regulated between 20 and 150 psi, water flush pressure was controlled when it was used, and lubricating oil was controlled.

The drill had the following specifications:

| | |
|---|---|
| Bore: | 2 5/8 in. |
| Stroke: | 2 in. |
| Rifle Bar Pitch: | 1/24, 1/30, and 1/40 |

Drill steel was $\frac{7}{8}$ in. hex and 4-ft long, and bits were tungsten carbide insert, single-chisel type. These were resharpened after each series of tests. Darley Dale sandstone contained angular quartz grains, strongly cemented with some mica with 2000 psi shear strength, and a medium-grained granite had a shear strength of 5250 psi. Rates of penetration were carefully measured, and drill cuttings were collected for energy balance studies: the blows per minute, rotation, and piston stroke, forward and backward, were also measured.

Cheetham and Inett (1953) found that for the test drill they used, the blows per minute varied as the square root of the operating pressure (absolute) measured at the inlet of the drill (Figure 3.36). The design of self-rotating percussive drills makes the rotation of the chuck and steel dependent on the piston stroke. However, the rotation is not directly proportional to the operating pressure but (under no load) is the sum of the rotation due to the piston stroke and rotation, and that due to the inertia of the moving parts of the drill. That due to inertia is the difference between the measured rotation and the calculated value from piston travel (Figure 3.37).

Where the thrust is too low, the bit is not in contact with the rock when the pressure pulse in the steel reaches the bit, there are excessive bounce and bit abrasion, and the drill is more likely to suffer mechanical breakdown (Figure 3.38). In the proper balanced phase of drill operation, maximum penetration is hypothet-

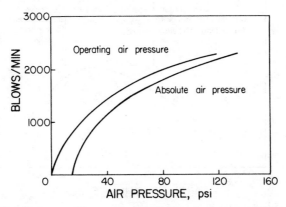

**Figure 3.36** Relationship between blow rate and (a) absolute air pressure and (b) operating air pressure (rifle bar: 1 in 30; applied thrust: zero) (Cheetham and Inett, 1953–55).

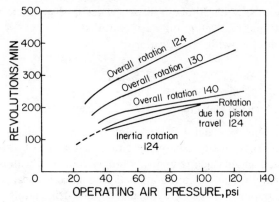

**Figure 3.37** Variation of rotation rates with operating air pressures (rifle bars: 1 in 24, 1 in 30, and 1 in 40; applied thrust: zero) (Cheetham and Inett, 1953–55).

ically obtained due to (1) elimination of bounce, (2) the bit's being in contact with rock when blow is delivered, and (3) less abrasive wear. Also, the bit edges travel through smaller angles between blows, the machine is more easily controlled, and the drill is thus working more efficiently.

If the thrust is too high, it requires higher turning torque and has limited rebound time for rotation, and the combined resistance of these factors then causes stalling. Nominal stroke occurs when the shank is in proper position. Reduced thrust does not return the shank to this optimum position, and overstroke occurs. On the other hand, increased thrust reduces and eliminates overstroke, and too much thrust

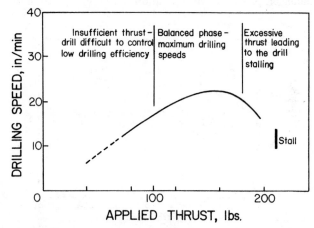

**Figure 3.38** Variation of drilling speed with applied thrust (rifle bar: 1 in 30; air pressure: 87 psi; bit: 1½-in.-diameter chisel) (Cheetham and Inett, 1953–55).

reduces the stroke below nominal (Tables 3.8). The drilling rate increased with both thrust and operating pressure, the balanced phase, a most efficient level, being between 100 and 175 psi operating pressure and 60 and 80 psi thrust for the drill tested. The higher thrusts reduced bounce of the drill steel and kept the cutting edge of the bit in contact with the rock; rotation speeds were lower, which reduced bit abrasion (Figures 3.38, 3.39); and higher drilling efficiency reduced drill break-down. Drilling rates beyond optimum resulted in excessive wear of the drill and breakage of steel.

At constant air pressure, the potential energy of the drill head was essentially constant, and the actual energy output was determined by the piston velocity and the nominal piston stroke, the latter being influenced by the thrust. The nominal stroke was defined as that distance traveled by the piston before striking the shank when the latter was in correct position. If the shank was not returned to this position, then overstrike occurred, it being limited by the crushing and the chuck.

Increase in thrust reduced overstroke (Table 3.8) and further decreased nominal stroke. The overstroke disappeared because of the position of the shank and resistance to rotation at high thrust which leads ultimately to stalling (compare with U.S. Bureau of Mines data; see Chapter 4).

The blows per revolution ($1\text{-}\frac{5}{8}$-in. single-chisel bit) increased with the thrust. In the first section of the curves (Figure 3.40), the thrust had a small effect on the revolutions per blow, indicating a bouncing phase. A further increase in thrust overcame rotation due to inertia. Peak drilling was obtained when rotation due to inertia ceased. Both the rotation due to inertia and that due to piston travel were calculated and observed (Figure 3.41) as a function of thrust. Observed values were much higher than calculated values.

**TABLE 3.8. Reduction in Length of Piston Overstroke as the Applied Thrust Is Increased**

| Air Pressure (psi) | Applied Thrust (lb) | Stroke Length (based on the average of 25 sucessive cycles) (in.) | Average Total Stroke Length (in.) |
|---|---|---|---|
| 70 | 0 | Full nominal stroke of 2.0 in. Maximum overstroke of 0.5 in. | 2.50 |
| 70 | 78 | Full nominal stroke of 2.0 in. Majority of strokes vary between zero and 0.5 in. overstroke | 2.21 |
| 70 | 90 | Full nominal stroke of 2.0 in. Majority of strokes show an overstroke but over a smaller range than at 78-lb thrust | 2.14 |
| 70 | 108 | Full nominal stroke of 2.0 in. Overstroking tending to be reduced and also very uniform in length | 2.03 |

[a]Source: Cheetham and Inett (1953–55).

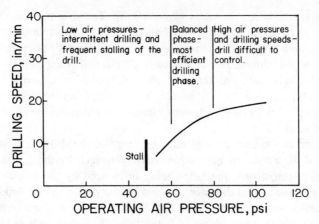

**Figure 3.39** Variation of drilling speed with air pressure (rifle bar: 1 in 30; applied thrust: 108 lb; bit: $1\frac{1}{2}$-in.-diameter chisel) (Cheetham and Inett, 1953–55).

As the thrust was increased above optimum, it also increased the resistance to rotation, the piston was unable to move, and the drill stalled. Further, more blows were required to complete one revolution of the chuck (Figure 3.40). The slope of the blows per minute versus thrust curves is initially very steep, showing little increase in blows per revolution, corresponding to the bouncing phase of the steel operation. As the thrust was increased, inertial rotation was eliminated at a critical thrust which itself increased with operating pressure. Stalling for the test drill occurred at lower blows per revolution as air pressure and thrust were increased,

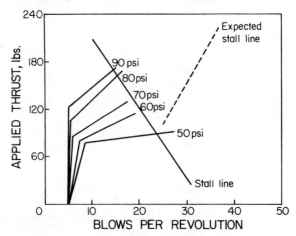

**Figure 3.40** Relationship between applied thrust and blows per revolution for a range of air pressures (rifle bar: 1 in 30; bit: $1\frac{5}{8}$-in.-diameter chisel) (Cheetham and Inett, 1953–55).

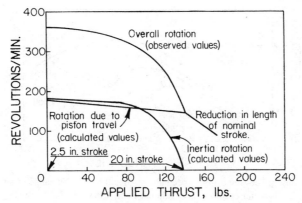

**Figure 3.41** Effect of applied thrust on rotation (rifle bar: 1 in 30; air pressure: 80 psi; bit: $1\frac{5}{8}$-in.-diameter chisel) (Cheetham and Inett, 1953–55).

giving a change of the parameters at optimum drilling conditions (Figure 3.38). That is, the peak drilling speed was obtained when the greater part of the inertial rotation was eliminated.

At a given thrust the resistance to rotation was constant, but the drill developed more power with the increase of air pressure, the stroke was lengthened, and the piston overcame the resistance which caused stalling (Figure 3.41). The fastest drilling speeds were obtained as well-defined optimums for certain pressure–thrust combinations (Figure 3.42) which are characteristics of a given drill.

Rotation of the bit (indexing) to furnish a new rock surface for each blow is essential to effective drilling. The required rotation between blows for complete rock removal depends on the energy per blow, shape of the bit, and the physical properties of the rock. Too little rotation results in repeated blows near the same position and does not give the proper indexing to yield maximum chip size, and it results in inefficient drilling, whereas too much rotation produces ridging and poor rock removal. Fast rotation with three different rifle pitches and different thrust and pressure shows that fast rotation in Darley Dale sandstone produces rapid drilling at lower thrust, slow rotation produces faster drilling at high thrust, and higher speeds occur for slow rotation. These factors may be different for other types of rock.

For the drill used in these tests in sandstone, the 1 : 40 pitch gave the highest penetration rate at 87-psi operating pressure. For higher operating pressure at 107-psi thrust, the penetration rate was higher for slow rotation (Figures 3.43, 3.44). Under these conditions, increased pitch of the rifle bar increased blow rate because of reduced energy absorption by the splines.

The division of rotation into that caused by the piston strokes and by inertia shows that initial rotation was less and decreased to zero at lower thrust for the slower rotation (1 : 40), with a somewhat similar reduction in stroke length (Figure

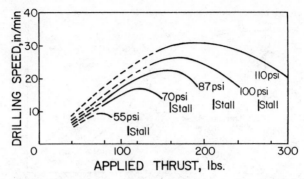

**Figure 3.42** Relationship between drilling speed and applied thrust with a range of air pressures (rifle bar: 1 in 30; bit: $1\frac{1}{2}$-in.-diameter chisel) (Cheetham and Inett, 1953–55).

3.45). In the tests by Cheetham and Inett (1953), drill cuttings were flushed both by air and by water. The results of drilling by the two means at given drilling conditions gave little difference for the drilling conditions shown for sandstone (Figure 3.46).

Penetration rates were measured for cross bits, continuously sharpened. Decrease in drilling speed was greatest for wet drilling, and gage loss was so severe in granite that the drill stalled (Figure 3.47). The reasons for the greater wear with wet drilling were given as (1) cuttings suspended in water were more abrasive than in air and (2) wet drilling allowed faster rotation, causing more abrasive wear.

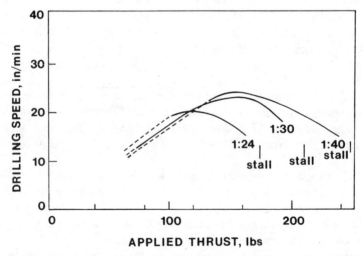

**Figure 3.43** Variation of drilling speed with applied thrust for a range of rifle bars (rifle bars: 1 in 24, 1 in 30, and 1 in 40; air pressure: 87 psi; bit: $1\frac{1}{2}$-in.-diameter chisel) (Cheetham and Inett, 1953–55).

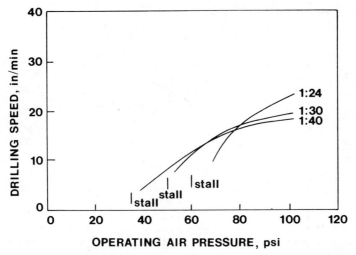

**Figure 3.44** Variation of drilling speed with air pressure for a range of rifle bars (rifle bars: 1 in 24, 1 in 30, and 1 in 40; applied thrust: 108 lb; bit: 1½-in.-diameter chisel) (Cheetham and Inett, 1953–55).

There was 50% more gage loss (measured across the cutting shoulder) in sandstone than in granite. The water required for the effective flushing should be above 40 gph for the particular drill and operating conditions used.

For no-load conditions, increased rifle pitch at constant pressure increased the blow rate because there was reduced energy absorbed by the spline and by inertia. The relationship among rpm, thrust, inertia, rotation, and reduction in stroke is

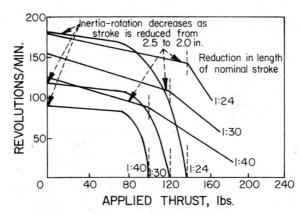

**Figure 3.45** Effect of applied thrust on rotation for a range of rifle bars (rifle bars: 1 in 24, 1 in 30, and 1 in 40; air pressure: 80 psi; bit: 1⅝-in.-diameter chisel) (Cheetham and Inett, 1953–55).

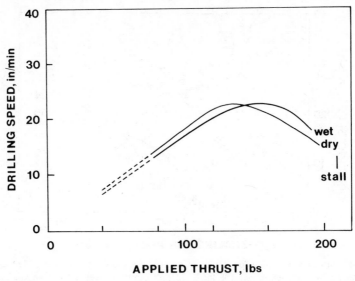

**Figure 3.46**  Variation of drilling speed with applied thrust with a wet and dry flush (rifle bar: 1 in 30; air pressure: 87 psi; bit: 1½-in.-diameter chisel) (Cheetham and Inett, 1953–55).

shown in Figure 3.45. Although a slow-rotation machine is stalled at higher thrust than a fast-rotation machine, the latter maintains a full stroke over a wider range of thrust. A short-pitch rifle bar produces a higher percentage inertial rotation than less pitch, but for long-pitch, stall results only in higher thrust. Thus, though the pitch of the rifle bar is important, a proper combination of rifle bar pitch, thrust, and air pressure within limited ranges for each will give efficient drilling.

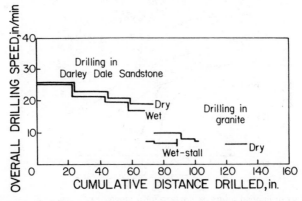

**Figure 3.47**  Decrease in drilling speed with overall distance drilled with a wet and dry flush (rifle bar: 1 in 30; applied thrust: 122 lb; air pressure: 85 psi; bit: 1½-in.-diameter steel cruciform) (Cheetham and Inett, 1953–55).

Cuttings may be flushed by air or water. For one type of rock drilled by Cheetham and Inett (1953), the particular rock–drill combination showed little advantage for one method of flushing over the other (Figure 3.46). For drilling in granite, however, the penetration rate with wet drilling decreased more rapidly with depth of hole drilled, the bit wear for drilling in granite being greater for wet flushing (Figure 3.48).

Theoretically, under no-load conditions, the relationship between blows per minute and operating pressure is

$$bpm \propto p^{1/2} \tag{3.26}$$

and the amount of energy in the compressed air available for work was calculated by means of the polytropic expansion equation with the ratio of the specific heat capacities (lambda) $\lambda = 1.3$. The amount of free air used was a linear function of the inlet pressure (curve 1, Figure 3.48a), whereas the energy curve per cubic foot was a parabolic (curve 2), increasing less rapidly as the pressure became larger. Under standard conditions, the drill speed increased more rapidly above about 60

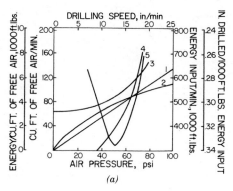

(a)

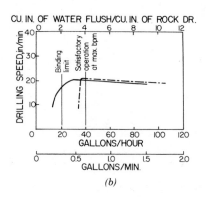

(b)

**Figure 3.48** (a) Relationships between energy input and drilling output. Curve 1: Relationship between inlet air pressure and quantity of free air used in unit time. Curve 2: Energy per cubic foot of input air at varying pressures. Curve 3: Relationship between drilling speed and air consumption under standard thrust conditions. Curve 4: Relationship between drilling speed and energy input per minute. Curve 5: Distance drilled (in.) per 1000 ft-lb energy (Cheetham and Inett, 1953–55). (b) Variation of drilling speed with water quantity passing (solid line) and cubic inches of water flush per cubic inch of rock drilled (broken line) (rifle bar: 1 in 30; applied thrust: 135 lb; air pressure: 87 psi; bit: $1\frac{1}{2}$-in.-diameter chisel) (Cheetham and Inett, 1953–55).

psi (curve 3), as did drilling speed with energy per minute (curve 4). The distance drilled per 1000 ft-lb of energy decreased rapidly to about 50 psi and increased rapidly thereafter (curve 5). Under the conditions indicated in Figure 3.48a, the most efficient drilling was 0.0336 in. × 1000 ft-lb of energy, corresponding to an air consumption of 80 cfm at 66 psi pressure. However, these data can be applied only for certain test conditions.

The quantity of air consumed was proportional to the air pressure and is capable of delivering a certain amount of energy in the extension cycle. The most efficient calculated operation condition was found to be (curve 5) 0.0336 in./1000 ft-lb of energy input, which corresponds to 90 cfm at 66 psi. Different test conditions gave different optimum results, which varied considerably from the calculated values.

# 4

# PERCUSSIVE DRILLING—
# RESEARCH AND TESTING

## 4.1   LABORATORY TESTS—USBM

As a second part of the U.S. Bureau of Mines investigation (Paone et al., 1969), laboratory studies were made of rock drillability with two self-rotating percussive drills (Table 3.1). The variables measured were rotational speed of the drill rod, operating air pressure, thrust, air consumption, and penetration rate. In the test procedure, each hole was collared with an old bit, and a new bit was used for the test. The operating air pressure and feed pressure (thrust) were set at predetermined values, and drilling was continued for 1 min or until the drill reached the limit of the carriage, whichever occurred first. Drilling was performed "dry" with air flushing.

For correlation analysis, the following physical properties were evaluated for nine test rocks: compressive strength, tensile strength (indirect method), Shore scleroscope hardness, density, longitudinal velocity, shear velocity, bar velocity, dynamic modulus of rigidity (shear modulus), static and dynamic Young's modulis, and Poisson's ratio (Table 4.1).

The average penetration rates for the nine types of rock (Figures 4.1–4.9) for the same levels of thrust and operating pressure have similar general patterns, but

**TABLE 4.1. Physical Properties of Rocks Drilled in the Laboratory**

| | Unit | Oneota Member, Prairie du Chien Formation Mankato Stone, MN | Oneota Member, Prairie du Chien Formation Kasota Stone, Kasota, MN | Rockville Quartz Monzonite Rockville Granite Rockville, MN | Morton Granite Gneiss Rainbow Granite Morton, MN |
|---|---|---|---|---|---|
| Geologic name / Commercial name / Locality | | | | | |
| Compressive strength | psi ($kg/cm^2$) | 7,700 (541) | 14,750 (1039) | 20,500 (1441) | 28,200 (1982) |
| Tensile strength | psi ($kg/cm^2$) | 1,350 (95) | 915 (64) | 1,540 (108) | 2,040 (143) |
| Shore hardness, scleroscope | units | 45 | 36 | 90 | 92 |
| Apparent density | $g/cm^3$ | 2.58 | 2.51 | 2.63 | 2.66 |
| Static Young's modulus | $10^6$ psi ($kg/cm^2$) | 4.1 (0.288) | 5.2 (0.366) | 9.1 (0.640) | 10.4 (0.731) |
| Longitudinal velocity | fps (m/s) | 18,300 (5578) | 16,900 (5151) | 20,600 (6279) | 18,400 (5608) |
| Bar velocity | fps (m/s) | 15,600 (4755) | 12,700 (3871) | 16,500 (5029) | 17,500 (5334) |
| Shear velocity | fps (m/s) | 9,900 (3018) | 5,600 (1707) | 10,600 (3231) | 11,200 (3414) |
| Dynamic Young's modulus | $10^6$ psi ($kg/cm^2$) | 8.8 (0.619) | 8.9 (0.626) | 10.9 (0.766) | 10.9 (0.766) |
| Poisson's ratio | | 0.29 | 0.44 | 0.32 | 0.20 |
| Shear modulus | $10^6$ psi ($kg/cm^2$) | 3.4 (0.239) | 1.1 (0.77) | 4.0 (0.281) | 0.5 (0.035) |

| | Unit | St. Cloud Gray Granodiorite | Sioux Quartzite | | Biwabik Iron Formation | |
|---|---|---|---|---|---|---|
| Geologic name | | | | | | |
| Commercial name | | Charcoal Granite | Dresser Basalt | Jasper Quartzite | Taconite A | Taconite B |
| Locality | | St. Cloud, MN | Dresser, WI | Jasper, MN | Aurora, MN | Babbitt, MN |
| Compressive strength | psi (kg/cm$^2$) | 33,320 (2342) | 44,500 (3128) | 56,400 (3965) | 64,200 (4513) | 67,350 (4735) |
| Tensile strength | psi (kg/cm$^2$) | 1,700 (120) | 2,485 (175) | 2,660 (187) | 4,420 (311) | 3,040 (214) |
| Shore hardness, scleroscope | units | 91 | 84 | 90 | 80 | 81 |
| Apparent density | g/cu cm | 2.73 | 2.99 | 2.62 | 3.23 | 3.84 |
| Static Young's modulus | 10$^6$ psi (kg/cm$^2$) | 9.8 (0.689) | 12.2 (0.858) | 10.1 (.710) | 15.0 (1.05) | 13.4 (0.942) |
| Longitudinal velocity | fps (m/s) | 18,800 (5730) | 21,700 (6614) | 16,500 (5029) | 19,300 (5883) | 20,700 (6309) |
| Bar velocity | fps (m/s) | 14,600 (4450) | 19,100 (5822) | 14,500 (4420) | 18,600 (5669) | — |
| Shear velocity | fps (m/s) | 10,000 (3048) | 11,900 (3627) | 10,900 (3322) | 12,200 (3719) | 11,700 (3566) |
| Dynamic Young's modulus | 10$^6$ psi (kg/cm$^2$) | 9.6 (0.675) | 14.5 (1.02) | 9.3 (0.654) | 14.7 (1.03) | 17.9 (1.26) |
| Poisson's ratio | | 0.33 | 0.29 | 0.28 | 0.24 | 0.26 |
| Shear modulus | 10$^6$ psi (kg/cm$^2$) | 3.7 (0.260) | 5.6 (0.394) | 4.2 (0.295) | 5.9 (0.415) | 7.1 (0.499) |

*Source:* Paone et al. (1969).

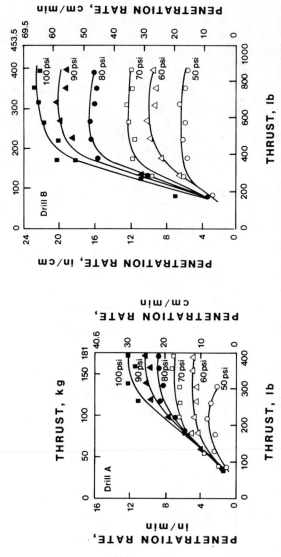

**Figure 4.1** Penetration rate versus thrust for operating pressures in a hard rock (taconite A), laboratory test (Paone et al., 1969).

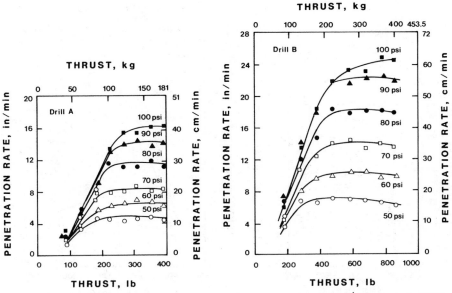

**Figure 4.2** Penetration rate versus thrust for operating pressures in taconite B (Paone et al., 1969).

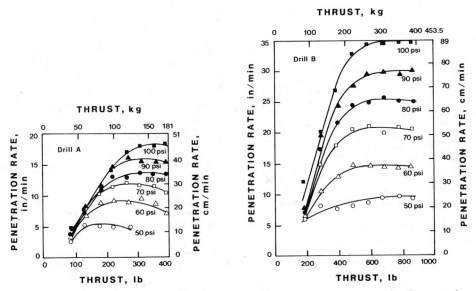

**Figure 4.3** Penetration rate versus thrust for operating pressures in jasper quartzite (Paone et al., 1969).

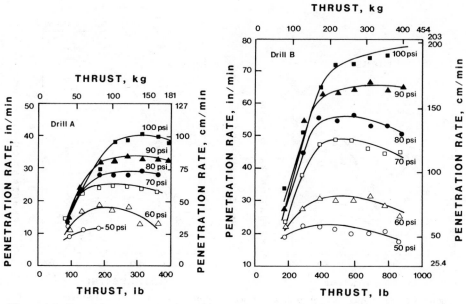

**Figure 4.4** Penetration rate versus thrust for operating pressures in Mankato stone (Paone et al., 1969).

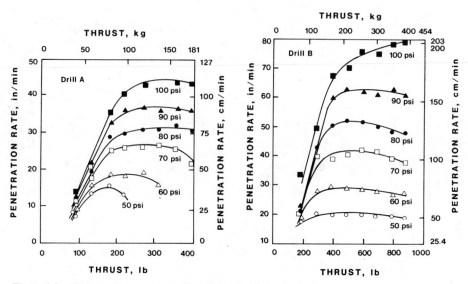

**Figure 4.5** Penetration rate versus thrust for operating pressures in Kasota stone (Paone et al., 1969).

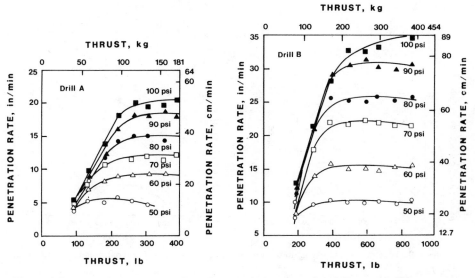

**Figure 4.6** Penetration rate versus thrust for operating pressures in rainbow granite (Paone et al., 1969).

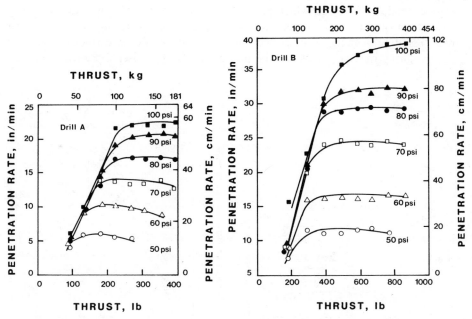

**Figure 4.7** Penetration rate versus thrust for operating pressures in Rockville granite (Paone et al., 1969).

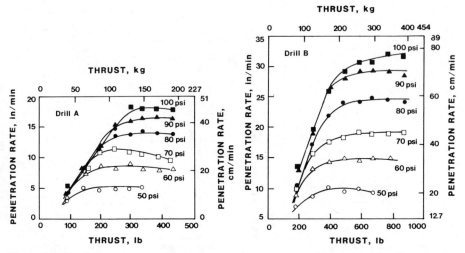

**Figure 4.8** Penetration rate versus thrust for operating pressures in Charcoal granite (Paone et al., 1969).

the rock properties affect the optimum values. Investigations of the effect of thrust on penetration rate (Hustrulid, 1968) showed that low thrust results in overtravel or free rotation of the bit, which produces overindexing and poor chip formation. Also, the piston may impact the drill rod while the bit is not in contact with the rock, resulting in energy loss. Higher thrust reduces rotation even though indexing occurs during the back stroke of the piston but also increases the torque required

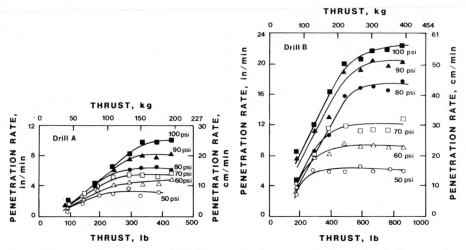

**Figure 4.9** Penetration rate versus thrust for operating air pressures in Dresser basalt (Paone et al., 1969).

for indexing, and, when thrust is increased to a critical value, the drill will stall. Each drill has a characteristic optimum thrust for maximum penetration that corresponds to good contact at the bit–rock interface at optimum indexing.

For both drills A and B, the optimum thrust increases with increasing operating pressure (Figures 4.1–4.9). After this thrust is reached, there is little increase in the penetration rate. The optimum thrust for operational drilling is just beyond the inflection point on the penetration rate curve, which also gives the maximum penetration rate for minimum bit wear, for optimum energy for rotation, and for minimum drill maintenance. The penetration rate increases with the operating pressure, so the best drilling efficiency is obtained at the highest pressure recommended by the manufacturer. This is governed largely by maintenance considerations; that is, excessively high operating pressure results in excessive wear or breakage of drill, rod, and bits.

The curves for all the types of rock drilled in the laboratory tests are similar in shape. They are steeper in the low thrust range for granite than those for sandstone but are quite similar for the data obtained for the same rocks drilled in the field, differences being accounted for largely by the *in situ* structure of the rock. Also, the curves for quartzite (see Section 4.2, Figures 4.16–4.25) are much the same, but those obtained with a drill with independent rotation differ, as do those obtained with button bits (see Section 4.2, Figure 4.12).

## 4.2 FIELD TESTS—USBM

Investigations of rock drillability in the field were made to complement laboratory research. Evaluations of drillability used classifications of rocks by their physical properties or by their response to a small-scale test such as measurement of indentation (or rebound) to drilling the rock with a miniature bit. The equation proposed by Hustrulid (1968) uses energy factors for determination of drilling rate:

$$PR = (12E_w)(f)(T_r)/(E_v)(A_H) \qquad (4.1)$$

The machine variables, blow energy ($E_w$) and blows per minute ($f$), can be obtained from the drill manufacturer, and the values of $T_r$ and $E_v$ can be determined from laboratory tests. Equation 4.1 in modified form was used as the basis for a prediction equation for drilling rock in place.

In laboratory research described earlier, Lundquist and Anderson (1969) measured the strain energy in the drill rod and found that an energy loss occurred in the conversion of kinetic energy of the piston to strain wave energy in the drill steel.

Selim and Bruce (1970) developed prediction equations by applying statistical regression analysis to drill parameters and physical properties, but the analysis involved a limited range of interrelated properties for only two pneumatic drill

**TABLE 4.2. Measured Physical Properties of Rocks Drilled in the Field**

| | Unit | Neagunee Iron Formation / Humboldt Iron Silicate / Humboldt, MI | Banded Gray Gneiss[a] / Hornblende Schist / Randville, MI | Granite Pegmatite / Randville, MI | Rib Hill Quartzite / Wausau Quartzite / Wausau, WI | Welded Tuff / Wausau Argillite / Wausau, WI |
|---|---|---|---|---|---|---|
| Geologic name | | Neagunee Iron Formation | Banded Gray Gneiss[a] | Granite Pegmatite | Rib Hill Quartzite Wausau Quartzite | Welded Tuff Wausau Argillite |
| Commercial name | | Humboldt Iron Silicate | Hornblende Schist | | | |
| Locality | | Humboldt, MI | Randville, MI | Randville, MI | Wausau, WI | Wausau, WI |
| Tensile strength | psi ($kg/cm^2$) | 2,080 (146) | 1,080 (76) | 1,230 (86) | 2,510 (176) | 2,620 (184) |
| Compressive strength | psi ($kg/cm^2$) | 59,550 (4186) | 29,600 (2081) | 12,750 (896) | 31,650 (2225) | 31,400 (2207) |
| Shore hardness, scleroscope | units | 76 | 76 | 88 | 100 | 72 |
| Density | $g/cm^3$ | 3.50 | 2.99 | 2.63 | 2.64 | 2.73 |
| Static Young's modulus | $10^6$ psi ($kg/cm^2$) | 11.1 (0.780) | 14.6 (1.03) | 5.9 (0.415) | 10.5 (0.738) | 7.6 (0.534) |
| Longitudinal velocity | fps (m/s) | 17,300 (5273) | 20,500 (6248) | 16,500 (5029) | 17,200 (5243) | 19,400 (5913) |
| Bar velocity | fps (m/s) | 16,100 (4907) | 18,800 (5730) | 12,000 (3658) | 16,400 (4999) | 18,000 (5486) |
| Shear velocity | fps (m/s) | 11,300 (3444) | 11,900 (3627) | 9,500 (2896) | 11,800 (3597) | 11,500 (3505) |
| Dynamic Young's modulus | $10^6$ psi ($kg/cm^2$) | 14.0 (0.984) | 14.0 (0.984) | 7.9 (0.555) | 10.5 (0.738) | 12.0 (0.844) |
| Shear modulus | $10^6$ psi ($kg/cm^2$) | 6.2 (0.436) | 5.6 (0.394) | 3.1 (0.218) | 4.9 (0.344) | 4.9 (0.344) |
| Poisson's ratio | | 0.13 | 0.24 | 0.07 | 0.07 | 0.23 |

[a]Not official name used by U.S. or Canadian Geological Survey.

| | Unit | St. Cloud Gray Granodiorite / Charcoal Granite / St. Cloud, MN | Warman Quartz Monzonite / Diamond Gray Granite / Isle, MN | Dresser Basalt / Dresser, WI | Oneota Member, Prairie du Chien Formation / Shiely Limestone / St. Paul Park, MN | Biwabik Iron Formation / Mt. Iron Taconite / Mt. Iron, MN |
|---|---|---|---|---|---|---|
| Geologic name | | St. Cloud Gray Granodiorite | Warman Quartz Monzonite | | Oneota Member, Prairie du Chien Formation | Biwabik Iron Formation |
| Commercial name | | Charcoal Granite | Diamond Gray Granite | Dresser Basalt | Shiely Limestone | Mt. Iron Taconite |
| Locality | | St. Cloud, MN | Isle, MN | Dresser, WI | St. Paul Park, MN | Mt. Iron, MN |
| Tensile strength | psi ($kg/cm^2$) | 1,850 (130) | 1,780 (125) | 4,020 (283) | 820 (58) | 4,330 (304) |
| Compressive strength | psi ($kg/cm^2$) | 28,950 (2035) | 24,350 (1712) | 40,800 (2868) | 14,200 (998) | 51,350 (3610) |

| | units | 87 | | 85 | | 81 | | 35 | | 80 | |
|---|---|---|---|---|---|---|---|---|---|---|---|
| Shore hardness scleroscope | units | 87 | | 85 | | 81 | | 35 | | 80 | |
| Density | g/cm³ | 2.66 | | 2.65 | | 2.99 | | 2.48 | | 3.36 | |
| Static Young's modulus | 10⁶ psi (kg/cm²) | 9.8 | (0.689) | 9.3 | (0.654) | 13.1 | (0.921) | 6.2 | (0.436) | 15.7 | (1.10) |
| Longitudinal velocity | fps (m/s) | 18,500 | (5639) | 17,800 | (5425) | 21,900 | (6675) | 16,700 | (5090) | 18,850 | (5745) |
| Bar velocity | fps (m/s) | 16,800 | (5121) | 15,900 | (4846) | 18,600 | (5669) | 14,700 | (4481) | 17,300 | (5273) |
| Shear velocity | fps (m/s) | 10,800 | (3292) | 10,400 | (3170) | 10,200 | (3190) | 9,200 | (2804) | 11,400 | (3975) |
| Dynamic Young's modulus | 10⁶ psi (kg/cm²) | 10.3 | (0.729) | 9.5 | (0.668) | 12.6 | (0.886) | 9.0 | (0.633) | 14.1 | (0.991) |
| Shear modulus | 10⁶ psi (kg/cm²) | 4.2 | (0.295) | 3.8 | (0.267) | 4.9 | (0.344) | 2.8 | (0.197) | 5.9 | (0.415) |
| Poisson's ratio | | 0.25 | | 0.23 | | 0.29 | | 0.28 | | 0.19 | |

| | Unit | Biwabik Iron Formation | | Babbitt Diabase | Duluth Gabbro | Trap Rock |
|---|---|---|---|---|---|---|
| Geologic name | | Biwabik Iron Formation | Biwabik Iron Formation | Babbitt Diabase | Duluth Gabbro | Trap Rock |
| Commercial name | | Aurora Taconite | Babbitt Taconite | Babbitt Diabase | Ely Gabbro | Trap Rock |
| Locality | | Hoyt Lakes, MN | Babbitt, MN | Babbitt, MN | Ely, MN | Tofte, MN |
| Tensile strength | psi (kg/cm²) | 3,160 (222) | 4,100 (289) | 3,550 (250) | 2,150 (151) | 730 (51) |
| Compressive strength | psi (kg/cm²) | 52,400 (3684) | 51,850 (3645) | 53,300 (3747) | 29,600 (2081) | 9,800 (689) |
| Shore hardness scleroscope | units | 83 | 86 | 90 | 89 | 43 |
| Density | g/cm³ | 3.07 | 3.12 | 2.99 | 2.85 | 2.68 |
| Static Young's modulus | 10⁶ psi (kg/cm²) | 13.3 (0.935) | 13.0 (0.914) | 11.7 (0.823) | 12.9 (0.907) | 8.5 (0.598) |
| Longitudinal velocity | fps (m/s) | 19,000 (5791) | 19,850 (6050) | 19,650 (5989) | 21,800 (6645) | 18,000 (5730) |
| Bar velocity | fps (m/s) | 17,800 (5425) | 18,500 (5639) | 17,500 (5334) | 18,900 (5761) | 16,500 (5029) |
| Shear velocity | fps (m/s) | 11,900 (3627) | 12,100 (1076) | 11,300 (3444) | 11,800 (3597) | 10,800 (3292) |
| Dynamic Young's modulus | 10⁶ psi (kg/cm²) | 13.6 (0.956) | 15.3 (1.076) | 12.4 (0.872) | 13.6 (0.956) | 10.6 (0.745) |
| Shear modulus | 10⁶ psi (kg/cm²) | 5.9 (0.415) | 6.2 (0.436) | 5.1 (0.359) | 5.3 (0.373) | 4.1 (0.288) |
| Poisson's ratio | | 0.16 | 0.22 | 0.24 | 0.28 | 0.28 |

**TABLE 4.2.** (*Continued*)

| | Unit | Oneota Member, Prairie du Chien Formation Winona Dolomite Winona, MN | Oneota Member, Prairie du Chien Formation Mankato Stone Mankato, MN | Sioux Quartzite New Ulm Quartzite New Ulm, MN | Sioux Quartzite Jasper Quartzite Jasper, MN | Rockville Quartz Monzonite Rockville Granite Cold Spring, MN |
|---|---|---|---|---|---|---|
| Geologic name | | | | | | |
| Commercial name | | | | | | |
| Locality | | | | | | |
| Tensile strength | psi (kg/cm²) | 600 (42) | 910 (64) | 2,250 (158) | 2,950 (207) | 1,300 (41) |
| Compressive strength | psi (kg/cm²) | 13,800 (970) | 17,800 (1251) | 22,250 (1564) | 43,700 (3072) | 22,000 (1547) |
| Shore hardness scleroscope | units | 52 | 49 | 66 | 92 | 91 |
| Density | g/cm³ | 2.62 | 2.60 | 2.61 | 2.63 | 2.65 |
| Static Young's modulus | $10^6$ psi (kg/cm²) | — | 7.4 (0.520) | 5.8 (0.408) | 9.4 (0.661) | 9.6 (0.675) |
| Longitudinal velocity | fps (m/s) | 17,300 (5273) | 18,200 (5547) | 17,200 (5243) | 16,600 (5060) | 18,000 (5486) |
| Bar velocity | fps (m/s) | — | 14,600 (4450) | 16,200 (8938) | 16,100 (4907) | 15,500 (4724) |
| Shear velocity | fps (m/s) | — | 9,600 (2926) | 11,500 (3505) | 11,600 (3537) | 10,100 (3078) |
| Dynamic Young's modulus | $10^6$ psi (kg/cm²) | — | 8.6 (0.605) | 10.0 (0.703) | 9.9 (0.696) | 9.2 (0.647) |
| Shear modulus | $10^6$ psi (kg/cm²) | — | 3.4 (0.239) | 4.4 (0.309) | 4.8 (0.337) | 3.7 (0.260) |
| Poisson's ratio | | — | 0.27 | 0.14 | 0.03 | 0.26 |

| Geologic name<br>Commercial name<br>Locality | Unit | Anorthosite<br>Tofte, MN | Duluth Gabbro<br>Ely Gabbro<br>Duluth, MN | Bad River<br>Dolomite<br>Marble<br>Grandview, WI | Gabbro<br>Primax Gabbro<br>Mellen, WI | Negaunee Iron<br>Formation<br>Iron Ore<br>Palmer, MI |
|---|---|---|---|---|---|---|
| Tensile strength | psi (kg/cm$^2$) | 1,500 (105) | 1,990 (140) | 1,010 (710) | 1,810 (127) | 1,680 (118) |
| Compressive strength | psi (kg/cm$^2$) | 18,700 (1315) | 26,500 (1863) | 18,150 (1276) | 25,050 (1761) | 32,050 (2253) |
| Shore hardness scleroscope | units | 91 | 75 | 52 | 82 | 65 |
| Density | g/cm$^3$ | 2.71 | 2.91 | 2.85 | 2.93 | 3.33 |
| Static Young's modulus | 10$^6$ psi (kg/cm$^2$) | 12.2 (0.858) | 9.2 (0.647) | 11.6 (0.815) | 14.8 (1.04) | 10.0 (0.070) |
| Longitudinal velocity | fps (m/s) | 21,800 (6645) | 19,200 (5852) | 19,100 (5822) | 21,000 (6401) | 14,800 (4511) |
| Bar velocity | fps (m/s) | 18,600 (5669) | 16,900 (5151) | 17,300 (5273) | 18,200 (5547) | 12,600 (3840) |
| Shear velocity | fps (m/s) | 11,500 (3505) | 10,600 (3231) | 10,900 (3322) | 11,400 (3975) | 8,800 (2682) |
| Dynamic Young's modulus | 10$^6$ psi (kg/cm$^2$) | 12.5 (0.879) | 11.2 (0.787) | 11.5 (0.808) | 13.0 (0.914) | 8.4 (0.591) |
| Shear modulus | 10$^6$ psi (kg/cm$^2$) | 4.8 (0.337) | 4.4 (0.309) | 4.6 (0.323) | 5.1 (0.359) | 3.4 (0.239) |
| Poisson's ratio | | 0.27 | 0.27 | 0.26 | 0.27 | 0.23 |

Source: Schmidt (1972).

**TABLE 4.3. Coefficients of Rock Strength**

| Rock Name | CRS | Rock Name | CRS |
|---|---|---|---|
| Humboldt iron silicate | 2.39 | Shiely limestone | 0.57 |
| Hornblende schist | 1.64 | Mt. Iron taconite | 1.47 |
| Granite pegmatite | 0.77 | Aurora taconite | 2.62 |
| Wausau quartzite | 0.78 | Babbitt taconite | 2.84 |
| Wausau argillite | 2.28 | Babbitt diabase | 2.44 |
| Winona dolomite | 0.47 | Ely gabbro | 1.21 |
| Mankato stone | 0.45 | Trap rock | 0.64 |
| New Ulm quartzite | 0.75 | Anorthosite | 0.73 |
| Jasper quartzite | 1.01 | Duluth basalt | 2.11 |
| Rockville granite | 0.84 | Marble | 0.68 |
| Charcoal granite | 1.21 | Primax gabbro | 1.02 |
| Diamond gray granite | 0.82 | Iron ore | 1.28 |
| Dresser basalt | 2.86 | | |

*Source:* Schmidt (1972).

sizes. For field tests (Schmidt, 1972), two drills (Table 3.1) used in laboratory tests were mounted on a mobile drill unit. The primary experiments were performed with $1\frac{1}{2}$-in. (3.81 cm), four-wing, cross-type bits with tungsten carbide inserts, and comparative experiments were made with 2-in. (5.08-cm) cross-type bits and with $1\frac{1}{2}$-in. (3.81-cm) button bits, the latter being used only with the smaller drill in soft to moderately hard rocks. Twenty-five rock types were drilled, holes were collared with old bits, and a new or resharpened bit was used for the first drill run. The

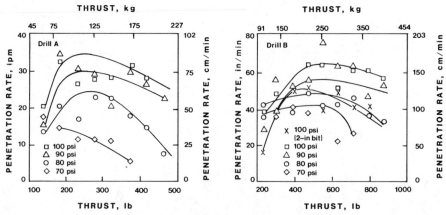

**Figure 4.10** Penetration rate versus thrust for operating pressures in Shiely limestone. (Unless otherwise indicated, bits are $1\frac{1}{2}$-in., four-insert type.) (Schmidt, 1972.)

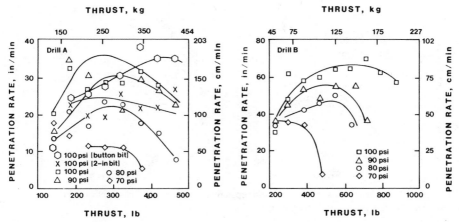

**Figure 4.11** Penetration rate versus thrust for operating pressures in Mankato stone. (Unless otherwise indicated, bits are $1\frac{1}{2}$-in., four-insert type.) (Schmidt, 1972.)

tests in each rock were made at thrust pressures of 20–90 psi in increments of 10 psi to the maximum or until the drill stalled. Tests were repeated as operating pressures increased in increments of 10 psi from 70 to 100 psi. The second drill was then tested with the same sequence of parameter changes, resulting in a 4 × 8 factorial experiment for each drill for each type of rock.

In soft to moderately hard rocks, samples were taken for property tests with an air-powered diamond core drill. For harder rocks, which the small diamond drill could not cut, hand samples were taken for subsequent coring in the laboratory.

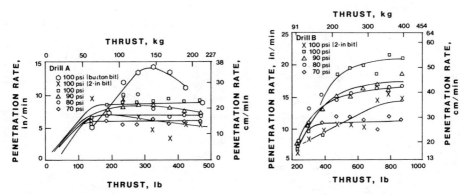

**Figure 4.12** Penetration rate versus thrust for operating pressures in Dresser basalt. (Unless otherwise indicated, bits are $1\frac{1}{2}$-in., four-insert type.) (Schmidt, 1972.)

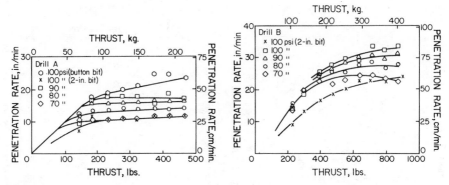

**Figure 4.13**   Penetration rate versus thrust for operating pressures in Primax gabbro. (Unless otherwise indicated, bits are $1\frac{1}{2}$-in., four-insert type.) (Schmidt, 1972.)

Physical properties measured were tensile strength, compressive strength, Shore scleroscope hardness, density, longitudinal velocity, shear velocity, bar velocity, dynamic Young's modulus, and Poisson's ratio (Table 4.2). The Coefficients of Rock Strength were determined using 10 irregularly shaped rocks averaging 7.5 cm³ each (Table 4.3).

As in the laboratory experimentation, the optimum thrust is defined as the axial force applied to the drill to keep the bit in contact with the rock for most efficient drilling. With insufficient thrust, the drill rod rebounds from the rock, energy is lost, and overindexing may occur. For each combination of air pressure, type of drill, and rock, there is an optimum thrust at which the maximum amount of rock is drilled per unit of energy and the maximum penetration rate is achieved. Exceeding the optimum thrust increases the rotation torque, and, for a machine with

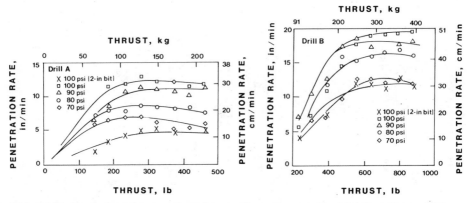

**Figure 4.14**   Penetration rate versus thrust for operating pressures in Humboldt iron silicate. (Unless otherwise indicated, bits are $1\frac{1}{2}$-in., four-insert type.) (Schmidt, 1972.)

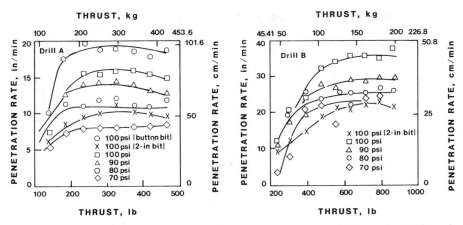

**Figure 4.15**  Penetration rate versus thrust for operating pressures in Rockville granite. (Unless otherwise indicated, bits are $1\frac{1}{2}$-in., four-insert type.) (Schmidt, 1972.)

rifle-bar rotation, the drill will eventually stall. Hence, a drill with independent rotation has an advantage for high thrust.

For the rocks in which they were used, button bits achieved higher penetration rates (Figures 4.10–4.15) than wedge bits. The reason may be related to the mechanics of chip formation (Lundquist, 1968; Paone and Tandanand, 1966). The optimum thrust is easily identified for softer rocks and, in some cases, occurs at lower operating pressure in harder rock. For some rocks, it was not reached in the thrust range investigated, or the optimum could not be detected.

## 4.3  FIELD TESTS—QUARTZITE

Hustrulid (1971b) reported the results of drilling experiments in two kinds of quartzite with three types of drills (Table 4.4) and four types of drill steel (Table 4.5). The rock in the Robinson deep gold mine has a uniaxial compressive strength of 45,000 lb/in.², and that in the Vlakfontein mine, 39,000 lb/in.². For most tests, the drills (without independent rotation) were mounted on horizontal slides, and the thrust was furnished with a coaxial compressed air cylinder. Measurements were made of operating air pressure, thrust, time to drill 24 in., air consumption, blows per minute, and rod rotation per minute.

As with results of the U.S. Bureau of Mines tests, for a given operating pressure, the penetration rate increases with the thrust to a maximum, decreases, and, finally, the drill stalls. Hustrulid (1971) defined the optimum thrust as that near-peak penetration at which bit wear is a minimum. In a broader sense, the best drilling conditions should also include optimum wear and maintenance factors for the machine, steel, and bits.

**TABLE 4.4. Rock Drills Used in Drilling Tests**

| Designation | A SECO S24 Steel Eng.Co. | B RH77 Delfos- Atlas Copco | C HC-53 Victoria Engineering |
|---|---|---|---|
| Piston Diameter in. (cm) | 3 (7.62) | 2 7/8 (7.30) | 2 5/8 (6.67) |
| Piston Weight (with rifle nut) lb (kg) | 6.688 (3.03) | 4.80 (2.18) | 4.907 (2.23) |
| Rifle Bar | 1:55 1:30 | 1:38 | 1:64 1:30 1:23 |
| Nominal Piston Stroke in. (cm) | 3 (7.62) | 2.75 (6.99) | 2.375 (6.03) |
| Drill Steel | 1 in. hex | 1 in. hex | 1 in. hex (2.54) |
| (Integral) | (2.54) | (2.54) | 7/8 in. hex (2.22) |
| Chisel Bit Diameter in. (mm) | 0.99 (39) | 0.99 (39) | 39 (0.99) 30 (0.76) |

*Source:* Hustrulid (1971b).

**TABLE 4.5. Description of Drill Steel Used**

| Designation | Description |
|---|---|
| 1 | Staved shank, 1 in. (2.54 cm), 78 in. (198 cm) long, 39 mm bit |
| 2 | Rubber collar, 1 in. (2.54 cm) hex, 74 in. (188 cm) long, 39 mm bit |
| 3 | Rubber collar, 7/8 in. (2.22 cm), 74 in. (188 cm) long, 39 mm bit |
| 4 | Staved shank, 1 in. (2.54 cm) hex, 74 in. (188 cm) long, 39 mm bit |
| 5 | Staved shank (Avesta steel), 1 in. (2.54 cm), 73 in. (185 cm) long, 39 mm bit |

*Source:* Hustrulid (1971b).

The shapes of the thrust–penetration curves for *in situ* quartzite are somewhat more uniform than for the rocks drilled in the field tests by the Bureau of Mines, probably because of the consistent properties of the quartzite at great depths below the earth's surface (Figures 4.16–4.25), even for different drills, steel, and rifle bars (Tables 4.5, 4.6). A straight line represents approximately the optimum thrust locus. From the equation $F_1 = \alpha pA$, the experimental results differ from the theoretical values (Table 4.7). The value of $\alpha$ required for agreement between theory and test results is approximately 0.5, and the prediction equation for quartzite then becomes

$$F_1 = 0.5 \, pA \qquad (4.2)$$

In conjunction with standard thrust tests, three different rifle bars were used.

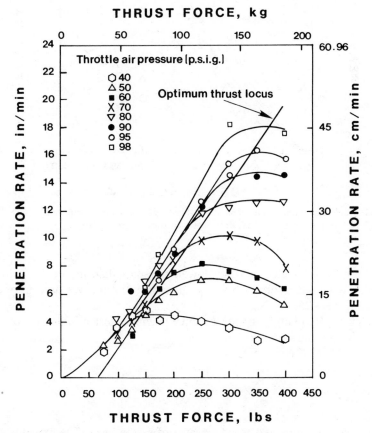

**Figure 4.16** Penetration rate as a function of thrust for various machine throttle air pressures, Robinson deep mine, rock drill A, rifle bar 1 : 55, drill steel 1 (Hustrulid, 1971).

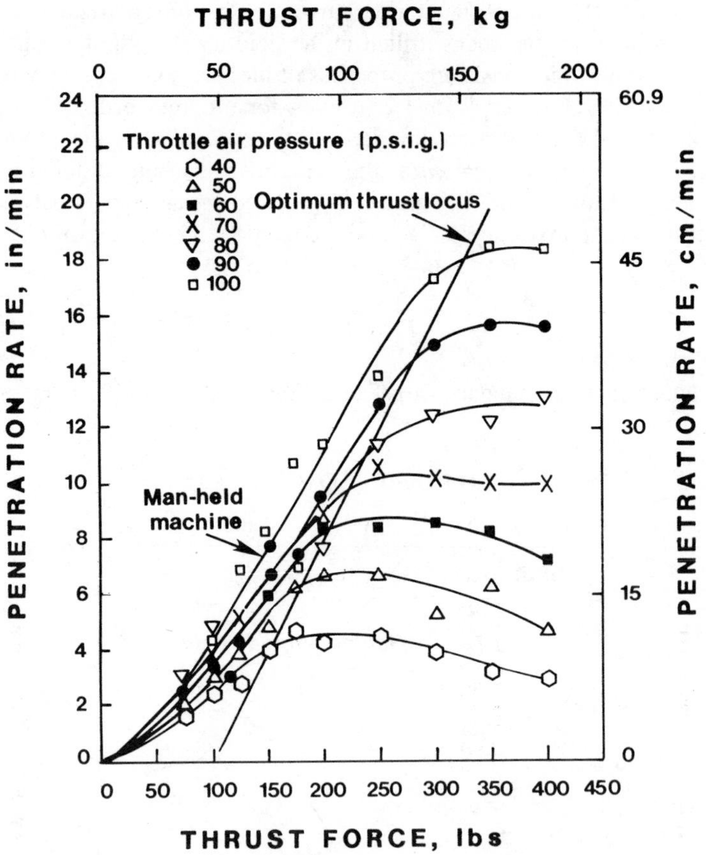

**THRUST FORCE, kg**

**Figure 4.17** Penetration rate as a function of thrust for various machine throttle air pressures, Robinson deep mine, rock drill B, rifle bar 1 : 38, drill steel 1 (Hustrulid, 1971).

The 1 : 64 rifle bar stalled at much lower pressures than the others, which was contrary to expected results. Also, a smaller bit (30 mm) was used in one series of tests (Figure 4.25).

The ratio of optimum thrust for the three machines tested in the quartzite in the Robinson deep mine was calculated from

$$\frac{F_{t1}}{F_{t2}} = \frac{A_1}{A_2} \tag{4.3}$$

where $F_{t1}$, $F_{t2}$ = thrust and $A_1$, $A_2$ = area of piston head. The experimental and theoretical thrust ratios agree within 10% (Table 4.8).

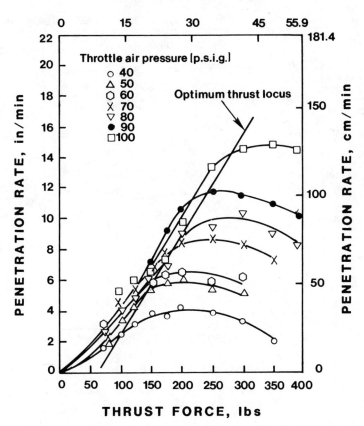

**Figure 4.18**  Penetration rate as a function of thrust for various machine throttle air pressures, Robinson deep mine, rock drill C, rifle bar 1:30, drill steel 1 (Hustrulid, 1971).

The measured data for blow frequency (Table 4.9) were used to calculate values for $K_0$ from the equation

$$\bar{f} = K_0(6pAg/SW)^{1/2} \tag{4.4}$$

for each of the drills (Table 4.10). Predicted values fell between 20 and 30, averaging 22, showing some dependence on air pressure. For $K_0 = 22$, a variation of less than 5% exists for predicted values for operating pressures above 60 lb/in.². 

Impact energy in the drill steel was not measured but was calculated (Table 4.11) by means of

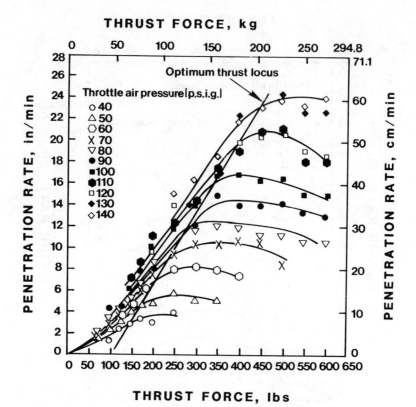

**Figure 4.19** Penetration rate as a function of thrust for various machine throttle air pressures. Vlakfontein mine, rock drill A, rifle bar $1:55$, drill steel 1 (Hustrulid, 1971).

$$E_i = \frac{pAS\beta_0^2}{12} \tag{4.5}$$

where $\beta_0$ = constant.

From this, the applicability of the penetration rate prediction equation

$$PR = \frac{12E_i \bar{f} T_r}{A_H \times E_v} \tag{4.6}$$

was determined (Table 4.12) using a value of 0.8 for $T_r$, the compressive strength of the quartzite (45,000 lb/in.²) for $E_v$ and 1.85 in.² for $A_H$. The predicted values checked very closely with measured values except at higher operating pressures. For the tests in the Vlakfontein mine (Table 4.13), a value of $E_v = 39,000$ lb/in.² was used, and the predicted rates were higher than the measured rates, indicating

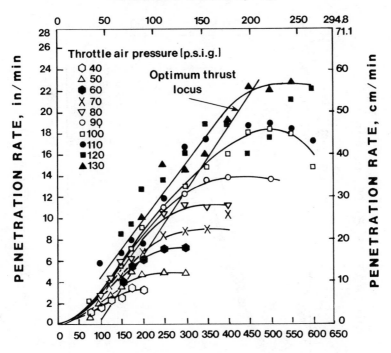

**THRUST FORCE, kg**

**Figure 4.20** Penetration rate as a function of thrust for various machine throttle air pressures. Vlakfontein mine, rock drill B, rifle bar 1:38, drill steel 1 (Hustrulid, 1971).

that the value of $T_r = 0.8$ does not hold or that the compressive strength is not the best measure of specific energy. The predicted and measured rates for the tests at both mines agree well and would be in better agreement for a higher value of $E_v$ for the Vlakfontein tests except for drill C.

Evaluation of penetration rates must also take into consideration the consumption of compressed air. The penetration rate as a function of free air consumption (Figures 4.26, 4.27) may be expressed for the rocks in the two mines in the form

$$PR = K_1 (\text{cfm})^2 \tag{4.7}$$

where the value of $K_1$ is, for the Robinson deep mine, $6.67 \times 10^{-4}$ and, for the Vlakfontein mine, $6.25 \times 10^{-4}$. The air consumption also varied as the square root of the operating pressure (Table 4.14).

To determine the effects of hole diameter, drill C was modified, and a 30-mm

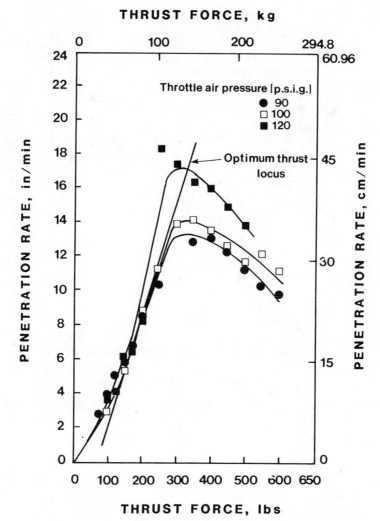

**Figure 4.21** Penetration rate as a function of thrust for various machine throttle air pressures. Vlak-fontein mine, rock drill C, rifle bar 1:30, drill steel 1 (Hustrulid, 1971).

chisel bit was used to compare drilling rates with those of a 39-mm bit (Table 4.15). The ratio of peak penetration rates was determined by the inverse of the area ratios:

$$\frac{PR_1}{PR_2} = \frac{\frac{\pi}{4}(39)^2}{\frac{\pi}{4}(30)^2} = 1.69 \tag{4.8}$$

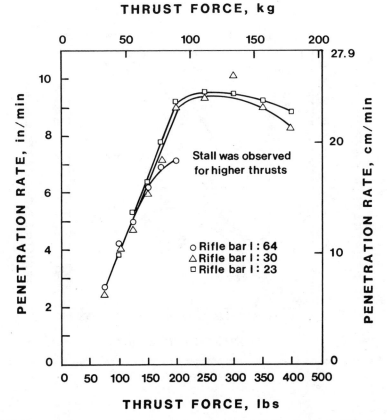

**Figure 4.22**  Penetration rate as a function of thrust for rock drill C, using three different rifle bars. Robinson deep mine, throttle air pressure 80 lb$^2$/in.$^2$g, drill steel 1 (Hustrulid, 1971).

where it is assumed that the specific energy is the same for both bit diameters. Poor removal of cuttings, variations in rock properties, and other factors could have caused the differences in theoretical and measured values. Also, the predicted penetration rates in equation 4.8 were much higher than the measured rates for high operating pressures (Table 4.16).

### 4.3.1  Thrust

As indicated previously, Hustrulid (1968) has shown that the thrust required to keep the bit in contact with the rock is given by

$$F_t = \frac{\bar{f}}{30} \left(1 + \beta\right) \int_0^\tau \sigma_i \, dt \qquad (4.9)$$

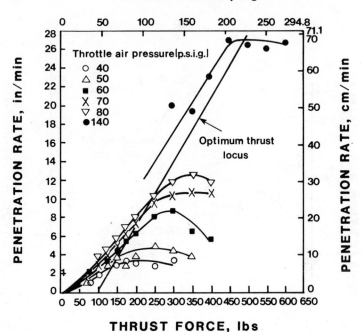

**Figure 4.23** Penetration rate as a function of thrust for various machine throttle air pressures. Vlakfontein mine, rock drill A, rifle bar 1:30, drill steel 2 (Hustrulid, 1971).

where $\sigma_i$ = incident stress as a function of time, $\tau$ = duration of incident pulse, $\beta$ = coefficient of momentum transfer from piston to drill steel, normally $0 \le \beta \le 0.2$, and $F_t$ = minimum thrust.
This is simplified to

$$F_t = \frac{\bar{f}}{30}\,(1 + \beta)\,\frac{WV_s}{g} = \frac{K_0}{30}\,\beta_0(1 + \beta)\,pA \qquad (4.10)$$

where, for convenience,

$$\alpha = \frac{K_0\beta_0}{30}\,(1 + \beta) \qquad (4.11)$$

The combination of equations 4.9, 4.10, and 4.11 gives the relationship of the penetration rate to the drill machine parameters:

$$PR = \frac{K_0 \times T_r}{A_H \times E_v} \left[ \frac{6Sg}{W}\,(pA)^3 \right]^{1/2} \beta_0^2 \qquad (4.12)$$

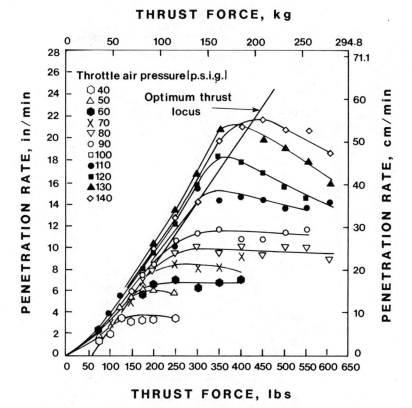

**Figure 4.24**   Penetration rate as a function of thrust for various machine throttle air pressures. Vlakfontein mine, rock drill C, rifle bar 1 : 30, drill steel 4, bit diameter 39 mm (Hustrulid, 1971).

The previous equation may then be used to predict penetration rates as they are related to thrust.

**Problem 1.** What is the ratio of penetration rates of two drills of different stroke, piston, weight, and piston area, assuming the same values of $K_0$, $T_r$, and $E_v$ for each machine?

**Problem 2.** What is the ratio of thrusts?

**Problem 3.** What are the penetration rates and required thrusts for the same machine in two different types of rock?

**Problem 4.** What are the rates when the same machine is used with two different diameter bits in the same rock?

Drill steel life is related to the peak stress and the number of times it is repeated. When a drill is operated at less than optimum thrust, the stress reflected from the

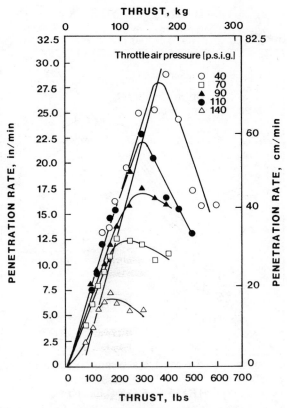

**Figure 4.25** Penetration rate as a function of thrust for various machine throttle air pressures. Vlakfontein mine, rock drill C, rifle bar 1:30, drill steel 3, bit diameter 30 mm (Hustrulid, 1971).

## TABLE 4.6. Penetration Rate Tests

| Figure No. | Drill | Mine | Rifle Bar | Drill Steel* | Comment |
|---|---|---|---|---|---|
| 4.16 | A | RD | 1:55 | 1 | |
| 4.17 | B | RD | 1:38 | 1 | |
| 4.18 | C | RD | 1:30 | 1 | |
| 4.19 | A | V | 1:55 | 1 | |
| 4.20 | B | V | 1:38 | 1 | |
| 4.21 | C | V | 1:30 | 1 | |
| 4.22 | C | RD | (3) | 1 | 3 rifle bars, 80 psi (5.6 kg/cm$^2$) |
| 4.23 | A | V | 1:30 | 2 | |
| 4.24 | C | V | 1:30 | 4 | 39 mm bit |
| 4.25 | C | V | 1:30 | 3 | 30 mm bit |

* See Table 4.5

*Source:* Hustrulid (1971).

**TABLE 4.7. Rock Drill Test Results**

| Rock Drill | Experimental | Theoretical (Eq. 14) |
|:---:|:---:|:---:|
| A | $F_t = 3.5p$ | $F_t = \alpha_1\ 7.06p*$ |
| B | $F_t = 3.5p$ | $F_t = \alpha_2\ 6.50p$ |
| C | $F_t = 2.75p$ | $F_t = \alpha_3\ 5.42p$ |

* $p$ (psi)

Drill A: 1 1/2 in. (3.81 cm), four-insert bits, laboratory rocks: 0.72
Drill B: 1 1/2 in. (3.81 cm), four-insert bits, laboratory rocks: 0.75
Drill A: 1 1/2 in. (3.81 cm), button bits, field rocks: 0.80
Both drills: 2 in. (5.08 cm), four-insert bits, field rocks: 0.62

*Source:* Hustrulid (1971).

bit end is almost equal to that in the incident wave (Figure 3.32*a*). However, if the bit and rock are in contact using the correct thrust, the reflected stress is diminished because of the energy used in drilling (Figure 3.32*b*). To maintain long drill steel life, which is reduced by fatigue, it is necessary to keep the number of high-stress reflected waves to a minimum. The piston delivers the same amount of energy to the drill steel for low and high thrust, and either the energy that is not transferred to the rock is trapped in the steel or part of it may be transmitted back to the piston. For this reason, drills are constructed so that when there is no thrust on the steel, the steel can move away from the piston, and the piston energy is absorbed by trapped air in the piston chamber and by the body of the drill machine. The primary effects of bit wear are to change the values of $E_v$, $T_r$, and $\beta$, and excessive gage wear reduces the clearance until the bit will become lodged in the drill hole.

For a simplified design of piston and drill (Figure 4.28), the peak stress is generated at the front of the pulse in the rod on impact. From the continuity of

**TABLE 4.8. Thrust Ratio**

| Rock Drill | Experimental | Theoretical |
|:---:|:---:|:---:|
| A/B | 1.00 | 1.08 |
| A/C | 1.27 | 1.30 |
| B/C | 1.27 | 1.20 |

*Source:* Hustrulid (1971).

**TABLE 4.9. Piston Impact Velocities and Energies as Functions of Machine Air Pressure for Rock Drills A, B, and C**

| Operating Pressure | | Blow Frequency | Piston Impact Velocity | | Impact Energy | |
|---|---|---|---|---|---|---|
| lb/in.$^2$ $g$ | (kg/cm$^2$) | (BPM) | (ft/s) | (m/s) | (ft-lb) | J |
| Rock Drill A | | | | | | |
| 40 | 2.8 | 1,200 | 17.8 | 5.43 | 32.8 | 44.4 |
| 50 | 3.5 | 1,290 | 19.8 | 6.04 | 40.8 | 55.3 |
| 60 | 4.2 | 1,380 | 21.8 | 6.64 | 49.3 | 66.8 |
| 70 | 4.9 | 1,450 | 23.4 | 7.13 | 57.2 | 77.5 |
| 80 | 5.6 | 1,500 | 25.1 | 7.65 | 65.5 | 88.8 |
| 90 | 6.3 | 1,550 | 26.6 | 8.10 | 73.6 | 99.8 |
| 95 | 6.65 | 1,600 | 27.4 | 8.35 | 78.0 | 105.8 |
| 100 | 7.0 | 1,660 | 28.1 | 8.56 | 81.1 | 110.0 |
| 110 | 7.7 | 1,720 | 29.4 | 8.96 | 90.5 | 122.7 |
| 120 | 8.4 | 1,800 | 30.8 | 9.39 | 98.8 | 134.0 |
| 130 | 9.1 | 1,850 | 32.0 | 9.75 | 106.6 | 144.5 |
| 140 | 9.8 | 1,900 | 33.2 | 10.12 | 114.3 | 155.0 |
| Rock Drill B | | | | | | |
| 40 | 2.8 | 1,500 | 19.3 | 5.88 | 27.6 | 37.4 |
| 50 | 3.5 | 1,640 | 21.4 | 6.52 | 34.3 | 46.5 |
| 60 | 4.2 | 1,740 | 23.6 | 7.19 | 41.5 | 56.3 |
| 70 | 4.9 | 1,830 | 25.4 | 7.74 | 48.0 | 65.1 |
| 80 | 5.6 | 1,910 | 27.2 | 8.29 | 55.2 | 74.9 |
| 90 | 6.3 | 1,980 | 28.8 | 8.79 | 61.9 | 83.9 |
| 100 | 7.0 | 2,080 | 30.4 | 9.26 | 69.0 | 93.6 |
| 120 | 8.4 | 2,200 | 33.3 | 10.15 | 82.7 | 112.1 |
| 130 | 9.1 | 2,275 | 34.7 | 10.58 | 89.5 | 121.4 |
| 140 | 9.8 | 2,475 | 36.0 | 10.97 | 97.0 | 131.5 |
| Rock Drill C | | | | | | |
| 40 | 2.8 | 1,520 | 16.2 | 4.93 | 19.9 | 27.0 |
| 50 | 3.5 | 1,600 | 18.0 | 5.49 | 24.8 | 33.6 |
| 60 | 4.2 | 1,670 | 19.8 | 6.04 | 29.8 | 40.4 |
| 70 | 4.9 | 1,750 | 21.3 | 6.49 | 34.7 | 42.1 |
| 80 | 5.6 | 1,900 | 22.8 | 6.95 | 39.8 | 53.6 |
| 90 | 6.3 | 2,000 | 24.2 | 7.38 | 44.6 | 60.5 |
| 100 | 7.0 | 2,100 | 25.5 | 7.77 | 49.6 | 67.3 |
| 110 | 7.7 | 2,180 | 26.8 | 8.17 | 55.0 | 74.6 |
| 120 | 8.4 | 2,250 | 28.0 | 8.53 | 59.6 | 80.8 |
| 130 | 9.1 | 2,300 | 29.1 | 8.87 | 64.6 | 82.6 |
| 140 | 9.8 | 2,350 | 30.2 | 9.20 | 69.5 | 94.2 |

*Source:* Hustrulid (1971).

**TABLE 4.10. Values of $K_o$**

| Operating Pressure | | Drill A | Drill B | Drill C |
|---|---|---|---|---|
| psi | kg/cm² | | | |
| 40 | 2.8 | 22.9 | 24.2 | 25.3 |
| 50 | 3.5 | 22.2 | 23.8 | 23.9 |
| 60 | 4.2 | 21.6 | 23.0 | 22.7 |
| 70 | 4.9 | 21.0 | 22.4 | 22.1 |
| 80 | 5.6 | 20.4 | 21.9 | 22.4 |
| 90 | 6.3 | 20.8 | 21.4 | 22.2 |
| 100 | 7.0 | — | 21.3 | 22.2 |

*Source:* Hustrulid (1971).

velocity and force equations, and for the same material in the piston and rod, it can be shown that the stress at the wave front is given by

$$\sigma_p = \sigma_i \left[ 1 + \frac{2A_3}{A_3 + A_1} \left( \frac{A_2 - A_1}{A_1 + A_2} \right) \right] \qquad (4.13)$$

where $\sigma_p$ = peak stress,

$$\sigma_i = 12 \frac{E}{C} \left( \frac{A_1}{A_1 + A_3} \right) V_s,$$

$E$ = Young's modulus, and $C$ = velocity of sound in steel.

**Problem:** Derive equation 4.12.

The peak stresses are given as functions of piston velocity for two types of drill steel (Table 4.17).

**TABLE 4.11. Impact Energy-Operating Pressure**

| Drill | Equation |
|---|---|
| A | $E_i = 0.816p$* |
| B | $E_i = 0.688p$ |
| C | $E_i = 0.496p$ |

* $p$ = psi

*Source:* Hustrulid (1971).

**TABLE 4.12. Predicted and Measured Penetration Rates—Robinson Deep Mine**

Penetration Rates, in./min (cm/min)

| Operating Pressure lb/in.² g | (kg/cm²) | Rock Drill A | | Rock Drill B | | Rock Drill C | |
|---|---|---|---|---|---|---|---|
| | | Actual | Predicted | Actual | Predicted | Actual | Predicted |
| 40 | 2.8 | 4.6 (11.7) | 4.6 (11.7) | 4.6 (11.7) | 4.8 (12.2) | 4.0 (10.2) | 3.5 ( 8.9) |
| 50 | 3.5 | 7.0 (17.8) | 6.1 (15.5) | 6.9 (20.1) | 6.5 (16.5) | 6.0 (15.2) | 4.6 (11.7) |
| 60 | 4.2 | 8.1 (20.6) | 7.9 | 8.7 (22.1) | 8.3 (21.1) | 6.6 (16.8) | 5.8 (14.7) |
| 70 | 4.9 | 10.2 (25.9) | 9.6 | 10.3 (26.2) | 10.2 (25.9) | 8.6 (21.8) | 7.0 (17.8) |
| 80 | 5.6 | 12.7 (32.3) | 11.4 | 12.8 (32.5) | 12.2 (31.0) | 10.0 (25.4) | 8.7 (22.1) |
| 90 | 6.3 | 14.7 (32.3) | 13.2 | 15.7 (39.9) | 14.1 (35.9) | 11.8 (30.0) | 10.3 (26.2) |
| 95 | 6.65 | 16.2 (41.1) | 14.4 | — (—) | — (—) | — (—) | — (—) |
| 100 | 7.0 | — (—) | 15.6 | 18.5 (47.0) | 16.6 (42.2) | 14.8 (37.6) | 12.0 (30.5) |

*Source:* Hustrulid (1971).

**TABLE 4.13. Predicted and Measured Penetration Rates—Vlakfontein Mine: Quartzite Composition Strength = 39,000 psi**

| Operating Pressure | | Penetration Rates, in./min (cm/min) | | | | | | | | |
|---|---|---|---|---|---|---|---|---|---|---|
| | | Drill A | | | Drill B | | | Drill C | | |
| lb/in.² g | (kg/cm²) | Predicted | Measured | Measured | Predicted | Measured | Measured | Predicted | Measured | Measured |
| 40 | 2.8 | 5.3 (13.5) | 3.8 ( 9.7) | 3.4 ( 8.6) | 5.5 (12.4) | 3.3 ( 8.4) | | 4.0 (10.2) | 3.8 ( 9.7) | |
| 50 | 3.5 | 7.0 (17.8) | 5.5 (14.0) | 4.8 (12.2) | 7.5 (19.1) | 4.9 (12.4) | | 5.3 (13.5) | 6.0 (15.2) | |
| 60 | 4.2 | 9.0 (23.1) | 8.2 (20.8) | 8.8 (22.4) | 9.6 (24.4) | 7.2 (18.3) | | 6.6 (16.8) | 6.8 (17.3) | |
| 70 | 4.9 | 11.1 (28.2) | 10.6 (26.9) | 10.7 (27.2) | 11.8 (30.0) | 9.0 (22.9) | | 8.1 (20.6) | 8.5 (21.6) | |
| 80 | 5.6 | 13.1 (33.3) | 12.4 (31.5) | 12.7 (32.3) | 14.1 (35.8) | 11.3 (28.7) | | 10.0 (25.4) | 10.0 (25.4) | |
| 90 | 6.3 | 15.2 (38.6) | 14.5 (36.8) | — (—) | 16.3 (41.4) | 13.9 (35.3) | | 11.9 (30.2) | 11.6 (29.5) | |
| 100 | 7.0 | 17.9 (45.7) | 16.9 (42.9) | — (—) | 19.2 (23.4) | 18.4 (46.7) | | 13.9 (35.3) | 14.0 (35.6) | |
| 110 | 7.2 | 20.8 (52.8) | 20.9 (53.1) | — (—) | | | | | | |
| 120 | 8.4 | 23.8 (60.5) | — | — (—) | | | | | | |
| 130 | 9.1 | 26.3 (66.8) | 24.1 (61.2) | — (—) | | | | | | |
| 140 | 9.8 | 29.0 (73.7) | — | — (—) | | | | | | |

*Source:* Hustrulid (1971).

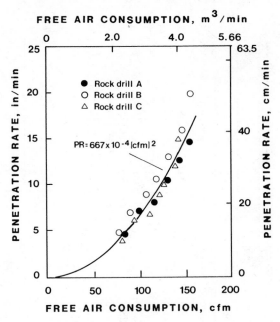

**Figure 4.26** Penetration rate as a function of the free air consumption for three rock drills at Robinson deep mine (Hustrulid, 1971).

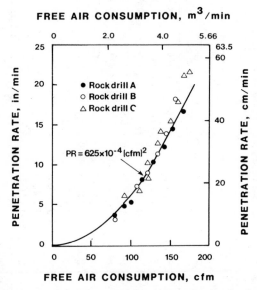

**Figure 4.27** Penetration rate as a function of the free air consumption for three rock drills at Vlak-fontein mine (Hustrulid, 1971).

**TABLE 4.14.  Air Consumption and Operating Pressure**

| Drill | Equation |
|-------|----------|
| A | cfm = 21.5 p 1/2 - 51.0 |
| B | cfm = 22.6 p 1/2 - 67.8 |
| C | cfm = 16.8 p 1/2 - 25.4 |

*Source:* Hustrulid (1971).

The manual thrust that can be applied to hand-held drills is about one half that for optimum penetration. An air cylinder will provide all of the thrust required (Hustrulid, 1971), or an alternative method is the use of a screw feed. Constant thrust with an air cylinder gives consistently better penetration rates than constant displacement with a screw feed (Table 4.18). For restricted working areas, an air leg is more practical (Figure 4.29) in which the pressure in the leg is adjusted to support the machine in line with the hole.

Assuming a person can exert a thrust equal to half their weight, and taking the weight of the drill into account, it requires an angle of 20° to obtain a thrust of 350 lb. If the person (150 lb) applies his weight to the drill, the weight of the person will increase the thrust. Also, the penetration rate with an air leg approaches that of an in-line thrust cylinder as the angle of inclination becomes 30° or less.

Thrust was applied manually to rock drill B by an operator and an assistant seated on the footwall, each facing the drill. From 12 to 80 holes 2 ft deep were drilled at each operating pressure. The penetration rates were consistently lower for manual thrust, especially at higher operating pressures, the men not being capable of applying the optimum thrust at high operating pressures. For example, at 80 lb/in.$^2$ operating pressure, the men were only able to apply 145 lb, whereas the optimum thrust is 250 lb.

**TABLE 4.15.  Penetration Rates: 30- and 39-mm Bits**

| Operating Pressure | | Penetration Rate (in./min) | | | | Ratio $R_1/R_2$ |
|---|---|---|---|---|---|---|
| lb/in.$^2$g | (kg/cm$^2$) | 30 mm Bit ($R_1$) | | 39 mm Bit ($R_2$) | | |
| 40 | 2.8 | 6.7 | (17.0) | 3.8 | ( 9.7) | 1.78 |
| 70 | 4.9 | 12.3 | (31.2) | 8.5 | (21.6) | 1.45 |
| 90 | 6.3 | 17.2 | (43.7) | 11.6 | (29.5) | 1.48 |
| 110 | 7.7 | 22.1 | (56.1) | 15.4 | (39.1) | 1.44 |
| 140 | 9.8 | 28.0 | (71.1) | 21.8 | (55.4) | 1.28 |

*Source:* Hustrulid (1971).

**TABLE 4.16. Predicted and Measured Penetration Rates**

| Operating Pressure | | Penetration Rate in./min (cm/min) | | | |
|---|---|---|---|---|---|
| lb/in.$^2$g | (kg/cm$^2$) | Predicted | | Measured | |
| 140 | 2.8 | 6.8 | (17.3) | 6.7 | (17.0) |
| 70 | 4.9 | 13.7 | (34.8) | 12.3 | (31.2) |
| 90 | 6.3 | 20.1 | (51.0) | 17.2 | (43.7) |
| 110 | 7.7 | 27.0 | (68.6) | 22.1 | (56.1) |
| 140 | 9.8 | 36.9 | (93.7) | 28.0 | (71.1) |

*Source:* Hustrulid (1971).

## 4.4   ROTARY-PERCUSSION DRILLING

In the previous two decades, according to Bullock (1974), percussive drilling productivity had risen by as much as 300% owing in part to the use of all hydraulic power for percussive drilling. The application of high-pressure hydraulic power to rock drills in the beginning was limited to rotary drills in rocks of low strength and abrasiveness. However, after the development of rotary-percussion drills, de-

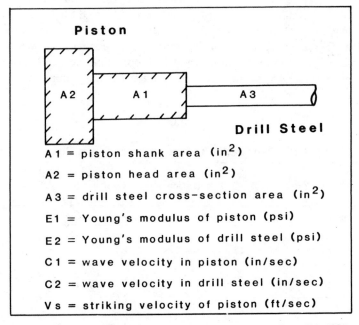

**Figure 4.28**  Diagram of a simplified piston and drill steel (Hustrulid, 1971).

**TABLE 4.17. Peak Stresses Calculated for Three Different Pistons**

1. Impacting in a 1-in. (2.54 cm) hexagonal drill steel

| Rock Drill Type | $A_1$ in.$^2$ (cm$^2$) | $A_2$ in.$^2$ (cm$^2$) | $A_3$* in.$^2$ (cm$^2$) | Peak Stress lb/in.$^2$ (kg/cm$^2$) |
|---|---|---|---|---|
| A | 3.06  (19.7) | 7.07  (45.6) | 0.83  (5.3) | $\sigma_p = 1\ 658.88\ V_s$ (116.1  $V_s$) |
| B | 1.77  (11.4) | 6.49  (41.9) | 0.83  (5.3) | $\sigma_p = 1\ 674.72\ V_s$ (117.2  $V_s$) |
| C | 1.93  (12.5) | 5.42  (35.0) | 0.83  (5.3) | $\sigma_p = 1\ 615.68\ V_s$ (113.1  $V_s$) |

*Corrected for water hole.

2. Impacting a ⅞-in. hexagonal drill steel

| Rock Drill Type | $A_1$ in.$^2$ (cm$^2$) | $A_2$ in.$^2$ (cm$^2$) | $A_3$* in.$^2$ (cm$^2$) | Peak Stress lb/in.$^2$ (kg/cm$^2$) |
|---|---|---|---|---|
| A | 3.06  (19.7) | 7.07  (45.6) | 0.55  (3.5) | $\sigma_p = 1\ 713.60\ V_s$ (120.0  $V_s$) |
| B | 1.77  (11.4) | 6.49  (41.9) | 0.55  (3.5) | $\sigma_p = 1\ 748.16\ V_s$ (122.4  $V_s$) |
| C | 1.93  (12.5) | 5.42  (35.0) | 0.55  (3.5) | $\sigma_p = 1\ 699.20\ V_s$ (118.9  $V_s$) |

* Corrected for water hole.

*Source:* Hustrulid (1971).

**TABLE 4.18. Penetration: Screw Versus Thrust Air Cylinder Feed**

| Operating Pressure | Rock Drill A* Penetration Rates in./min (cm/min) | | Rock Drill B* Penetration Rates in./min (cm/min) | |
|---|---|---|---|---|
| lb/in.$^2$g (kg/cm$^2$) | Screw Feed | Thrust Cylinder** | Screw Feed | Thrust Cylinder ** |
| 40  (2.8) | 3.9  (9.9) | 4.5  (11.4) | 3.5  (8.9) | 3.3  (8.4) |
| 50  (3.5) | 4.4  (11.2) | 7.0  (17.8) | 5.5  (14.0) | 4.8  (12.2) |
| 60  (4.2) | 7.3  (18.5) | 8.1  (20.6) | 7.8  (19.8) | 7.2  (18.3) |
| 70  (4.9) | 9.2  (23.4) | 10.2  (25.9) | 9.8  (24.9) | 9.0  (22.9) |
| 80  (5.6) | 9.9  (25.1) | 12.7  (32.3) | 10.1  (25.7) | 11.2  (28.4) |
| 90  (6.3) | 14.0  (35.6) | 14.7  (37.3) | 13.0  (33.0) | 14.0  (35.6) |
| 95  (6.7) | 12.4  (31.5) | 16.2  (41.1) | —  (—) | —  (—) |
| 100  (7.0) | —  (—) | —  (—) | 15.7  (39.9) | 18.4  (46.7) |

*Peak penetration rate used.
**Drill steel 1 used in all tests.

*Source:* Hustrulid (1971b).

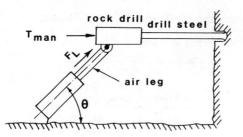

**Figure 4.29** Diagram of an air-leg and rock drill (Hustrulid, 1971).

velopments were also made to supply the power for percussion energy by hydraulic fluid pressures:

1. Rotation is continuous with a high torque of the same magnitude as that of a small rotary drill.

2. High thrust holds or quickly returns the cutting edges of the bit against the rock between blows. Thrust levels normally exceed those that the torque capabilities of either of the other types of percussion drills may overcome.

3. Since the drilling system employs both percussive and rotary drilling action, the bit is designed to use both types of comminution. That is, the bit has a proper rake angle to allow rock cutting by raking, a clearance angle to allow penetration by both the percussive force and the thrust, and an included angle sufficient to withstand the resulting high stress in the bit.

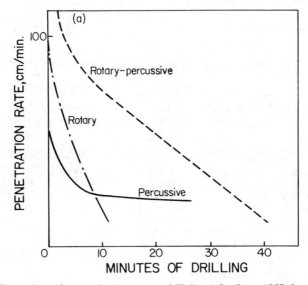

**Figure 4.30** Comparison of penetration rate versus drill time (after Inett, 1957; Larsen-Basse, 1973).

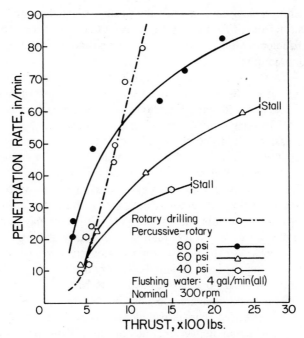

**Figure 4.31** Thrust penetration rate curves in shale (Fish, 1956–57).

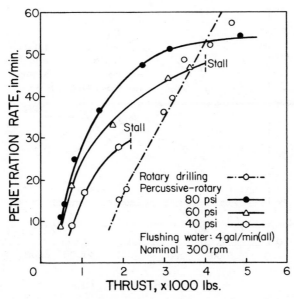

**Figure 4.32** Thrust penetration rate curves in Pennant sandstone (Fish, 1956–57).

**4.** The flushing system must remove the cuttings when they are formed, because clogging in front of the bit is more critical than for either rotary or percussive drills. Efficient cutting removal is also much more critical to prevent excessive comminution.

Rotary-percussive drills operate with chisel- (single or cross) shaped bits or fixed or roller bits implanted with carbide buttons, usually hemispherical in shape. Published work in 1973 related mostly to drilling with chisel-shaped bits.

Rotary-percussive drilling combines rotary and percussive techniques with the tool held constantly against the rock surface under high axial (thrust) load as it is turned between impacts. The bit insert wedge angle is usually 90°; however, its axis makes an oblique angle with the rock face to give a forward rake angle for the cutting action. Bit impact removes rock by crushing and fracture and forms cracks and chips in the rotation part of the cycle. Crushing may reduce the cutting friction on the clearance face to less than that for rotary drilling. This factor may vary for harder rocks. As for power, rotary-percussive drilling is more efficient than percussive drilling and less efficient than rotary drilling (Figure 4.30), although the bit wear rate for rotary-percussive drilling is between the other two and is therefore useful for rocks that are too abrasive for rotary drilling but in which percussive action is effective. The bit is under compression most of the time, which extends its life, and higher impact energies are used than for straight percussive drilling (Fairhurst and Lacabanne, 1957).

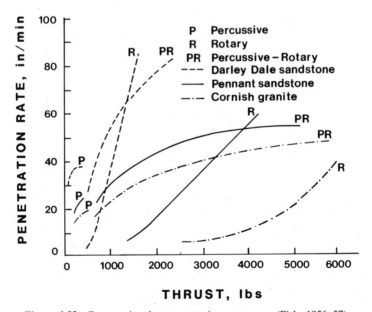

**Figure 4.33** Comparative thrust penetration rate curves (Fish, 1956–57).

**TABLE 4.19. Comparative Air Consumption (80 lb/in.$^2$ working pressure)**

| Type of Rock | Rotary-percussive | | | Rotary | | | Percussive | | |
|---|---|---|---|---|---|---|---|---|---|
| | Penetration Rate (in./min) | Air Consumption (ft$^3$/min) | ft$^3$ Air/in.$^3$ Rock Broken | Penetration Rate (in./min) | Air Consumption (ft$^3$/min) | ft$^3$ Air/in.$^3$ Rock Broken | Penetration Rate (in./min) | Air Consumption (ft$^3$/min) | ft$^3$ Air/in.$^3$ Rock Broken |
| Darley Dale Sandstone | 70 | 485 | 3.1 | 70 | 260 | 1.7 | 36 | 185 | 2.3 |
| Pennant Sandstone | 50 | 475 | 4.3 | 50 | 285 | 2.6 | 23 | 165 | 3.2 |

*Source:* Bullock (1974).

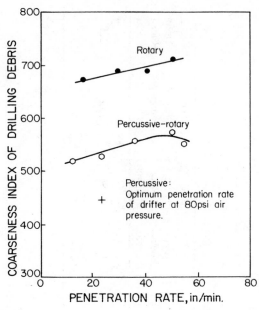

**Figure 4.34** The variation of size of drillings with rate of penetration in Pennant sandstone (Fish, 1956–57).

### TABLE 4.20. Rocks Used in the Experimental Work

| Rock Type | Description | Shear Strength lb per sq in. | Crushing Strength lb per sq in. |
|---|---|---|---|
| Chislet Shale | A typical coal measure shale | 1,700* | 5,400* |
| Darley Dale Sandstone | A rock typical of many in the Carboniferous series; fairly soft but abrasive | 1,500 | 8,000 |
| Pennant Sandstone | A very hard and abrasive coal measure sandstone | 2,900 | 12,500 |
| Cornish Granite | Fairly coarse-grained | 3,100 | 15,800 |

\* Across the bedding planes.

*Source:* Bullock (1974).

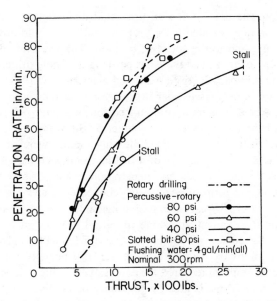

**Figure 4.35** Thrust penetration rate curves in Darley Dale sandstone (Fish, 1957a).

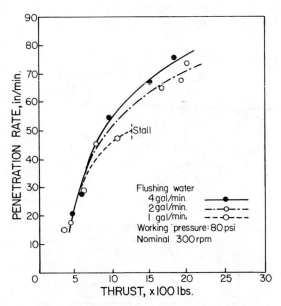

**Figure 4.36** Effect of variation of flushing water quantity in Darley Dale sandstone (Fish, 1957a).

The penetration rate decreases with time of drilling because of bit wear. The average penetration rate is stated to be proportional to the hold-down force, to the blow impact energy, and to the speed of rotation. The wear rate during operation depends on the type of rock, and the wear per length of hole drilled decreases with hold-down load and impact energy (Wahl and Kantenwein, 1961). Increased penetration rate results in decreased wear in sandstone but increased wear for granite (Fish, 1957a–c). Abrasion is more pronounced for the quartz-rich sandstone during the cutting part of the cycle. (See also Chapter 5, Bit Wear.)

In research by Fish (1957a–c), the following factors were proposed as being important in the technical suitability of a drilling method for a given application: (1) drilling characteristics: forces and penetration rates; (2) rate of bit wear; (3) power consumption and form of energy; (4) convenience and safety.

For a Hausherr DK7ES rotary-percussive machine, the PR in shale was higher (Figure 4.31) than for Pennant sandstone (Figure 4.32), the shale requiring lower thrust. The thrust–PR curves are similar to those for Pennant sandstone. Very high thrust is required for rotary drilling (Figure 4.33), especially for harder rock, C granite. The air consumption required for percussive drilling is lower (Table 4.19).

Air consumption at 80 psi is greatest for rotary-percussive and least for percussive drilling for the two sandstones, in terms of both cmf and cmf/cu in. of rock broken (Table 4.19). The coarseness of the drillings is greater for the rotary drilling (Figure 4.34).

Fish (1957a, b) reported the results of rotary-percussive drilling and comparative studies of percussive (single-insert), rotary, and rotary-percussive (single-insert) drilling. The first series of rotary-percussive tests was made with the Hausherr DK7ES machine with some tests being run with a separate rotary drive, a third

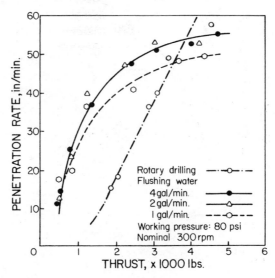

**Figure 4.37**   Effect of variation of flushing water quantity in Pennant sandstone (Fish, 1957a).

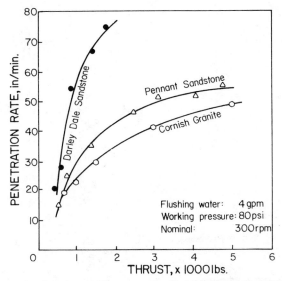

**Figure 4.38** Comparative thrust/penetration rate curves for rotary-percussive drilling in various rocks (Fish, 1957a).

series being run for bit wear tests. Percussive tests were made with a $3\frac{1}{2}$-in. drifter. Rotary-percussive bits were of normal and slotted types with a 9% cobalt and coarse-grained WC. The rocks tested were those tested earlier: shale, two sandstones, and granite (Table 4.20).

The slotted bit design for use in softer rocks gave the greatest penetration rate

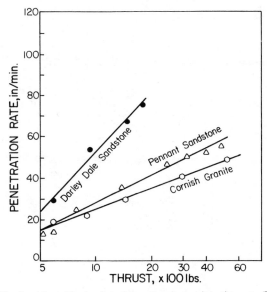

**Figure 4.39** Semi-logarithmic plot of thrust against penetration rate (Fish, 1957a).

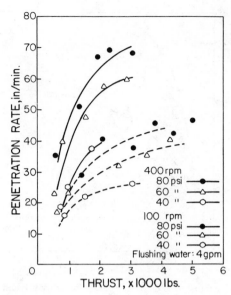

**Figure 4.40** Thrust/penetration rate curves in Pennant sandstone at 100 and 400 rpm (Fish, 1957b).

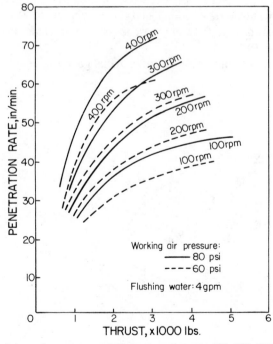

**Figure 4.41** Thrust/penetration rate curves in Pennant sandstone at 100, 200, 300, and 400 rpm (Fish, 1957b).

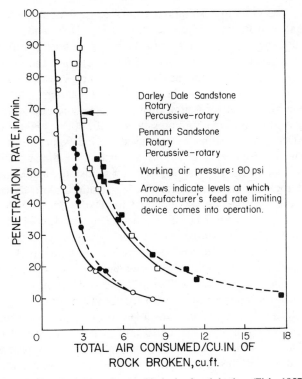

**Figure 4.42**  Air consumed per cubic inch of rock broken (Fish, 1957b).

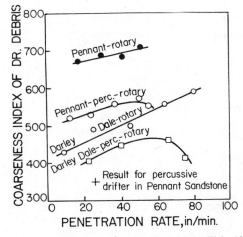

**Figure 4.43**  Coarseness index of drilling debris (Fish, 1957b).

**135**

(Figure 4.35). A reduction in flushing rate from 4 to 1 gal/min resulted in reduction of penetration rate and stalling (Figure 4.36) in Darley Dale sandstone and Pennant sandstone, the PR increasing linearly with the thrust for rotary drilling.

As noted, the Hausherr DK7ES (rotary-percussive) was used with independent rotation for some tests. The frequency and blow energy were controlled by the operating pressure, with a frequency in the range of 6000 bpm. The slotted bit (for soft rock) gave a higher PR for 4 gpm flushing water, and rates were higher than those for rotary drilling for operating pressures of 40, 60, and 80 psi (Figure 4.37) at low thrusts. Decreased flushing water caused a decrease in PR for both Pennant and Darley Dale sandstones. The PR as a function of thrust decreases in the following order: Darley Dale sandstone, Pennant sandstone, and C granite (Figure 4.38). For soft rocks, rotary-percussive drilling required more thrust than rotary drilling. When the rotation rate was doubled for Darley Dale sandstone, the PR was almost double for a given operating pressure (Figures 4.39–4.41). The results were similar for Pennant sandstone for rotation speeds ranging from 100 to 400 rpm, although the increase in PR was not so great in the harder rock (Figure 4.41).

The compressed air consumed for a unit volume of rock drilled was lower for high penetration rates (Figure 4.42). The coarseness of cuttings for the two sandstones was greatest for rotary and increased for both types of drilling with increase of PR (Figure 4.43). This indicates a decrease of specific energy of drilling for this range of drill operation with the increase of penetration rate for both types of drilling in sandstone, but to a certain optimum value for rotary-percussion only.

In studies of drill hole flushing media, Fish (1957a–c) investigated rotary, percussive, and rotary-percussive drilling. Two laboratory tests were run wet and dry

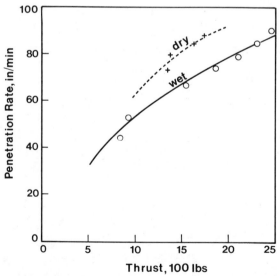

**Figure 4.44** Thrust/penetration rate curves for wet and dry rotary-percussive drilling in Darley Dale sandstone (Fish, 1957).

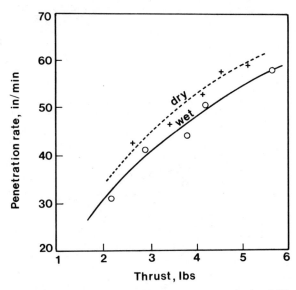

**Figure 4.45**   Thrust/penetration rate curves for wet and dry rotary-percussive drilling in Pennant sandstone (Fish, 1957).

on Pennant and Darley Dale sandstone to determine thrust versus penetration rate (Figures 4.44–4.46). The Darley Dale sandstone was more difficult to drill, and the penetration rates were higher for rotary drilling.

### 4.4.1  Hydraulic Drills

The physical characteristics and dimensional data of available hydraulic drills (Bullock, 1974) were obtained from manufacturers (Table 4.21), and some performance characteristics were calculated (Tables 4.22, 4.23).*

Advantages of the hydraulic drills are (1) reduction of machine noise because of no escaping compressed air; (2) no fog from oil and compressed air; (3) low steel breakage; (4) higher drill rates; and (5) increased productivity of broken rock or ore.

The number of blows per minute varies from 1800 to 9600 with corresponding ft-lb per blow of about 250–500 (Table 4.21). Typical penetration rates for five types of drills vary from 1.1 to 15 ft/min depending on the type of rock (Tables 4.22, 4.23), and costs per foot are lower for hydraulic drilling (Table 4.24).

A further comparison of drilling capabilities using a measure of the specific energy, which is defined by Bullock (1974, 1975, 1976) as the total amount of work required to remove a unit volume of rock from the hole and to reduce it to a specific size, includes (1) all of the work, (2) a particular drilling system, (3) removal of rock, and (4) reduction to specific size.

*Data in this section in English units only.

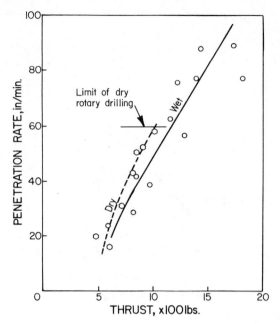

**Figure 4.46**  Thrust/penetration rate curves for wet and dry rotary drilling in Darley Dale sandstone (Fish, 1957).

The work output is then calculated from

$$\Sigma W = 12fF_p(T_p) + 24\pi F_{Tq}N(T_T) + F_{Th}PR \qquad (4.14)$$

where $f$ = blows per minute; $F_p$ = energy per blow, ft-lb; $T_p$ = empirical energy transfer ratio; $F_{Tq}$ = torque, ft-lb; $N$ = revolutions per minute; $T_T$ = torque energy transfer ratio; $F_{Th}$ = thrust, lb; and $PR$ = penetration rate, in./min.

All hydraulic drills can drill at 50–100% greater penetration rate than pneumatic drills, and it has been shown that there is a correlation between chip size and the penetration rate. The rock–bit interaction, including the rebound between blows and the flushing of cuttings, is a critical factor.

Equation 4.14 was used to calculate the work rate of pneumatic drills, independent rotation pneumatic, all-hydraulic rotary-percussion, and all-hydraulic drills for comparison purposes (Figures 4.47, 4.48). For the ratio of output work to installed horsepower—that is, the mechanical efficiency—the pneumatic drills all rate the lowest, with the rotary hydraulic drill being fifth from the top. In general, the efficiency of hydraulic drills is greater than that of the other types. Also, the specific energy of drilling is lower for the hydraulic drills, where the specific energy is defined as the total amount of work of the drilling system to remove a unit volume of rock.

TABLE 4.21. Specification Describing the Various All-Hydraulic Rock Drills (information furnished by the manufacturer except where indicated)

| | | | | | | | | Rock Drill | | | | | Recommended Drill Water | |
|---|---|---|---|---|---|---|---|---|---|---|---|---|---|---|
| Manufacturer | Drill | Blows/Minute | Ft-lb Blow | Maximum RPM | Maximum Torque (ft-lb) | Operating Torque[b] (ft-lb) | Maximum Thrust (lb) | Weight (lb) | Length (in.) | Hgt. Above Feed (in.) | Width or Diam. (in.) | Installed HP/Drill | Pressure (psi) | Flow (gpm) |
| Atlas–Copco | COP 1038 HD | 2,500–3,800 | ≈250[a] | 300 | 300 | 30 | 1,320 | 298 | 38.7 | 8.82 | 10.2 | 60 | 87–145 | 9.2–13.2 |
| Ingersoll–Rand | HARD II | 9,600 | 50 | 200 | 140 | 14 | 2,900 | N.R. | 31.5 | 4.0 | N.R. | 75 | N.R. | N.R. |
| Ingersoll–Rand | HARD III | 9,300 | 63 | 225 | 139 | 14 | 2,000 | 475 | 38¾ | 10¾ | 9⅝ diam. | 62.5 | 100 | 35 |
| Joy | JH-2 | 12,000 | 90 | 450 | 800 | 35 | 1,400 | 340 | 26 | 9 | 9 | 75 | 100 | N.R. |
| Krupp | HB-100-B | 1,650–1,800 | N.R. | 320 | 210 | 21 | 2,200 | 260 | 44.5 | 7.3 | 11.1 | 60 | N.R. | N.R. |
| Krupp | HB-100-S | 1,650–1,800 | N.R. | 150 | 500 | 50 | 2,200 | 310 | 47.8 | 15.7 | 13.0 | 60 | N.R. | N.R. |
| Krupp | HB-100-O | 1,650–1,800 | N.R. | 40 | 1,700 | 170 | 2,200 | 310 | 47.8 | 15.7 | 13.0 | 75 | N.R. | N.R. |
| Krupp | HB-50 | 2,300–2,800 | N.R. | 250 | 120 | 12 | 880 | 180 | 26.0 | 11.3 | 10.6 | 50 | N.R. | N.R. |
| LeRoi | LHD155 | 2,000 | 250 | 150 | 50 | 5 | 2,000 | 405 | 34 | 5 | 11 | 50 | N.R. | N.R. |
| Montabert | H-40 | 3,200 | | 200 | 470 | 47 | N.R. | 220 | 33 | 6⅛ | N.R. | 30 | 60 | 3¼ |
| Montabert | H-50 | 2,700 | 145 | 200 | 687 | 69 | N.R. | 220 | 33 | 6⅛ | N.R. | 30 | 60 | 3¼ |
| Montabert | H-60 | 2,200 | 220 | 100 | 870 | 87 | N.R. | 330 | 43¼ | 6⅛ | N.R. | 30 | 60 | 4½ |
| Montabert | H-100 | 1,400 | 365 | 200 | 220 | 22 | 2,650 | 331 | 31.2 | 3.35 | 13.1 | 40 | 88–147 | 4 |
| Secoma | RPH 35 | 2,600 | 145 | 200 | 148 | 15 | 3,600 | 243 | 33.0 | 8.1 | 10.6 | 40 | 73–147 | 4 |
| Tamrock | HE425 | | | 200 | | | | | | | | | | |
| Tamrock | HL432 | 3,100 | 133 | 200 | 148 | 15 | 3,600 | 243 | 31.3 | | | | 73–147 | 6.6 |
| Tamrock | HL438 | | | 300 | 148 | 15 | 3,600 | 247 | 31.9 | | | | 88–147 | 8 |

[a] Calculated Information.
[b] The 10% torque was used to calculate the amount of energy used for torque during normal drilling.
N.R., no report.
*Source:* Bullock (1974).

**TABLE 4.22. Mines and Quarries Using Hydraulic Drills in North America**

| Hydraulic Drill User | | | Type | | | | | Drill Bits | | | Drill Steel | | | | | |
|---|---|---|---|---|---|---|---|---|---|---|---|---|---|---|---|---|
| Name | Place | Manufacturer | Drill | Jumbo | No. of Jumbos | No. of Drill Jumbos | Rock Type or Strength (psi) | Diameter (in.) | Life (ft) | Type | Diameter (in.) | Shape | Type | Length (ft) | Life (ft) | Penetration Rate (ft/min) |
| Occidental Petroleum | Grand Junction | Atlas–Copco | COP 1038 | Boomer H132 | 1 | 2 | Oil Shale | $2\frac{1}{4}$ | 20,000 | Cross | | Round | Carborized | 14 | 5,000 | 10.0 |
| St. Joe Minerals | Fletcher | Ingersoll–Rand | Hard III | 96 RMH | 2 | 2 | Dolomite 5,000–40,000 | $1\frac{1}{2}$ | 350 | RPD Single Pass | $1\frac{1}{4}$ | Round | Alloy | 14 | N.R. | 12 |
| Dravo | Maysville | Ingersoll–Rand | Hard III | 96 RMH | 2 | 2 | 22,000 | $1\frac{3}{4}$ | N.R. | Cross | $1\frac{1}{4}$ | Round | Alloy | 14 | N.R. | 5.1 |
| | | | | | | | | $1\frac{3}{4}$ | N.R. | RPD | $1\frac{1}{4}$ | | | 14 | N.R. | 7.8 |
| | | | | | | | | 2 | N.R. | RPD | $1\frac{1}{4}$ | Round | Alloy | 14 | N.R. | 8.0 |
| Cooper Range | White Pine | Joy | JH-Z | EPM-2R | Test | 2 | | $1\frac{3}{4}$ | 2,000 | RPD | $1\frac{1}{4}$ | Round | Alloy | 13.5 | 8,000 | 15.5 |
| | | | | | | | | $1\frac{3}{4}$ | 2,500 | Button | $1\frac{1}{4}$ | Round | Alloy | 13.5 | N.R. | 9.0 |
| | | | | | | | | 2 | 2,800 | RPD | $1\frac{1}{4}$ | Round | Alloy | 13.5 | N.R. | 13.0 |
| | | | | | | | | $2\frac{1}{2}$ | 2,000 | Cross | $1\frac{1}{4}$ | Round | Alloy | 13.5 | N.R. | 6.5 |
| Mt. Bruno Quarries | St. Bruno de Montarville, Quebec | Montabert | H-100 | J.C. Hydrofore | 1 | 1 | Hornfels 33,000 | $2\frac{1}{2}$ | | | $1\frac{1}{2}$ | Hex | T-38 | 10 Ext. | N.R. | 2.7 |
| | | | | | | | | $2\frac{1}{2}$ | | | $1\frac{1}{2}$ | Hex | T-38 | 10 Ext. | N.R. | 2.6 |
| | | | | | | | | $3\frac{1}{2}$ | | | $1\frac{1}{2}$ | Hex | T-38 | 10 Ext. | N.R. | 2.0 |
| | | | | | | | | $3\frac{1}{2}$ | | | $1\frac{1}{2}$ | Hex | T-38 | 10 Ext. | N.R. | 1.7 |
| | | | | | | | | 4 | | | $1\frac{1}{2}$ | Hex | T-38 | 10 Ext. | N.R. | 1.2 |
| International Nickel | Creighton Mine | Montabert | H-50 | J.C. MJM-21H | 3 | 3 | N.R. | $1\frac{11}{16}$ | 750 | Cross | $1\frac{1}{4}$ | Hex | Carborized | 12 | 3,500 | 2.3 |
| Agnew Lake Mine | Espanola, Ontario | Montabert | H-70 | Juro Fan Drill | 1 | 1 | Arkose & Polymictic Conglomerate | $2\frac{1}{8}$ | | Cross | $1\frac{1}{2}$ | Hex | N.R. | 6 Ext. | N.R. | 2.2 |
| | | | | | | | | 3 | | Cross | | | | | | 1.1 |
| Amax | Buick Mine | Montabert | H-45 | Crawler | Test | 1 | Dolomite | $1\frac{5}{8}$ | N.R. | Cross | $1\frac{1}{4}$ | Round | Alloy | 12 | N.R. | 8.0 |

*Source:* Bullock (1974).

TABLE 4.23. Partial List of Mines and Quarries Using Hydraulic Drills in South Africa and Europe

| Date of Use | Hydraulic Drill User | Manufacturer | Type: Drill | Type: Jumbo Unit | No. of Units | No. of Drills | Rock Type or Strength (psi) | Drill Bits: Diameter (in.) | Drill Bits: Life (ft) | Drill Bits: Type | Drill Steel: Diameter (in.) | Drill Steel: Shape | Drill Steel: Type | Drill Steel: Length (ft) | Drill Steel: Life (ft) | Penetration Rate (ft/min) |
|---|---|---|---|---|---|---|---|---|---|---|---|---|---|---|---|---|
| 73–74 | Norcem Brevik, Norway | Atlas Copco | COP 1038 | Boomer | 1 | 2 | 15,640 | $1\frac{7}{8}$ | 1650 | Cross | $1\frac{1}{2}$ | Hex. | N.R. | 14 | 4000 | 4.7 |
| 73–74 | Norcem Brevik, Norway | Atlas Copco | COP 1038 | H132 | 1 | 2 | 15,640 | $2\frac{1}{4}$ | N.R. | N.R. | N.R. | N.R. | N.R. | 14 | N.R. | 4.4 |
| 73–74 | Norcem Brevik, Norway | Atlas Copco | COP 1038 | H132 | 1 | 2 | 15,640 | $2\frac{1}{2}$ | N.R. | N.R. | N.R. | N.R. | N.R. | 14 | N.R. | 4.2 |
| 74–75 | Anglo-Amer. Welcom | Atlas Copco | COP 1038 | Promec T260 | 18 | 2 | O.F.S. Quartzite | 1.69 | 722 | Cross | $1\frac{1}{8}$ | Hex. | Carburized | 15.8 | 2000 | 4.3 |
| 72 | Boliden Liasvalle | | COP 1038 | Promec 296 | 2 | 2 | Sandstone | 2 | 394 | Cross | N.R. | N.R. | N.R. | N.R. | 3900 | 4.6 |
| | | | | 297 | | 4 | Sandstone | | | | | | | | | |
| 76 | Preussag Erzbergwerk Bad Grund | Krupp | KIS HB-50 | Alimak | 1 | 1 | Iron Ore | 1.57 | 919 | Chisel | N.R. | Hex. | Alloy | 9.8 | 1150 | 6.1 |
| 76 | | Krupp | HB-50 | NCB | 6 | 1 | Limestone 15,950 | $1\frac{3}{4}$ | N.R. | Cross | 1 | N.R. | N.R. | N.R. | N.R. | 8.0 |
| 75 | Trierer Kalk Waaerbillig Luxemberg | Krupp | HB-50 | DHM-50 | 1 | 1 | Limestone | 1.57 | 2100 | Intregal | 1 | Hex. | Alloy | 9.8 | 2100 | 7.7 |
| 76 | Knauf Anhydrite-Heidenheim | Krupp | HB-51 | Knauf | 1 | 1 | Basalt 55,000 | 1.61 | N.R. | Cross | $1\frac{1}{4}$ | N.R. | Carburized | N.R. | N.R. | 7.6 |
| 75 | Franzefoss Quarry, Norway | Krupp | HB-101 | DHR-75 | 1 | 1 | Granite 35,000 | 3.0 | N.R. | Cross | $1\frac{1}{4}$ | Round | T-38 | 10 | N.R. | 2.1 |
| | | | | | | | Hornfels 40,000 | 3.0 | N.R. | Button | $1\frac{1}{4}$ | Round | T-38 | 10 | N.R. | 3.1 |
| | | | | | | | | 3.0 | N.R. | Button | $1\frac{1}{2}$ | Round | T-38 | 10 | N.R. | 2.9 |
| 75 | Diabase Quarry, Wolfshagen | Krupp | HB-103 | DHR-75 | 1 | 1 | Diabase 40,000 | $3\frac{1}{2}$ | N.R. | Button | $1\frac{1}{2}$ | Round | T-38 | 10 | N.R. | 2.5 |
| 75 | Tele M. Ebbers Quarry, Hunneberg/Harz Area | Krupp | HB-101 | DHR-75 | Test | 1 | Diabase 55,000 | $3\frac{1}{2}$ | N.R. | Button | $1\frac{1}{2}$ | Round | T-38 | 10 | N.R. | 2.0 |
| | | | | | | | | $3\frac{1}{2}$ | N.R. | Cross | $1\frac{1}{2}$ | Round | T-38 | 10 | N.R. | 1.6 |

**TABLE 4.23.** (*Continued*)

| Date of Use | Hydraulic Drill User | Type | | | No. of Units | No. of Drills | Rock Type or Strength (psi) | Drill Bits | | | Drill Steel | | | | | Penetration Rate (ft/min) |
|---|---|---|---|---|---|---|---|---|---|---|---|---|---|---|---|---|
| | | Manufacturer | Drill | Jumbo Unit | | | | Diameter (in.) | Life (ft) | Type | Diameter (in.) | Shape | Type | Length (ft) | Life (ft) | |
| 75 | Boliden Renstom | Tamrock | HE-425 | Minimatic H | 1 | 2 | 40,000 | 1.65 | 525 | Cross | 1 | Hex. | Carburized | 15.8 | 591 | 4.4 |
| 76 | Barbara Rohstoff-Namen-Beltriebe, W. Ger. Wohlverwahrt | Tamrock | HE-425 | Minimatic H | 1 | 1 | 18,000 | 1.77 | N.R. | Cross | N.R. | Round | Carburized | 13.1 | N.R. | 6.7 |
| | | | | | | | | 3.50 | N.R. | Cross | N.R. | Round | Carburized | 13.1 | N.R. | 2.6 |
| 75–76 | Universal, Arge KW Solk, Austria Stein e.d. Enns | Tamrock | HE-425 | PV-2 ZR-650H | 1 | 2 | 30,000 | 1.40 | 525 | Intregal | 1 | Hex. | Carburized | 11.5 | 525 | 4.9 |
| 75 | Ammi Sarda Masua Mine, s.p.A., Italy Sardinia | Tamrock | HE-425 | Minimatic H | 1 | 2 | 15,000–25,000 | 1.5 | 1312 | Intregal | 1 | Hex. | Carburized | 13.1 | 1312 | 5.9 |
| | | | | | | | | 3.0 | N.R. | Cross | 1 | Hex. | Carburized | 13.1 | N.R. | 2.6 |
| 74–76 | Outokumpu Oy, Hammaslahti Finland Mine | Tamrock | HE-425 | Minimatic H | 1 | 2 | 45,000 | 1.65 | 656 | Intregal | 1 | Hex. | Carburized | 10.4 | 656 | 4.3 |
| | | | | | | | | 3.00 | 1312 | Cross | 1 | Hex. | Carburized | 10.4 | 1312 | 2.0 |
| 76 | Anglo Amer. Waal Reefs So. Africa | Tamrock | HE-426 | Minimatic H | 1 | 2 | O.F.S. Quartzite | 1.69 | N.R. | Cross | $1\frac{1}{2}$ | Hex. | Carburized | 15.8 | N.R. | N.R. |
| 76 | LKAB Kiruna & Malmber Get Mine | Tamrock | HL-438 | Maxomatic H3R | 2 | 3 | N.R. | 1.69 | N.R. | Cross | N.R. | Hex. | Carburized | N.R. | N.R. | N.R. |
| | | | | | | | | 1.77 | N.R. | Cross | N.R. | Hex. | Carburized | N.R. | N.R. | N.R. |
| | | | | | | | | 4.0 | N.R. | Cross | N.R. | Round | Carburized | N.R. | N.R. | N.R. |
| | | | | | | | | 5.0 | N.R. | Cross | N.R. | Round | Carburized | N.R. | N.R. | N.R. |

*Source:* Bullock (1974).

**TABLE 4.24. Sample Comparison of Drilling Costs Between Pneumatic and Hydraulic Drilling**

| Item of Cost | Mine A Jumbo | | Mine B Jumbo | |
|---|---|---|---|---|
| | Pneumatic | Hydraulic | Pneumatic | Hydraulic |
| Capital expenditures | $0.071/ft | $0.080 | N.R. | N.R. |
| Operating wages | 0.085 | 0.071 | 0.107 | $0.063 |
| Compressed air | 0.015 | — | N.R. | — |
| Compressed air dist. system | 0.037 | — | N.R. | — |
| Electrical power | — | 0.002 | N.R. | 0.003 |
| Electrical dist. system | — | 0.017 | N.R. | N.R. |
| Job maintenance | | | | |
|   Wages | 0.055 | 0.057 | N.R. | N.R. |
|   Parts | 0.058 | 0.013 | N.R. | N.R. |
| Drill maintenance | | | | |
|   Wages | 0.004 | 0.004 | N.R. | 0.013 |
|   Parts and supplies | 0.015 | 0.026 | N.R. | 0.016 |
| Drilling accessories | 0.166 | 0.155 | N.R. | 0.101 |
| Total cost | $0.506 | $0.455 | | N.R. |

*Source:* Bullock (1974).

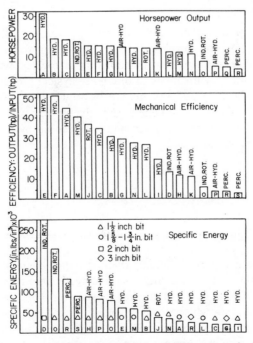

**Figure 4.47** Comparison of specific energy for different drilling systems in the Bonneterre dolomite (Bullock, 1974).

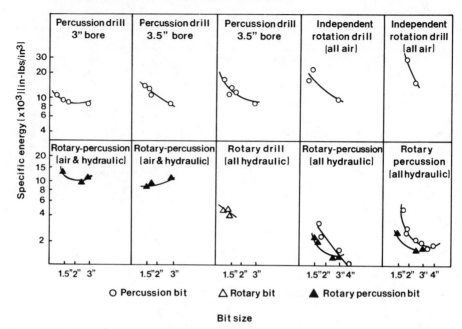

**Figure 4.48** Comparison of specific energy for different drilling systems in Bonneterre dolomite (Bullock, 1974).

For tests in dolomite (Figure 4.48), the rotary-percussion drill with a rotary-percussion-type bit gave the lower specific energy except for a 4-in. (10.16-cm) diameter bit. However, the optimum combination of rock properties and bit size and type would be different for rocks of different physical properties.

Some of the disadvantages of hydraulic drills are (1) the capital cost (which is offset by advantages), (2) high heat generated by the power pack, (3) space required by the power pack, and (4) maintenance problems caused by operators not experienced with this type of drill.

## 4.5  OIL SHALE—LABORATORY TESTS

The commercial oil shales of western Colorado have been found to be difficult to drill with pneumatic percussive drills for blasting purposes. Research on the drillability of oil shale was designed (Brune, 1983) to define some of its drilling characteristics as a function of physical properties and hydrocarbon content to determine in turn the best drill operation parameters. Tests were also made with a larger hydraulic drill to ascertain whether it is adaptable to drilling of oil shale.

The standard physical properties such as strengths and moduli have been measured for shale, but, as is the case for most rocks, there is not a consistent relation between either standard properties or special properties such as CRS, oil content,

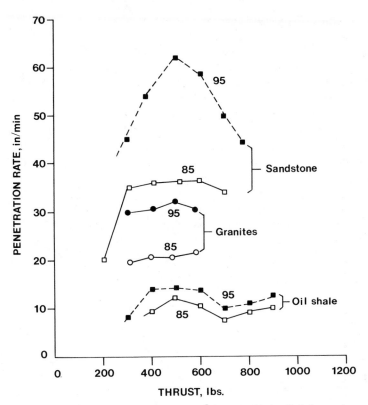

**Figure 4.49** Comparative penetration rates, drill A, $1\frac{5}{8}$-in. cross bit in oil shale, sandstone, and two granites (Brune, 1969).

**TABLE 4.25. Drills Used for Oil Shale.**

| Drill | A | B | C |
|---|---|---|---|
| Type | Pneumatic Rifle Bar 1 : 30 | Pneumatic Independent Rot. | Hydraulic Percussion Independent Rot. |
| Maximum operating pressure and flow Piston diameter,[a] length and weight | $p$ = 95 psi $V$ = 200 cfm $D$ = 170.4 mm $L$ = 202 mm $W$ = 2700 g | $p$ = 85 psi $V$ = 270 cfm $D$ = 91.2 mm $L$ = 447 mm $W$ = 12600 g | $p$ = 3150 psi $V$ = 25 gpm $D$ = 53.9 mm $L$ = 407 mm $W$ = 4922 g |
| Nominal stroke[a] | 64 mm | n.a. | 15.6 mm |
| Blow rate | 1800/min @ 95 psi | 1650/min @ 85 psi | 9000/min @ 3750 psi |
| Blow energy | 50 ft-lb @ 95 psi | 140 ft-lb @ 85 psi | 45 ft-lb @ 3300 psi/130 F |
| Power output (ft-lb/min) | 90,000 | 235,000 | 405,000 |

[a]Data are only approximate; details are proprietary.
*Source:* Mfg's. data.

and drillability. The kerogen content in the shale does affect drillability but not in proportion to the percentage present.

The rate of penetration was measured for pneumatic and hydraulic drills with both cross and button bits. In addition to standard properties, tests were made of the energy absorption characteristics by means of a split Hopkinson bar apparatus, and the kerogen content was determined as part of the test procedure.

One of the qualitative properties of rock that affects drillability is the brittleness of the minerals in the matrix of the rock. Oil shales are more plastic in their response to impact, because mineral grains are crushed less than in harder rocks and are not as easily fragmented and removed from the face of the drill hole. Button bits may be more effective in some types of oil shale. The most effective level of impact energy per blow appears to be less than that for harder rock. It has been suggested

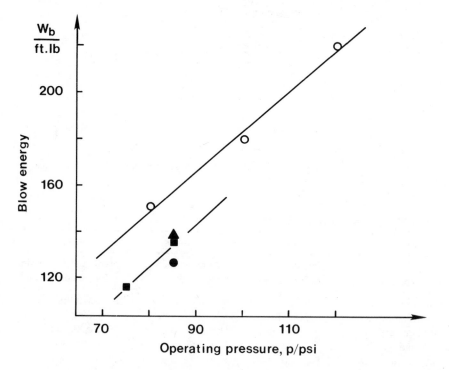

O manufacturer's data, W bl – 1.750p – 8.33, r 2 – 0.993

● measured, average for oil shale A1

▲ measured, average for Barre Granite

■ measured, average for oil shale A2

All measured data: $W_{bl}$ = 1.87 p – 2.40, $r^2$ = 0.75

**Figure 4.50**  Blow energy versus operating pressure, drill B, pneumatic (Brune, 1969).

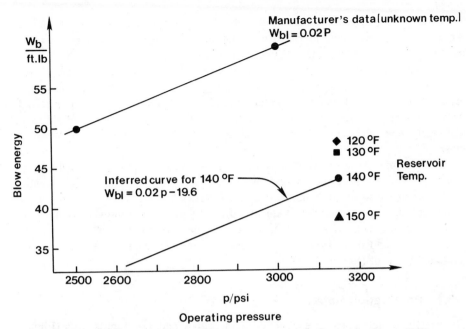

**Figure 4.51**  Blow energy versus operating pressure, drill C, hydraulic (Brune, 1969).

that the best method for drilling oil shale may be a combination of a rotary drill action with a light impact imposed. This method has apparently not been thoroughly investigated or reported in the literature.

A comparison of transmitted impact energy measured in a split Hopkinson bar for sandstone and oil shale showed that 67% of the energy in sandstone is utilized

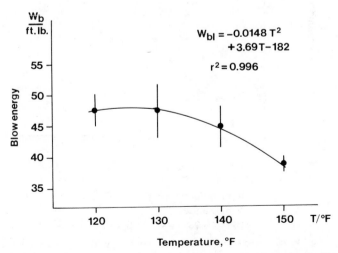

**Figure 4.52**  Blow energy versus hydraulic fluid temperature, drill C, hydraulic (Brune, 1969).

in breakage, whereas in oil shale much of the energy is reflected from the end of the specimen. This offers a partial explanation of the greater penetration rate of percussive drills in sandstone and granite (Figure 4.49).

In a series of tests, three types of percussive drills (Table 4.25) were used for the drillability of oil shale, granite, and sandstone. Meaningful comparisons of results from other drills were not directly possible. Two of the drills were powered with compressed air, and one with hydraulic fluid. Cross and button bits of different diameters were selected on the recommendations of drill operators in the oil shale mines.

Blow energies were found to be typical for pneumatic and hydraulic drills (Figures 4.50, 4.51). The blow energy for the hydraulic drill decreased significantly with an increase in oil temperature (Figure 4.52), which affected the consistency of the drillability data obtained with this drill. Likewise, the blow frequency increased rapidly with an increase in operation pressure for the pneumatic drills (Figure 4.53) and more slowly with thrust (Figure 4.54) and oil temperature for the hydraulic drill (Figure 4.55).

### 4.5.1 Penetration Rates

The most complete data were obtained with drill A and $1\frac{5}{8}$-in. cross bits (Figure 4.49) (Brune, 1983). Drilling rates as high as 62 in./min were obtained in sand-

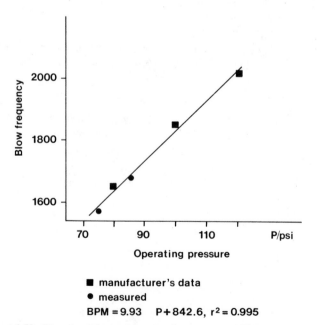

■ manufacturer's data
● measured
BPM = 9.93    P + 842.6, $r^2$ = 0.995

**Figure 4.53** Blow frequency versus operating pressure, drill B (Brune, 1969).

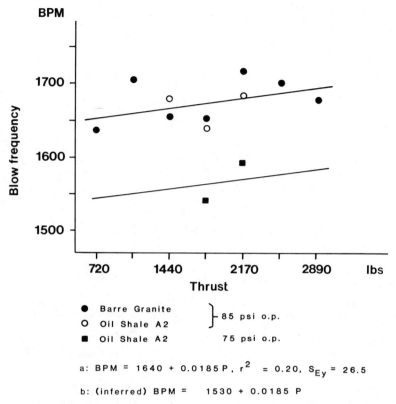

**Figure 4.54** Blow frequency versus thrust, drill B, pneumatic (Brune, 1969).

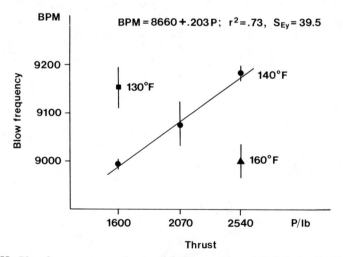

**Figure 4.55** Blow frequency versus thrust and fluid temperature, drill C, hydraulic (Brune, 1969).

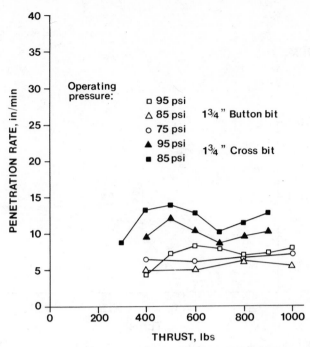

**Figure 4.56** Penetration rate versus thrust, drill A, oil shale A2 (Brune, 1969).

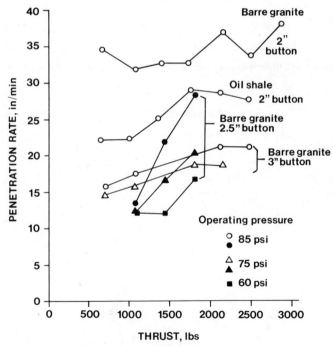

**Figure 4.57** Penetration rate versus thrust, drill B, Barre granite and oil shale, button bit (Brune, 1969).

**150**

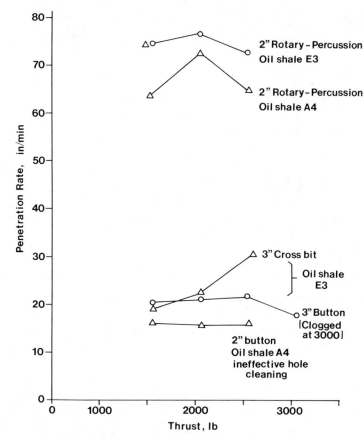

**Figure 4.58**  Penetration rates versus thrust, oil shale, drill C, hydraulic, button, cross and rotary-percussion bits (Brune, 1969).

stone, 30 in./min in granite, and only about 12 in./min in oil shale. This emphasizes the field observation that oil shale absorbs impact energy without crushing, failure, or separation of the mineral grains, which does occur in sandstones and granites. Further, button bits were less effective in oil shale than were cross bits (Figure 4.56), even though button bits drill faster in oil shale than in granite at lower thrust. They also gave slightly faster rates than cross bits in oil shale as long as they did not clog (Figure 4.57).

The hydraulic drill was the most effective with rotary-percussion bits where the teeth have a forward rake (Figure 4.58). The button bits, as with pneumatic drills, were less effective and clogged badly at high thrust. The large amount of energy reflected in the drill rod in tension (26.8%) for oil shale and much smaller (1.2%) in granite (Figures 4.59, 4.60) indicates one marked difference in energy transfer at the rock–bit interface for oil shale and granite.

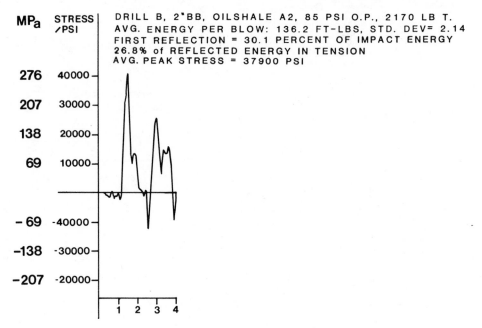

**Figure 4.59** First impact and reflected pulse in drill steel (Brune, 1969).

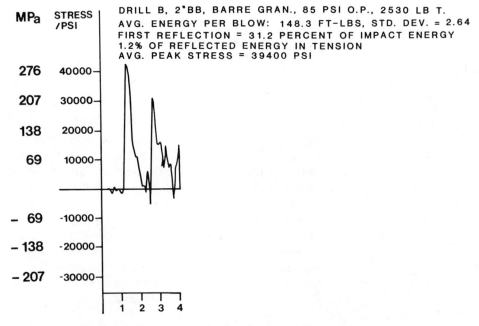

**Figure 4.60** First impact and reflected pulse in drill steel (Brune, 1969).

## 4.6   FORCE-PENETRATION CURVES

The dynamic force–penetration relation at the rock–bit interface has been used with physical property data and specific energy values to evaluate percussive drill performance. Although this information is valuable in yielding some basic data, unfortunately most of the research has been done at or near zero thrust. As indicated by the available data for operating drills, a force penetration curve is also dependent upon the length, diameter, weight, and velocity of the piston, the geometry of the rock and bit, and the structural and physical properties of the rock. One experimental method used to evaluate the functional elements of a percussive drill system is a drop tester with which the dimensions of the piston can be varied and the velocity of impact measured. The piston velocity in this apparatus can be easily measured and the energy calculated, whereas this is extremely difficult to do with an operating drill.

Experiments with drop testing on granite and sandstone with zero thrust by Miaioang and Johnson (1983) show that the pulses in the steel display different energy transfer effectiveness for granite and sandstone. Theoretical and calculated load-displacement curves indicate both elastic and plastic behavior. While medium-length (400-mm) pistons were found to be more efficient (the most effective velocity was about 7.5 m/s), the lowest specific energy did not correspond to the highest efficiency. Tests at or near zero thrust do not correspond with operating conditions, where the thrust would create different test conditions.

# 5

## DRILL BITS—DESIGN AND WEAR

There have been numerous investigations of the properties of various grades of tungsten–carbide cobalt alloys to determine their behavior in tools used for the rotary, percussive, and rotary-percussive drilling of rocks of different drillability characteristics. Several factors influence the wear and microscopic fracture of WC inserts or cutting edges used for different types of bits.

Some correlations have been established between standard physical properties of alloys and rocks, such as strengths, elastic constants, resistance to creep, fatigue, and so on, and the wear characteristics of WC alloys. The percentage of cobalt and the grain size of the WC have a marked effect on its properties and wear resistance. Impact strength and rupture strength increase with the percentage of cobalt, and the hardness decreases. Hardness also decreases rapidly with temperature above 400°K.

Several wear mechanisms have been proposed for WC, including micro-machining, impact, fatigue, and thermal stresses, while the detailed wear processes

154

may be due to erosion of the cobalt, pitting, fracturing of the cobalt and WC, and the tearing out of the WC grains. The wear is different for wet and dry drilling for sandstone and similar abrasive rocks.

Bits are designed with different percentages of cobalt and of WC grain sizes, depending on the use of different types of bits for rotary, percussive, or rotary percussive drilling. Wear is a function of several factors, including the WC–Co properties, rock properties, bit design, type of drill, and the various drilling process factors. Analysis of bit wear tests in quartzite shows that certain machine operating conditions are more favorable for good drilling rates and minimum wear, both for the cutting edge and for the gage.

In the three or more decades of the use of tungsten–carbide alloys for cutting tools, including rock drill bits, there has been considerable research on WC–Co properties related to wear and impact. Much of this dealt with properties of sintered carbide as a function of carbide grain size and percent cobalt. Test results are reported mostly in terms of adapted standard properties, such as tensile, flexure tensile, compressive, and shear strengths; elastic constants, creep, torsion, hardness, and fatigue. Few of the effects of temperature on several of these properties have been investigated.

In relation to property tests, the interest of the mining industry is centered on the response of WC–Co alloys in their use in percussive, rotary-percussive, and rotary drilling. Bit wear in rock drilling is a major factor in determining the cost of drilling and may determine the drilling method for a given rock. Wear decreases penetration rates and increases drilling forces, which may also cause major fracture of inserts. The type and degree of wear depends on the strength and abrasiveness of the rock and on the properties of the carbide inserts.

Wear processes have been classified by Larsen-Basse (1973) as surface impact spalling, surface impact fatigue spalling, thermal fatigue, and abrasion. Other researchers have used similar classifications. These factors may cause various degrees of wear, depending on WC–Co material variables and rock types. For hard rocks of medium abrasiveness such as granite, frontal wear by impact and impact fatigue causes wear. For highly abrasive rocks such as quartzite, gage wear by abrasion is a major factor. For rotary drilling of soft rocks, abrasive wear dominates for highly abrasive rocks, such as sandstone; wear by thermal fatigue is critical for nonabrasive rocks such as calcite.

Thus, different metals for drill tools must be chosen for different rocks. However, hard, abrasion-resistant carbides have poor toughness, and tough, spall-resistant alloys have relatively poor abrasion resistance, and improvement has been made by WC grain-size control. The carbide porosity also affects its strength.

## 5.1  TUNGSTEN–CARBIDE ALLOYS

Tungsten–carbide alloys consist of two phases: WC in small particles or crystals, and cobalt, often with minor amounts of W (tungsten) and C (carbon) dissolved in it, which increases the hardness and lowers the ductility of the Co. The size of the carbide grains usually varies between 1 and 5 $\mu$m, the proportion Co from 1 to 12%. Standard properties and wear resistance to friction and impact are markedly influenced by these two factors.

The average diameter of the WC grains and the mean free path between the grains are thus important characteristics and may be measured metallographically by means of the relations (Gurland and Bardzil, 1955) as follows:

$$d = \frac{4}{\pi} \frac{N_L}{N_S} \tag{5.1}$$

$$p = \frac{1 - f}{N_L} \tag{5.2}$$

where $d$ = average diameter of grains, $N_L$ = number of noncontiguous grains intersected on a metallographic plane by a line of unit length, $N_s$ = number of noncontiguous grains per unit area, $p$ = mean free path between grains, and $f$ = volume fraction of dispersed phase.

In addition to the determination of the mean free path, several types of tests are used to define the properties of WC–Co alloys. These include the following:

1. Transverse rupture test, ambient and high temperatures;
2. Charpy impact test, notched and unnotched specimens, ambient and high temperatures;
3. Vickers (Rockwell A) hardness, ambient and high temperatures;

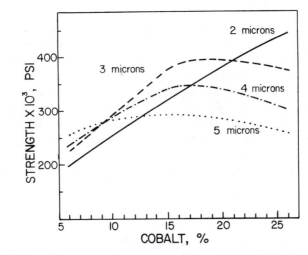

**Figure 5.1** Variation of transverse-rupture strength with composition (Gurland and Bardzil, 1955).

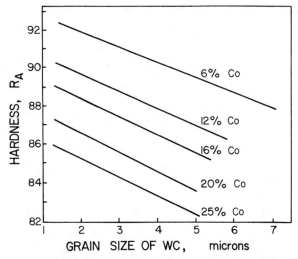

**Figure 5.2** Rockwell hardness of WC–Co alloys (Gurland and Bardzil, 1955).

4. Bending creep;
5. Fatigue limit—cyclic bending;
6. Compressive strength;
7. Deformation;
8. Tensile strength;

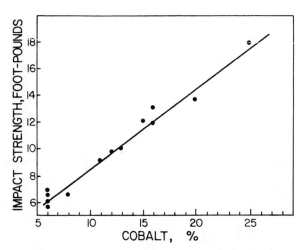

**Figure 5.3** Impact strength of WC–Co alloys of compacts having a WC grain size from 1.4 to 3.1 (Gurland and Bardzil, 1955).

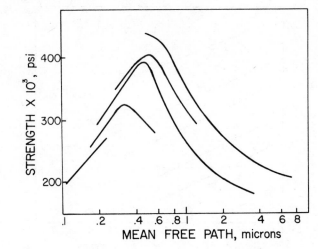

**Figure 5.4** Plot of the transverse-rupture strength as a function of mean free path (Gurland and Bardzil, 1955).

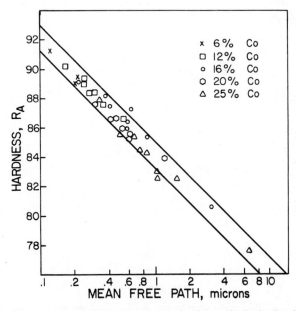

**Figure 5.5** Rockwell hardness as a function of the mean free path (Gurland and Bardzil, 1955).

9. Torsional strength;
10. Elastic constants, $E$ and $\mu$;
11. Wear resistance:
    a. standard test,
    b. rotary drilling in rock,
    c. percussive and rotary-percussive drilling in rock;
12. Thermal shock resistance.

Some temperature effects include those on hardness, transverse rupture strength, compressive strength, and impact strength (Charpy unnotched), and qualitative tests of temperatures generated in the WC in bits in rotary drilling.

Gurland and Bardzil (1955) found that transverse strength, impact (Charpy) strength, and hardness vary with grain size (Figure 5.1) and percent cobalt (Figures 5.2, 5.3) and that these properties may also be related to mean free path (Figures 5.4, 5.5).

## 5.2  TEMPERATURE EFFECTS

### 5.2.1  Tranverse Rupture Strength

The effects of grain size and percent Co on WC alloy properties have been studied extensively, and those combinations that are most advantageous for a given drilling operation are usually selected by drill bit manufacturers. According to Gurland and Bardzil (1955), the greatest strength appears to be in the 3- to 4-$\mu$m range. The transverse rupture strength is not affected by increased temperature from 270 to about 673°K but drops rapidly as the temperature is increased above 673°K.

### 5.2.2  Compressive Strength

In the Co composition range to about 8%, the compressive strength decreases rapidly as the temperature is increased above 290°K, showing approximately the same type of decrease as the transverse rupture strength (Kreimer, 1968), although the decrease is linear.

### 5.2.3  Vickers Hardness

Whereas the compressive strength decreases by a factor of about 2 with increase in temperature from 300 to 1000°K, the Vickers hardness decreases by a factor of 2–3 over this same temperature range (Figure 5.6) (Tretyakov, 1962). There has been little research done on the relation of the effects of temperature on the impact or wear properties of WC–Co, although temperature has been monitored in rotary bits. A study of thermal fatigue cracks in WC–Co induced by friction led Lagerquist

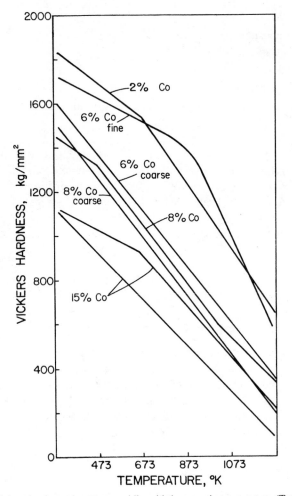

**Figure 5.6**   Vickers hardness decreases rapidly with increase in temperature (Tretyakov, 1962).

(1975) to the following conclusions: (1) the number of thermal fatigue cracks increases with cobalt content; (2) propagation rate decreases with cobalt content and WC grain size; (3) thermal fatigue cracks propagate preferentially at WC–Co boundaries, rates being controlled by interparticle distance.

Adamson et al. (1980) point out that, aside from hardness and wear resistance, WC has limitations due to (1) brittleness with high compressive but low bending (tensile) strength, (2) coefficient of thermal expansion, 50% that of steel, and (3) low coefficient of heat conduction. The brittleness and low tensile strength are related to the fracture characteristics as reported by Montgomery, Latin, and others. Adamson et al. (1980) also report the results of earlier work by Latin (1961) on

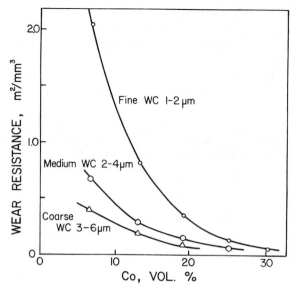

**Figure 5.7**  Wear resistance/percent Co (after Latin, 1961).

the wear of WC–Co as a function of percent Co and grain size of WC (Figure 5.7).

The surface of drill bits may reach high temperatures during drilling, even with water flushing, but no quantitative data are available. When WC–Co sliders were run dry against sandstone wheels (simulating rotary drilling), a glassy layer could have been formed on the wheel, indicating temperatures in excess of 1250°C (Rae, 1966). Temperatures of 450–500°C were found near hard metal inserts in dry rotary-percussive drilling of 5-ft holes in sandstone (Fish, 1957b). Whitbread (1960a, b), in measuring the temperature on the back side of hard metal inserts in dry rotary drilling of sandstone, found that a temperature equilibrium was reached after 5 min of drilling and that the average was 150°C. The highest temperature was 490°C for severe conditions, and temperatures above 320°C were common. The temperature increases with thrust and rock hardness and with the size of the wear scar on the bit. It is a minimum for rake angles around zero, and high temperatures are created for low penetration rates because of the tool friction. As the penetration rate is increased and chip formation takes place, the temperature initially falls and then increases.

## 5.3  ABRASIVE WEAR—MECHANISMS

Blomberry et al. (1974) showed that anomalies arising from reported observations of two abrasive mechanisms in rotary drilling of sandstone with sintered tungsten–

carbide tools were due to (1) selective cobalt removal and (2) micro-fracture of the carbide skeleton. Initially, fine abrasive particles of rock remove cobalt, forming small pits. This removal of cobalt may lower the fracture strength of the surface layers, and cracks propagate from the pits, followed by micro-fracturing of carbide grains and micro-spalling of the surface layer. Cobalt content and its distribution, the size of abraded particles (from the abrasive material, rock), and other factors control the abrasive wear (Figure 5.7).

Several systems have been used to classify different types of wear. Applications of terms such as "abrasion," erosion," or "impact" result from macroscopic or (low-power) microscopic examination of the wear surface.

Five micro-mechanisms of wear have been proposed:

1. Micro-machining: the abrasion cuts and removes chips.
2. Impact: localized impact causes crushing.
3. Adhesion: high temperature causes local welding, and tearing occurs.
4. Thermal stresses: cyclic stresses cause fatigue fractures.
5. Chemical: reactions with chemically active species in drilling processes occur.

Terms such as "abrasive wear" may indicate only the apparent nature of the wear rather than the actual mechanisms. Several major wear factors may operate simultaneously, such as the chemical environment, which may influence adhesion, and lubrication in metal–metal or metal–rock contact.

The importance of fundamental abrasive mechanisms, rather than the gross types of wear, became clearer as a result of recent microscopic investigations in the effects of rotary drilling of sandstone. Two apparently distinct types of wear were found: (1) fracture of surface layers, and (2) the removal of the cobalt binder, which was followed by the pull-out of tungsten–carbide grains. No correlation was found with either cobalt content or wear rate. A possible explanation may be found in the simultaneous action of two wear modes where cobalt removal affected the fracture resistance.

Examinations with a scanning electron microscope (SEM) have shown that heavily deformed surfaces displayed all states of wear, and the sequence of changes could not be identified. Studies carried out on surfaces that were abraded in steps make it possible to examine the surface at successive stages of wear.

The experimental studies of abrasion on sandstone were made with WC–Co of composition 8–12% cobalt. Polished surfaces were abraded on blocks of Stawell sandstone with a pressure of about 60 kPa, which is a small fraction of that used in rock drilling, which resulted in low wear rate.

The conclusions drawn from the abrasion experiments were the following:

1. Removal of small amounts of cobalt (0.2 $\mu$m depth) occurring by chemical etching makes the surface susceptible to intergranular cracking and micro-

spalling by abrasion. The depth of the spalled layer (3–4 $\mu$m) is much greater than that of the etching.

2.  The rate of wear is controlled by the cobalt removal.
3.  Alloys with more than the average amount of cobalt are most susceptible to fracture.

The study by Blomberry et al. (1974) confirmed earlier observations that (1) selective removal of cobalt followed by pull-out of loose tungsten–carbide grains and (2) the fracture of the carbide skeleton are two possible wear mechanisms. Analyses indicated a need for further study of the stress states in the WC, effect of pH and impurities in the water, the temperature of operation, and possible chemical reactions. Carbide grains themselves were found to undergo little wear, and, hence, removal of cobalt is the controlling wear factor.

## 5.4   ROTARY DRILLING—WEAR

Fish et al. (1959) conducted full-scale rotary drilling tests in sandstones and limestones. The WC was 9% coarse-grained. Wear was measured as milligrams of loss per 1000 ft of peripheral travel distance. The results are given graphically in terms of width of wear flat as a "medium" penetration rate (Figure 5.8) and a "high" penetration rate (Figure 5.9). The rate of wear increases with rpm but decreases with penetration rate (Figure 5.10) for sandstone.

In tests with rotary bits in a fine-grained sandstone, 95% quartz, with a grain size of 0.23 mm, cemented carbides with grain sizes of 1 and 5 $\mu$m were used by Stjernberg et al. (1975), who found that the rate of wear was a function of cutting speed (Figure 5.11) and is markedly less for coarse-grained alloy. The hardness of the alloy varies with depth below its surface (Figure 5.12).

It has been stated that high temperatures induced in the carbide may have a marked effect on its hardness and wear resistance. An increase of rotary speed increases the temperature, followed by deformation, reduction of cobalt content, and possible increase of hardness below the surface. Thus, at low temperatures the wear is by abrasive carbide-grain removal, and at high temperatures by plastic deformation and grain boundary sliding, which result in greater wear of coarse-grained carbide.

Rotary drilling is most commonly done with drag bits, whose action resembles that of a tool for machining metals. The cutting is done by a WC–Co insert brazed to a steel cylindrical body. Bits may have a two-pronged fork for drilling holes, mounted on cylinders for drilling cores, on large heads for tunneling excavation, or on endless chains as picks for coal or overburden. For smaller boring bits, speeds of rotation usually vary between 100 rpm for rock and 1500 rpm for coal (Fish, 1959). The clearance angle is usually 15–20°, and the rake angle about zero (Fairhurst and Lacabanne, 1957). The chips removed by rotary drilling are coarser

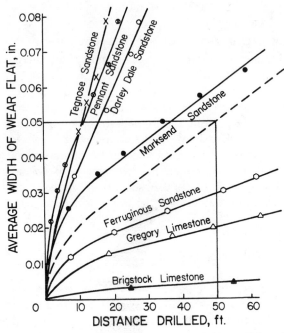

**Figure 5.8** Bit wear results at ''medium'' penetration rate (Fish et al., 1959).

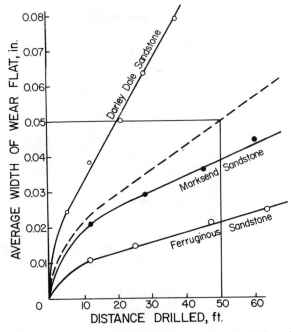

**Figure 5.9** Bit wear results at ''very high'' penetration rate (Fish et al., 1959).

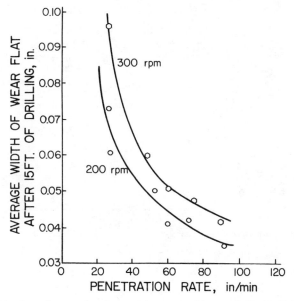

**Figure 5.10** Effect of rotation speed and penetration rate on bit wear in Darley Dale sandstone (Fish et al., 1959).

than those created by percussive drilling, and rotary drilling is more efficient for certain softer rocks. However, tool wear and fracture are severe in hard abrasive rock.

The hypothetical (static) mechanism of rock material removal by this bit (Figure 5.13) includes a force buildup from the applied torque, creating fracture of the rock

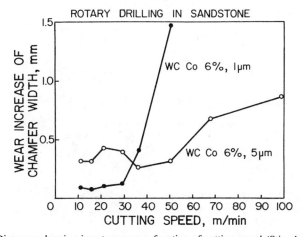

**Figure 5.11** Diagrams showing insert wear as a function of cutting speed (Stjernberg et al., 1975).

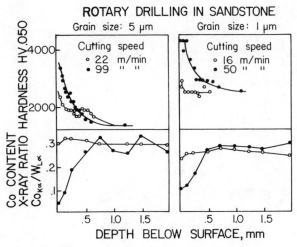

**Figure 5.12** Variation of microhardness and Co content after rotary drilling in sandstone (Stjernberg, 1975).

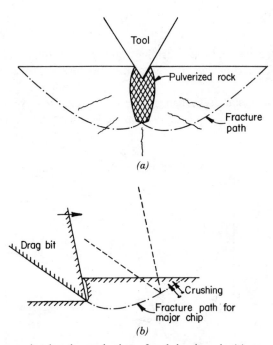

**Figure 5.13** Diagrams showing the mechanism of rock breakage in (*a*) percussive and (*b*) rotary drilling (after Maurer, 1969; Larsen-Basse, 1973).

initiated at or near the tool tip. The fracture usually extends beneath the plane of cut, then curves upward and connects with the free surface. The tool then loses contact with the rock and, again, impacts the rock. The later chipping and crushing of the rock at the tool rake and clearance faces takes place, resulting in some bit material removal. Tool impact with oscillations of frequencies up to 2000/min (Jackson and Hartman, 1962) causes fine ridges on the rock and chip surface. The thrust and torque on the bit increase rapidly with depth of cut and with decreasing rake angle, and the penetration rate increases with increased clearance angle (Nevill and Cron, 1962).

Bit wear markedly affects the rotary drilling parameters. The major wear is due to abrasion of the rake face, but some wear loss also occurs on the clearance face. Gage wear is important for rotary drilling (Nevill and Cron, 1962; Appl and Rowley, 1963) in that the contact area between the rock and the rake face is smaller than the depth of cut, and wear takes place only in rounding of the tool nose. Rock fracture in chipping begins at the tool tip, and, hence, the rounding of the tip requires that the downward thrust must be increased to maintain the penetration rate. This creates an increase in the friction over the clearance face and also in the torque, but the increase of torque is not as pronounced as the increase in thrust. The wear is proportional to the linear distance traversed by the cutting edge (Fish, 1958; Nevill and Cron, 1962; Jackson and Hartman, 1962), and the wear per foot drilled is larger at low penetration rates and at high speeds of rotation. At low penetration rates, there may be a higher degree of rock crushing instead of chip formation, which also increases the wear. Wear appears to be independent of rake angle, disregarding breakage of sharp tips and thrust effects.

Bit life lengthens with its hardness, but at higher hardness tools tend to fracture. The rate of wear is caused more by the abrasiveness of rock than by its strength.

In a way, Fish et al. (1959) found a general qualitative correlation between bit wear in rocks and their abrasiveness, which was measured in sliding with loads proportional to the crushing strength of the rock. A rough correlation between the abrasiveness and the quartz content of the rocks tested, all sedimentary, was also indicated.

Nevill and Cron (1962) noted for granite some initial flaking of the insert rake face at the start of the test with each bit, partially due to the sharp edges of the tool. While fine chipping of the cutting edge took place, it diminished for very low or very high bit speeds. In these cases, the tool may have been in more constant contact with the rock with fewer impacts.

The stresses in the cutting edge can be severe, as those that have been calculated by Appl and Rowley (1963), where a large chip is about to be formed. The stress is maximum at the clearance face and the curvature of the worn cutting edge, where tool fractures tend to be initiated. The compressive stress is normal to the clearance face and may have values up to 2210 N/mm$^2$ (320,000 lbf/in.$^2$) in hard rocks at low penetration rates. If the penetration rate is increased, it is believed that the rock is removed by brittle shear rather than by plastic deformation, and the maximum stress is lower and roughly proportional to the radius of the rounded cutting edge, which emphasizes the importance of edge wear. The wear in rotary drilling then is said to be largely due to abrasion and impact fatigue, possibly augmented by thermal fatigue.

Studies have been made of the wear mechanisms in rotary drilling of abrasive rock with tungsten carbide–cobalt tools of different abrasion and fracture resistance. Larsen-Basse et al. (1974) show that rock drilling grades containing from 8–12% cobalt wear by both cobalt erosion and fracture of the carbide skeleton. Although fracture may account for most of the weight loss, the wear rate is larger for the WC–Co alloys having the greatest fracture resistance and the lowest abrasion resistance. Tools used in rock-cutting operations of rotary, percussive, or rotary-percussive methods are subjected to different types of wear. However, the influence of metal property variables—such as grain size, cobalt content, and cobalt mean free path—on the physical properties is important in all types of drilling.

From the studies made to determine both (1) the wear mechanisms that are operative in rotary drilling of abrasive rocks and (2) the influence of physical properties on the rate of wear, it was concluded that the main abrasive constituents in rocks are quartz grains, which are soft abrasives compared with tungsten carbide. Also, wear may occur by (1) removal of small lumps of the composite due to brittle fracture of the tungsten–carbide skeleton, (2) the preferential removal of cobalt followed by "uprooting" and fracture of the tungsten–carbide grains, or (3) both mechanisms. The work by Larsen-Basse et al. (1974) gives experimental evidence of the wear mechanisms for rotary drilling of abrasive Stawell sandstone.

The tungsten carbide–cobalt composites used in their series of tests were four samples containing 8–12% cobalt. Sintered integral drill inserts had been prepared from tungsten carbide of a nominal 6-$\mu$m particle size and cobalt of 99.98% purity. An abrasive index for each composite was measured, which is the weight loss of metal per unit time with the specimen subjected to controlled abrasion under a constant applied load. For this test, a 3.18 × 1.90 × 0.48 cm sample was held by a weight of 20 kg against a steel sheet of 16-cm diameter upon which is fed a slurry of 30-mesh alumina. The wheel was rotated at 100 rpm. This test gave an empirical rating of abrasion resistance, the reciprocal of the abrasive index. Although there was a decrease in abrasion resistance and hardness with increasing cobalt content, the index values of 8.5 and 11.0% cobalt alloys are similar (Figure 5.14).

On the other hand, the transverse rupture strength does not show a correlation with cobalt content (Table 5.1, Figure 5.14). These departures from reported behavior (Larsen-Basse, 1973; Kreimer, 1968) are associated with the metal structure. That is, the average tungsten–carbide grain size varied from the nominal 6 $\mu$m, being 4–5 $\mu$m for 8.0–12.0%, 5–6 $\mu$m for the 8.5%, and 3–4 $\mu$m for the 11.0% cobalt alloys. This cobalt content is below that at which a continuous thick film of cobalt surrounds each tungsten–carbide grain. Thus the microscopic structure consisted of a tungsten–carbide skeleton enclosing pockets of cobalt, but the existence of thin surface films of cobalt was possible. The distribution of cobalt throughout

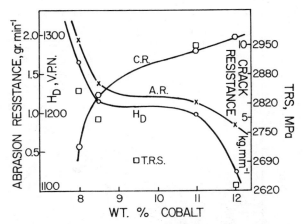

**Figure 5.14** Some physical properties of the tungsten carbide–cobalt composites used in the drilling experiments as a function of the cobalt content (Larsen-Basse, 1973).

the structure varies from grade to grade, the cobalt mean free path, as well as the grain size, being larger in the 8.5% than in the 11.0% composite. This explains the relatively low hardness and abrasion resistance of the 8.5% cobalt composite compared with those of 8.0 and 11.0% (Larsen-Basse, 1973; Exner and Gurland, 1970). The results also show that anomalies may be encountered in classification by cobalt content alone.

The fracture resistance of the composites was determined by the Palmqvist method, the crack length at the corners of a Vickers-hardness indentation being measured as a function of the applied load (Palmqvist, 1962, 1963; Exner and Gurland, 1970; Dawihl and Altmeyer, 1964) (Figure 5.14). The curves do not pass through the origin, indicating that some surface strains were present (Exner, 1969). However, linear relationships occurred between total crack length and the applied load (Figure 5.15), the reciprocal of the slopes of the curves being taken as a

**TABLE 5.1. Physical Properties of the Tungsten Carbide–Cobalt Composites Used in the Drilling Experiments**

| Composition (Wt% Co) | Cohesive force (Oe) | Density (g/cm$^3$) | Hardness (V.P.N.) | Abrasion Index (mg/min) | Transverse Rupture str (MPa)[a] |
|---|---|---|---|---|---|
| 8.0 | 118 | 14.69 | 1270 | 500 | 2840 |
| 8.5 | 106 | 14.70 | 1218 | 715 | 2780 |
| 11.0 | 86 | 14.36 | 1200 | 866 | 2950 |
| 12.0 | 82 | 14.32 | 1125 | 1185 | 2620 |

[a]S.I. units: $10^3$ psi $= 6.895$ MPa.
*Source:* Larsen-Basse (1973).

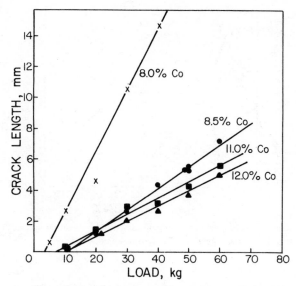

**Figure 5.15**  The sum of the crack lengths at the corners of a Vickers hardness impression as a function of the applied load for the tungsten carbide–cobalt composites used in drilling. The reciprocal of the slope gives a measure of the fracture resistance (Larsen-Basse et al., 1974).

measure of the fracture resistance. Differing from transverse rupture strength, the fracture resistance varies smoothly with cobalt content (Figure 5.14), indicating that crack-length data give physical properties that are obtainable from strength or fracture measurements (Dawihl and Altmeyer, 1964). The fracture resistance of 8.5% cobalt composite was higher than that of 8.0%, possibly because of the larger mean free path in the former.

Thus abrasion and fracture resistance of WC–Co are two of the important parameters related to the wear of tools in rotary drilling of abrasive rocks. For low-cobalt contents, the abrasion resistance is high and the crack resistance is low, and the reverse holds at high-cobalt contents (Figure 5.16). Hence, laboratory tests predict that the rates of wear and wear mechanism vary with cobalt content. Stawell sandstone was used for tests because of its abrasiveness and uniformity of structure. It is a fairly even-grained, dense, silica-bonded sandstone consisting principally of quartz with a few grains of water-worn zircon and rounded tourmaline. The size of quartz grains ranges from 90 to 500 $\mu$m, and the transverse rupture strength is 7030 kPa.

In rotary drilling tests reviewed by Larsen-Basse (1973), tool geometry was designed to create a rapid rate of wear to avoid the drilling of deep holes. Drill bits 13 mm long, 6.5 mm wide, and 1.5 mm thick were spark-machined, and the cuts were taken so that the drilling axis of the bit was parallel to that of the original insert. The bits were mounted in stainless-steel shanks and ground to give a 6.5-mm-wide chisel with two equally inclined faces with an included angle of 110°.

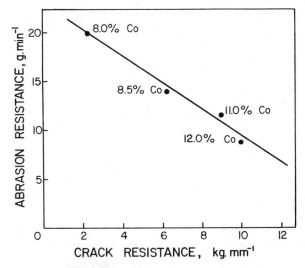

**Figure 5.16** The relation between crack resistance and abrasion resistance for the 8.0, 8.5, 11.0, and 12.0% cobalt composites (Larsen-Basse et al., 1974).

The faces of the chisel were polished metallographically to a 0.5-$\mu$m diamond finish. The drilling tests were carried out on a Joy machine drilling vertically upward at 60 rpm into a rock sample clamped onto a floating crosshead. Crosshead loads of 23 and 35 kg were used without lubrication or cooling. For all cobalt contents, a substantial wear flat was produced on the tools after 1600 revolutions, this being sufficient to produce a hole approximately 15 mm deep in the sandstone.

Tool weight loss measurements were made at regular intervals, and for some tests the cuttings were collected and weighed. The cutting edges were examined by means of optical microscopy and SEM. The weight of rock cuttings as a function of the weight loss from the tool showed a rapid decrease of cutting power after a critical loss of tool weight (Figure 5.17). A wear flat had usually formed by 200 revolutions, and after 1600 revolutions all tools showed excessive weight loss.

The rate of weight loss from all cobalt composites increased rapidly during the initial "running-in" period, reaching an approximate constant weight loss per revolution. The bit weight loss was smaller for the 8.0% cobalt composite. The low value for the 8.5% cobalt composite after 1600 revolutions (23-kg load) was related to the slipping of the tool on the rock surface. The increase of the load from 23 to 36 kg increased the tool weight loss, but not proportionately to load (Figure 5.18).

### 5.4.1 Structure of Worn Surfaces

The tests produced a wear flat, some edge or gage wear, and some scouring or grooving of the leading edge, with the trailing edge remaining in a polished condition. The gage wear was not extensive, and most of the weight loss was due to

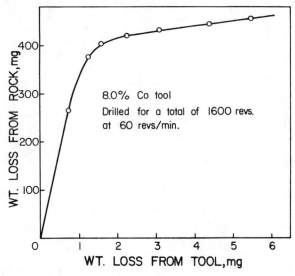

**Figure 5.17** The weight loss from the rock as a function of the weight loss from the tool during a typical drilling test (Larsen-Basse et al., 1974).

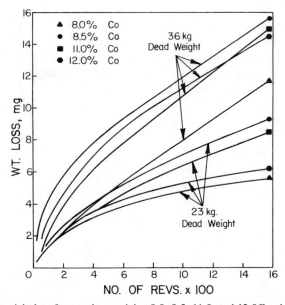

**Figure 5.18** The weight loss from tools containing 8.0, 8.5, 11.0, and 12.0% cobalt as a function of the number of revolutions during drilling under loads of 23 and 35 kg (Larsen-Basse et al., 1974).

the erosion of the wear flat. This was pitted with regions of micro-fracturing, leading to metal removal with some preferential removal of cobalt, which had resulted in the protrusion of tungsten–carbide grains. Increase in load resulted in an increased incidence of pitting and micro-fracture, with gross grooving also occurring. Changes in the actual wear mechanisms with changes in the cobalt content of the tool could not be detected.

The character of the bottom and sides of the drill hole indicated that the sandstone was "machined" by crushing and fracturing and some tearing out of the quartz grains. This produced a generally smooth but microscopically irregular surface at the bottom of the hole. Heating, melting, and slipping were observed with the 8.5% cobalt tool after 1600 revolutions under a load of 23 kg.

Despite the fact that the tungsten–carbide samples had nearly the same grain sizes and composition, they exhibited variable abrasion and fracture resistances. However, this variation of operating properties demonstrates the influence of carbide grain size, cobalt content, and cobalt mean free path (cobalt distribution) on the physical properties of WC–Co. A qualitative correlation only was found between hardness, abrasion resistance, and weight loss. For example, the lowest weight loss occurred with a composite of 8.0% cobalt of highest hardness, abrasion resistance, and a relatively high transverse rupture strength. The weight loss from the 8.5% cobalt alloy was excessive. This might be due to the slightly larger carbide grain size and larger cobalt mean free path. Accurate evaluation, however, will require further analysis of the micro-structural details than provided by usual laboratory measurements; that is, valid analysis of wear data requires knowledge of the wear rate as a function of the wear process and geometry of the wear surface.

The form of an applicable function gives a linear fit for plots of wear rate $dW/dt$ versus wear $W$, or a relationship of the type

$$\frac{dW}{dt} = KW^A \tag{5.3}$$

The values of the "constants" $A$ and $K$ are sensitive to the scar geometry and their dependence on drilling load and tool type. The parameters $A$ and $K$ are related to the metallurgical parameters measured, but a direct interpretation of the results was not possible.

One of the most important factors in defining wear mechanisms is the distribution of cobalt throughout the tungsten–carbide skeleton. Wear of all the rotary tools tested appeared to take place by a combination of (1) erosion of cobalt and (2) micro-fracture of the carbide skeleton. The relative extent of each did not vary with cobalt content, abrasion, or fracture resistance of the composite. Cobalt erosion and the loosening of carbide grains make only a small contribution to the total weight loss. The major mechanism of weight loss in all cases is probably related to the micro-fracture of the carbide skeleton; thus the fracture resistance is a most important parameter.

It was concluded by Larsen-Basse et al. (1974) that drilling tests and wear tests, in general, give reasonable correlations between the wear and the physical properties of the WC–Co but give limited information on the wear mechanisms. As noted earlier, Blomberry et al. (1974) state that the abrasive wear of WC in the rotary drilling of sandstone may occur by any of these mechanisms: micro-machining, impact (other than percussive), adhesion, local welding, tearing, thermal stresses, fracturing, and/or chemical reactions. Further study was recommended of (1) stress states and preferential cobalt removal, (2) effects of pH, (3) temperatures in bits, and (4) chemical reactions affecting WC under operating conditions.

The following conclusions were drawn by Blomberry et al. (1974):

1. Abrasive wear by soft abrasives is by preferential removal of cobalt, with most weight loss by fracture of the carbide;
2. The first stage is formation of pits in the cobalt;
3. This alters the stess and lowers fracture resistance;
4. Micro-fractures in the pits propagate, causing spalling;
5. Cobalt removal is greater by loose abrasive particles;
6. This model is consistent with earlier results;
7. Chemical etching has an effect similar to mechanical erosion of cobalt.

In tests on rotary drill bits conducted by Nevill and Cron (1962) to determine the effects of clearance angle, it was found that in basalt and granodiorite the wear was independent of thrust and varied linearly with the (rotary) distance traversed by the cutting edge, except for initial penetration. For experimental drilling of slate (Jackson and Hartman, 1962), the wear related somewhat to the thermal conductivity and the abrasion resistance of the alloys, the latter being deformed by a standard abrasion test. The weight loss/time curves (Figure 5.19) are for a running-in period followed by a steady-state condition. For these carbides, the wear for running in was large. The wear rates during running-in and the steady state are functions of the cobalt content. Alloys with large amounts of TiC and TaC, often used in metal machining, have a poorer performance than the straight WC–Co alloys for drilling slate.

Latin (1961) studied wear in simulated rotary drilling in a relatively complete experiment that allowed an evaluation of the influence of hardness, cobalt content, tungsten–carbide grain size, and mean free cobalt path. For these tests, cutting tools with WC–Co tips were used to cut sandstone in a lathe. Two nonsimulated conditions of this method were that the cutting velocity increases with outward distance and that the load on the tool increases with wear. However, the weight loss was found to be proportional to the area of rock traversed, and, hence, the effect of these two factors is small.

The wear rate for these rotary tests increased with both cobalt content and grain size (Figures 5.20, 5.21). For a given cobalt content, fine WC grains caused the greatest hardness and minimum wear. In terms of the volume resistance, the inverse of the volume wear rate (Figure 5.22) gives the volume–wear resistance in terms

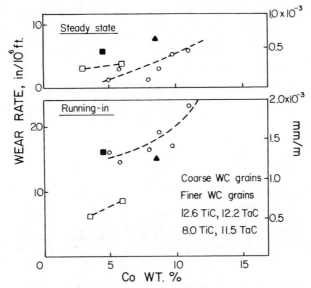

**Figure 5.19** Average wear rate during running-in and steady-state wear rate versus cobalt content for rotary drilling of slate (data from Jackson and Hartman, 1962).

of the volume percentage of cobalt and, in Figure 5.23, wear versus hardness. For the cobalt percentages for inserts used in rock drilling, the fine-grained metals have a better wear resistance for rotary drilling than do the coarse-grained.

As with percussive wear, the relation between wear resistance and hardness is different from that for more ductile materials and hard abrasives. Two factors could

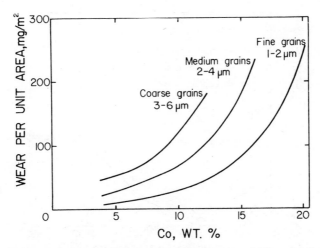

**Figure 5.20** General curves for abrasive wear in simulated rotary drilling as a function of cobalt content and WC grain size (after Latin, 1961).

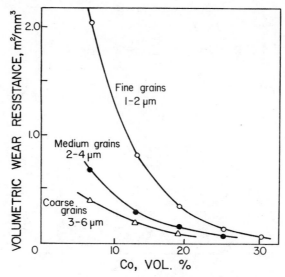

**Figure 5.21** Volume wear resistance versus volume percent cobalt (data derived from Fig. 5.20).

account for observed results: (1) the rock abrasive particles, such as quartz, are not hard abrasives, and (2) the abrasive mechanism is different for the WC–Co inserts from the cutting type of mechanism of ductile metal. Also, hardness in itself cannot be used as a single measure of wear resistance but must include grain size and other properties.

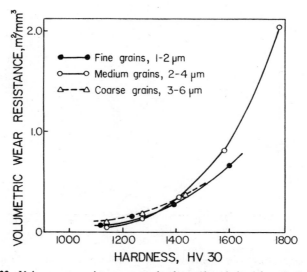

**Figure 5.22** Volume wear resistance versus hardness (data derived from Latin, 1961).

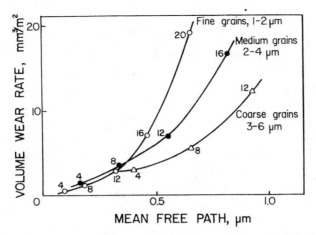

**Figure 5.23**  Volume wear rate as a function of mean free path. Numerals indicate weight–percent cobalt (data derived from Latin, 1961).

Important properties of WC–Co are related to the mean free path of the cobalt phase (Kreimer, 1968). The relation between wear resistance and mean free path in "typical" values listed in Latin's paper (1961) for the mean free path are given as a function of carbide grain size and cobalt content (Figure 5.23). The volume–wear rate is plotted versus this parameter and, like hardness (Fischmeister and Exner, 1966), the wear rate is a function of mean free path as well as grain size and cobalt content. For small mean free path, the coarse-grained WC–Co metals show the greatest wear and, for large values of mean free path, the least wear, but they are the same below WC–Co 0.27–0.31-$\mu$m mean free path.

In studies of the microstructure of alloys rubbed against calcite cylinders, Pons et al. (1962) found that wear occurs by three different mechanisms: (1) uniform wear of both carbide and cobalt, similar owing to a polishing action; (2) removal of the cobalt followed by uprooting of the carbide grains, which may subsequently scratch the metal surface; and (3) cracking of large carbide grains with pieces being torn out. Calcite is almost nonabrasive, which would lead to the expectation that thermal effects are of importance (a thesis supported by test results).

Carboni (1969) employed laboratory abrasion tests on several commercial WC–Co alloys. In these tests, metal specimens were pressed against a rotating steel wheel with a water-based slurry of corundum grains fed to the contact area. Hardness, abrasion resistance, and compressive proof stress are all dependent on the volume fraction of cobalt binder (Figure 5.24). This dependence of hardness and wear resistance is similar to the curves derived from Latin's data (Figure 5.25a, b). The abrasion resistance is proprotional to the inverse initial tungsten–carbide grain size (Fisher subsieve), and for three specimens the wear rate is proportional to the fraction of the wear–scar length taken up by straight grooves (Figures 5.26, 5.27). For this experiment straight grooves were made on the metal surface by abrasive grains, which retain their cutting position for some distance before being

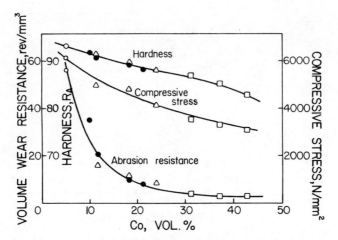

**Figure 5.24** Abrasive wear resistance, hardness, and compressive proof stress as functions of the cobalt content. The different notations indicate different producer grades (data from Carboni, 1969; Larsen-Basse, 1973).

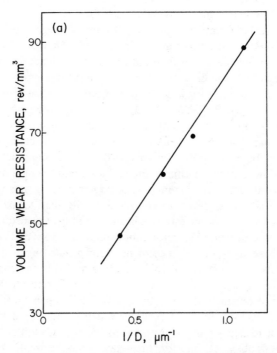

**Figure 5.25** (a) Abrasion resistance versus inverse initial WC powder size (Fisher subsieve) for WC alloys containing 95.5% WC, 0.5% TiC–TaC, and 4% Co (data drom Carboni, 1969). (b) Wear rate as a function of the fraction of total wear–scar length taken up by distinct abrasion grooves for various WC-base alloys (measured in photos from Carboni, 1969; Larsen-Basse, 1973).

**178**

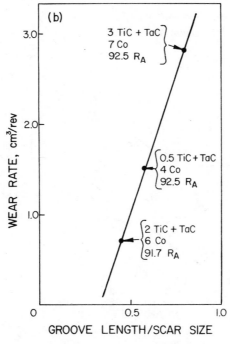

**Figure 5.25** (*Continued*)

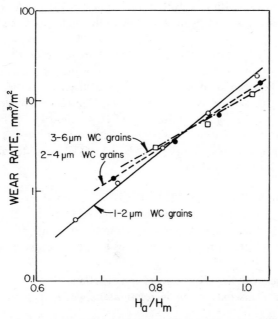

**Figure 5.26** Abrasion wear rates versus $H_a/H_m$ (data from Latin, 1961; Larsen-Basse, 1973).

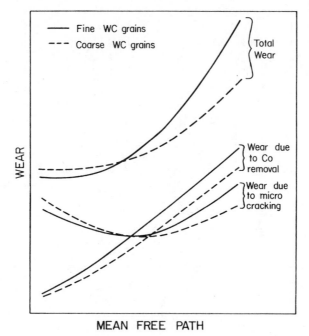

**Figure 5.27** Proposed general dependence of wear by the two cooperating mechanisms and total wear on the mean free cobalt path (Larsen-Basse, 1973).

crushed or blunted or starting to roll. From Figure 5.27 it can be seen that the action of the abrasive, when it is in contact with the metal, is more important than metal hardness.

It was concluded by Larsen-Basse (1973) that more experimental rotary wear data are required to give a detailed understanding of the mechanisms of wear and the relationships between wear rate and rock and insert properties. He discusses the wear behavior of several types of materials. The room temperature hardness values of the WC–Co alloys are in the range 1000–1800 HV. The sliding wear against rock is primarily due to the quartz content of the rock (Fish et al, 1959). The hardness of quartz at room temperature is ~ 1500 H. He states also that several different types of wear may take place:

1. Abrasion by hard abrasives. This occurs when the abrasive has a hardness at least 1.6 times the metal hardness (Kurschov et al., 1970). The metal surface may attain high temperatures during drilling, whereas the newly exposed quartz grains may have a lower temperature and possibly a higher hardness.

2. Adhesion between rock constituents and the composite, especially at elevated temperature. This may result in relations between wear rate and distance,

load, and hardness similar to those observed. However, quartz is not the only rock constituent causing adhesive wear, and this mechanism of wear is of minor importance.

3. Abrasion by relatively soft abrasives. Latin's 1961 results show the ratio between the hardness of the abrasive and the bulk hardness of the insert metal, $H_a/H_m$. The wear rate is not linear with $H_a/H_m$ as proposed by Nathan and Jones (1966–67). Instead, a plot of $H_a/H_m$ in log-log coordinates is linear (Figure 5.26). This, incidentally, shows that there is also a linear relationship between log abrasion resistance and log bulk hardness. The slopes of the lines are 8.5 for fine, 6.9 for medium, and 5 for coarse WC grains. The difference may be due to differences in work hardening.

Thus, microscopic wear of WC–Co alloys by quartz grains in rocks is more complex than "conventional wear," for which two different mechanisms are postulated by Larsen-Basse (1973):

1. Wear by removal of small fragments of the material by brittle fracture, as is known to occur in the abrasion of brittle materials like Si, Ge, NaCl, and glass (Dobson and Wilham, 1963).
2. Wear by preferential erosion of cobalt grains. Doeg (1959–60) proposed this as a mechanism for rock-drilling conditions; it has also been shown to be responsible for wear in the cavitation erosion of WC–Co alloys (Dawihl and Frisch, 1968).

Brittle micro-fracture depends on both crack initiation and crack propagation. It is postulated that for the hard, brittle alloys in which crack propagation is easy, crack initiation may be the controlling factor. Cracks may initiate both at the carbide–binder interface and in the carbide grains. Where the mean free cobalt path is constant, the coarser-grained alloys may wear more because of fracture initiation in the large WC grains. The wear may decrease with increased transverse rupture strength (TRS) of the alloy, that is, with increased mean free path up to 0.3–0.4 m. Here the TRS increases because thicker cobalt layers are stressed less and may inhibit advancing cracks (Kreimer, 1968). For larger mean free paths, the TRS decreases because the increased cobalt content results in decreased strength and also increased ductility. Crack propagation is also more difficult for the coarse-grained alloys (Fischmeister and Exner, 1966), and they may consequently wear less than the fine-grained alloys. If this theory is correct, wear due to brittle micro-fracture would vary with mean free path (Figure 5.27). The wear by preferential erosion of cobalt results from (1) the hardness of the cobalt, which decreases with length of mean free path; (2) the number and size of the cutting asperites compared to the exposed cobalt (i.e., relative to the mean free path); (3) the total area of cobalt exposed; and (4) the resistance of the WC grains to "uprooting" and fracture.

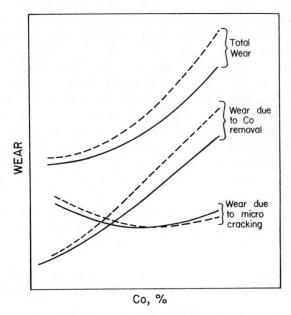

**Figure 5.28** Proposed general dependence of wear by the two cooperating mechanisms and total wear on cobalt content (Larsen-Basse, 1973).

Quartz grains fracture easily during abrasion, and one could expect cutting asperities of sufficiently small size to be available in rocks containing free quartz. When the WC grain size is increased at constant mean free path, the rate of wear would tend to fall because of the decrease in cobalt content. If grain size is increased at constant cobalt content, internal stresses in the cobalt decrease with greater mean free path; the wear rate, therefore, rises with increased grain size. The two sets of curves (Figure 5.28) give total wear rate/mean free path curves of shape similar to the experimentally determined curves (Figure 5.23). Similar results are obtained with the cobalt content as the variable. Yielding curves are of a shape similar to the experimental curves (Figure 5.20).

This discussion indicates that the rotary wear probably takes place by both micro-fracture and preferential cobalt removal and that the change in wear behavior with WC grain size, cobalt content, and mean free path changes the effectiveness of these two mechanisms.

## 5.5 PERCUSSIVE DRILLING—WEAR

In a study of surface spalling due to impact (Latin, 1961) WC–Co inserts with a 70° included angle were positioned with the cutting edge on sandstone and impacted with a 163-J (120 ft-lb)/blow in an impact-testing machine. Spalling of the cutting edge was evaluated qualitatively by visual examination. Harder inserts formed thin, definite spalls, and the more ductile ones burred on the edge with the appearance of micro-chipping. Latin termed this "impact abrasion." The burring was possibly due to plastic deformation of the cutting edge or the formation of very small spalls.

The change between erosion due to spalls and to edge burring occurred at ~12% Co for materials with fine WC grains (1–2 $\mu$m) and at ~8% Co for a WC grain size of 2–4 $\mu$m. It has also been found that for large specimens impact strength increases with cobalt content and with WC grain size (Kreimer, 1968). This transition also occurs for both medium and fine WC grains at hardness of 1400–1450 HV and a mean free path of 0.35–0.1 $\mu$m, which again indicates that the mean free path is an important parameter.

In percussive wear research by Cheetham and Inett (1953–55), the factors that affect bit wear in sandstone and granite rates in penetration were most effective for wet drilling, and gage loss was severe enough in granite to cause stalling. The reasons for greater wear and wet drilling (Table 5.2) are that (1) cuttings suspended in water are more abrasive than in air, and (2) wet drilling allows faster rotation, causing more abrasive wear. There was 50% more gage loss (measured across the cutting shoulder) in sandstone than in granite. The water required for effective flushing should be more than 40 gph for the particular drill and operating conditions used (Figure 3.47).

Fish (1957a) performed percussive drilling tests with cross bits and wet and dry drilling in connection with dust supression studies. The rates of wear were much more rapid for wet drilling (Figure 5.29). Cutting-edge profiles of the carbide inserts were measured, and bit life was determined (Table 5.3) to be greater for dry drilling.

**TABLE 5.2. Bit Wear When Drilling with a $1\frac{1}{2}$-in.-Diameter Steel Crucifrom Bit with a Wet and Dry Flush**

| Rock Type | Flush | Bit diam. across cutting edges (in.) | Loss in diam. (in.) | Bit diam. across base of reaming angle angle (in.) | Loss in diam. (in.) | Distance drilled (in.) | Cumu- lative distance drilled (in.) |
|---|---|---|---|---|---|---|---|
| Darley Dale | Water | 1.524 | | 1.498 | | | |
| Sandstone | | 1.363 | 0.161 | 1.465 | 0.033 | 69.75 | 69.75 |
| Grey Granite | Water | 1.275 | 0.088 | 1.445 | 0.020 | 18.00 | 87.75 |
| Darley Dale | Air | 1.520 | | 1.500 | | | |
| Sandstone | | 1.420 | 0.100 | 1.472 | 0.028 | 74.00 | 74.00 |
| Grey Granite | Air | 1.342 | 0.078 | 1.458 | 0.014 | 59.50 | 133.50 |

*Source*: Cheetham and Inett (1953–55).

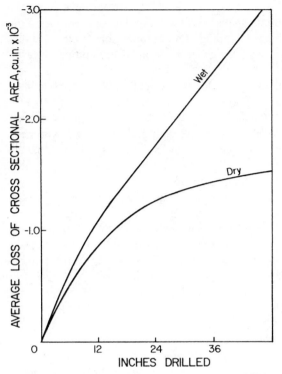

**Figure 5.29** Wet and dry bit wear curves (Fish, 1959).

Percussive drilling in magnetite (Stjernberg et al., 1975), a relatively nonabrasive rock, created a network of cracks known as "snake skin." Carbides used are shown in Table 5.4. The drill was an Atlas Copco RH658, actuating pressure, 0.7 $N/mm^2$, and hold-down pressure, 0.15 $N/mm^2$. Drill holes were 17 mm in diameter and 0.23 cm deep, requiring about 500 impacts. Corners of carbide inserts were studied microscopically. Networks of cracks were formed initially with more cracks being formed in further drilling, these being 6–100 $\mu$m deep. The first cracks, which appeared after 1000 blows, are not produced by fatigue but are thought to be created by tensile stress, possibly caused by thermal expansion.

**TABLE 5.3. Life of Bits**

| Dry Bit<br>(in. drilled) | Wet Bit<br>(in. drilled) | "Life" Ratio<br>(Dry/Wet) |
|---|---|---|
| 24 | 15 | 1.6/1.0 |
| 36 | 18 | 2.0/1.0 |
| 48 | 21 | 2.3/1.0 |

*Source*: Fish (1965).

**TABLE 5.4. Cobalt Content and Grain Size**

| Grade | Cobalt Content (wt %) | Grain Size ($\mu$m) | Compressive Strength (N/mm$^2$) |
|:-----:|:---------------------:|:-------------------:|:-------------------------------:|
| A | 11 | 4 | 4,120 |
| B | 9.5 | 3 | 4,200 |
| C | 9.5 | 5 | 3,900 |
| D | 9.5 | 5 | 4,050 |

*Source:* Stjernberg et al. (1975).

## 5.6  REVIEW OF PERCUSSIVE BIT WEAR

Tungsten–carbide inserts for drill bits of many types had replaced the hard steel formerly used, and over half of the world's supply of cemented tungsten carbide was used by the mining industry because of its excellent strength and wear resistance properties (Lardner, 1974). Sintered carbides of tungsten, titanium, or chromium consist of finely divided hard particles of the carbide sintered in iron, nickel, or cobalt. The carbides exist as individual grains of a finely divided network of dissolved grains resulting from precipitation or cooling.

The performance of carbides usually results from their high hardness and compressive strength. The hardness is 1000 to 1500 on the Knoop scale, density of sintered carbides varies from 12.0 to 14.7 g/cm$^3$, and compressive strength is about 633 ksi, depending on the cobalt percentage present. The strength is also considerably affected by the porosity.

Larsen-Basse (1973) reviewed the known factors affecting bit wear in rock-drilling processes and found that they include shock impact, impact fatigue, spalling, abrasion by quartz, and thermal fatigue. He reported that rock is attacked by the bit fragments in three stages: (1) elastic deformation, (2) pulverization and creation of radial cracks, and (3) fracture along path of maximum shear stress. Bit wear depends on the hardness and abrasiveness of the rock. Although the wear on chisel and button bits occurs in a somewhat different manner, it has some features in common. The frontal wear on a chisel bit may consist of impact fatigue at points of high-impact stress and of abrasion where the bit slides over rock or rock fragments. The wear on the outside or gage surface is predominantly by means of abrasion. Both frontal and gage wear are critical in bit performance and vary with rock type and other related factors, most of which have been only partially evaluated.

The carbide erosion depends on the hardness and abrasiveness of the rock. In hard rocks, the frontal wear is more important, and metal is removed by flaking or chipping (Dahlin, 1962; Montgomery, 1968). Montgomery et al. (1970) found that the larger wear spalls produced by drilling in granite vary in diameter between

25 and 200 $\mu$m, depending on the bit material, and have a thickness of 10–14 $\mu$m. The spalls are initiated from cracks in the worn surface and follow the Co/WC or WC/WC interfaces. The metal fragment size is independent of the rock properties, but the frequency of particle removal from the bit increases with rock hardness. This is affirmed by the work of Dahlin (1962), which shows that greater blow energy and higher rock hardness result in more severe wear and an increasingly rough metal surface on a macroscopic scale. Softer and more abrasive rocks create increasing numbers of abrasive grooves, especially near the corners of the bit and on the gage faces. For soft rocks of low abrasiveness, a ''snake skin'' pattern may form on the bit surface, which is a ''craze cracking'' due to thermal fatigue of the surface metals.

The bit wear in rock types that cause mainly gage wear is approximately proportional to the footage drilled (Bloemsma et al., 1947) but appears to be more directly related to drill time (Holliday, 1968). The latter corresponds more closely to the surface of rock traversed in both indexing and forward advance, and it confirms the theory that gage wear is largely abrasive in nature.

As stated earlier, the bit-wedge angle is about 90°, and its axis makes an oblique angle to give a forward rake angle. Rotary-percussive drilling is used in rocks that are too abrasive for rotary drilling but where percussion is effective (Figure 5.30). The bit is in compression more, which extends its life, and impact energies are greater than for percussive drilling (Fairhurst and Lacabanne, 1957).

The penetration rate is decreased because of tool wear, and the average penetration rate is proportional to hold-down force, to blow-impact energy, and to speed of rotation. The unit wear rate depends on the rock type and decreases with hold-down load and impact energy (Wahl and Kantenwein, 1961). Increased penetration rate results in decreased wear in sandstone but increased wear for granite (Fish,

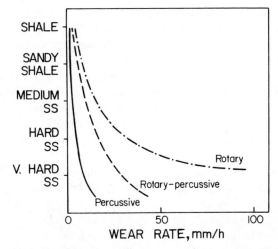

**Figure 5.30** Comparison of rock hardness versus wear rate curves for the three major drilling processes (after Inett, 1957; Larsen-Basse, 1973).

1958). For bit wear due to both abrasion and impact-failure micro-spalling, one would expect these results, and abrasion is more marked for the quartz-rich sandstone. In harder, less abrasive granite, it is postulated that impact fatigue increases if the impact energy is increased for higher penetration rates. The formation of spalls in the impact cycle of rotary-percussive drilling has been observed (Montgomery, 1969; Wahl and Kantenwein, 1961).

### 5.6.1 Rotary-Percussive Drilling Bit Wear (Laboratory)

Montgomery (1968) did research on wear in simulated rotary-percussive drilling, using high hold-down forces to give test conditions that correspond somewhat to drilling conditions. Single-button bits were slid against granite rock surfaces at speeds of 6.1–55 m/min (21–180 ft/min) under hold-down loads of 222–667 N (50–150 lbf) impacted with blow energies in the range of 5.4–19 J (4–14 ft-lb). The wear rate was proportional to the number of blows and the blow energy (Figure 5.31). A plot of wear rate proportional to the inverse of the blow energy gives more scatter in the data but fits a theory of activation of spall sites.

The abrasive wear per blow is theoretically proportional to the hold-down load, $L$, and to the distance of sliding between blows, $u/F$, where $u$ = sliding velocity and $F$ = frequency of impact. According to Montgomery (1968), experimental scatter is responsible for the deviation of his data points from a straight line relation between wear rate and the inverse blow energy (Figure 5.31). He estimates the

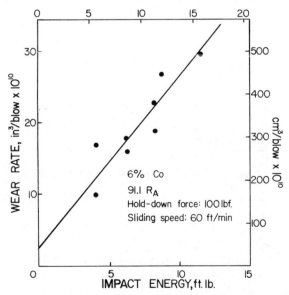

**Figure 5.31** Wear rate as a function of impact energy in simulated rotary-percussive drilling of granite. Tool material: 6% Co, $R_A$ = 91.1 (data from Montgomery, 1968; Larsen-Basse, 1973).

contribution from abrasion to be 4%. Larsen-Basse (1973) assesses the contribution of abrasion by plotting the total wear rate for constant blow energy versus $Lu/F$ (Figure 5.32). Apart from scatter of experimental origin, the data represent linearly increasing wear with blow energy. The lines indicate the rate of abrasive wear and the intercept wear due to impact only. Thus the total wear is made up of two parts: impact or impact–fatigue, and wear and abrasive wear. The impact wear, furthermore, is proportional to the impact blow energy, and the abrasive wear is proportional to the product of the hold-down load and the distance of sliding.

The data of Figure 5.31 are for different values of $Lu/F$. These are normalized as follows: The broken line in Figure 5.32 represents the wear for abrasion only. The data of Figure 5.31 are made to correspond to the $Lu/F$ value of 3.47 J (2.56 ft-lb) per blow, and the results of Figure 5.33 are obtained. This now gives a better linear function and the intercept wear rate of 41 cm$^3$ (2.5 in.$^3$)/blow $\times$ 10$^{10}$, which is the abrasive wear at $Lu/F$ = 3.47 J. Thus, the wear rate is given by

$$R = W/F = AE + BLZ \tag{5.4}$$

where $R$ = weight loss per blow, $W$ = total weight loss per minute, $F$ = impact frequency, $E$ = blow energy, $L$ = hold-down load, $Z$ = distance of sliding between blows, and $A$ and $B$ are constants.

For this example, the contribution from abrasive wear is small, an average about 10% of the total wear. The rock drilled was granite, hard but not highly abrasive, and the bit insert had a high hardness with a good resistance to abrasion.

Montgomery used the same technique to investigate (Figures 5.34, 5.35) the wear resistance (inverse wear rate) versus hardness and cobalt content, the only

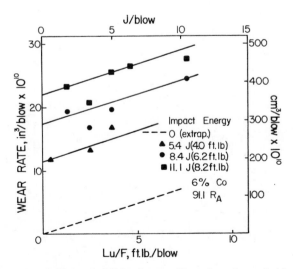

**Figure 5.32** Wear rate versus distance of sliding between blows times the applied load. Same material as in Figure 5.45 (data from Montgomery, 1968; Larsen-Basse, 1973).

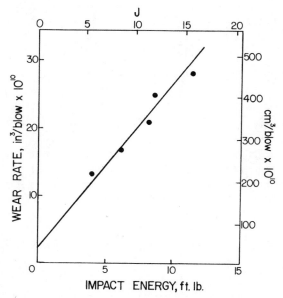

**Figure 5.33** Data from Figure 5.53 corrected to correspond to the same value of $Lu/F$, 3.47 J (2.56 ft-lb)/blow (Larsen-Basse, 1973).

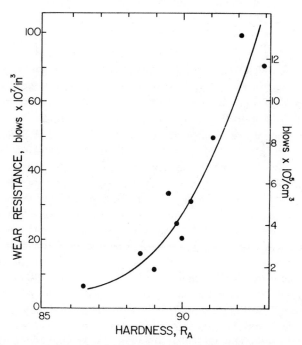

**Figure 5.34** Total wear resistance at 8.15 J (6.0 ft-lb) blow energy and constant $Lu/F$ 5.7 J (4.2 ft-lb)/blow versus hardness for a number of different WC–Co alloys (data from Montgomery, 1968; Larsen-Basse, 1973).

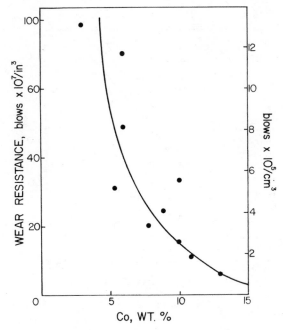

**Figure 5.35**   Total wear resistance versus cobalt content. Same conditions as in Figure 5.48 (data from Montgomery, 1968; Larsen-Basse, 1973).

WC–Co properties given. The data are for blow energy 8.15 J (6.0 ft-lb) and a $Lu/F$ value 5.7 J (4.2 ft-lb) per blow. Here, wear resistance increases with hardness and with decrease of cobalt content. Scatter of data is due to variations in grain size, porosity, and impurities, and to experimental conditions.

The total wear rate was also measured for 8.15- and 15.6-J (6.0- and 11.5-ft-lb) blow energies; for a value of $Lu/F$ some of the hardest samples of inserts fractured at higher blow energies. Most of the materials exhibited a high wear rate at the high blow energy, but three inserts containing 10–12% cobalt showed the inverse. One insert tested at zero blow energy in pure sliding gave the highest wear rate. Work hardening of the surface layers may increase the resistance to abrasive wear. However, the data were found to be insufficient to reach definitive conclusions.

For inserts showing increased wear rate and blow energy, the contributions due to abrasion and impact were identified. Although only one insert was tested for each of the two blow energies and the results are thus not statistically representative, Montgomery's 1969 results represent the only published detailed studies of wear in rotary-percussive drilling; an analysis was made accordingly. For wear rate versus blow energy, straight lines were drawn for each bit material for the plotted values for the two blow energies. The intercept gives the abrasive wear, and the

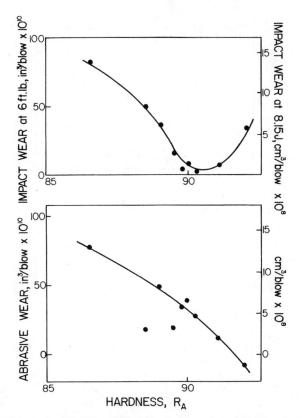

**Figure 5.36** Contributions of impact wear and abrasive wear in rotary-percussive drilling at 8.15 J (6.0 ft-lb) blow energy versus specimen hardness (derived from Montgomery, 1968; Larsen-Basse, 1973).

slope times the blow energy gives the impact wear. The results give wear as a function of hardness and of cobalt content (Figures 5.36, 5.37). A small negative value for the abrasive wear was obtained for one value of hardness, the reason for which was not determined.

The abrasive wear decreases with lower cobalt content and increased hardness. The relation between abrasion resistance and hardness is not linear as it often is for single-phase metals and for those with smaller amounts of hard constituents which are eroded by hard abrasives. The impact wear resistance for those test conditions is a maximum at 90–91 $R_A$ and 6–8% cobalt. The same behavior occurs for the compressive strength of WC–Co alloys versus percent Co and may occur also for transverse rupture strength and impact strength (Kreimer, 1968). Again, insufficient data were available to permit a comparison with fatigue strength.

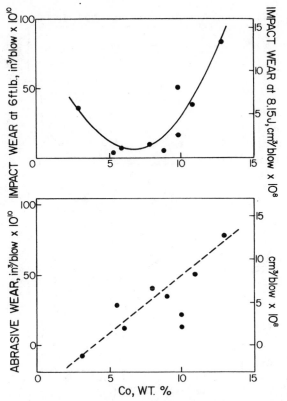

**Figure 5.37** Contributions of impact wear and abrasive wear in rotary-percussive drilling at 8.16 J (6.0 ft-lb) blow energy versus cobalt content (derived from Montgomery, 1968; Larsen-Basse, 1973).

### 5.6.2   Wear for Rotary-Percussive Field Tests

In research reported by Fish (1956, 1958), he proposed the following factors as being important in the technical suitability of a drilling method for a given application:

1. Drilling characteristics: forces and penetration rates;
2. Rate of bit wear;
3. Power consumption and form of energy; and
4. Convenience and safety.

Bit wear as a function of PR and distance drilled was again evaluated by means of the width of the wear "flat." The general shapes of the wear curves for different types of drills cutting in Darley Dale sandstone and Pennant sandstone are similar (Figures 5.38–5.40), the wear for rotary-percussive being about halfway between

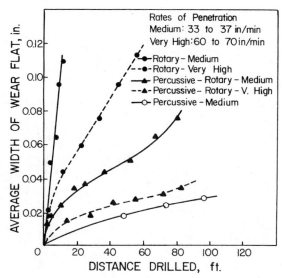

**Figure 5.38** A comparison of rotary, rotary-percussive, and percussive bit wear in Darley Dale sandstone (Fish, 1956–57).

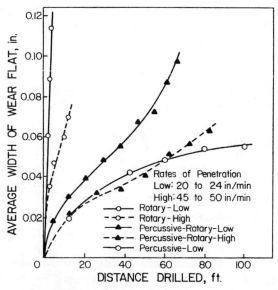

**Figure 5.39** A comparison of rotary, rotary-percussive, and percussive bit wear (low and high) in Pennant sandstone (Fish, 1956–57).

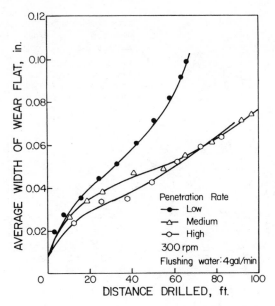

**Figure 5.40**  Rotary-percussive bit wear curves at various penetration rates in Pennant sandstone (Fish, 1957b).

the rotary and the percussive. The wear rate was more favorable in the harder Pennant sandstone.

Fish (1957a, b) reported the results of rotary-percussive drilling and comparative studies of percussive (single-insert), rotary, and rotary-percussive (single-insert) drilling. The first series of rotary-percussive tests was made with a Hausherr DK7ES machine, with some tests being run with a separate rotary drive, a third series being run for bit wear tests. Straight percussive tests were made with a $3\frac{1}{2}$-in. drifter. Rotary-percussive bits were of normal and slotted types with a 9% cobalt and coarse-grained WC. The rocks tested were shale, two sandstones, and a granite.

Bit wear measurements were made primarily in the two sandstones; only a limited number were made in granite. The width of the flat eroded on the cutting edge was again taken as a measure of the wear. For comparison of bit wear, four general levels of initial penetration rate were used: low, medium, high, and very high (Table 5.5).

Low was for Pennant sandstone (with $3\frac{1}{2}$ in. drifter), medium for Darley Dale sandstone percussive, high rotary-percussive, and very high rates in Darley Dale sandstone with rotary and rotary-percussive.

Bit wear per distance drilled in Pennant sandstone was greatest at low PR for the rotary-percussive drills, had the same order in rotary drills, but was greater for the rotary drill (Figures 5.40, 5.41). The rate wear (width of flat) decreases with

**TABLE 5.5. Levels of Penetration Rate in Bit Wear Studies**

| | |
|---|---|
| Low | 20 to 24 in. per min |
| Medium | 33 to 37 in. per min |
| High | 45 to 50 in. per min |
| Very High | 60 to 70 in. per min |

*Source*: Fish (1957).

penetration rate (Figure 5.42) in Darley Dale sandstone. The rotary-percussive wear was greatest for high PR in C granite (Figure 5.43); it was greatest for rotary-percussive low PR and about the same for rotary-percussive high PR and percussive low PR (Figure 5.44). The rate of wear at high PR increased with rpm (Figure 5.45) for Pennant sandstone and Darley Dale sandstone (Figure 5.46), but at low rpm the bit failed in the latter. The air consumed per unit volume of rock drilled was much lower for high PR. The coarseness of the cuttings for the two sandstones was greatest for rotary and increased for both types of drilling with increase in PR. This indicates a decrease of specific energy of drilling with increase of penetration

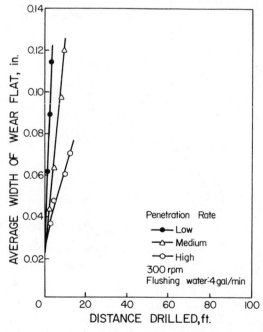

**Figure 5.41**  Rotary bit wear curves at various penetration rates in Pennant sandstone (Fish, 1957b).

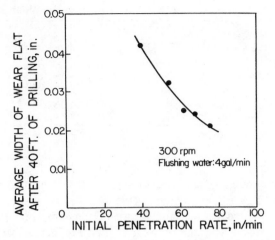

**Figure 5.42** Variation of rate of rotary-percussive bit wear with rate of penetration in Darley Dale sandstone (Fish, 1957b).

rate for two types of drilling in sandstone, but to a certain optimum value only for rotary-percussive drilling.

Fish (1958) also made bit wear studies for rotary-percussive drilling in Pennant sandstone. The wear increased rapidly in the first 10 ft of drilling, and the width of wear flat increased at almost a constant rate thereafter (Figure 5.47).

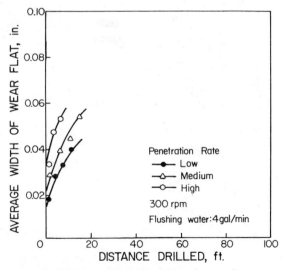

**Figure 5.43** Rotary-percussive bit wear curves in Cornish granite at various rates of penetration (Fish, 1957b).

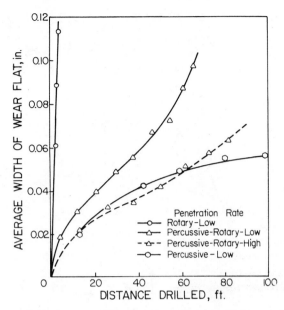

**Figure 5.44** A comparison of rotary, rotary-percussive, and percussive bit wear in Pennant sandstone (Fish, 1957b).

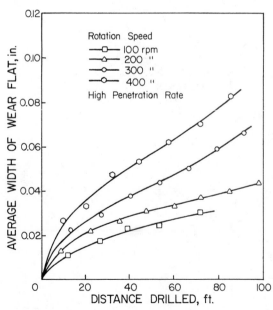

**Figure 5.45** Rotary-percussive bit wear curves at various rotation speeds in Pennant sandstone (Fish, 1957b).

**197**

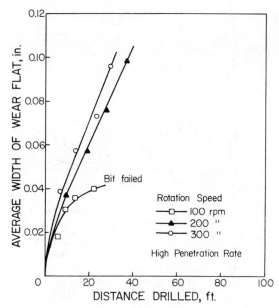

**Figure 5.46** Rotary bit wear curves at various rotation speeds in Darley Dale sandstone (Fish, 1957b).

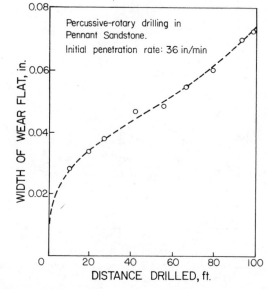

**Figure 5.47** Variation of width of wear flat with distance drilled. Rotary-percussive drilling in Pennant sandstone initial penetration rate: 36 in./min (Fish, 1958).

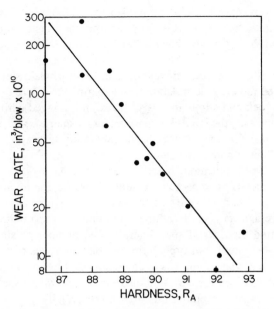

**Figure 5.48** Wear rate for $\frac{7}{16}$-in. buttons drilling Barre granite at 6.0 ft-lb impact energy as a function of hardness (Montgomery, 1969).

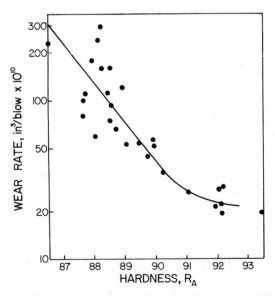

**Figure 5.49** Wear rate for $\frac{7}{16}$-in. buttons drilling Barre granite at 11.5 ft-lb impact energy as a function of hardness (Montgomery, 1969).

**199**

Montgomery et al. (1969, 1970) concluded from laboratory percussive drilling tests on Barre granite that hardness is the most important factor affecting wear rate of button bits, and WC grain size has a secondary effect (Figures 5.48–5.50; Table 5.6).

Montgomery (1968) also studied wear of tungsten–carbide inserts from bits used in field drilling and those tested in laboratory experiments where the WC specimen was held against rock and simultaneously struck and dragged across the rock surface. In the latter case, the hold-down force, blow energy, and sliding velocity were controlled. The WC inserts used in the laboratory tests were $\frac{7}{16}$-in.-diameter, hemispherically shaped button inserts.

Montgomery (1968) indicated, based on experimental evidence obtained, that wear occurred by means of impact that resulted from fatigue micro-spalling of the carbide. Sliding wear effects were minor. The fatigue wear rate was equal to the product of the average rock spall volume, the number of spall sites, and a probability factor relating the critical spall energy. The number of sites found in the laboratory tests were small, but probability of spallation was near unity. This led to the conclusion that when a site has been conditioned by impact, little additional energy is required to cause a spall. WC bit inserts used were usually 10% cobalt, both single and X chisel, or button bits used commercially.

Observations of wear of bits used in field drilling was by means of microscopic studies of wear surfaces and examination of wear particles recovered from drillings. Studies of spalled particles indicated sizes and shapes that were consistent with the postulation of the formation of spalls.

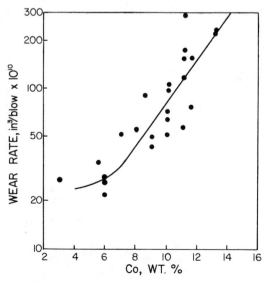

**Figure 5.50** Wear rate for $\frac{7}{16}$-in. buttons drilling Barre granite at 11.5 ft-lb impact energy as a function of nominal weight percent cobalt (Montgomery, 1969).

## TABLE 5.6. Wear Rate of Igetalloy Rock Bit Inserts Drilling Ibaraki Granite

| Grade | Hardness RA** | Cobalt Content wt % | Mean Grain Size, $\mu$ | Mean Cobalt Thickness, $\mu$ | Wear Rate $\times 10^6$ lb per ft Drilled |
|-------|---------------|---------------------|------------------------|------------------------------|-------------------------------------------|
| GR 35 | 88.1 | 7.78  | 2.53 | 0.118 | 27.7, 34.3 |
| GR 35 | 87.8 | 7.90  | 2.40 | 0.114 | 33.3, 37.0 |
| GR 40 | 87.2 | 9.45  | 2.35 | 0.135 | 43.5, 44.8 |
| GR 50 | 86.9 | 10.55 | 2.38 | 0.153 | 45.3, 47.8 |
| GR 2  | 89.4 | 7.91  | 1.60 | 0.076 | 31.6 |
| GR 2  | 89.6 | 7.91  | 1.60 | 0.076 | 23.0 |
| GR 3  | 88.2 | 9.80  | 1.74 | 0.104 | 47.2, 52.0 |
| GR 5  | 88.0 | 11.78 | 1.69 | 0.118 | 51.6 |
| GR 5  | 87.6 | 11.78 | 1.69 | 0.118 | 54.5 |
|       | 86.8 | 7.76  | 5.05 | 0.235 | 40.4, 44.8 |

*Source:* Montgomery et al. (1969).

## TABLE 5.7. Wear Rate Data of Various Carbide Grades

| Grade | Hardness $(R_A)$ | Nominal Composition (%) | Wear Rate at 6.0 ft-lb (in.³/blow × 10¹⁰) | Wear Rate at 11.5 ft-lb (in.³/blow × 10¹⁰) |
|-------|------------------|-------------------------|-------------------------------------------|--------------------------------------------|
| Carboloy 895 | 92.9 | WC 94.0 / Co 6.0 | 14 | * |
| Carboloy 999 | 92.1 | WC 97.0 / Co 3.0 | 10 | 27 |
| Carboloy 860 | 92.0 | WC 91.0 / TaC 4.0 / Co 5.0 | 8 | 21 |
| Carboloy 44A | 91.1 | WC 94.0 / Co 6.0 | 20 | 26 |
| Kennametal 3404 | 90.3 | WC 94.5 / Co 5.5 | 32 | 35 |
| Kennametal 3406* | 90.0 | WC 92.0 / Co 8.0 | 49 | 57 |
| Carboloy 779 | 89.8 | WC 91.0 / Co 9.0 | 40 | 44 |
| Tungaloy PR-2 | 89.5 | WC 90.0 / Co 10.0 | 38 | 53 |
| Carboloy 248 | 89.0 | WC 89.0 / Co 11.0 | 87 | 120 |
| Carmet 2102 | 88.6 | WC 90.0 / Co 10.0 | 140 | 74 |
| Adamas 569 | 88.5 | WC 90.0 / Co 10.0 | 63 | 110 |
| Carboloy 241 | 87.7 | WC 90.0 / Co 10.0 | 130 | 100 |
| Tungaloy PR-45 | 87.7 | WC 88.5 / Co 11.5 | 280 | 80 |
| Tungaloy PR-50 | 86.6 | WC 87.0 / Co 13.0 | 160 | 230 |

* This grade failed to survive at the higher blow energy.

*Source*: Montgomery (1968).

As described previously, the laboratory tests consisted of impacting a simulator (1200–3600 bpm) held down against the surface of Barre granite and measuring of the button surface with a stylus-type profilimeter. The hold-down forces were 50–150 lb, sliding velocities 20–180 ft/min, and blow energies 4.0–13.8 ft-lb. Several grades of WC were tested (Table 5.7). Wear occurred by the formation of thin spalls. Often, no wear occurred until after 400–800 blows. The spalls in Carboloy grade 241 (10% cobalt) appear to have diameters of about 100 $\mu$m. Results were similar for Carmet 2102 (10% cobalt), but for Carboloy 44A (6% cobalt) wear occurred by smaller uniform spalls.

It was found that the wear of button bits was proportional to the number of blows. Although there is considerable scatter in data, the plot of wear rate, cubic inches per blow, against the reciprocal of blow energy gives a straight line on a log–log plot (Figure 5.51). The effect of wear rate due to sliding was small (Figure 5.52) on Carboloy 241. Wear rate was again found to be a function of carbide hardness. Some soft carbides wore faster at low blow energies, with a possibly different wear mechanism.

In usual unlubricated sliding wear, the wear is proportional to the sliding distance and is almost proportional to the bearing load if the mechanism of wear is constant. This is not the case for carbide rock–bit inserts, where the wear is not proportional to the load. However, the wear rate is proportional to the number of blows and is a function of blow energy.

Rozeneau (1963) has suggested that fatigue wear is a rate process for sliding surfaces; that is, wear rate is a function of energy input analogous to the theory of absolute reaction rates for predicting effects of temperature on chemical reaction

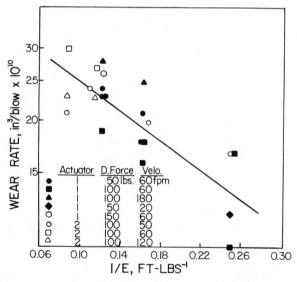

**Figure 5.51**  Wear rate of Carboloy 44A button insert drilling Barre granite as a function of the reciprocal of the blow energy (Montgomery, 1968).

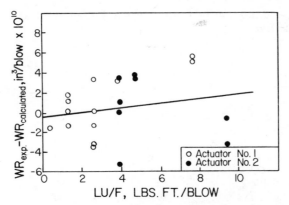

**Figure 5.52** Wear rate of Carboloy 44A button insert drilling Barre granite attributable to sliding (Montgomery, 1968).

rates. The number of favorable events (individual chemical reactions) is equal to the product of a "pre-exponential" term giving the total number of atomic or molecular collisions, and an exponential term, $e^{-E^*/kT}$, which is the proportion of the collisions that result in a reaction. $E^*$ is the energy of activation of the reaction, $k$ is Boltzmann's constant, and $T$ is the absolute temperature. Montgomery (1968) used Rozeneau's concept to formulate an expression for wear of carbide as a function of blow energy.

It was assumed that the energy distribution over the carbide surface is not temperature-dependent. The rate equation is modified for a measure of the energy population of surface points. Thus,

$$n = Nce^{-E^*\sqrt{\epsilon}} \tag{5.5}$$

where $n$ is the number of favorable events (formation of micro-spalls), $N$ the number of impacts, $c$ the number of sites of possible spall formation per impact, $E^*$ the effective activation energy for the formation of a micro-spall, and $\epsilon$ the average energy at the individual sites of possible spall formation. The size of the micro-spalls was found to be approximately constant for the energy range investigated; therefore, total wear is proportional to the number of micro-spalls formed. (The size of the micro-spalls is characteristic of a particular grade of carbide.) The number of sites of possible spall formation, $c$, is determined by the properties of the rock drilled, the geometry of the rock–carbide contact, the properties of the particular carbide, and so on, but it should be constant for any particular system. The average energy $\epsilon$ is also a function of blow energy and the number of possible spall sites. Equation 5.5 is then rewritten

$$W = \overline{V}n = \overline{V}Nce^{-E^*/(\nu E/c)} \tag{5.6}$$

**TABLE 5.8. Evaluation of the Parameters in Equation 5.6 for Various Commercial Carbide Grades**

| Grade | Hardness $(R_A)$ | Vc $(in.^3/blow \times 10^{10})$ | $cE^*/\nu$ (ft-lb) |
|---|---|---|---|
| Carboloy 999 | 92.1 | 54 | 2.30 |
| Carboloy 860 | 92.0 | 42 | 2.34 |
| Carboloy 44A | 91.1 | 30 | 0.50 |
| Kennametal 3404 | 90.3 | 37 | 0.22 |
| Kennametal 3406q | 90.0 | 63 | 0.35 |
| Carboloy 779 | 89.8 | 46 | 0.22 |
| Tungaloy PR-2 | 89.5 | 68 | 0.83 |
| Carboloy 348 | 89.0 | 150 | 1.30 |
| Adamas 569 | 88.5 | 160 | 1.30 |
| Tungaloy PR-50 | 86.6 | 290 | 0.83 |

*Source*: Rozeneau (1963).

where $W$ is the total wear, $\overline{V}$ the average volume of a spall, $E$ the blow energy, and $\nu$ the fraction of the blow energy causing stresses in the carbide surface.

The wear rate $WR$ (defined as $dW/dN$) is then written

$$WR = \overline{V}ce^{-(c\nu)(E^*/E)} \tag{5.7}$$

This requires that effects of abrasion be negligible. Abrasion effects can be taken into account by addition of the term for sliding wear. The wear rate obtained with Carboloy 44A correlated with the reciprocal of the blow energy indicates that equation 5.2 may describe percussive wear.

The data in Table 5.8 were used to evaluate the "pre-exponential" term, $\overline{V}_c$, and the term in the exponent, $cE^*\nu$.

The pre-exponential term relates number and size of possible micro-spalls. The number $c$ was evaluated where an estimate of the volume of an average micro-spall was made.

One micro-spall for Carboloy 860, used for drilling in Barre granite (Table 5.9), was produced in about 50 blows, a micro-spall from Adamas 569 produced almost every blow, and about 30 micro-spalls produced at every blow from Carboloy 44A, which emphasizes the fatigue nature of impact wear.

The exponential term relates to the proportion of events (stress pulses) that result in micro-spalls. Its value is almost 1, so all but the low-energy stress pulses may cause micro-spalls in metal that has been critically fatigued. Wear rate increase is less than 1 in relation to blow energy, whereas chemical reaction rates generally increase more rapidly with temperature. In a chemical reaction, only a small fraction of colliding atoms or molecules have the critical energy necessary to result in reaction.

**TABLE 5.9. Tungsten Carbide Micro-spalls**

| Grade | Estimated Average Dimensions ($\mu$) | Estimated Volume (in.$^3$) | Sites of Possible Spall Formation Per Blow Per Button |
|---|---|---|---|
| Carboloy 860 | $250 \times 400 \times 10$ | $2000 \times 10^{-10}$ | 0.02 |
| Carboloy 44A | $12 \times 16 \times 10$ | $1 \times 10^{-10}$ | 30.0 |
| Adamas 569 | $50 \times 150 \times 10$ | $200 \times 10^{-10}$ | 0.8 |

*Source*: Rozeneau (1963)

This value of the exponential term agrees with the pre-fatigue mechanism of micro-spallation; that is, fatigue damage accumulates over many blows so that observed spall formation requires little additional energy. If the effective activation energy approaches zero, the exponential term approaches unity, in which case the wear rate becomes independent of blow energy. $E^*/\nu$ was estimated for the three grades of carbide for which estimates of $c$ were made. The values (Table 5.10) are consistent with the observation that harder grades of WC are more brittle and therefore would be spalled more easily.

For a more general equation of wear, the contribution of abrasive wear is taken into account by adding a term for sliding wear to equation 5.7 which is a function of the product of hold-down force and velocity. Thus,

$$WR = \overline{V}ce^{-(c/\nu)(E^*/E)} + \phi(Lu/F) \tag{5.8}$$

where $L$ is the hold-down force, $u$ the sliding velocity, and $F$ the flow frequency. The correction for sliding wear for Carboloy 44A and Barre granite is small, amounting to less than 4% of the total wear. The sliding wear may be larger for softer, less abrasion-resistant, carbide grades, but it is not a major factor for any of the commercial carbides for drilling conditions.

**TABLE 5.10. Activation Energy**

| Grade | $cE^*/\nu$ (ft-lb) | $E^*/\nu$ (ft-lb) | Area of Average Spall (in.) | Effective Activation Energy, Energy per in.$^2$ of Spall (ft-lb)/in.$^2$ |
|---|---|---|---|---|
| Carboloy 860 | 2.34 | 120. | $500 \times 10^{-6}$ | $0.02 \times 10^6$ |
| Carboloy 44A | 0.56 | 0.019 | $0.25 - 10^{-6}$ | $0.08 \times 10^6$ |
| Adamas 569 | 1.30 | 1.6 | $50 \times 10^{-6}$ | $0.03 \times 10^6$ |

*Source*: Rozeneau (1963).

Montgomery (1968) concluded that

**1.** The chief wear mechanism of cemented carbide inserts in rotary-percussive drilling is fatigue micro-spalling caused by stresses generated by the blows of the bit on the rock. The effect of sliding wear hold-down force is small.

**2.** The rate of production of micro-spalls depends on the hardness of the rock, and the micro-spall size depends on the grade of carbide. Large numbers of spalls are produced in hard rock, but only occasional spalls are produced in soft limestone.

**3.** Such surface fatigue wear can be described by an equation similar to the absolute reaction rate expression for chemical reactions.

In studies of drill-hole flushing media, Fish (1957c) investigated rotary, percussive, and rotary-percussive drilling. Two laboratory tests were run wet and dry on Pennant and Darley Dale sandstones to determine thrust versus penetration rate. The Darley Dale sandstone is more difficult to drill, and the penetration rates were higher for rotary drilling.

It was found that for wet rotary-percussive drilling, the wear in Pennant sandstone was greatest (Figure 5.53) for high penetration rates, but the opposite for dry drilling (Figure 5.54). The same order of wear was found for Darley Dale sandstone (Figures 5.55, 5.56). The wear rates were the same for rotary drilling both wet and dry for Darley Dale sandstone (Figure 5.57) but were greater for wet drilling in Pennant sandstone (Figure 5.58). It was also stated that in rotary-percussive drilling bit temperatures of the order of 450–500°C were reached (as shown by

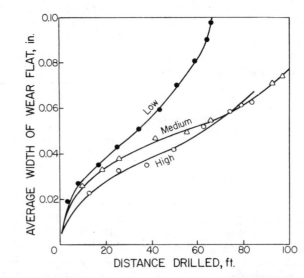

**Figure 5.53**  Rotary-percussive bit wear curves for wet drilling in Pennant sandstone (Fish, 1957a).

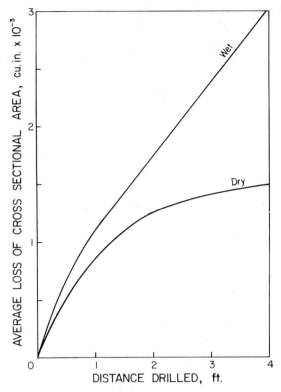

**Figure 5.58** Percussive bit wear curves for wet and dry drilling in Pennant sandstone (Fish, 1957a).

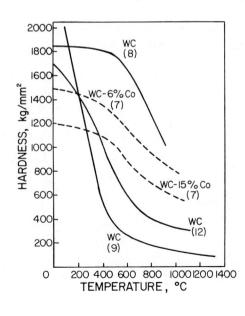

**Figure 5.59** Hardness versus temperature curves for tungsten carbide, from several authors (Osborn, 1969).

**209**

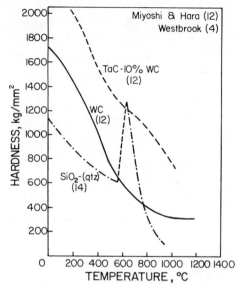

**Figure 5.60**   Hardness versus temperature curves for TaC–10% WC, WC, and quartz (Osborn, 1969).

the carbides tested and reported by Osborn (1969) exhibit a rapid decrease of hardness between 400 and 800°C. Similar temperature characteristics were obtained for TaC, WC, and quartz (Figure 5.60).

## 5.7   BIT WEAR—QUARTZITE

Hustrulid (1971b) has given an analysis of the principles of bit wear for single-edge bits and also data for bit wear for drilling in quartzite. The effects of wear are stated to change (1) the values of $T_R$, the energy transfer from the drill steel to the rock; (2) $E_r$, the specific energy for a given steel–rock combination; and (3) the factor $\beta$, the coefficient of momentum transfer from the drill steel to the piston.

In this analysis, it was assumed that the volume of bit material removed from the bit by drilling is proportional to the volume of rock removed by drilling; that is,

$$\frac{V_c}{V_{\text{rock}}} = \gamma \tag{5.9}$$

where $V_c$ = volume bit material removed (in.$^3$), $V_{\text{rock}}$ = volume of rock removed (in.$^3$), and $\gamma$ = constant. The value of $\gamma$ is determined by the properties of the bit material, the rock, the bit configuration, and other factors.

Differentiation of equation 5.9 gives

$$\frac{dV_c}{dt} = \gamma \frac{dV_{\text{rock}}}{dt} \tag{5.10}$$

but

$$V_{\text{rock}} = A_H \times L \tag{5.11}$$

where $A_H$ = hole area (in.$^2$) and $L$ = total length of hole drilled (in.).
Substitution of equation 5.11 into equation 5.10 gives

$$\frac{dV_c}{dt} = \gamma A_H \times PR \tag{5.12}$$

where $PR$ = penetration rate $(dL/dt)$ and $(dV_c)/(dt)$ = rate at which bit material wears (in.$^3$/min). The material removed in a hole of depth L is, therefore,

$$V_c = \gamma A_H L \tag{5.13}$$

For chisel bit with an included angle of $2\theta$ (Figure 5.61), the flat width, $W_o$, in terms of $L$ is

$$W_o = \left(\gamma \pi d_h L \tan \theta\right)^{1/2} \tag{5.14}$$

where $d_h$ = hole diameter.
Detailed data are not available to describe the effects on the values of $E_r$, $T_R$, and $\beta$ of bit wear. The gage wear of bits is critical in percussion drilling and is analyzed similarly to tip wear. Thus, equation 6.13 gives

$$V_c = \gamma V_{\text{edge}} \tag{5.15}$$

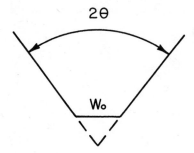

**Figure 5.61** Chisel bit configuration (Hustrulid, 1971a).

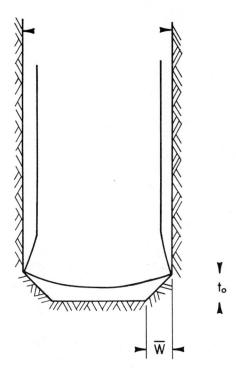

**Figure 5.62**   Diagram showing the effects of gage wear of a bit leaving a ring of rock with a triangular cross section (Hustrulid, 1971a).

When the volume of rock is drilled from a circular ring by a worn bit, it leaves a triangular cross section around the bottom edge of the hole (Figure 5.62).

More energy is needed to drill the outer rock than that near the center of the hole because of confining effects, and before the entire bit cutting edge contacts the hole bottom, the edges must cut through this ring of rock. It has been observed that the width and length of the triangular cross section are very similar, $\overline{W} \approx t_o$. Furthermore, $\overline{W}$ is found to be nearly independent of hole diameter but is related to the penetration per blow. It is assumed that the bit edge makes an angle of $\theta$ with the axis of the drill steel (Figure 5.61). The volume of bit material removed as a function of the decrease in gage $\overline{d}$ is given by (Figure 5.63)

$$V_c = \frac{1}{2} \frac{\overline{d}^3 \tan \theta}{\cos \phi \, \sin^2 \phi} \tag{5.16}$$

The percent of the hole volume removed by the bit edges is

$$\frac{V_{\text{edge}}}{V_{\text{hole}}} = (d_h \overline{W} - \overline{W}^2)/d_h^2 \tag{5.17}$$

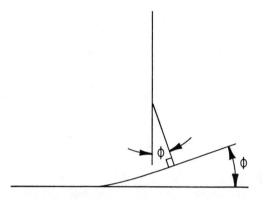

**Figure 5.63** Diagram showing the angles between the edge of the bit and axis of the drill steel and the bottom of the hole (Hustrulid, 1971a).

because

$$V_{\text{hole}} = \frac{\pi d_h^2 L}{4} \tag{5.18}$$

and

$$V_{\text{edge}} = \pi (d_h \overline{W} - \overline{W}^2) L \tag{5.19}$$

A combination of equations 5.16, 5.17, and 5.18 gives

$$\frac{\overline{d}^3 \tan \theta}{\cos \phi \sin^2 \phi} = \gamma \pi (d_h \overline{W} - \overline{W}^2) L \tag{5.20}$$

or

$$\overline{d} = 2\pi\gamma \frac{\cos \phi \sin^2 \phi}{\tan \theta} (d_h \overline{W} - \overline{W}^2)^{1/3} (L)^{1/3} \tag{5.21}$$

and bit gage in terms of $L$ is

$$\text{bit gage} = d_h - 4 \left[ \frac{\pi\gamma \cos \phi \sin^2 \phi}{\tan \theta} (d_h \overline{W} - \overline{W}^2) \right]^{1/3} L^{1/3} \tag{5.22}$$

Equation 5.22 can be expressed as

$$\text{bit gage} = d_h - K' L^{1/3} \tag{5.23}$$

where

$$K' = 4\left[\frac{\pi\gamma \cos\phi \sin^2\phi}{\tan\theta}(d_h\overline{W} - \overline{W^2})\right]^{1/3}$$

Research by Cook et al. (1968) indicates that for tungsten carbide drilling quartzite $\gamma$ is $4 \times 10^{-5}$. Hence

$$\text{bit gage} = d_h - K'L^{1/3} \tag{5.24}$$

Te Water and Mihulka (unpublished) measure bit gage wear in the drilling of quartzite at West Rand Consolidated Mines with chisel bits having gages of 40 and 30 mm ($\frac{7}{8}$-in. hexagonal drill, 48-in.-long, RH 67 rock drill). Their results are described by the following:

for $d = 40$-mm bit:

$$\text{bit gage} = d_h - 4.5(L)^{1/3} + 17.5L > 59 \text{ ft} \tag{5.25}$$

for $d = 30$-mm bit:

$$\text{bit gate} = d_h - 3.2(L)^{1/3} + 13.5L > 75 \text{ ft} \tag{5.26}$$

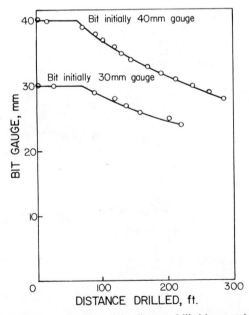

**Figure 5.64** Bit gage wear as a function of the distance drilled in quartzite (Hustrulid, 1971).

Except for the constants, these equations are of the same form as equation 5.23 and show that very little gage wear occurs (loss of bit gage) until a certain depth is reached. This indicates that the breakage of the corners of the hole is related to the bit sharpness. When the cutting edges dull, more rock is left in the corners, and the constant $K'$ in equation 5.24 becomes $K' = 3.61$ for the 40-mm bit and $K' = 2.46$ for the 30-mm bit for $\overline{W} = 0.2$ in.; $\gamma = 3 \times 10^{-5}$, $\phi = 5°$, $\theta = 45°$.

The experimental and theoretical values of $K'$ do not agree well. The ratio between the gage losses for the two bit sizes is

$$R_o = \frac{(d_{h1} - \text{bit gage})_{40}}{(d_{h2} - \text{bit gage})_{30}} = \left( \frac{d_{h1}\overline{W}_1 - \overline{W}_1^2}{d_{h2}\overline{W}_2 - \overline{W}_2^2} \right)^{1/3} \qquad (5.27)$$

Based on the assumption that $\overline{W} = k_o d_h$, $R_o = 1.21$, and, from experimental results, $R_o = 1.41$ (slope ratio), and $R_o' = 75/79 = 1.27$ (intercept ratio). This agreement is fair for the assumptions made.

Tests were made in quartzite to find the relationship between rate of penetration and the footage drilled. A sharpened bit was used to drill holes 2 ft in length, no resharpening was done during the tests, and drilling was normally continued until either the drill steel or the bit failed. The decrease in penetration rate show qualitatively the effect of bit wear on the energy transfer and on the specific energy (Figure 5.64). When the operating pressure was increased to 120 lb/in.$^2$ and the applied thrust was increased to 600 lb, a staved shank drill steel drilled only 64 ft

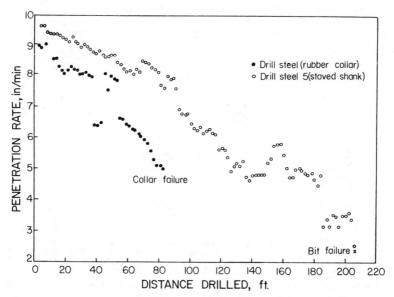

**Figure 5.65** Penetration rate as a function of the distance drilled using an initially sharpened bit in quartzite (Hustrulid, 1971b).

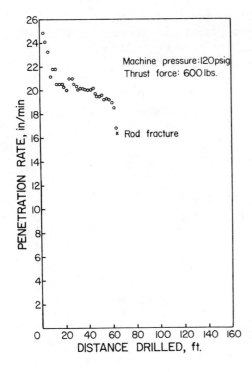

Figure 5.66 Penetration rate as a function of distance drilled in quartzite using an initially sharpened bit (Hustrulid, 1971a).

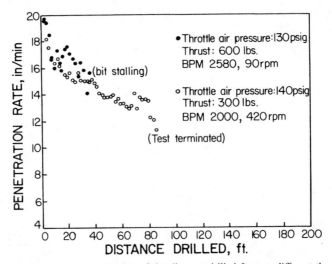

Figure 5.67 Penetration rate as a function of the distance drilled for two different throttle pressure–thrust conditions (Hustrulid, 1971a).

before failing (Figure 5.65) as compared with 206 ft for the conditions in Figure 5.66. A large decrease occurs, because the peak stresses in the steel existing at 120 lb/in.$^2$ $g$ are approximately 1.4 times as large as those at 60 lb/in.$^2$ $g$, and because the applied thrust (600 lb) is too high for an operating pressure of 120 lb/in.$^2$ $g$. Hence, bit rotation probably occurs while the bit is against the rock. The torsional stresses and the longitudinal stresses superimposed may exceed the strength of the drill steel.

The initial and final penetration rates and the distance drilled show that at an air pressure of 60 lb/in.$^2$ $g$ the penetration rate decreases by approximately 0.36%, and at 120 lb/in.$^2$ $g$, by 0.39%. This would affirm the assumption that the volume of bit removed is proportional to the volume of rock drilled.

The penetration rate is a function of the footage drilled for two machine pressure-thrust conditions ($Mp$ = 130 lb/in.$^2$ $g$), thrust = 600 lb, ($Mp$ = 140 lb/in.$^2$ $g$), thrust = 300 lb (Figure 5.67). The penetration rate/distance drilling curves for these operating conditions are quire similar.

The machine operating at 130 lb/in.$^2$ $g$ was very much overthrusted, and it stalled after only 36 ft had been drilled, the penetration rate decreasing more rapidly (roughly 1.2% per foot drilled) than with the machine operating at 140 lb/in.$^2$ $g$ with 300 lb thrust (0.35% per foot drilled). Thus, improper operating conditions resulted in a decrease of almost a factor of 4 in penetration rate.

# 6

## ROTARY DRILLING

### 6.1 DIAMOND DRILLING

Experimental evaluation of the drillability of rock with diamond drills has been based on (1) the hardness and toughness of rocks, (2) the power input per volume of hole, and (3) the rate of penetration related to resistive forces of the rock and the applied penetrative forces. Diamond drilling is a major cost item in exploration operations, and, hence, it is important to be able to predict the drillability of rock for calculation of accurate cost estimates of drilling.

Experimental curves were obtained (Paone and Bruce, 1963; Paone and Madson, 1965) for AX size ($1\frac{1}{8}$-in. diameter) (1) surface-set and (2) impregnated bit performance in the laboratory for rock types ranging from dense hard taconite to soft limestone, and a theory developed. The compressive strength and other rock properties were shown to have some usefulness for the prediction of drilling rates. These studies evaluated pertinent drilling variables as well as the relationship of penetration to the physical properties of rock.

Drillability by diamond drills, as it is related to rock properties, depends on the method of energy application used to fragment the rock. For example, effects of abrasive hardness cannot be based on the process of comminution by means of impact energy, and hardness is not an adequate criterion for drillability, because other factors, such as modulus of elasticity, strength, bonding, and so on, affect rock fracture processes.

**218**

Diamond drilling rates of advance based on specific energy involve all types of energy consumed, such as energy of plastic deformation, friction, kinetic energy of chips, heat losses, and energy of comminution. The energy partitioning involved in these processes varies with rock type and is difficult to measure. Also, the processes of energy transfer from the bit to the rock must be known for the correct evaluation of the resistive forces. A rock drillability index that uses the length of hole drilled per unit time, and wear factors, is useful for application to field drilling, if other related operation parameters are known.

Compressive, tensile, or shear strengths, or a simple index relating them, would offer a useful drillability index, if these strength characteristics of rock are related to the forces developed by the bit at the bit–rock interface.

The difference between specific energy of drilling (energy expended at the bit–rock interface) and specific power (overall power required to drill a unit volume of hole) are related to drillability. Three combinations of factors related to rock drillability have been used: (1) drilling rate as related to hardness and toughness, which are difficult to measure; (2) specific drill power, which does not include drill power losses; and (3) penetration as a function of the resistance to the drilling forces. The last relationship also considers the forces (torque and thrust) in relation to the energy at the bit instead of the power input into the drill. Drilling rates are related to compressive, tensile, and shear strength. Studies have shown that compressive strength is useful because (1) computed "drilling strength" for many rocks approximates compressive strength, (2) compressive strength is easily determined, and (3) compressive strength is a direct function of tensile and shear strength.

## 6.1.1 Theory of Abrasive Drilling

The forces on a diamond core drill bit cause (1) surface failure or crushing due to the thrust and (2) rock removal by the plowing action of the diamonds. Rock resistance to drilling can, therefore, be resolved into (1) the reaction against thrust, and (2) the frictional (plowing) resistance that opposes the torque of the drill friction being the dominant factor. The mechanism of plowing by the diamonds is somewhat similar to the mechanism of friction developed between metal surfaces. However, the abrasion of the rock surface results in a variable resistance because of intermittent shearing of asperities and variable properties due to the granular structure of rock.

The coefficient of friction is defined as

$$\mu = \frac{\text{shear strength } (\tau)}{\text{yield strength } (\sigma_c)} \qquad (6.1)$$

The shear strength of the rock is assumed to approximate its bulk shear strength, and the yield strength $\sigma_c$ under the diamond points is defined as

$$\sigma_c = \frac{dP}{dA} \qquad (6.2)$$

where $dP$ is the average of the applied force for each diamond and $dA$ is the mean cross-sectional area of each crater formed in the rock.

The total thrust $F_v$ is equal to that for the number of the diamonds $n$ contacting the rock, or

$$F_v = n \cdot dP \qquad (6.3)$$

The coefficient of friction $\mu$ from equations 6.1 through 6.3 is

$$\mu = \frac{\tau}{\sigma_c} = \frac{\tau}{\dfrac{dP}{dA}} = \frac{\tau \cdot dA}{dP} = \frac{n\tau \cdot dA}{F_v} \qquad (6.4)$$

Thus, the coefficient of friction is a function of the shear strength of the rock, the number of diamonds in contact with rock surface, the mean area of the indented craters, and the thrust. Since the number of diamonds performing abrasion and $dA$ cannot be evaluated, the coefficient of friction likewise cannot be calculated directly. Its value also varies with the frictional behavior of the diamond surface, the properties of rock, fluid lubrication, and so forth. An average value of 0.4 has been assumed, because it is in close agreement with values obtained from experimental penetration rates under optimum conditions.

Friction studies (Bowden and Tabor, 1950) show that surface damage due to a pressure element sliding on a surface consists of microscopic fragmentation and groove formation due to shear. Similar phenomena have been observed (Blake, 1951) in diamond-cutting studies. Penetration by a diamond bit is, therefore, due to the action of the tangential force and the thrust on each diamond point that exceeds the failure strength of the rock plus the friction.

The resistance to bit rotation due to friction $R$ with thrust $F_v$ is

$$R = \mu F_v \qquad (6.5)$$

When a tangential force $F_t$ is also applied to the bit, the force required to abrade the rock—that is, overcome the drilling strength $S$—is the difference of these two forces; that is, the useful work performed per revolution $W_1$ at the bit face is

$$W_1 = 2\pi r (F_t - R) = 2\pi r (F_t - \mu F_v) \qquad (6.6)$$

where

$$r = \frac{D_1 + D_2}{4} = \text{mean radius of the cutting edge}$$

and $D_1, D_2$ = outer and inner diameters of core bit.

Since

$$F_t r = T = \text{applied torque} \tag{6.7}$$

then

$$W_1 = 2\pi(T - \mu F_v r) \tag{6.8}$$

The work done by the thrust per revolution is

$$W_2 = F_v \delta \tag{6.9}$$

where $\delta$ = advance per revolution.

If the effective drilling strength of the rock is $S$, then the work per revolution is

$$W_R = (SA) \cdot \delta \tag{6.10}$$

where $A$ is the cross-sectional area of the bit:

$$A = \frac{\pi}{4}(D_2^2 - D_1^2) \tag{6.11}$$

The work done on the rock (equation 6.10) in terms of rock strength is

$$W_R = W_1 + W_2 \tag{6.12}$$

Therefore,

$$\delta(SA) = 2\pi(T - \mu F_v r) + F_v \delta \tag{6.13}$$

from which

$$\delta = \frac{2\pi(T - \mu F_v r)}{SA - F_v} \tag{6.14}$$

Equation 6.14 predicts the penetration rate in terms of two unknowns: the coefficient of friction (approximately 0.4) and the drilling strength of rock $S$, which may be approximated by the compressive strength.

### 6.1.2 Experimental Procedures

Laboratory drilling tests with surface set and impregnated $AX$ bits were performed on blocks of rock of measured properties with an electrohydraulic diamond drill. Instrumentation on the drill included a tachometer, a torque gage, and a thrust gage. Rocks were drilled with controlled thrust and speed with the water flow constant.

For each test, a new bit was run at 500 rpm, a second bit at 700 rpm, a third at 900 rpm, and the fourth at 1100 rpm, with the thrust being increased in 1000-lb increments for each speed from 1000 to 4000 lb. The independent test variables were rock types, thrust, speed of rotation, and water flow, and the dependent variables were penetration rate, torque, power consumption, and bit wear, although the latter could not be measured precisely.

The boring capability and wear characteristics of similar diamond bits varied considerably. The surface-set bits were of 10-carat, 332-diamond size, with four waterways and the same matrix metal. The amount of exposed diamond varied around the periphery of the bit, and only part of the bit face appeared to be worn during each test. Also, the rock properties were not uniform, resulting in some penetration rate anomalies.

### 6.1.3   Penetration and Rock Properties—Surface-Set Bits

Analysis of drilling results obtained with 2000- and 1000-lb thrust versus compressive strength shows that thrust has a greater effect on penetration than the speed of rotation, the effect being more pronounced for rocks with lower compressive strength. Calculation of the penetration rate for a given rock type based on compressive strength of the rock ($S$ in equation 6.14) is in good agreement with experimental data for an assumed value $\mu$ of 0.4 (Figure 6.1). The relationships among penetration rate and tensile strength, static Young's modulus, and shear modulus all exhibit curves similar to that for compressive strength.

Correlations with the other properties were analyzed, and, though some were similar to those for compressive and tensile strength, others showed no apparent correlation. The shapes of these curves, which are similar to that for compressive strength (Figure 6.1), have the disadvantage, as with pneumatic drilling and diamond drilling with impregnated bits, that the penetration rate is sensitive to changes in strength at low values of the latter and insensitive at high values.

The specific energy $E_v$ of drilling, not including the work done against friction, is obtained by dividing the work by the volume of rock removed per revolution, or

$$S = \frac{2\pi}{SA}(T - \mu F_v r) + \frac{F_v}{A} = E_v \tag{6.15}$$

or the effective drilling strength $S$ is equal to the specific energy of drilling. The torque in this case is

$$T = \frac{\delta(SA - F_v)}{2\pi} + \mu F_v r \tag{6.16}$$

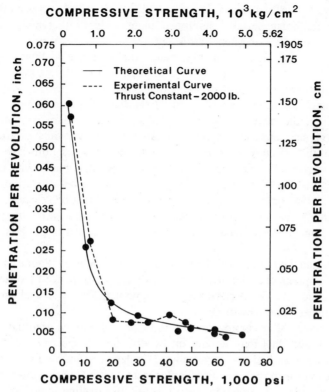

**Figure 6.1** Relationship of theoretical and experimental curves for diamond drilling drillability. Based on compressive strength (Paone and Bruce, 1963).

If the specific energy $E_v$ is to include the work done against friction, it is given by

$$E_v' = \frac{F_v}{A} + \frac{2\pi}{A} \cdot \frac{T}{S} \qquad (6.17)$$

which is the form of the specific energy equation developed by Teale (1965). With the work against friction included in $E_v'$, the specific energy is not equal to the drilling strength of the rock.

## 6.1.4  Penetration and Rock Properties—Impregnated Bits

In this series of experiments (Palone and Madson, 1965), impregnated diamond bits were tested on seven rock types in the laboratory and 21 types in the field. Fundamentally, the rotational speed, thrust, and torque must produce sufficiently high stress levels in the rock at the bit face to cause continuous rock failure and thus achieve effective penetration rates. Although higher speeds of rotation are more effective for drilling relatively harder rocks, high speed causes more bit wear. An increase in thrust causes a nonlinear increase in drilling rate, whereas the relation between torque and penetration rate for a given rock type is linear up to a critical torque value. Penetration rate trends for impregnated bits are similar to those for surface-set bits for rock with compressive strength less than 25,000 psi.

In an impregnated diamond bit, a sintered, powder metal matrix crown contains uniformly distributed fragmented bort or whole diamonds of selected screen sizes. During drilling, the matrix wears away, exposing new, sharp diamonds; consequently, the bit sharpens itself by exposing larger diamonds, which protrude beyond the surface of the crown. Both impregnated bits and surface-set bits have been used for taconite exploration, and one factor that favors impregnated bits is that they stand more abuse by inexperienced drillers. Two important operation conditions (McWilliams, 1957) are that the correct feed rate or thrust be selected, so that the bit can cut freely, and that the rock be abrasive enough to wear the matrix. Hence, factors affecting bit performance are properties of the matrix material and abrasiveness of the rock.

The field and laboratory drillability tests were made by Paone and Madson (1965) with impregnated bits to compare with the performance of surface-set bits (Paone and Bruce, 1963). Performance of impregnated bits was evaluated in terms of rotational speed and penetration rate.

The advantages of an impregnated bit are that it (1) can be used until the crown is worn away and thus may remain in the drill hole longer, (2) is stronger and more usable by unskilled operators, and (3) is less susceptible to damage or loss of diamonds in broken rock. Its disadvantages are that it (1) has a lower penetration rate and (2) requires resharpening when it is used for drilling in nonabrasive rocks. Also, the diamonds are small, cannot be reset, and, hence, cannot be salvaged.

Laboratory drilling experiments were carried out with AX-size impregnated bits with a medium-hard matrix on seven types of rock, and holes were drilled with an electrohydraulic diamond drill with instrumentation for measuring thrust, rotary speed, and torque. Holes were first collared with an old bit; then a new impregnated bit was used to drill at 500, 700, 900, and 1100 rpm with a thrust of 1000 and 2000 lb.

Field drilling experiments were performed with the trailer-mounted diamond drill machine equipped with hydraulic thrust. Twenty-one rock types were drilled in operating mines and quarries in Minnesota, Wisconsin, and Michigan. Thrust, revolutions per minute, and water pressure were measured. Experimental rotary speeds were 600, 1100, and 1600 rpm at thrusts of 1000 and 2000 lb, as in the laboratory experiments.

Standard property tests were made on core obtained from laboratory and field drilling. These included tensile strength, compressive strength, Shore scleroscope hardness, density, longitudinal wave velocity, shear wave velocity, bar velocity, dynamic modulus of rigidity (shear modulus), static and dynamic Young's modulus, Poisson's ratio, relative abrasiveness, and quartz content. The abrasiveness was determined in a steel drum rotation at 74 rpm with a steel paddle rotating at 627 rpm within the drum rotating in the same direction. A 400-g sample of $+\frac{3}{8}$ in. $\times$

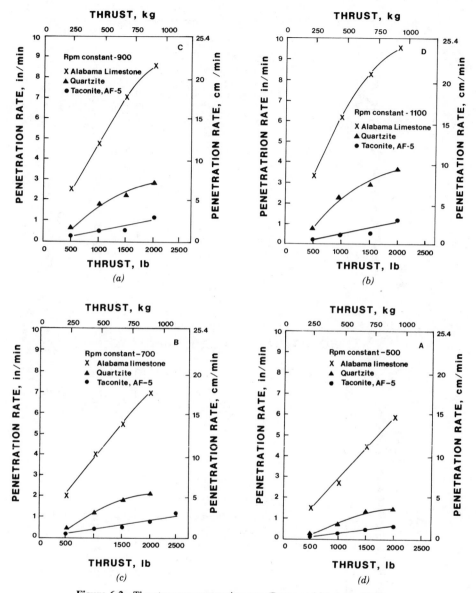

**Figure 6.2** Thrust versus penetration rate (Paone and Madson, 1965).

$-\frac{3}{4}$ in. material was tumbled in the drum, and the abrasivness of rock was determined by the weight loss of the paddle. Only the compressive and tensile strengths, relative abrasiveness, and penetration rates showed significant relationships to the penetration rates in the field.

The highest penetration rates achieved (Figure 6.2) were at the highest thrust and speed attainable by the drill. For most of the rocks, the increase in penetration

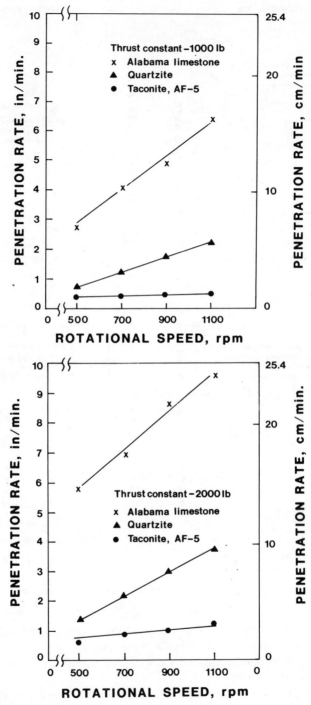

**Figure 6.3** Revolutions per minute versus penetration rate (Paone and Madson, 1965).

rate with thrust is not linear, the slope of the penetration curve decreasing at higher rpm. The increase of penetration rate with thrust is greatest for softer rocks at low rotary speeds and is more pronounced for harder rocks at higher speeds. The increase of penetration rate with rotational speed at constant thrust is linear; however, both high rotational speeds and thrust are required for effective drilling of hard rocks, with greater bit wear occurring at higher rotational speeds.

The test results of torque and penetration rate for limestone, quartzite, and taconite (Figures 6.3, 6.4) indicate that, up to a critical value, a linear relation exists between these two parameters. As pointed out earlier, the torque is a measure of resistive forces opposing rotation and is a function of the friction and abrasion at the bit–rock interface. Hence, the frictional resistance is increased, and the required torque is greater, being dependent upon strength of the rock, number of exposed diamond chips, amount of matrix in contact with the rock, applied thrust, fluid lubrication, and rotational speed. Penetration does not commence below a threshold value of torque (Figure 6.3) partly because limited diamond exposure permits only slight penetration. As indicated earlier, rotational speed, thrust, and torque must be increased until stress and continuous failure are induced in the rock, which also cause wear of the bit matrix. If the diamond points wear down to the matrix in nonabrasive rock, the bit must be resharpened by sandblasting or by altering the drill operation so that the rock abrades the matrix.

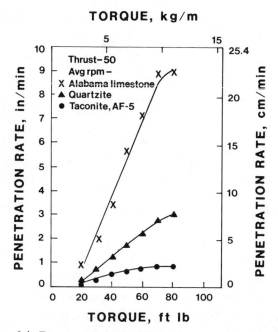

**Figure 6.4** Torque versus penetration rate (Paone and Madson, 1965).

The compressive and tensile strengths, relative abrasiveness, and penetration rate of rocks drilled in field tests have a wide range of values. The field experiments · were conducted with the same types of bits that were used in laboratory tests, and results showed that penetration rates were lower than those for surface-set bits.

In some rocks, the bits became polished and would not penetrate until they were resharpened in the hole by increasing speed and thrust or by sandblasting. Abrading the matrix by increasing the thrust and speed was successful in laboratory drilling but not in the field, because the smaller drill used in the field produced insufficient torque.

From the research with surface-set diamond bits, equation 6.15 was checked utilizing compressive strength as the effective rock strength. While the penetration rate and compressive strength (Figure 6.5) for impegnated bits show the same general trends, equation 6.17 does not apply for impregnated bits because of the difference in the coefficient of friction and the abrasive action of large and small diamonds. Penetration rates of impregnated bits do not change significantly with rock strength for values greater than 25,000 psi, and relationships with tensile strength (Figure 6.6) are similar to those for compressive strength.

For impregnated bits, rock hardness affects the penetration rate, but no strong

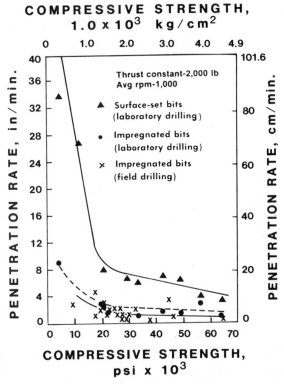

**Figure 6.5** Penetration rate versus compressive strength (Paone and Madson, 1965).

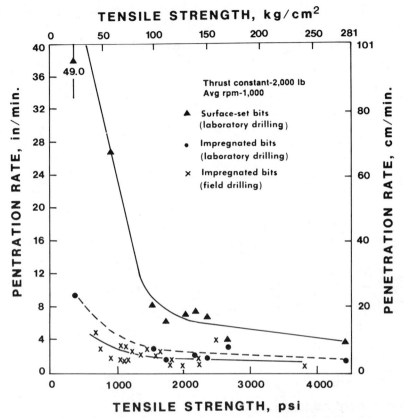

**Figure 6.6**  Penetration rate versus tensile strength (Paone and Madson, 1965).

correlation was found with Shore scleroscope hardness. Abrasion versus penetration rates (Figure 6.7) showed that taconite AF-5 and quartzite have approximately the same abrasive number, but the bit life in taconite was three times that in quartzite (Figure 6.8), whereas the penetration rate in taconite was about one-third of that in quartzite. The percent free quartz is much smaller in taconite than in quartzite, which may account for the lower rate of wear in taconite. Another factor affecting penetration rates in quartzite and taconite, both having high compressive strength, may be the grain size, quartzite having uniform large grains compared to the dense, fine-grained taconite.

The total bit life (footage) measured in the laboratory varies almost in the same order as the penetration rates (Table 6.1). The footages given are minimum values, because the rock was drilled with maximum (not optimum) thrust and speed. For less severe conditions, the bit life of the abrasive rocks might be greater, but with correspondingly lower penetration rates.

In all of the correlations of physical properties with drill performance, for low values of strength, abrasiveness, or quartz content, small changes in a property

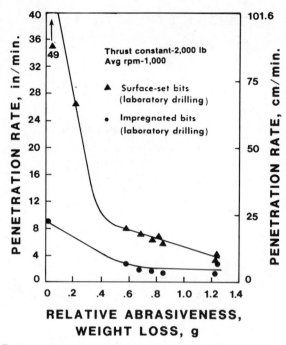

**Figure 6.7** Penetration rate versus relative abrasiveness (Paone and Madson, 1965).

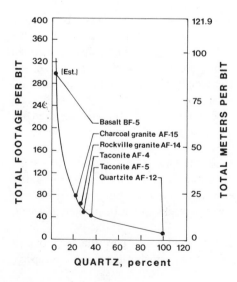

**Figure 6.8** Total footage per bit versus percentage quartz (Paone and Madson, 1965).

**TABLE 6.1. Total Length per Bit at Maximum Penetration Rates**

| Rock Type | Penetration Rate in./min (cm/min) | | Bit Life (ft) (meters) | |
|---|---|---|---|---|
| Alabama limestone | 9.60 | (24.4) | 5,000 * | (1524) |
| Basalt BF-5 | 2.45 | (6.22) | 300-400 † | (91-122) |
| Charcoal granite AF-15 | 2.40 | (6.10) | 80 | (24.4) |
| Rockville granite AF-14 | 3.20 | (8.13) | 65 | (19.8) |
| Taconite AF-4 | 2.50 | (6.35) | 42 | (12.8) |
| Taconite AF-5 | .89 | (2.26) | 44 | (13.4) |
| Sioux quartzite AF-12 | 2.57 | (6.53) | 16 | (4.88) |

* Estimate

† Estimate based on the amount of wear measured after drilling approximately 25 to 30 ft.

*Source:* Paone and Madson (1965).

result in a large change in penetration rate or bit life. Hence, as with surface-set bits in this low-value range, the change in drilling rate is oversensitive to small changes in properties, whereas in the higher range of properties, the drill parameters are insensitive to large changes in properties. It was this trend, which is similar to that found in later pneumatic drilling tests, that led to the use of the Protodiakov Coefficient of Rock Strength and specific energy as parameters for evaluation of penetration rates.

Paone et al. (1966) used the rock property and penetration rate data for surface-set and impregnated bits as the basis for a statistical regression analysis. The analysis showed that no single property is consistently a best indicator of penetration rate and that all of the standard rock properties are highly intercorrelated.

## 6.2 OTHER ROTARY DRILLING

Rotary drilling other than diamond drilling consists of two functional types: (1) drag bit and (2) roller core drilling (Tandanand, 1973).

### 6.2.1 Drag Bits

A drag bit (Figure 6.9) cuts in a manner similar to that in a metal cutting, the edge of a tooth moving in a helix with inclination $\omega$ along a radius $r$ and penetration $\delta$:

$$\omega = \tan^{-1} (\delta / 2\pi r) \qquad (6.18)$$

The mathematical relationship between pertinent parameters is based on the following assumptions:

1. The cutting edge of the tooth is normal to the motion and its width is greater than the depth of cut.

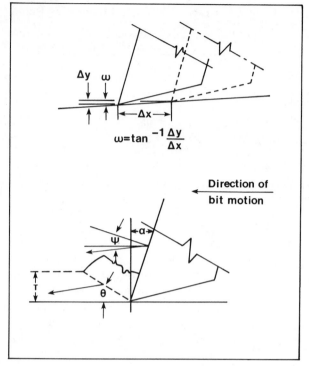

**Figure 6.9**  Geometry of drag bit and cutting force (Tandanand, 1973).

**2.** There is no side flow, and, hence, the problem is two-dimensional.

**3.** The shear plane is perpendicular to the direction of cutting.

**4.** The tooth is sharp and makes contact only at the cutting edge.

Based on the Mohr–Coulomb criterion of failure, the cutting force per unit width of the tooth is given by

$$F = \frac{2t}{n+1} \cdot \tau_o \frac{\cos \phi}{1 - \sin (\phi - \alpha + \psi)} \tag{6.19}$$

where $F$ = cutting force per unit width of tooth, $t$ = depth of cut, $n$ = stress distribution factor, $\alpha$ = rake angle, $\tau_o$ = cohesive strength of rock, $\phi$ = internal angle of friction or rock, and $\psi$ = angle of friction of cutting face and rock; that is, $\psi$ is the angle between the direction of the resultant cutting force and the normal to the rake face of the tool. The cutting force has vertical (thrust) and horizontal (tangential) components

$$F_v = F \sin (\psi - \alpha)$$

$$F_h = F \cos (\psi - \alpha) \tag{6.20}$$

The angle of friction $\psi$ and the stress distribution factor $n$ are functions of the rake angle. The forces fluctuate with time, depending on the rate of formation of chips, whose depth usually exceeds the depth of cut because of curvature of fracture surfaces. Increase of depth of cut increases the thrust and the rake angle, each requiring more torque for the same depth of cut.

## 6.2.2   Roller Cones

One of the techniques used to determine a property of rock that could be correlated with drillability by roller cones is the Schreiner test used by Gstalder and Raynal (1966). It uses a punch test which consists of loading a rock surface to failure with a flat-ended punch. The displacement-load curve is recorded, and the volume of broken rock is measured. The quantities determined are the following:

Hardness:

$$n_r = F_r/A \quad \text{kg/mm}^2 \tag{6.21}$$

where $F_r$ = load at failure and $A$ = cross-sectional area punch, mm$^2$.

Specific disintegration:

$$v_s = V/W_r \tag{6.22}$$

where $V$ = volume of rock broken, mm$^3$, and $W_r$ = total work required to break rock, kg/m.

Plasticity coefficient:

$$K = W_r/W_e$$

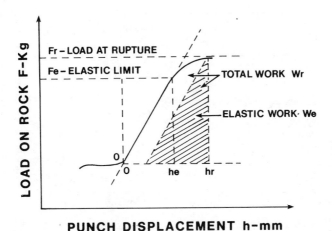

**PUNCH DISPLACEMENT h-mm**

**Figure 6.10**   Load displacement curve from hardness test (Gstalder and Raynal, 1966).

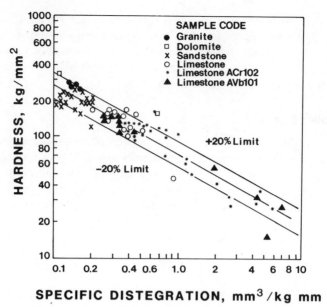

**Figure 6.11** Correlation of hardness and specific disintegration (Somerton et al., 1969).

The plastic coefficient is a measure of the crushing; that is, the elastic work is recovered and does not represent breakage, where $W_e$ = electric work (Figure 6.10), kg/mm.

Young's modulus:

$$E = (1 - \mu) F_e/Dh_e \quad \text{kg/mm}^2 \qquad (6.24)$$

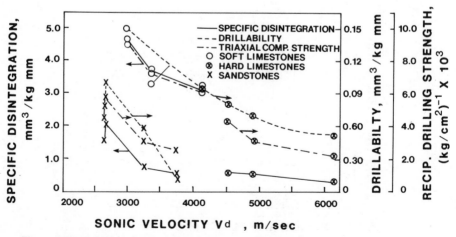

**Figure 6.12** Relation between sonic velocity and drilling quantities (Somerton et al., 1969).

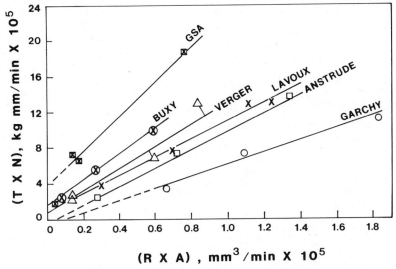

**Figure 6.13**   Drillability determination for limestones (Somerton et al., 1969).

where $\mu$ = Poisson's ratio; $F_e$ = load at elastic limit, kg; $D$ = diameter of punch, mm; and $h_e$ = punch displacement at elastic limit, mm.

The best correlations were obtained between hardness and specific disintegration (Figure 6.11), the best squares fit giving

$$n_r = 70.5 \times v_s^{-0.58} \tag{6.25}$$

The scatter of data is $\pm 20\%$, but the correlation was felt to be valid. Also, it permitted the measurement of the specific disintegration $v_s$ from an easily measured quantity $n_r$.

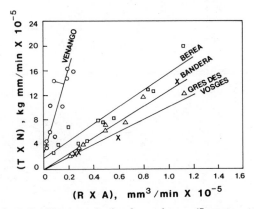

**Figure 6.14**   Drillability determination for sandstones (Somerton et al., 1969).

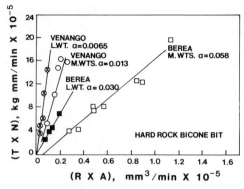

**Figure 6.15**  Effect of bit load on drillability (Somerton et al., 1969).

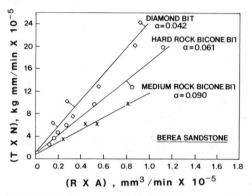

**Figure 6.16**  Effect of bit type on drillability (Somerton et al., 1969).

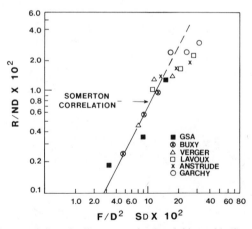

**Figure 6.17**  Drilling correlations for limestones, hard rock bicone bit (Somerton et al., 1969).

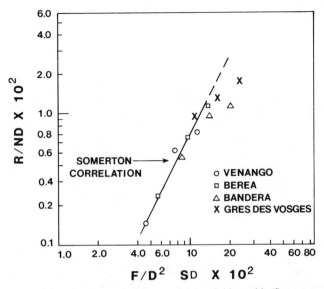

**Figure 6.18**  Drilling correlations for sandstones, hard rock bicone bit (Somerton et al., 1969).

The correlation with Young's modulus was not so favorable, but that of hardness with sonic velocity was well grouped. It was hoped that sonic velocity would give a good measurement of drillability (Figure 6.12), but good correlations were obtained for a soft rock bit only. One reason for possible differences was that hardness tests were conducted under atmospheric pressures. Hardness tests were also con-

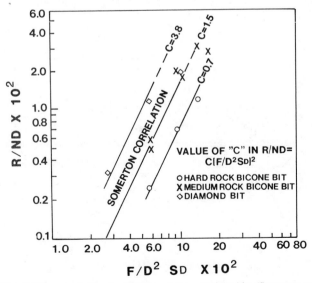

**Figure 6.19**  Drilling correlation for Berea sandstone, various bits (Somerton et al., 1969).

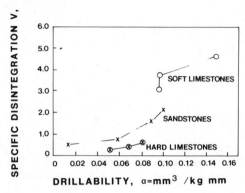

**Figure 6.20** Relation between drillability and specific distintegration (Somerton et al., 1969).

ducted under higher pressures, but sudden rupture of the rock no longer occurred when the confining pressure was above 300 psi (20 bars).

Roller cone bits are made with hard steel or carbide buttons for cutting or crushing the rock. As with pneumatic drills, one of the more meaningful measures of drillability is the specific energy of rock removal. Somerton et al. (1969) used the inverse or the volume drilled per unit energy:

$$\alpha = \frac{1}{E_v} = \frac{R \times A}{T \times N} \tag{6.26}$$

where $\alpha$ = volume drilled per unit of energy, $E_v$ = specific energy in ft-lb/in.$^3$, $A$ = cross-sectional area of hole, $R$ = drilling rate, $T$ = torque on bit, and $N$ = rate of rotation.

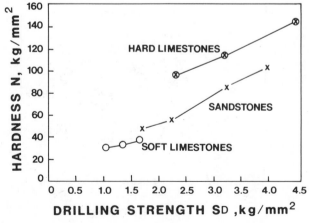

**Figure 6.21** Relation between drilling strength and hardness (Somerton et al., 1969).

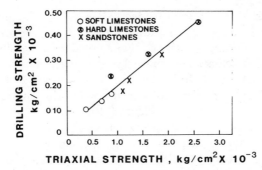

**Figure 6.22** Triaxial strength versus drilling strength (Somerton et al., 1969).

The analysis of drilling parameters may also employ dimensionless quantities:

$$\frac{R}{ND} = C\left[\frac{F}{D^2 S_d}\right] \tag{6.27}$$

where $D$ = bit diameter; $F$ = load on bit; $S_d$ = a strength parameter, psi, drilling strength; and $C$ = correlation constant.

A group of sandstones and limestones were tested for hardness, drilling rate, sonic velocity, triaxial strength, and elastic moduli. The results of laboratory tests indicated that drillability varied with bit type and that drilling efficiency was better for soft rocks.

The values of energy ($T \times N$) and volume ($R \times A$) (Figures 6.13–6.16) indicate a good correlation between energy and rock removed by drilling for different rocks and for different bit types. Consistent relationships were also found for Somerton's correlation (equation 6.27; Figures 6.17–6.19), between specific dis-

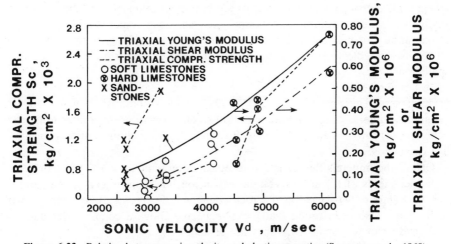

**Figure 6.23** Relation between sonic velocity and elastic properties (Somerton et al., 1969).

integration determined by Schreiner hardness (Figure 6.10) and drillability $\alpha$ (Figure 6.20), between hardness $n_r$ and drilling strength $S_d$ (Figure 6.21), and between triaxial strength and drilling strength $S_d$ (Figure 6.22). The results of the research on sandstones and limestones also showed a good correlation among these properties: specific disintegration, drillability, reciprocal of drilling strength, and sonic velocity (Figure 6.23). Comparisons with diamond drills showed that the roller cone bit had faster rates of advance (Figure 6.16).

Hence, for the rocks tested and for low bit weights, the rock drillability $\alpha$ is a good indicator of drilling efficiency. Drilling strength as defined above, biaxial strength, and hardness also gave reliable correlations. Hard rocks or high bit weights were not tested.

## 6.3 JET PIERCING

The jet piercing process uses a high-temperature flame emitted from a water-cooled combustion chamber to heat rocks that are susceptible to spallation, causing them to flake off. The spalled particles are then swept away by the combustion gases and steam from the cooling water. Such flame jets are used for drilling holes in certain hard rocks, for cutting slots to free blocks of monument stone, and for scaling spallable stone (Calaman and Rolseth, 1968).

The properties affecting spallability are complex, but a general formula for spallability is

$$\text{Spallability} \propto \frac{\text{Thermal diffusivity} \times \text{expansion (at } T_o) \times \text{grain size}}{\text{Compression strength (at } T_o)} \qquad (6.28)$$

where $T_o$ is the temperature at which rock becomes sufficiently plastic to prevent spalling. Inasmuch as differential thermal expansion is an important factor, if grains at the surface of the rock are heated by the flame to a higher temperature, then those just underneath will expand more, creating tensile stresses that result in spallation.

In terms of equation 6.28, the properties that are favorable to spallability are (1) high thermal expansion below 700°C; (2) high thermal diffusivity below 400°C; (3) granular interlocking structure with no fine clay, mica, or other alteration products; and (4) low percentages of soft, low-temperature melting minerals, or those that decompose.

One of the minerals that usually cause spallation is quartz, which not only has high coefficients of linear and volume expansion but changes crystalline form at 573°C (Figure 6.24). Usually, if a rock contains 20% free quartz or more, it will spall. It has also been found that rocks that have above a certain percentage of contained water will spall when they are heated.

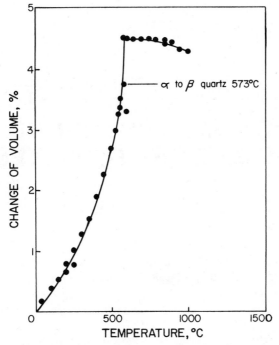

**Figure 6.24** Volumetric thermal expansion of quartz.

Advantages of jet piercing are (1) it will pierce hard formations that are difficult to drill, (2) the jet can be used to change or enlarge a drill hole by adjusting the rate of advance, (3) the jet channeling device obviates the use of drilling and shooting in dimension stone quarries, and (4) rates of drilling are high in most spallable rocks. Disadvantages are (1) capital cost of equipment is high, (2) energy costs are likewise high (the amount of heat of the jet entering the rock is estimated at 6%), and (3) jets are noisy and in quarrying operations create relatively large amounts of dust.

# 7

## ROCK BORING—PRINCIPLES AND LABORATORY TESTS

7.1 Boring Parameters
7.2 Rock Cutting—Laboratory
    7.2.1 Early Linear Cutter Research
    7.2.2 Linear Cutting on Sawn Surfaces
    7.2.3 Laboratory Tests—Preconditioned Surfaces

The mechanical cutting and boring of large openings have been developed and made practical for application in many types of rock formations in recent decades, and these techniques have an important place in processes that involve the fragmentation of rock for underground excavation. Although some tunnel boring had been done previously, an early significant project was carried out with a Robbins machine in 1954 at Oahe Dam, South Dakota, in a 25-ft-9-in.-diameter tunnel through faulted shale; similar projects were later completed with good rates of advance in softer rocks.

Most tunnel-boring machines (TBMs), raise borers, and vertical rotary drills, the latter having been used for many decades for drilling of petroleum, employ roller cutters mounted on a drilling head. Modern cutters for rock are of three types: disc, carbide button, or mill tooth. Some TBMs have used undercutters or drag cutters, but these usually have a high rate of wear. Mill tooth cutters are likewise subject to a high rate of abrasion and wear in harder rocks. On the TBM, only the cutting head, not the stem, rotates, which gives the machine limited steerability, and the body of the machine remains fixed during each stage of drill-head advance. This control is important in bore guidance and avoidance of deviation. Buckets on the cutting head pick up the broken muck and deposit it on a conveyor system which transports it to muck cars in the rear of the machine (Stack, 1982; Bickel and Kuesel, 1982).

Much of the testing of TBM roller cutters, particularly the disc type, has been made on both small and large linear cutting machines. In these tests the cutter is moved in a straight line relative to a block of rock. This simulates the curvilinear path of a cutter on a machine only to a limited extent but does yield data of fundamental value. On the other hand, a 6-ft-diameter rotating laboratory cutting head was tested (Ozdemir et al., 1983) and has a more realistic cutting geometry. However, this machine has no method provided for the removal of cuttings in the invert if the machine is operated in a horizontal or down-slope position. Consequently an abnormal amount of regrinding of cuttings takes place, and the data obtained for the cutter forces do not represent the conditions for a cutterhead operating under field conditions where the cuttings are continuously removed.

## 7.1  BORING PARAMETERS

The performance of rotary boring machines is basically controlled by the same types of parameters as pneumatic drilling, the relation of the optimum levels of performance being determined largely by the particular characteristics of the boring machine and the rocks excavated. In the case of a TBM and similar boring machines these are (1) thrust, (2) torque, (3) specific energy, (4) cutter type and geometry, (5) cutting radius and cutter forces, (6) cutting velocity, (7) cutter sharpness, (8) cutter spacing, and (9) rock properties and geologic conditions. These factors are all interrelated and, in turn, determine the penetration rate, which, with the down time, fixes the overall advance rate in the excavation of the opening, which may be a tunnel, shaft, raise, drill hole, or incline. The emphasis in this book is largely on the excavation processes in mining operations, although rotary drilling of smaller vertical holes is of major interest in petroleum exploration and production.

The failure of the rock in boring is cyclic with cutter penetration (Cook, 1968), and crushing occurs immediately beneath the cutters at a critical load when the stresses also cause chipping. The processes of crushing and chipping are constantly repeated as the cutter moves over the rock surface, as are the load–penetration relationships for any rock–bit combination. These are also determined by the rock properties and the bit geometry. An approximate average linear relationship exists between thrust and penetration for roller bits (Teale, 1964), for disc roller-type cutters (Bruce and Morrell, 1969a), and for field machines (Gaye, 1972). Peak and average values of thrust and cutter forces are extremely variable, as found by measurements of cutter forces.

The indexing, or spacing of cuts, affects the value of the thrust and other operation parameters of penetration for a given rock. With disc roller cutters of different cutting-edge angles, indexing has a pronounced effect on cutter thrust, increase of thrust with spacing being nearly linear (Miller, 1974; Ozdemir, 1975). There is a critical spacing for a given penetration beyond which interaction between the cuts ceases, and forces on the cutters increase in a curvilinear manner. For tricone roller bits (Teale, 1964), the relationship between torque and penetration is essentially linear, as is that for a small two-winged bit.

The thrust and the rolling force, which determine the torque, are affected by increased spacing (Roxborough, 1969; Roxborough and Rispen, 1972) and also have a nearly linear relation between them.

Specific energy, or the energy required to excavate a unit volume of rock, has been used as one measure of the efficiency of rock cutting. However, the power for tunnel boring, for example, is usually about 5% of the total cost of excavation, whereas the advance rate, wear, and failure of the cutters, as well as maintenance expenses are more dominant cost factors.

For rock-cutting systems, the specific energy is a variable function of the rock properties, type of cutting tool, penetration, spacing (indexing) of cuts, velocity of cutting, and virtually all of the other machine parameters. For most cutter–rock combinations, it decreases rapidly with increased indexing at a given penetration to a certain critical or optimum indexing distance, but beyond this point the specific energy increases, and increased penetration at a given spacing may also cause a reduction in specific energy. Types of cutters have been designed to break rocks of different hardness and strength; this breakage is accomplished by a combined crushing and chipping action, the type of failure being influenced by the geometry of the cutting and the mechanical and geological properties of the rock. For hard to very hard rock, tungsten–carbide button, disc, or cone type cutters act primarily by crushing, although chipping occurs when high stresses are induced in the rock. Because crushing consumes excessive energy, these tools are relatively inefficient, but they are more resistant to wear. Disc roller-type cutters will cut hard rock, although they are more subject to wear. However, they are the most efficient; that is, they require the lowest specific energy and are usually preferable for cutting, except for the wear factor. Hence, wear must be considered in the choice of cutters for a given rock formation.

The important geometric factors of insert cutters are their shape and spacing. For harder rock, the inserts are more rounded and more closely spaced. The geometric factors of a disc cutter are the diameter and the included edge angle. Diameter has a minor influence on cutter forces, and increase of edge angle causes the forces to increase rapidly.

The cutters on a rotary boring machine travel on circular paths mostly of different radii, whereas those on standard laboratory linear cutting machines move in a straight line. Along a short-radius circular path the breakage is more concentrated on the inner side of the cutter, which suggests that the side forces are directed outward. Linear cutting closely simulates that along larger-radius paths, especially close to the gage. The force on the cutter can be resolved into rolling, vertical, and side components, and these three components may be measured on appropriately designed experimental linear or rotary machines. The cutters are placed at calculated, fixed radial and angular positions on the cutterhead for proper spacing and force distribution, and they traverse the rock surface at velocities proportional to their radial distances. The loading rate may have some effect on the ultimate strengths of rocks; that is, it has been observed that the higher the rate of load, the greater the effective strength of the given rock.

Crack-growth propagation rates and strain rates affect the effective strength or drilling resistance of the rock. For loading rates on a 90° wedge impacting Tennessee marble with velocities to 250 ft/s, the specific crushing energy increased with a velocity increase of 10 ft/s, beyond which no significant effect was found (Haimson and Fairhurst, 1970). For commercial-size cutters with linear cutting tests at velocities of 40 in./s, the cutter forces also increased with cutting velocity to approximately 10 in./s. For greater velocities, the cutter forces were independent of the velocity. Most field cutters traverse the rock at velocities greater than this.

Cutter wear may be critical, especially in hard rock, where cutter costs are, consequently, high. Rock hardness and abrasivity affect wear as well as penetration rate, and effective traverse velocity and quartz content of the rock mass have been used in attempts to predict wear. Laboratory studies (Sasaki et al., 1972) showed that wear, which increases but less rapidly than the penetration rate, also increases with the velocity of cutting.

Rock properties, the compressive, shear, and tensile strengths, as well as the deformation modulus are related to the borability (Hustrulid, 1970), and a limited correlation of compressive strength was found with cutting performance. For one rock a good relationship between the compressive strength and the specific energy was obtained (Muirhead and Glossop, 1957), and there was a close correlation among specific energy, muck volume, and compressive strength (Rad and Olson, 1974a,b). Borability and compressive strength are not always directly related; a Schmidt–Hammer survey (Ross, 1970) of the Mountains Tunnel project rocks showed a poor correlation with rates. Other results of field tests are described later (Chapter 8).

Regression analysis of data (Bruce and Morrell, 1969a,b) obtained for smooth-surface blocks under laboratory conditions indicates that the depth of penetration of a disc cutter can be approximately determined from the vertical force, Shore scleroscope hardness, and rock density with a predictor equation. The effects of anisotropy (Benjumea and Sikarskie, 1965), that is, of bedding plane orientation (normal and parallel), on specific energy by wedge penetration into Indiana Limestone showed almost a doubling of specific energy for cutting perpendicular bedding planes. The anisotropy of a foliated rock cut with discs (Ozdemir, 1975) also affected the specific energy.

Rock weaknesses facilitate most processes of rock fragmentation, and artificial joints (Miller, 1974) introduced by sawing in rock blocks gave the lowest linear cutting forces when the planes were parallel to the cutting surface. The frequency of joints also affected the cutting efficiency, but only for close spacing of joint planes. In the Nast Tunnel, the machine advance rate increased with decreasing dip of joints (Ozdemir, 1975), but no effect was found of the frequency of occurrence of joints. Though joints may increase rock borability, they also increase the probability of roof falls.

Several other methods have been developed in attempts to predict or relate borability to measureable mechanical properties of rock. Tarkoy and Hendron (1975) used empirical relationships among simple rock rebound, abrasion properties, and penetration rates, but found much scatter in the data. Miller (1974) and Roxborough and Phillips (1975) used a predictor equation involving compressive and shear strengths.

In brief summary, thrust increases linearly with penetration rate for various cutting tools, but the rolling force increases curvilinearly with penetration rate. Specific energy is one measure of cutting efficiency, but it is not a critical factor in the determination of costs. Disc rollers are the most efficient roller cutters, but they are subject to more wear. Linear cutting in the laboratory effectively simulates the action of field cutters. The cutting velocity affects cutter forces only under 10 in./s, but most field cutter velocities are higher than this. Secondary cutting and grinding are also critically affected by efficiency of the removal of cuttings.

Cutter wear in hard rock causes high excavation costs. Some rock properties affect wear, but more experimental field data are needed. Rock properties and accompanying geological structures, such as joints and rock weaknesses, significantly affect boring performance.

## 7.2 ROCK CUTTING—LABORATORY

As with other types of drilling, boring, and rock fragmentation, it would be desirable to have a simple test available that could be used to determine the drillability or borability, or fragmentation resistance, of rock for the prediction of work effort and excavation cost estimation for tunnel boring. Unconfined compressive strength has been used more than any other standard property and may serve as a first approximation, but it is limited in its reliability for complete prediction purposes. Another example is the CRS, which has been used for percussive drilling, but it also has marked limitations as a fragmentation index.

The results of both linear and rotary roller cutter tests have been employed to establish predictor equations for the purpose of estimating machine penetration rates and performance. Most of this research has been done for estimation of TBM performance.

### 7.2.1 Early Linear Cutter Research

Early research was conducted with small sawn rock blocks mounted on a dynamometer table in which the three components of force on the cutter were measured for a mill-tooth cutter on soft rock only, the total force capacity being limited to 3000 lb (Peterson, 1969). One major research discovery was that rock cutting is influenced by prior cuts due to rock removal and consequent subsurface damage to the rock, and, thus, a steady-state condition is reached only after the removal of three or four layers of rock. This was also found to be an important factor in determining the conditions for linear cutter tests, which are described below. The disc-cutting tests reported by Morrell et al. (1970), which were carried out at the U.S. Bureau of Mines, were on smooth sawn rock surfaces, and, hence, the results differ from those obtained from a surface preconditioned by cutting. Although the use of a sawn surface does permit measurement of crater dimensions and results in smoother cutting, it does not include the effects of previous cuts such as those that occur on a tunnel face. These laboratory results are useful, however, for comparison purposes and for the basic data they furnish. Test conditions may include (1) pre-set cutter forces or (2) pre-set penetration per cut, and the machine may be flexible or rigid. These factors must be considered in the analysis and application of test results to field conditions.

## 7.2.2 Linear Cutting on Sawn Surfaces

For the test results from experiments on smooth block surfaces (Morrell et al., 1970) a linear cutting machine utilized 7-in. diameter cutters with 60 and 90° angles, an available 14,000-lb cutting force, a pre-set cutting force, and a cutting velocity of 3 in./s. These values were measured together with crater dimensions, and the specific energy of cutting was calculated. Particle-size distribution determinations were made of the fragments for some tests (Rad and Olson, 1974b). The first series of tests was made on five sedimentary rocks, three limestones, a dolomite, and a marble (Table 7.1). A second series of tests was made with metamorphic and igneous rocks (Table 7.2), two granites, a basalt, and a quartzite. The vertical $(F_v)$ and horizontal $(F_h)$ rolling forces were measured as outputs of strain gages from load transducers, the average being determined by measuring the area under the force–distance traces and dividing by the distance. The widths and depths of the craters were measured at 1-in. intervals, and the crater volume was calculated from the weight of the cutter fragments—that is, the chips and crushed material.

The analysis of the test results for both groups of rock included the following factors: crater width–depth, crater volume/length–$F_v$, depth–$F_v$, $F_h$–$F_v$, energy–crater volume, and specific energy–$F_v$. From these, prediction equations were determined, as well as some other parameters. Craters were cut with spacing large enough that they did not break into adjacent craters. Hence, the important effects of spacing for these tests were not included in the testing or analyses in these two series of experiments.

For crater width–depth studies of softer rock, only the results for marble were plotted, all typically showing a greater width for a 90° cutter (Figures 7.1, 7.2), and the results for igneous and metamorphic rocks were all approximately the same but changed somewhat for rocks of different properties or structure. This relation for most rocks was found to be linear and is expressed by

$$W = W_o + KD \tag{7.1}$$

where $W$ = average crater width (in.), $W_o$ = intercept (in.), $K$ = the slope of a straight line (in./in.), and $D$ = average crater depth (in.).

For hard rocks with a density of 2.6–3.0 and a Shore hardness between 85 and 105, the following predictor equation was found to apply:

$$W = 0.017 + D\left(\frac{2.88}{SH} + \frac{7.88}{\rho}\right) \tag{7.2}$$

where $W$ = predicted crater width (in.), $D$ = crater depth (in.), $SH$ = Shore scleroscope hardness (scleroscope units), and $\rho$ = rock density, (g/cm$^3$).

Another predictor equation for both hard and soft rock was proposed for rocks with densities between 2.3 and 3.0 g/cm$^3$ and compressive strengths between 9000 and 67,000 psi:

**TABLE 7.1. Physical Properties of Rocks Tested**

| | Unit | Salem Limestone | Salem Limestone | Oneota Member, Prairie du Chien Formation | Holston Limestone | Cordell Dolomite Member, Manistique Formation |
|---|---|---|---|---|---|---|
| Geologic name | | | | | | |
| Commercial name | | Indiana Limestone, Type 2 | Indiana Limestone, Type 1 | Kasota Stone | Tennessee Marble | Valders White rock |
| Locality | | Bedford, IN | Bedford, IN | Kasota, Mn | Knoxville, TN | Valders, WI |
| Compressive strength | psi | 9,126 | 9,991 | 13,184 | 16,809 | 27,230 |
| Tensile strength | psi | 679 | 502 | 792 | 1,219 | 793 |
| Shore hardness | Scleroscope units | 27 | 32 | 37 | 55 | 68 |
| Apparent density | Slugs/ft$^3$ | 4.455 | 4.635 | 4.818 | 5.186 | 5.056 |
| Apparent density | g/cm$^3$ | 2.302 | 2.395 | 2.487 | 2.681 | 2.613 |
| Static Young's modulus | 10$^6$ psi | 3.5 | 4.4 | 5.7 | 9.0 | 5.7 |
| Longitudinal velocity | fps | 14,570 | 14,610 | 17,119 | 20,058 | 12,815 |
| Bar velocity | fps | 13,062 | 12,007 | 14,708 | 16,845 | 12,118 |
| Shear velocity | fps | 11,482 | 8,489 | 9,360 | 10,590 | 8,513 |
| Dynamic Young's modulus | 10$^6$ psi | 5.29 | 4.65 | 7.42 | 10.29 | 5.17 |
| Poisson's ratio | | 0.27 | 0.33 | 0.28 | 0.32 | 0.20 |
| Shear modulus | 10$^6$ psi | 2.09 | 2.32 | 2.90 | 4.07 | 2.55 |

*Source:* Morrell et al. (1970).

**TABLE 7.2. Physical Properties of Rocks Tested**

| | Unit | Lac Du Bonnet Quartz Monzonite | St. Cloud Gray Grano-diorite | Dresser Basalt | Sioux Quartzite |
|---|---|---|---|---|---|
| Geologic name | | | | | |
| Commercial name | | Lac du Bonnet Granite | Charcoal Granite | Basalt | Jasper Quartzite |
| Rock type | | Quartz Monozonite | Hornblende–Biotite Granodiorite | | Quartzite |
| Locality | | Lac du Bonnet, Manitoba, Canada | St. Cloud MN | Dresser, WI | Jasper, MN |
| Compressive strength | psi | 38,300 | 39,110 | 63,610 | 67,470 |
| Tensile strength | psi | 1,133 | 1,376 | 1,982 | 2,057 |
| Shore hardness | | | | | |
| Scleroscope units | | 95.6 | 95.9 | 85.8 | 105.7 |
| Apparent density | Slugs/ft$^3$ | 5.091 | 5.266 | 5.879 | 5.117 |
| Apparent density | g/cm$^3$ | 2.624 | 2.714 | 3.029 | 2.637 |
| Static Young's modulus | 10$^6$ psi | 8.069 | 7.748 | 14.25 | 10.63 |
| Longitudinal velocity | fps | 15,951 | 18,838 | 21,873 | 17,254 |
| Bar velocity | fps | 14,255 | 17,231 | 19,514 | 16,988 |
| Shear velocity | fps | 9,895 | 11,119 | 12,264 | 11,821 |
| Dynamic Young's modulus | 10$^6$ psi | 7.18 | 10.86 | 15.54 | 10.25 |
| Poisson's ratio | | 0.2706 | 0.2479 | 0.2725 | 0.1152 |
| Shear modulus | 10$^6$ psi | 3.46 | 4.52 | 6.14 | 4.96 |

*Source:* Morrell et al. (1970).

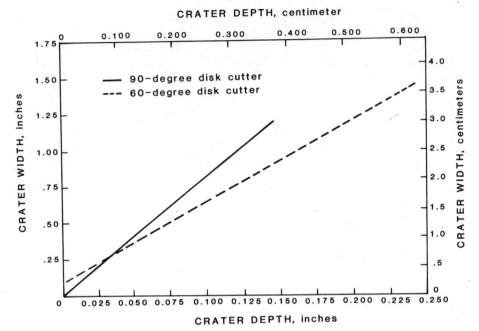

**Figure 7.1** Crater width–crater depth relation for Tennessee marble (Morrell et al., 1970).

$$W = 0.025 + D\left(\frac{1.85 \times 10^4}{\sigma_c} + \frac{7.88}{\rho}\right) \tag{7.3}$$

where $W$ = predicted crater width (in.), $D$ = crater depth (in.), $SH$ = Shore scleroscope hardness (scleroscope units) $\sigma_c$ = compressive strength, and $\rho$ = rock density ($g/cm^3$).

The volumes removed were greater for marble, the 60° cutter being the most effective, and the values for the ig-met rocks were consistently lower (Figures 7.3, 7.4). The equations of these curves are in the form

$$V/L = KF_v^x \tag{7.4}$$

where $V/L$ = crater volume per unit length (in.$^3$/ft), $K$ = a constant (in.$^3$/ft-lb), $F_v$ = average vertical cutter force (lb), and $x$ = an exponent (varying from 1.83 for soft rock to 2.93 for hard rock).

Practically, the efficiency of disc cutters increases with load but is limited primarily by the strength of the cutter edge and bearings. In addition, the volume broken is critically dependent on the cut spacing, as in the tests on preconditioned surfaces described below.

For crater depth-$F_v$ the 60° cutter was again most effective in marble (Figure 7.4), whereas for a 90° cutter the crater depth was consistently greater for softer

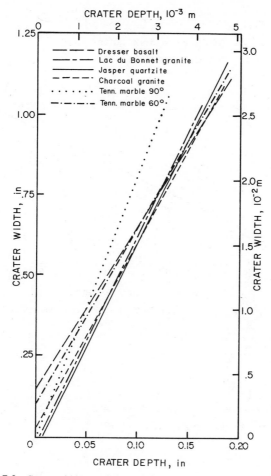

**Figure 7.2** Crater width as a function of crater depth (Morrell et al., 1970).

rocks (Figures 7.5–7.7). For softer rocks it was assumed that the curves (as for crater width) were of the form:

$$D = D_o + KF_v^x \tag{7.5}$$

where $D$ = crater depth (in.), $D_o$ = intercept (in.), $K$ = constant (in./lb), and $F_v$ = average vertical cutter force (lb).

For harder rocks,

$$D = KF_v^x \tag{7.6}$$

where $D$ = crater depth (in.), $K$ = constant depending on rock properties, $F_v$ = average vertical cutter force (lb) and $x$ = an exponent that differs for each rock (from 1.1 to 1.2).

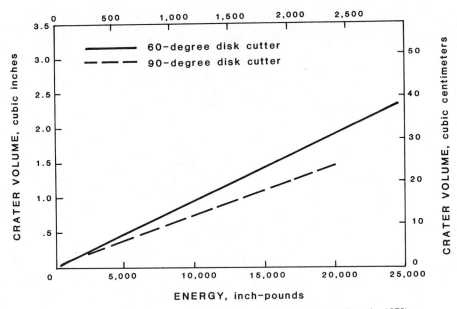

**Figure 7.3**  Volume–energy relation for Tennessee marble (Morrell et al., 1970).

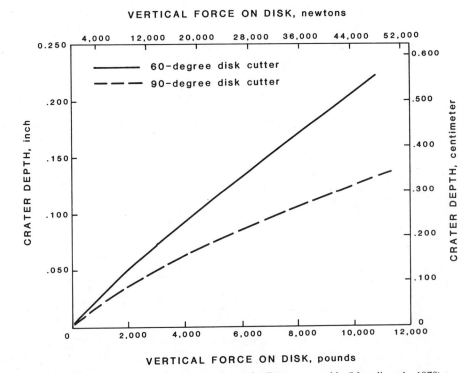

**Figure 7.4**  Crater depth–vertical force relation for Tennessee marble (Morrell et al., 1970).

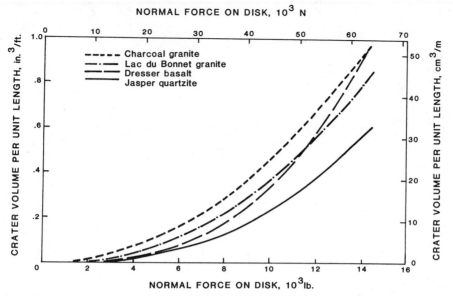

**Figure 7.5** Crater volume per unit length as a function of normal cutter force (Morrell et al., 1970).

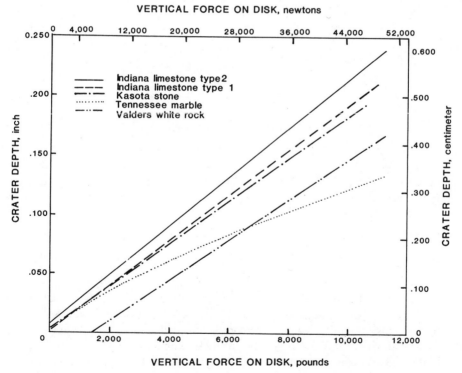

**Figure 7.6** Crater depth–vertical force relation for all rocks tested (Morrell et al., 1970).

**253**

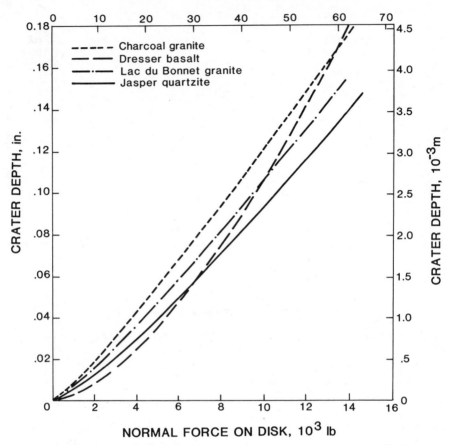

**Figure 7.7**   Crater depth as a function of normal cutter force (Morrell et al., 1970).

For depth of penetration for a disc cutter in rocks with approximately the same physical properties, the prediction equation was developed using stepwise multiple linear regression:

$$D = 0.004 + F_v^{1.2}\left(\frac{1.25 \times 10^{-4}}{SH} + \frac{3.50}{E_s}\right) \qquad (7.7)$$

where $D$ = estimated crater depth (in.), $F_v$ = average vertical cutter force, $SH$ = Shore scleroscope hardness (scleroscope units), and $E_s$ = static Young's modulus (lb/in$^2$). This equation has a multiple correlation coefficient of 0.957 and a standard error of estimate of 0.013 and is valid for rocks with a Shore hardness between 85 and 105 and a Young's modulus between $7.7 \times 10^6$ and $14 \times 10^6$ psi. The constant 0.004 can be dropped without loss of accuracy.

The horizontal or tangential force determines the cutterhead torque and is greater

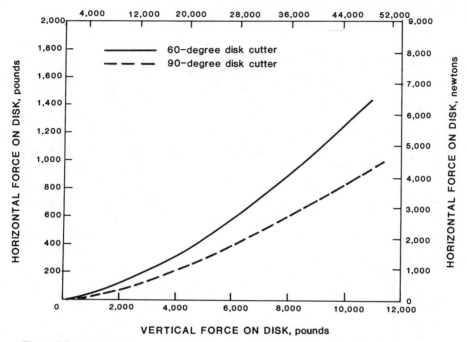

**Figure 7.8** Horizontal force–vertical force relation for Tennessee marble (Morrell et al., 1970).

for a 60° disc for marble (Figures 7.8, 7.9) and much the same for all of the types of rock tested for a 90° cutter, related to the vertical force. The tangential and the horizontal, or rolling, forces are required to move the disc cutter across a rock surface. The equation that best fits the data is

$$F_t = KF_v^x \tag{7.8}$$

where $F_t$ = tangential cutter force (lb), $F_v$ = average vertical cutter force (lb), $K$ = a constant that differs for each rock (lb), and $x$ = an exponent ranging from 1.7 to 2.1. The data show that the tangential force increases approximately as the square of the vertical force.

The volume per unit energy is also greater for a 60° cutter for marble (Figure 7.3), whereas the volumes for softer rocks are consistently much higher than for harder ig-met rocks (Figures 7.10, 7.11). The cratering energy of a disc cutter is the sum of the vertical and horizontal work. The vertical work is the average horizontal force times the length of the run and was found to be a small fraction of the horizontal work, usually from 6 to 10%. The crater volume per unit energy for Tennessee marble is typical of all of the soft rocks tested and has the form

$$V = V_o + KE^x \tag{7.9}$$

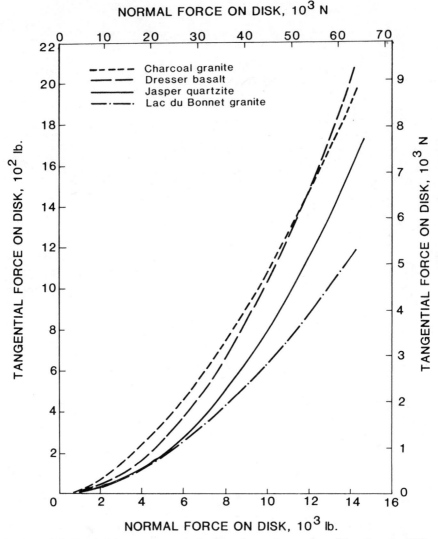

**Figure 7.9** Tangential cutter force as a function of normal cutter force (Morrell et al., 1970).

where $V$ = crater volume (in.$^3$), $E$ = energy (in.-lb), $K$ = a constant different for each rock, (in.$^3$/in.-lb), and $x$ = an exponent that varies between 1.2 and 1.4.

The crater volume predictor equation for rocks with a Shore hardness between 85 and 105, a tensile strength between 1100 and 2000 psi, and a dynamic Young's modulus between $7 \times 10^6$ and $15 \times 10^6$ psi was developed:

$$V = -0.054 + E^{1.1}\left(\frac{1.48 \times 10^{-4}}{SH} + \frac{1.41 \times 10^{-2}}{\sigma_t} + \frac{3.74 \times 10^1}{YM_d}\right) \quad (7.10)$$

where $V$ = crater volume (in.$^3$), $E$ = energy (in.-lb), $SH$ = Shore scleroscope

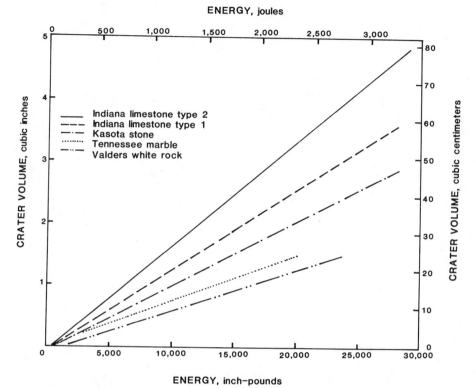

**Figure 7.10** Crater volume–energy relation for sedimentary rocks tested (Morrell et al., 1970).

hardness (scleroscope units), $\sigma_t$ = uniaxial tensile strength (lb/in.$^2$), and $YM_d$ = dynamic Young's modulus (lb/in.$^2$).

A similar equation was developed for rocks of different properties with compressive instead of tensile strength and shear modulus instead of Young's modulus. This predicts crater volume for rocks whose properties fall between the soft and the hard rocks:

$$V = -0.090 + E\left(\frac{8.43 \times 10^{-1}}{\sigma_t} + \frac{8.48 \times 10^{-4}}{SH} + \frac{6.67 \times 10^{1}}{SM}\right) \quad (7.11)$$

This applies to rocks with a Shore scleroscope hardness between 27 and 105, with compressive strengths between 9000 and 67,000 psi, and with shear modulus ($SM$) between 2.1 and 6.14 × 10$^6$ psi. The specific energy (energy per unit volume of rock broken) decreases rapidly for the ig-met rocks (Figure 7.3) until a vertical force of 12,000–14,000 lb is reached. The curves are described by

$$E_s = KF_v^{-x} \quad (7.12)$$

where $E_s$ = specific energy (in.-lb/in.$^3$), $F_v$ = average vertical cutter force (lbs), $K$ = a constant different for each rock (in.-lb/in.$^3$), and $x$ = an exponent.

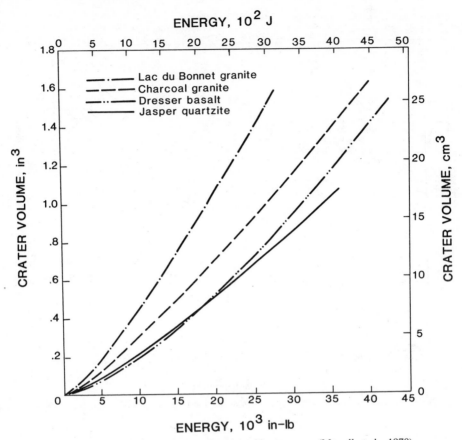

**Figure 7.11**   Crater volume as a function of input energy (Morrell et al., 1970).

The vertical force should be large enough that the specific energy for a given rock is on the flat portion of the curve. The ratio of specific energy to compressive strength is sometimes taken as a measure of the efficiency of rock fragmentation.

The method of development of prediction equations for disc cutter performance used by the Bureau of Mines is of interest for comparison with other methods which utilize conditioned versus smooth or ground surfaces. Morrell and Larsen (1974) used a stepwise linear regression analysis, which entered one independent variable at a time, to produce a series of equations with each equation having one more variable. The significance of the regression coefficients was determined by F-testing at a 95% level, using the following model for both 60 and 90° cutters:

$$Y_{90} = \beta_0 + X\left[\frac{\beta_1}{RP_1} + \frac{\beta_2}{RP_2} + \cdots + \frac{\beta_n}{RP_n}\right] \qquad (7.13)$$

where $Y$ = dependent variable to be estimated, either crater depth or crater volume; $X$ = independent variable, vertical force or energy; $\beta_i$ = the desired regression coefficients from regression analysis; and $RP_i$ = physical properties of the rock.

These physical properties were entered in as reciprocals, because resistance to penetration or fragmentation is an inverse function of their values. Thus, values for standard properties were effective in establishing the prediction equations for rocks whose property values fall within a given range—that is, for those of the rocks tested. Although the most important predictor equation is that for the crater depth, which gives the penetration rate, those for the crater width, crater volume, and the tangential or rolling cutter force are also of interest. The rock properties found by the regression analysis to be related to machine operation parameters were Shore hardness, density, static Young's modulus, compressive strength, tensile strength, and dynamic Young's modulus.

However, these equations have two important deficiencies: (1) they are based on data for cutting tests on smooth, solid, sawn surfaces or rock, which does not represent the condition of a tunnel face, and (2) they do not account for the effect of cutter spacing.

## 7.2.3 Laboratory Tests—Preconditioned Surfaces

As a result of partial factorial experiments on conditioned rock faces (Latin square design), it was found (Ozdemir et al., 1976b) that for sharp disc cutters excavating a red granite (Colorado), the side forces and vertical forces all increased with cutter angle, diameter, spacing, and penetration, as did the vertical and rolling forces.

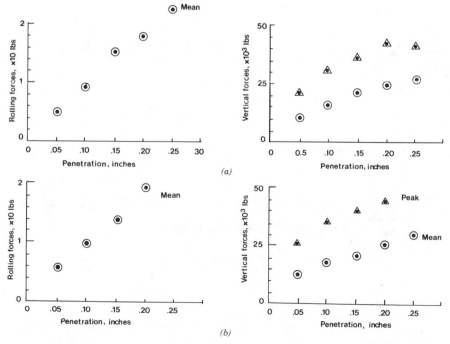

**Figure 7.12** Rolling and vertical cutter forces as a function of penetration in (a) a red granite from Colorado and (b) Charcoal Gray granite (Ozdemir, 1977).

## TABLE 7.3. Red Granite—Colorado

Petrographic analysis:

    Color: dark red-brown

    Texture: massive, holocrystalline, hypidiomorphic, porphyritic, medium grained

    Mineralogical composition:

| | |
|---|---|
| Quartz | 50% |
| Microline | 30% |
| Plagioclase | 8% |
| Biotite | 4–5% |
| Pyroxene | 3% |

    Classification: Holocrystalline, hypidiomorphic, porphyritic, medium grained biotite granite

    Physical properties:

| | |
|---|---|
| Compressive strength | 20,000 psi |
| Shear strength | 2,650 psi |
| Tensile strength | 1,700 psi |
| Density | 2.7 g/cm$^3$ |
| Young's modulus | 7 × 10$^6$ psi |

*Source:* Ozdemir et al. (1976b).

## TABLE 7.4. Charcoal Gray Granite (St. Cloud Gray Granodiorite)

Petrographic analysis:

    Color: gray

    Texture: massive, crystalline, medium grained

    Mineralogical composition:

| | |
|---|---|
| Quartz | 17% |
| Microline | 20% |
| Plagwelase | 41% |
| Biotite–chlorite | 9% |
| Hornblende | 12% |
| Magnetite | 1% |
| Rutile-apatite | 0.1% |

    Classification: crystalline, medium-grained granodiorite

    Physical properties:

| | |
|---|---|
| Compressive strength | 32,000 psi |
| Shear strength | 3,200 psi |
| Tensile strength | 1,900 psi |
| Density | 2.7 g/cm$^3$ |
| Young's modulus: | 10.3 × 10$^6$ psi |

*Source:* Ozdemir et al. (1976b).

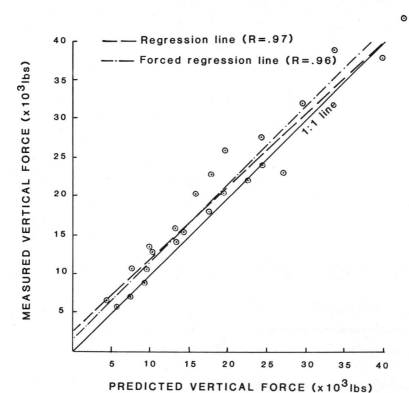

**Figure 7.13**  Comparison of predicted vertical forces to those measured from tests with sharp disc cutters in Colorado red granite (Ozdemir, 1977).

The latter two are the more important in determining machine torque and thrust, and both increase with penetration (Figure 7.12).

Only two granites of 50 and 17% quartz content (Tables 7.3, 7.4) were tested with sharp disc cutters. However, the derived prediction equation for vertical force gave reasonable agreement for both types of rock, with a line of slope of 1:1, a line determined by standard regression, and a force regression line with an imposed slope of 1.

Tests with disc cutters were made using a small milling machine that had been adapted to cut small blocks of rock (Miller and Wang, 1975; Ozdemir et al., 1976a) and a larger linear cutter developed by Hustrulid (1970) to accommodate larger blocks and full-size TBM cutters. As a result of these tests and those from other investigators, it was found that meaningful experimental results and their interpretation could be obtained for disc cutting only by consideration of all of the pertinent variables.

Research with sharp disc roller-type cutters has generally been limited to the study of one or two selected variables, while the remaining variables were kept

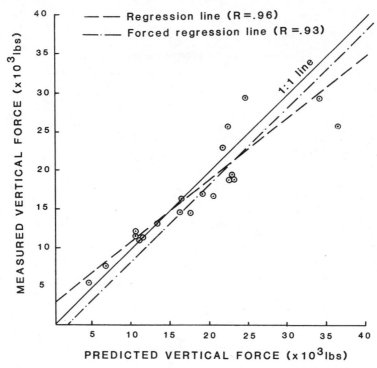

**Figure 7.14** Comparison of predicted vertical forces to those measured from tests with sharp disc cutters in Charcoal Gray granite (Ozdemir, 1977).

constant. The four essential machine performance variables are (1) cutter-edge angle, (2) diameter, (3) spacing, and (4) penetration, with five levels to cover the range of values usually found on a TBM.

A full factorial experiment of four variables at five levels requires 625 tests, which is impractical. However, effective fractional factorial design considers only the second-order interactions, of which there are eight. With the assumption of minimal interaction a Latin square experimental program on the linear cutter was used.

The force components were

MSF—mean side force (lb)

MPSF—mean peak side force (lb)

MVF—mean vertical force (lb)

MPVF—mean peak vertical force (lb)

MRF—mean rolling force (lb)

which represent the averages for five cuts over the rock surface. Experimental data

**TABLE 7.5. Rock Properties**

| Rock type | Origin | Description | Bulk Density, g/cm$^3$ | Uniaxial Compressive, $\sigma_c$ MPa | Tensile Strength, $\sigma_t$ MPa | Shear Strength,[a] $\tau$ MPa |
|---|---|---|---|---|---|---|
| Gregory sandstone | Mansfield, Notts. | Massive, medium-grained, highly porous yellow sandstone | 2.35 | 50.0 | 3.53 | 6.64 |
| Merrivale granite | Merrivale, Devon | Coarse-grained, unaltered, typical granite | 2.62 | 174.2 | 9.96 | 20.83 |
| Dolerite | Belford, Northumberland | Coarsely jointed, unaltered, fine-grained dyke | 2.91 | 339.8 | 27.53 | 48.36 |
| Plas Gwilym limestone | Colwyn Bay, Clwyd | Massive, medium- to fine-grained, crystalline, carboniferous limestone | 2.65 | 155.0 | 13.72 | 23.05 |

[a]Shear strength calculated from $\tau = 0.5\sqrt{\sigma_c \cdot \sigma_t}$.
*Source:* Ozdemir et al. (1976b).

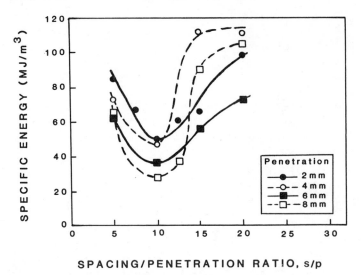

**Figure 7.15**   Effect of $s/p$ on specific energy in Merrivale granite (Snowden et al., 1982).

were analyzed by (1) determination of significance of each variable by analysis of variance techniques, and (2) derivation of relationships between the variable properties and the measured forces.

Dimensional analysis requires the simultaneous inclusion of those independent parameters that affect the behavior of the model. The following were felt to be

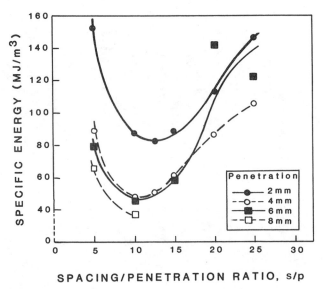

**Figure 7.16**   Effect of $s/p$ on specific energy in dolerite (Snowden et al., 1982).

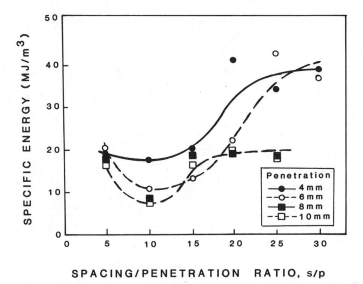

**Figure 7.17**  Effect of $s/p$ on specific energy in Gregory sandstone (Snowden et al., 1982).

necessary in view of previous analysis of cutter testing. *Rock properties: C =* uniaxial compressive strength (psi); $\tau$ = unconfined shear strength (psi). *Cutter geometry: R* = cutter radius or *D* = cutter diameter (in.); $\alpha$ = cutter edge angle (degrees). *Operating parameters: s* = spacing of cuts (in.); *p* = depth (penetration) of cuts (in.).

The cutting speed was not included because it has little effect on cutter performance in the range of velocities used. Other nondimensional variables such as geologic or mechanical faults in the rock mass may affect the cutting but are difficult to assess quantitatively.

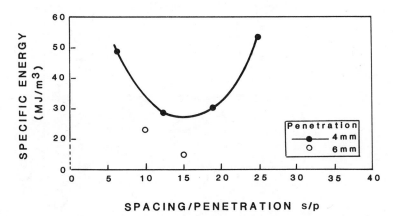

**Figure 7.18**  Effect of $s/p$ on specific energy in Plas Gwilym limestone (Snowden et al., 1982).

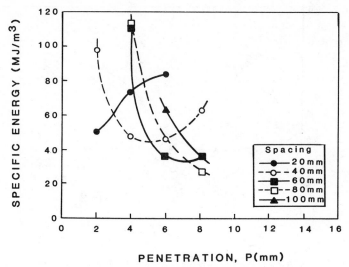

**Figure 7.19**  Effect of penetration on specific energy in Merrivale granite (Snowden et al., 1982).

For a dimensional analysis the vertical force can be written

$$VF = f(R, \alpha, C, \tau, s, p) \tag{7.14}$$

or

$$VF/CR^2 = f(\alpha, x/R, R/p, \tau/C). \tag{7.15}$$

That is, the vertical force is equal to the quantity $CR^2$ multiplied by a function of the $\pi$ terms.

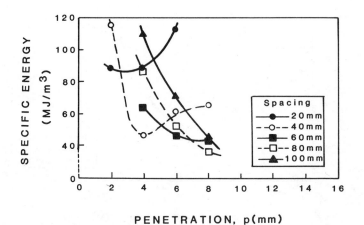

**Figure 7.20**  Effect of penetration of specific energy in dolerite (Snowden et al., 1982).

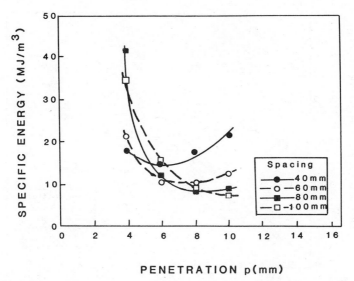

Figure 7.21 Effect of penetration on specific energy in Gregory sandstone (Snowden et al., 1982).

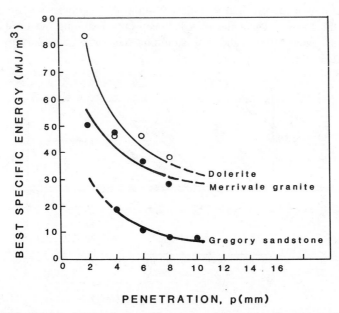

Figure 7.22 Relations between best specific energy and penetration (Snowden et al., 1982).

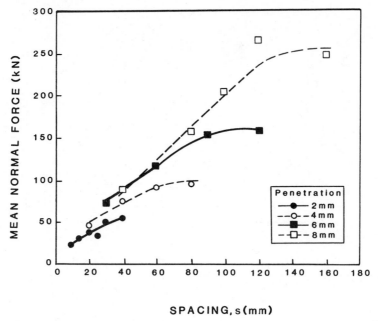

**Figure 7.23** Effect of spacing and penetration on the normal force in Merrivale granite (Snowden et al., 1982).

The equation for vertical cutter force is derived by finding the area of contact of a disc cutter or the crushed volume involving the compressive strength of the rock, and the force to shear the rock between cuts assuming the rock is chipped out owing to shear failure, and then summing the forces necessary to crush and shear the rock:

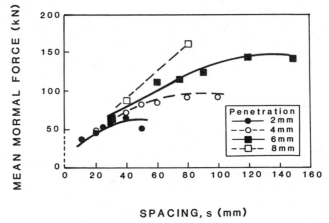

**Figure 7.24** Effect of spacing and penetration on the normal force in dolerite (Snowden et al., 1982).

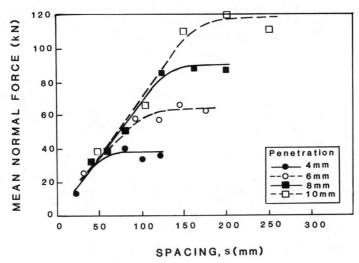

**Figure 7.25** Effect of spacing and penetration on the normal force in Gregory sandstone (Snowden et al., 1982).

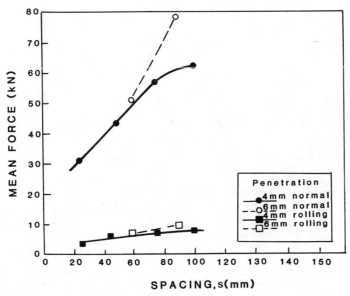

**Figure 7.26** Effect of spacing and penetration on the normal and rolling forces in Plas Gwilym limestone (Snowden et al., 1982).

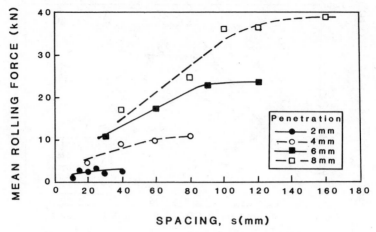

**Figure 7.27** Effect of spacing and penetration on the rolling force in Merrivale granite (Snowden et al., 1982).

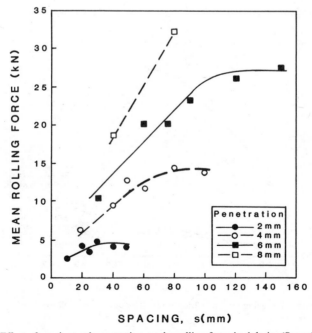

**Figure 7.28** Effect of spacing and penetration on the rolling force in dolerite (Snowden et al., 1982).

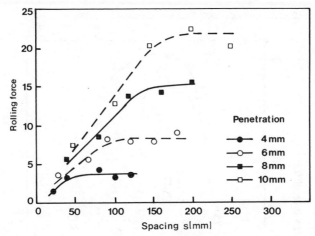

**Figure 7.29** Effect of spacing and penetration on the rolling force in Gregory sandstone (Snowden et al., 1982).

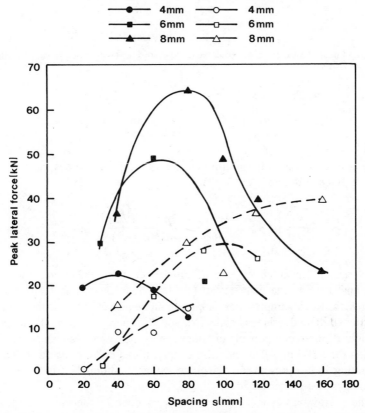

**Figure 7.30** Effect of spacing and penetration on the peak lateral force in Merrivale granite (Snowden et al., 1982).

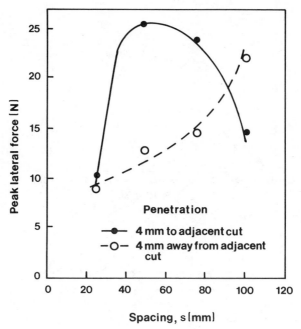

**Figure 7.31** Effect of spacing on the peak lateral force in Plas Gwilym limestone (Snowden et al., 1982).

$$VF = D^{1/2}p^{3/2}\left[\frac{4}{3}C + 2\tau\left(\frac{s}{p} - \tan\alpha/2\right)\right]\tan\alpha/2 \qquad (7.16)$$

The rolling force is found by multiplying the vertical force times the cutting coefficient that gives (Figures 7.13, 7.14):

$$RF = \left[Cp^2 + \frac{4\tau\phi(s - 2p\tan\alpha/2)}{D(\phi - \sin\phi\cos\phi)}\right]\tan\alpha/2 \qquad (7.17)$$

In a series of linear cutter tests (Snowden et al., 1982), sandstone granite, dolerite, and a limestone were tested (Table 7.5) to determine their disc-cutting characteristics. Rock samples were approximately 1-m cubes with the cutting face preconditioned. The debris was collected, specific energy calculated, and peak and mean values of three components of cutter forces measured, as well as spacing and penetration for a 200-mm-diameter disc cutter with an 80° cutting edge. The forces on the cutting tool represent the "average values," although it is not stated how these averages were determined.

Specific energy was calculated from the work done by the rolling force divided by the volume of broken material from the cut. The ratio of the cut spacings (s) to the penetration ($p$) is taken as a meaningful performance parameter. The optimum value of $s/p$ for granite, dolerite, and sandstone was found to be about 10, whereas that for the limestone tested was approximately 15, and the optimum penetration is about 10 mm for the first three rocks listed previously (Figures 7.15–7.31).

The normal and rolling forces follow the same general pattern, increasing to a plateau with spacing and higher plateaus with increase in penetration. The pattern for limestone (Figure 7.21) may have been due to the fact that the rock was more fractured or fissured.

# 8
## ROCK BORING—FIELD EVALUATION

8.1 Field Tests
    8.1.1 Penetration Rate
    8.1.2 Rock Index Properties
    8.1.3 Penetration Rate and Index Properties
8.2 Cutter Wear

## 8.1 FIELD TESTS

Tunnel-boring machine field performance is usually evaluated in terms of utilization, penetration rate, advance rate, and cutter wear (Nelson et al., 1984). The penetration rate is generally defined as penetration distance per unit time or penetration per cutterhead revolution, utilization is the percentage of available shift time during which machine excavation takes place, and advance rate is the product of the penetration rate and machine utilization. The total shift time for purposes of analysis may be taken as the sum of the machine utilization time and the down time due to (1) maintenance and repair of the TBM, (2) repair of the backup system, (3) effect of ground conditions, and (4) miscellaneous causes. In an extended study of several tunnels, Nelson et al. (1984) found the average percentage of delays for maintenance and repair, backup system repair, and ground conditions to be about 21, 20, and 14%, respectively; miscellaneous, 13%. Excavation time was 32%. These performance factors can be only partially evaluated by laboratory tests.

Delays due to the cutter changes are significant. Failure of cutterhead motors caused only moderate delays but decreased the overall penetration rate. It is estimated that more than 30% of the distance excavated on projects studied was completed with less than the full complement of cutterhead motors. The effect of wear on penetration rate is usually only discernible in abrasive rock.

Correlations between rock index properties and penetration rate normalized with respect to thrust improved the in statistical significance relative to those based on penetration rate only. In this study penetration rate is related to thrust by a field penetration index, $R_f$, which is defined as the ratio of the average thrust per cutter

274

to the penetration rate. Linear regressions of $R_f$ and total hardness, or a linear combination of abrasion and rebound hardness, produced the highest degree of statistical significance compared to correlations with single index properties, including rebound hardness, abrasion hardness, uniaxial compressive strength, point load index, and Brazilian tensile strength. Other properties have also been used for formulation of prediction equations.

The penetration rate, thrust, and rolling forces are interrelated for different rock types and have been proposed as a critical surface of three variables (Figure 8.5). The cutting process has been analyzed in terms of the penetration rate versus average thrust and rolling force. In the group of tunnels studied, the ratio of the average rolling force to average thrust per cutter at maximum torque levels was found to vary from approximately 0.15 for high-strength rock to 0.25 for low-strength rock. These values can be used to estimate average optimum thrust and the penetration rate by means of either the total hardness or linear combinations of abrasion and rebound hardness.

The relationship of penetration rates and field penetration indices with the fracture toughness, $K_{Ic}$, and critical energy release rate, $G_{Ic}$, were evaluated for three relatively brittle and high-strength rocks. The penetration rate and field penetration index varied linearly with $G_{Ic}$, and no relation between $K_{Ic}$ and performance was identified. For massive, brittle materials, it appears that $G_{Ic}$, which includes effects of rock strength and stiffness, may have potential for prediction of penetration rate and $R_f$. TBM and drill-and-blast methods of excavation are often compared, and one or the other may be more applicable to a given site because of capital costs, operating costs, rock properties, support problems, or other related factors.

The cutterhead is usually driven forward by hydraulic cylinders and is operated in a single cycle of forward thrust generally of 4 to 5 ft (1.2 to 1.5 m) of movement. The approximated average thrust per cutter, $T$, as compared to those measured on laboratory cutters, may be calculated by

$$T = \frac{N_t(p_c - p_o)\,\pi d^2}{4n} \tag{8.1}$$

where $N_t$ is the number of thrust cylinders in use, $d$ is a cylinder diameter, $n$ is the total number of cutters, $p_c$ is the total pressure for a given forward stroke, and $p_o$ is the pressure required to advance the TBM and trailing gear without face contact.

The torque, $T_o$, is determined from

$$T_o = \frac{N_m P e}{2\pi s} \tag{8.2}$$

where $N_m$ is the number of motors in use, $s$ is the rotational speed of the cutterhead, $P$ is the power consumed by each motor, and $e$ is the efficiency of the motor and drive train.

The average rolling force per cutter, $t$, is estimated by equating the torque, $T_o$, with the product of $t$ and the weighted average radial cutter distance, $R$, which gives

$$t = \frac{N_m Pe}{n 2\pi s R} \tag{8.3}$$

This equation, however, does not account for resistance to rotation from sources such as the weight of muck-filled buckets, friction along the periphery of the cutting head, resistance to rotation at the disc cutters, and friction caused by muck accumulations.

The most common TBM cutting tool is the single disc cutter, which is composed essentially of a hardened ring or disc on a hub mounted on a central shaft so it can rotate about its longitudinal bearing on a steel saddle bolted to the cutterhead face. Common disc cutters are 15.5 in. (394 mm) in diameter and are capable of transmitting up to 45 kips (200 kN) of thrust. The cutter ring is composed of a high alloy and heat-treated steel and is attached to the hub by means of heat shrinking. As noted earlier, cutters are laid out to give the proper distribution of thrust over the cutterhead.

The operating parameters of the TBM and similar boring machines are interrelated, such as machine thrust, the three components of force on the cutters, the penetration per revolution, cutter spacing, and wear or sharpness of cutters. These are in turn related to the mechanical properties and other characteristics of the rock. That is, the machine performance in any given section of rock in a tunnel is affected in varying degrees by a large number of factors. Nelson et al. (1984) have made one of the more detailed quantitative studies of the effects of these factors on TBM operations in the field, and their relatively comprehensive treatment and analysis would require a detailed computer program to produce meaningful results—that is, engineering and cost studies. Other computer programs such as COSTUN (Wheby and Cikanek, 1974) have incorporated some of these factors and procedures.

### 8.1.1  Penetration Rate

The analysis by Nelson et al. (1984) shows that relationships between penetration rate and rock properties vary with the location (e.g., Morgan et al., 1979) as do penetration rate and uniaxial compressive strength, point load indices, and the National Coal Board cone indenter test.

A comparison of rock hardness indices, uniaxial compressive strength, and penetration rates for TBM projects showed that some correlations with penetration rates were significant (Tarkoy, 1975; Tarkoy and Hendron, 1975). Correlations have also been made with combinations of rock property indices (Blindheim, 1976, 1979)—that is, brittleness, abrasion, and a drilling rate index, as well as total hardness (McFeat-Smith and Tarkoy, 1979) which is a combination of abrasion and rebound hardness indices and (Jenni and Balissat, 1979) a combined index of rebound hardness, point load strength, abrasion hardness, and mineral hardness. Such correlations for different machines have shown appreciable scatter (Korbin, 1979) where correlations could be demonstrated for different machines on the Keilder Aqueduct but were different for each machine for similar rock units. Correlations of average hardness indices for 13 TBM projects (Tarkoy and Hendron, 1975) for machines with different types of cutters operating at a variety of torque and thrust levels indicate general trends but give accurate predictions only taking into account the variations in machine design, operating levels, and rock mass properties.

An index called specific penetration (Wanner, 1975; Aeberli and Wanner, 1978; Wanner and Aeberli, 1979), which is the ratio of penetration per machine revolution to the average thrust per cutter, normalizes TBM performance for cutterhead rotation rate and operating thrust. Correlations were demonstrated with indentation tests, but the effects of discontinuity frequency and rock texture varied with rock type. The field penetration index, $R_f$ (Hamilton and Dollinger, 1971), is calculated as the average thrust per cutter divided by the penetration per cutterhead revolution (the inverse of specific penetration) and may be more useful than penetration rate alone in developing correlations (Dollinger, 1984).

However, rock property indices are independent of parameters of machine design, operating levels, conditions of cutters, and rock mass characteristics (Korbin, 1979). The factors that affect TBM performance change not only from site to site but on a given job. Also, contractor practices vary, and the mucking, support, supply, and maintenance systems affect TBM performance. The penetration rate is determined by well-defined machine operating factors. The six tunnels studied by Nelson et al. (1984) were excavated by similar TBMs with single disc cutters. The tunnel diameters varied from 18.5 to 21.3 ft (5.6 to 6.5 m), only the last four factors (Table 8.1) showing significant variation for these projects.

Nelson et al. (1984), in the analysis of the records of six tunnel operations, found that the penetration rates and average thrust per cutter showed significant scatter that varied with rock type and with machine operator. Data screened according to operator and other factors showed a more regular trend. The average penetration rate varied from 0.28 in./rev. (7.1 mm/rev.) to 0.42 in./rev. (10.7 mm/rev.) for sandstone, limestone, and shale.

## 8.1.2  Rock Index Properties

Nelson et al. (1984) reported several types of index rock tests where properties have been correlated with TBM penetration rates to make prediction evaluations (Tarkoy, 1973, 1975, 1979; Descoeudres and Rechsteiner, 1973; Tarkoy and Hendron, 1975; Barendsen and Cadden, 1976; McFeat-Smith and Tarkoy, 1979; Ozdemir, 1977) as related to both laboratory and field cutting of rock. The index tests for the study of Nelson et al. (1984) are among the most definitive for field correlation (Tables 8.2, 8.3).

## TABLE 8.1. Factors Affecting TBM Performance

| Factors Affecting TBM Performance | Elements Contributing to Machine Use and Rock Excavation |
|---|---|
| Machine configuration | Thrust, torque, and gripper reaction capacities, gripper configurations, mechanism for steering, machine stiffness |
| Cutterhead design | Cutterhead profile, spatial array of cutters, cutter spacing, muck removal |
| Cutter design | Type of cutter, cutter composition, cutter geometry, capacity of roller bearing |
| Contractor practices | Material and utility supply systems, maintenance schedule, muck removal, operator technique, coordination of machine excavation with erection of support, experience of tunneling crew |
| Contract restrictions | Tunnel alignment, line, and grade; access to tunnel, length of tunnel |
| Ground conditions | Lithology, discontinuities, weathering, ground water, *in situ* state of stress |
| Intact rock properties | Material properties and index parameters of strength, deformability, abrasivity, and resistance to fracture |

*Source:* Nelson et al. (1984).

## TABLE 8.2. Summary of Index Property Tests, Procedures, and Additional Sources of Test Results[a]

| Index Property Test | Referenced Procedure | Additional Sources[b] of Test Results | | |
|---|---|---|---|---|
| | | Buffalo | Rochester | Chicago |
| Schmidt Hammer rebound hardness | ISRM | | B, D | |
| Modified Taber | Tarkoy and Hendron (1975) | | B, D | |
| Total hardness | Tarkoy and Hendron (1975) | | B, D | |
| Uniaxial compression | ISRM | A | B, D, E | |
| Point load index | ISRM | A | B, C, D | |
| Brazilian tensile strength | ISRM | | | |

[a] Nelson et al., 1984.
[b] Additional data sources: A-Goldberg-Soino, 1978; B-Haley and Aldrich, 1976; C-Critchfield, 1984; D-Haley and Aldrich, 1978; E-Keifer and Associates, Inc. 1976a & b.
Source: Nelson et al. 1984

Rock from each of the lithologically distinct zones in excavated tunnels were tested for Schmidt Hammer (L-type) rebound hardness, modified Taber abrasion hardness, uniaxial compression, diametrical point load, and Brazilian tensile strength tests (Table 8.2).

## 8.1.3  Penetration Rate and Index Properties

In correlations of penetration rates and average index properties (Table 8.3), data for penetration rates used were limited to tunnel sections where the rock unit was relatively unweathered and unjointed; the data were screened to remove shifts during which all motors were not operating, cutters were changed, steel sets were erected, or mechanical problems were encountered. Work shifts of less than 1 h were discounted as well as shifts in tunneling in the abrasive Grimsby sandstone if the cutter rolling distance was greater than $5 \times 10^5$ ft ($1.5 \times 10^5$ m) to eliminate the effect of extreme cutter wear (Table 8.4).

For the screened data for average penetration rates, thrust per cutter, and field penetration, indices for full-face rock units indicated the standard deviation for the penetration rates and thrusts varied only between 5 and 15%.

These data were used to formulate predictive equations (in./rev) by linear regression techniques (Table 8.5). (Index properties for saturated rock were available only for Grimsby sandstone; data were for air-dry core only.) The values of the coefficient of determination, $r^2$, were determined according to the number of degrees of freedom in each data set and can be interpreted as the percentage of the variation in penetration rate that can be anticipated by a single-parameter linear equation. As noted earlier, the highest degree of correlation was found between penetration rate and a linear combination of abrasion and rebound hardness.

Penetration rate and abrasion hardness show the least variation; in contrast, the data for axial compressive strength show wide scatter and do not correlate well (Figure 8.1). The total hardness, $H_T$, which is the product of the rebound hardness and square root of the abrasion hardness (Figure 8.2), gives the 95% confidence and prediction intervals for the estimation of the mean of all penetration rates in rock of a given total hardness. Thus, for the $H_T$ equation, predictions may exceed 40% in error 5% of the time.

The field penetration index, $R_f$, ratio of the average thrust per cutter to penetration rate as a function of index properties, provides a measure of the average penetration rate normalized with respect to average thrust per cutter. Predictive regression equations relating $R_f$ (kips/in.) with various index properties (Table 8.6) and correlations with abrasion hardness and a combination of abrasion and rebound hardness have the highest statistical significance. The $r^2$ values are substantially larger than those for penetration rate. For the data for $R_f$ plotted as a function of the total hardness (Figure 8.3), the best linear fit of the data shows that predictions should be less than 25% in error for rocks with a total hardness of 50. Accuracy of prediction depends on the number of data points used. Thus, the correlations with rock index properties, which are calculated using penetration rate normalized with respect to thrust, have a higher statistical significance than those based only

**TABLE 8.3. Summary of Index Property Test Results**

| | Schmidt Hammer Rebound Hardness | | | Taber Abrasion Hardness | | | Total Hardness | | |
|---|---|---|---|---|---|---|---|---|---|
| Location/Rock Unit | Average | SD | No. of Tests | Average | SD | No. of Tests | Average | SD | No. of Tests |
| Buffalo | | | | | | | | | |
| Bertie Formation | | | | | | | | | |
| Falkirk Member | 45.5 | 4.7 | 10 | 1.53 | 0.71 | 7 | 55.1 | 18.6 | 7 |
| Oatka Member | 39.4 | 4.5 | 10 | 0.66 | 0.13 | 4 | 34.7 | 9.4 | 4 |
| Rochester | | | | | | | | | |
| Williamson Shale | 33.7 | 5.2 | 4 | 0.64 | 0.15 | 5 | 27.7 | 6.7 | 4 |
| Lower Sodus Shale | 27.8 | 8.2 | 5 | 0.51 | 0.09 | 7 | 21.5 | 1.3 | 3 |
| Reynales Limestone | 46.8 | 6.0 | 13 | 3.27 | 2.47 | 10 | 82.5 | 30.0 | 10 |
| Maplewood Shale | 23.5 | — | 1 | 0.58 | 0.06 | 4 | 19.3 | — | 1 |
| Grimsby Sandstone—air dry | | | | | | | | | |
| All tests | 38.1 | 7.4 | 24 | 2.07 | 2.72 | 18 | 43.7 | 34.5 | 15 |
| Middle Grimsby | 39.9 | 2.7 | 6 | 0.36 | 0.01 | 2 | 23.8 | — | 1 |
| Grimsby Sandstone—saturated | | | | | | | | | |
| All tests | 29.7 | 6.7 | 5 | — | — | — | — | — | — |
| Middle Grimsby | 33.7 | 4.1 | 3 | — | — | — | — | — | — |
| Chicago | | | | | | | | | |
| Joliet Formation | | | | | | | | | |
| Romeo Member | 49.0 | — | 1 | 2.06 | — | 1 | 70.3 | — | 1 |
| Markgraf Member | 43.1 | — | 1 | 0.82 | — | 1 | 49.1 | — | 1 |

on the penetration rate. The field penetration index, $R_f$, correlated with total hardness shows the highest degree of statistical significance from among those with the index properties: rebound hardness, abrasion hardness, uniaxial compressive strength, point load index, and Brazilian tensile strength.

The normalization of penetration rate with respect to the average thrust per cutter helps define the influence of rock properties on the force required for penetration and on the interaction between cutter and rock. Not only are rock properties variable, but the thrust per cutter varies almost in the extreme. The effective relationship between the average cutter thrust and penetration is not linear (Ozdemir et al., 1976a; Hamilton and Dollinger, 1971; Howarth, 1981). For a given rock, $R_f$ may increase at a high rate when the thrust exceeds a critical level. Cutter penetration is usually proportional to the square of the thrust (Dubrignon and Janach, 1981), and the rolling, or tangential, force varies directly as the penetration.

It is useful to show graphically the interrelationships of thrust, cutter forces, spacing, penetration, and so on, but this type of display is limited to two or three variables. Nelson et al. (1984) have shown (Figure 8.4) that the thrust and rolling force behave in a similar manner with respect to both penetration and each other.

The laboratory data from Snowden et al. (1982) provide for visualizing the relationship among penetration rate, thrust, and rolling force. In three dimensions (Figure 8.5) the data for each rock define a line in the three-dimensional space,

**TABLE 8.3. (Cont'd)**

| Uniaxial Compressive Strength Ksi (MPa) | | | Point Load Index Ksi (MPa) | | | Brazile Tensile Strength Ksi (MPa) | | |
|---|---|---|---|---|---|---|---|---|
| Average | SD | No. of Tests | Average | SD | No. of Tests | Average | SD | No. of Tests |
| 27.3 (188) | 7.5 (52) | 34 | 1.06 (7.3) | 0.22 (2.5) | 17 | 1.93 (13.3) | 0.55 (3.8) | 19 |
| 20.2 (139) | 4.1 (28) | 17 | 0.37 (2.6) | 0.19 (1.3) | 15 | 1.89 (13.0) | 0.48 (3.3) | 19 |
| 11.9 ( 82) | 0.6 ( 4) | 3 | 0.09 (0.6) | 0.03 (0.2) | 9 | — | — | — |
| 11.2 ( 77) | 1.4 (10) | 6 | 0.12 (0.8) | 0.04 (0.3) | 10 | — | — | — |
| 18.6 (128) | 6.2 (43) | 13 | 0.64 (4.4) | 0.36 (2.5) | 16 | 2.17 (15.0) | 0.34 (2.3) | 3 |
| 9.8 ( 68) | 2.2 (15) | 3 | 0.13 (0.9) | 0.03 (0.2) | 9 | — | — | — |
| 18.9 (130) | 11.3 (78) | 8 | 0.52 (3.6) | 0.29 (2.0) | 50 | 1.47 (10.1) | 0.69 (4.8) | 6 |
| 13.5 ( 93) | 2.3 (16) | 3 | 0.99 (6.8) | 0.14 (1.0) | 3 | 1.15 ( 7.9) | 0.25 (1.7) | 3 |
| 15.7 (108) | 13.1 (90) | 9 | 0.42 (2.9) | 0.30 (2.1) | 25 | 0.88 ( 6.1) | 0.88 (6.1) | 10 |
| 6.6 ( 46) | 3.4 (23) | 3 | 0.30 (2.1) | 0.25 (1.7) | 3 | 0.47 ( 3.2) | 0.23 (1.6) | 3 |
| 34.4 (237) | 9.6 (66) | 17 | 1.33 (9.2) | 0.43 (3.0) | 3 | 2.47 (17.0) | 0.39 (2.7) | 3 |
| 24.4 (168) | 5.3 (37) | 38 | 1.32 (9.1) | 0.11 (0.8) | 3 | 1.75 (12.1) | 0.09 (0.6) | 3 |

*Source:* Nelson et al. (1984).

and data for several rock types therefore represent a surface with two-dimensional projections for two rock types of the penetration rate versus rolling force plane. The surface shows the variation in cutting performance as the cutter on rocks of greater competence and strength and stiffness. This model was developed primarily for conceptual purposes.

The curves (Figure 8.4) represent a transition in rock types from the Maplewood Shale to the Reynales Limestone in the Culver–Goodman Tunnel, from weak, noncrystalline material to a relatively hard and stiff, crystalline rock. The curves for the penetration rate versus thrust and rolling force for excavation in the Culver-Goodman Tunnel in both the Maplewood Shale and Reynales Limestone (Figure 8.6) were drawn consistent with the curvilinear trends of the linear cutter data of Snowden et al. (1982). The points A, A', and B are recorded field values of the penetration rate, thrust, and rolling force. The new thrust capacity of the machine (dashed line—right) is an upper limit of the available thrust, while the net torque capacity is shown by the dashed line on the left. This total rolling force represents the maximum average rolling force per cutter delivered to the rock.

Nelson et al. (1984) state that point A' represents the rock conditions shortly after the second change of the main bearing seal in the Maplewood Shale. Resistance caused by packed muck buckets reduced the torque delivered to the cutters. Cleaning the muck buckets and stopping the dust abatement spray caused additional

**TABLE 8.4. Summary of Average Penetration Rate, Thrust per Cutter, and Field Penetration**

| Location/Rock Unit | Tunnel | Tunnel Station | Penetration Rate | | Average Cutter Thrust Load kips (kN) | | $R_f$ kips/in. (kN/mm) | |
|---|---|---|---|---|---|---|---|---|
| | | | Average | SD | Average | SD | Average | SD |
| Buffalo | | | | | | | | |
| Bertie Formation | | | | | | | | |
| Falkirk Member | C11 outbound | 46+43 to 53+02 | 0.299 (7.60) | 0.037 (0.94) | 30.2 (134) | 1.3 (6) | 100.4 (17.6) | 14.9 (2.6) |
| Oatka Member | C11 outbound | 26+76 to 32+59 | 0.411 (10.44) | 0.069 (1.76) | 24.2 (108) | 1.9 (8) | 58.8 (10.3) | 9.5 (1.7) |
| | C11 inbound | 28+18 to 32+81 | 0.423 (10.74) | 0.075 (1.91) | 24.9 (111) | 0.8 (4) | 58.8 (10.3) | 10.5 (1.8) |
| Rochester | | | | | | | | |
| Williamson and | | | | | | | | |
| Lower Sodus Shales | | 142+77 to 148+68 | 0.395 (10.03) | 0.012 (0.31) | 22.2 (99) | 1.0 (4) | 56.0 (9.8) | 3.6 (0.6) |
| Renales Limestone | | 115+17 to 117+17 | 0.267 (6.78) | 0.016 (0.41) | 31.8 (141) | 1.2 (5) | 119.2 (20.9) | 9.3 (1.6) |
| Maplewood Shale | | 96+89 to 100+27 | 0.411 (10.44) | 0.007 (0.18) | 22.0 (98) | 1.0 (4) | 53.2 (9.3) | 2.8 (0.5) |
| Grimsby Sandstone | | | | | | | | |
| All tests | | 27+55 to 59+40 | 0.310 (7.87) | 0.033 (0.84) | 25.1 (112) | 1.8 (8) | 82.4 (14.4) | 11.0 (1.9) |
| Middle Grimsby | | 30+87 to 41+62 | 0.325 (8.26) | 0.016 (0.41) | 25.3 (113) | 2.6 (12) | 74.5 (13.0) | 12.8 (2.2) |
| Chicago | | | | | | | | |
| Joliet Formation | | | | | | | | |
| Romeo Member | | 260+07 to 267+41 | 0.316 (8.03) | 0.028 (0.71) | 32.6 (145) | 0.8 (4) | 103.0 (18.0) | 9.9 (1.7) |
| Markgraf Member | | 231+45 to 235+24 | 0.366 (9.30) | 0.033 (0.84) | 30.9 (137) | 1.0 (4) | 84.6 (14.8) | 7.5 (1.3) |

*Source:* Nelson et al. (1984).

## TABLE 8.5. Penetration Rate Prediction Equations

| Rock Index Property | Units | Equation[a] | $r^2$ |
|---|---|---|---|
| Rebound hardness | | $PR = 0.5412 - 0.0048\ H_R$ | 35.9 |
| Abrasion hardness | | $PR = 0.4112 - 0.0468\ H_A$ | 58.8 |
| Total hardness | | $PR = 0.4383 - 0.00196\ H_T$ | 48.5 |
| Uniaxial compressive strength | ksi | $PR = 0.4093 - 0.00287\ q_u$ | 4.1 |
| Point load index | ksi | $PR = 0.4001 - 0.0715\ I_{s(50)}$ | 17.9 |
| Brazilian tensile strength | ksi | $PR = 0.3656 - 0.0141\ \sigma_t$ | 0.0 |
| Rebound hardness and abrasion hardness | | $PR = 0.4831 - 0.00214\ H_R - 0.0368\ H_A$ | 60.3 |

[a] Penetration rate, PR, predicted by these equations is in units of in./rev.
*Source:* Nelson et al. (1984).

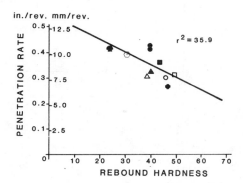

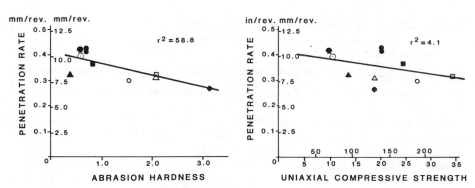

**Figure 8.1** Penetration rate versus rebound hardness, abrasion hardness, and uniaxial compressive strength (Nelson et al., 1984).

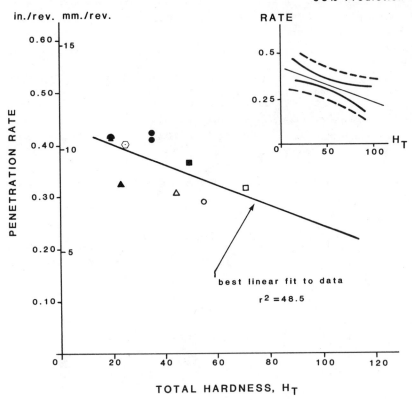

**Figure 8.2** Plot of penetration rate versus total hardness (Nelson et al., 1984).

## TABLE 8.6. Field Penetration Index Prediction Equations

| Rock Index Property | Units | Equation[a] | $r^2$ |
|---|---|---|---|
| Rebound hardness | | $R_f = 2.4327\,H_R - 17.12$ | 61.7 |
| Abrasion hardness | | $R_f = 20.79\,H_A + 52.92$ | 70.5 |
| Total hardness | | $R_f = 1.021\,H_T + 34.32$ | 85.1 |
| Uniaxial compressive strength | ksi | $R_f = 1.884\,C_o + 41.64$ | 29.8 |
| Point load index | ksi | $R_f = 31.89\,I_{s(50)} + 56.91$ | 27.1 |
| Brazilian tensile strength | ksi | $R_f = 26.00\,\sigma_T + 37.38$ | 11.3 |
| Rebound hardness and abrasion hardness | | $R_f = 1.415\,(H_A + 0.1\,H_R) + 5.18$ | 85.4 |

[a] $R_f$ predicted by these equations is in units of kips/in.
*Source:* Nelson et al. (1984).

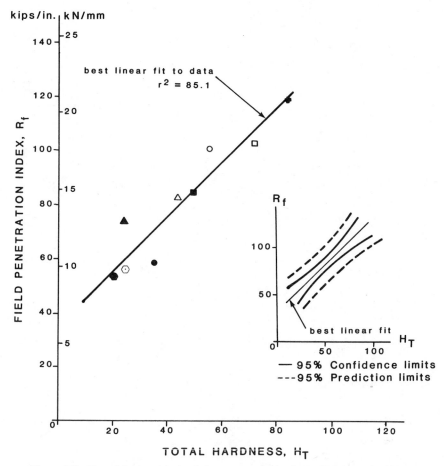

**Figure 8.3**  Plot of field penetration index versus total hardness (Nelson et al., 1984).

rolling force which changed the cutting conditions from points A′ and A, as indicated by the arrow.

In excavation of rock from the Maplewood Shale to larger tunnel sections of Reynales Limestone at constant amperage at maximum capacity, the rolling force followed a vertical path from point A to point B at net torque capacity. At the same time, the penetration rate decreased, and the thrust increased, that is, from point A to point B on the thrust curve (right). Thus, on the surface representing rock response, the relationships among penetration rate, thrust, and rolling force represent a critical surface.

## 8.2  CUTTER WEAR

Cutter wear may be caused by factors such as abrasion of the disc ring, loss of fit between the ring and hub, and bearing failure. The type of failure may vary with

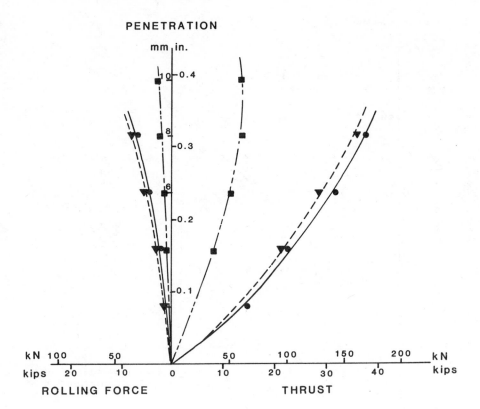

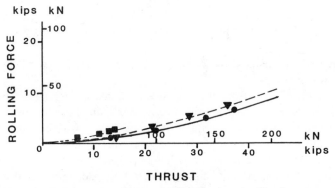

**Figure 8.4** Plots of penetration versus thrust and rolling force, and rolling force versus thrust (based on data from Snowden et al., 1982).

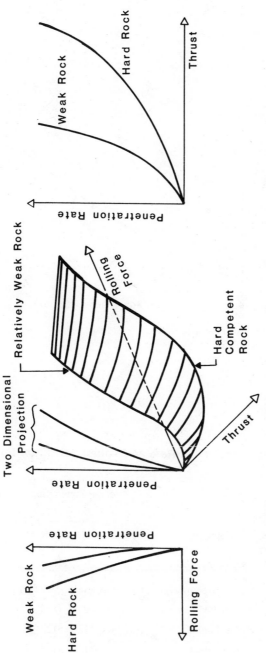

**Figure 8.5** Three-dimensional plot showing interaction among penetration, thrust, and rolling force (Nelson et al., 1984).

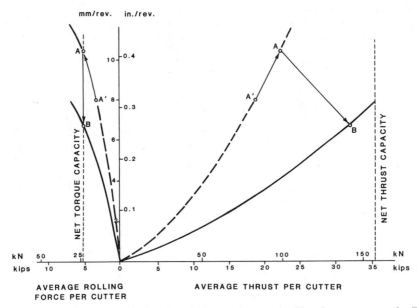

**Figure 8.6** Penetration rate as a function of average thrust and rolling force per cutter for TBM performance in the Culver–Goodman tunnel (Nelson et al., 1984).

**TABLE 8.7. Average Penetration Rate and Thrust per Cutter for the Culver–Goodman, C11, and Chicago Tarp Tunnels**

| Tunnel | Stations | Average Penetration Rate in./rev. (mm/rev.) | Average Thrust per Cutter kips (kN) |
|---|---|---|---|
| Culver–Goodman | 3+98 to 28+00 | 0.29 (7.4) | 24.0 (107) |
| | 28+00 to 85+71 | 0.29 (7.3) | 25.8 (115) |
| | 85+71 to 140+00 | 0.30 (7.6) | 21.8 ( 97) |
| | Entire tunnel | 0.30 (7.7) | 24.5 (109) |
| C11 outbound | Entire tunnel | 0.31 (7.9) | 25.9 (115) |
| C11 inbound | Entire tunnel | 0.31 (7.8) | 24.7 (110) |
| Chicago TARP | Entire tunnel | 0.26 (6.5) | 30.3 (135) |

*Source:* Nelson et al. (1984).

**TABLE 8.8. Summary of Average Number of Cutter Replacements per 1000 ft (305 m) for the Culver–Goodman, C11, and Chicago TARP Tunnels**

| Tunnel | Stations | Average Number of Cutter Replacements per 1000 ft (305 m) of Tunnel per Number of Positions in Each Cutter Group | | | |
|---|---|---|---|---|---|
| | | Center | Face | Gage | All Cutters |
| Culver–Goodman | 3+98 to 28+00 | 2.39 | 1.46 | 3.60 | 2.51 |
| | 28+00 to 85+71 | 0.86 | 1.27 | 3.02 | 2.08 |
| | 85+71 to 140+00 | 0.61 | 0.69 | 1.16 | 0.89 |
| | Entire tunnel | 1.35 | 1.20 | 2.80 | 1.95 |
| C11 outbound | Entire tunnel | 0.21 | 0.26 | 0.25 | 0.25 |
| C11 inbound | Entire tunnel | 0.21 | 0.25 | 0.28 | 0.26 |
| Chicago TARP | Entire tunnel | 0.87 | 0.25 | 0.46 | 0.45 |

*Source:* Nelson et al. (1984).

the cutter position on the head. The center cutters travel in smaller circular paths and may be abraded by scuffing. Those at the gage apply their thrust at an angle to the line of action to the cutter which develops an oblique set of loads. The higher velocity of travel of the gage cutters subjects their bearings to higher temperatures and decreases the bearing life. Also, in a horizontal tunnel the gage elements must regrind the muck that accumulates at the invert. Thus, the hub and saddle as well as the disc of a gage cutter are abraded more than those of the components of the others.

The records of cutter replacement on tunnel projects (Nelson et al., 1984) give data for the frequency of cutter replacements and their position on the cutter head as well as detailed information for wear parameters for one tunnel. For sections of the tunnels of different lithology or major change in construction procedures, the average rates of penetration range from 0.21 to 0.31 in/rev, while the average thrust varies only 20% (Table 8.7). From the average number of replacements per 1000 f (305 m) of tunnel for the center, face, and gage positions (Table 8.8), it was found that for a relatively constant penetration rate and thrust the number of replacements in the Culver–Goodman Tunnel was substantially higher. For rock type the replacement rate was higher in Grimsby Sandstone than in the other rock units.

A summary of both rolling distance and operating hours for positions (Tables 8.9, 8.10) shows that rolling distances varied by a multiple of 2 depending on rock type and that the face cutters rolled twice as far in one tunnel as in another. On the other hand, the operating time for the gage cutters was about half that for the

**TABLE 8.9. Summary of Average Rolling Distances at Time of Cutter Replacement for the Culver–Goodman, C11, and Chicago TARP Tunnels**

| Tunnel | Stations | Predominant Rock Units | Average Rolling Distances for Cutter Position Groups [ft (m) × 10⁶] | | | | |
|---|---|---|---|---|---|---|---|
| | | | Center | Face | Gage | All Cutters |
| Culver–Goodman | 3+98 to 28+00 | Queenston Shale and Lower Grimsby Sandstone | 0.06 (0.02) | 0.93 (0.28) | 0.64 (0.20) | 0.73 (0.22) |
| | 28+00 to 85+71 | Grimsby and Thorold Sandstones | 0.17 (0.05) | 1.05 (0.32) | 0.80 (0.24) | 0.89 (0.27) |
| | 85+71 to 140+00 | Maplewood Shale, Reynales Limestone, Lower Sodus and Williamson Shales | 0.21 (0.06) | 1.93 (0.59) | 1.99 (0.61) | 1.83 (0.56) |
| C11 outbound | Entire tunnel | Falkirk and Oatka Members of the Bertie Dolostone Formation | 0.14 (0.04) | 1.20 (0.37) | 1.00 (0.30) | 1.04 (0.32) |
| | Entire tunnel | | 0.68 (0.21) | 4.89 (1.49) | 8.44 (2.57) | 5.53 (1.69) |
| C11 inbound | Entire tunnel | Falkirk and Oatka Members of the Bertie Dolostone Formation | 0.66 (0.20) | 4.70 (1.43) | 7.77 (2.37) | 5.43 (1.66) |
| Chicago TARP | Entire tunnel | Romeo and Markgraf Members of the Joilet Formation | 0.15 (0.05) | 4.66 (1.42) | 4.82 (1.47) | 3.73 (1.14) |
| | 185+84 to 305+00 (abrasion failure only) | Romeo and Markgraf Members of the Joilet Formation | | 4.80 (1.46) | 4.41 (1.34) | 4.64 (1.41) |

*Source:* Nelson et al. (1984).

**TABLE 8.10. Summary of Average Cutter Life for the Culver–Goodman, C11, and Chicago TARP Tunnels**

| Tunnels | Stations | Average Clock Hours Before Replacement for Cutter Position Groups | | | |
|---|---|---|---|---|---|
| | | Center | Face | Gage | All Cutters |
| Culver–Goodman | 3+98 to  28+00 | 64 | 106 | 43 | 74 |
| | 28+00 to  85+71 | 178 | 119 | 51 | 90 |
| | 85+71 to 140+00 | 238 | 216 | 122 | 176 |
| | Entire tunnel | 151 | 136 | 63 | 103 |
| C11 outbound | Entire tunnel | 567 | 447 | 432 | 451 |
| C11 inbound | Entire tunnel | 551 | 403 | 403 | 413 |
| Chicago TARP | Entire tunnel | 150 | 509 | 270 | 353 |
| | 185+84 to 305+00 (abrasion failure only) | — | 488 | 256 | 392 |

*Source:* Nelson et al. (1984).

face position. The rolling distance varied for different tunnels and cutter positions (Figures 8.7, 8.8).

Inspection of some worn cutters indicated that abrasion was a primary cause of replacement, and a number of truncated discs occurred in another owing to bearing failure. Most abrasion that occurs on the disc ring is due to the abrasiveness of the

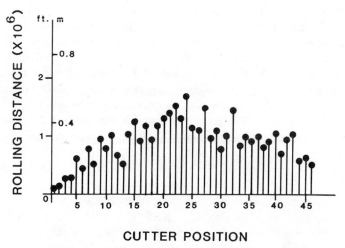

**CUTTER POSITION**

**Figure 8.7**  Average rolling distance per cutter position, station 28+00 to 85+71, Culver–Goodman Tunnel (Nelson et al., 1984).

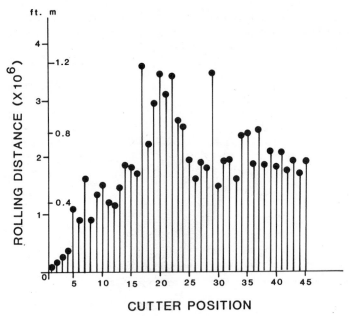

**Figure 8.8**   Average rolling distance per cutter position, station 85+71 to 140+00, Culver–Goodman Tunnel (Nelson et al., 1984).

rock. On the other hand, bearing failures result from the high forces and temperatures generated at the bearings. Bearing failures may be increased by fine-grained minerals that penetrate the bearing seals, and the life of a bearing depends on the time that high levels of thrust are carried by them. The effective thrust required for excavation is always related to the abrasiveness of the rock.

Cutter wear caused by abrasion was recorded by the measurement of diameters of the three outer gage cutters operating in three stratigraphic horizons—a sandstone, a limestone, and shales. This decrease in diameter plotted against rolling distance (Figures 8.8, 8.9; Table 8.11) showed that the sandstone is more abrasive than the limestone or shales. Lithographically, the lower Grimsby is medium- to coarse-grained, often weakly cemented; its quartz content is less than 50%, and it has an argillacious matrix. The middle Grimsby is fine-grained, thin-bedded, and about 30% quartz in an argillacious matrix. The top Grimsby is thick-bedded, well cemented and has a quartz content as high as 90%. Thus the quartz present is not a direct measure of abrasiveness, but it does cause more abrasion than other minerals.

A rock abrasiveness factor, $A_R$, was determined in the laboratory as the reciprocal of the weight loss of an abrasion wheel. Thus a low value indicates a high weight loss and means the rock is abrasive. The values given (Table 8.10) are combined averages for the lower and middle Grimsby sandstones.

The rate of cutter wear is given as the rolling distance per unit diameter loss or the reciprocal of the slope of the lines representing the wear data. The quartz

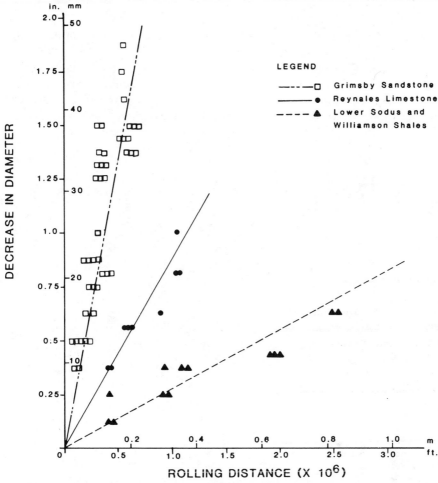

**Figure 8.9** Decrease in diameter versus rolling distance for gage cutters, Culver–Goodman Tunnel (Nelson et al., 1984).

**TABLE 8.11. Rock Properties and Rates of Cutter Disc Wear for Five Rock Units in the Culver–Goodman Tunnel**

| Rock Unit | Rolling Distance per Unit Disc Diameter Loss ft/in. (m/mm) × 10⁶ | Estimated Quartz Content (%) | Uniaxial Compressive Strength[a] ksi (MPa) | Rock Abrasiveness[b] |
|---|---|---|---|---|
| Lower Grimsby sandstone | 0.27 (0.003) | 60 | 13.5 (93) | 3 |
| Middle Grimsby sandstone | 0.32 (0.004) | 70 | 8.3 (57) | 3 |
| Upper Grimsby sandstone | 0.41 (0.005) | 90 | 29.4 (203) | 6 |
| Reynales limestone | 1.13 (0.014) | 10 | 15.0 (103) | 12 |
| Lower Sodus and Williamson shales | 3.65 (0.044) | 5 | 11.6 (80) | 42 |

[a] Evaluated by Cornell personnel.
[b] Evaluated by Haley and Aldrich (1978) (low index number = high abrasiveness).
*Source:* Nelson et al. (1984).

content of the rocks was estimated from petrographic sections, and the uniaxial compressive strengths were evaluated by standard tests.

The only rock property that correlates linearly with the rate of cutter wear is $A_R$ (Table 8.11); uniaxial compressive strength and quartz content show little correlation. The upper Grimsby sandstone with the highest quartz percent and compressive strength was the least abrasive of three rock types.

# 9

## WATER JET CUTTING OF ROCK

9.1 Rock Cutting
9.2 Crow Theory
9.3 Rock Failure under Jet Impact
9.4 Rehbinder Theory
9.5 Hole Drilling with Water Jets
9.6 Coal—Jet Slotting
9.7 Coal Mining and Coal Cutting
9.8 Water Jet Applications

Large-diameter water jets propelled by hydraulic ''giants'' were used for the mining of certain types of placer deposits where the soil, gravel, and boulder materials deposited with valuable minerals, usually gold, were washed by erosion, the water being used to remove the material from *in situ*, transport it, and wash it and separate the gold by sluices or other means. The nozzles of hydraulic giants ranged in size from 1.25 to 10 in. in diameter (Peele and Church, 1941), and the pressures at the intake of the giants varied from 10 to 400 ft head (42–173 psi). The quantities of water consumed varied from 60 to 2800 miner's inches (1.5 ft$^3$/min). The duty of a miner's inch is the number of cubic yards of gravel broken down and sent through sluices in 24 h.

### 9.1 ROCK CUTTING

Crow's (1973) rock-cutting nomenclature includes the following:

$c$ = intrinsic speed of hydraulic rock cutting
$c_D$ = drag coefficient
$d_o$ = initial jet diameter
$D$ = drag of a sphere
$f$ = rock porosity

$F$ = universal function of $(v/c)$

$g$ = surface-averaged grain diameter

$G$ = acceleration of gravity

$h$ = depth of slot

$H$ = slot depth accumulated over $N$ passes

$k$ = permeability

$L$ = length of permeability sample

$m$ = mass of grains with radii less than $r$

$M$ = total mass of grains

$n$ = coordinate normal to the cutting surface

$N$ = number of jet passes

$p$ = pore pressure

$P_a$ = atmospheric pressure

$P_o$ = driving pressure in a permeability test or total initial pressure of jet

$P_s$ = pressure at cutting surface

$P_v$ = vapor pressure

$P_c$ = critical pressure for the inception of cutting

$r$ = grain radius

$Re$ = Reynolds number, $2\rho_w v_g r / \eta$

$u$ = volume flux of permeating fluid

$u_a$ = volume flux at atmospheric pressure

$v$ = feed rate of rock

$v_g$ = settling speed of a grain

$\eta$ = fluid viscosity

$\theta$ = angle of jet stream with respect to the horizontal

$\theta_o$ = incidence angle of the jet

$\mu_r$ = coefficient of internal friction

$\mu_w$ = friction coefficient for cavitation drag

$\rho_g$ = grain density

$\rho_r$ = bulk density of rock

$\rho_w$ = density of water

$\sigma$ = stress normal to a plane of failure

$\tau$ = shear stress at failure

$\tau_o$ = inherent shear strength

At the present time, small-diameter, high-pressure water jets have been the subject of intensive experimentation, and those pressures and diameters that have found practical application in rock fragmentation and mining of coal are used primarily for drilling or cutting of relatively soft rocks and coal. Experimentation has been carried out for drilling and cutting of harder rocks for fragmentation,

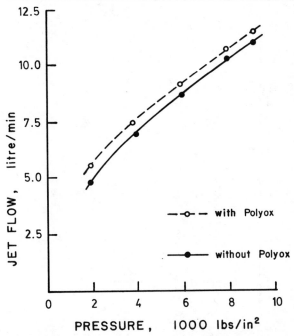

**Figure 9.1** Jet flow versus pressure with and without Polyox (Brook and Summers, 1969).

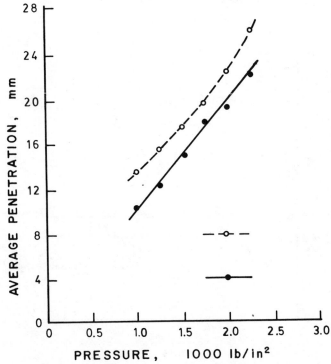

**Figure 9.2** Average penetration versus pressure for continuous jet and static target (Brook and Summers, 1969).

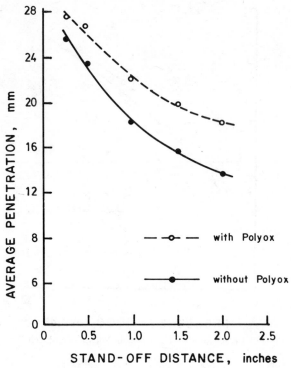

**Figure 9.3** Average penetration versus standoff distance for continuous jet and static target (Brook and Summers, 1969).

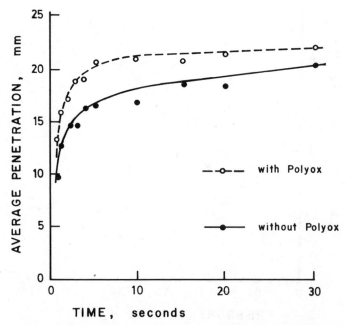

**Figure 9.4** Average penetration versus time for continuous jet and static target (Brook and Summers, 1969).

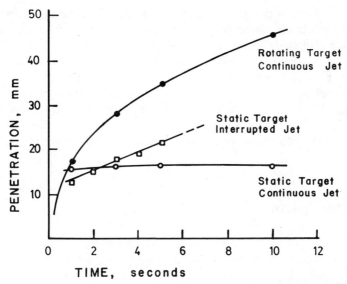

**Figure 9.5** Penetration versus time for various jet uses (Brook and Summers, 1969).

primarily for making holes, cutting slots, and assisting other methods of breakage such as drag-bit drilling and tunnel boring (Maurer, 1979). High-pressure water jets have also been used for other major industrial purposes such as cutting various materials, scaling corroded metals, and other cleaning processes.

While small-diameter water jets generated by pumps or intensifiers with pressures up to 50,000 psi have been investigated for various rock fragmentation purposes (Maurer, 1979; Symposia on Jet Cutting Technology, 1st through 7th), researchers have found most successful applications of pressures to 15,000 psi in the drilling of small holes in soft rock and cutting coal, and they show promise of application in areas such as assisting TBM cutters and drag bits.

The cutting efficiency of water jets for softer rocks is low compared to mechanical cutting. Crow and Hurlburt (1974) point out that a TBM with 40 MN thrust might deliver only 700 kW to a tunnel face whereas a water jet might deliver as much as $10,000 \text{ kw/cm}^2$. However, water jets also produce a fine material whose specific energy is very high. Their greatest practical use obviously is for specialized processes such as drilling small holes or assisting other types of rock cutting. The basic theories of jet cutting are based primarily on erosion of porous rock.

In a study of fixed versus traversing jets (sandstone), although in practice a water jet must be moving laterally with respect to the rock surface to be most effective, some of the mechanisms of water jet impact in the pressure–velocity regime of interest in rock cutting are defined where the target is not moving laterally (Brook and Summers, 1969). Factorial experiments were designed to measure effects of nozzle standoff, water pressure, and penetration time, six pressures being in the range of 4000–9000 psi (27.5–62 $\text{MN/m}^2$), five standoff values of 0.3–2.0 in. (7.62–50.8 mm), and 10 time periods 0.5–30 s. Tests were performed with and without an oil mixture (0.01% Polyox), the latter to make the jet more coherent

(Figure 9.1). Average penetration, taken as the best measure of jet performance, was found to increase almost linearly with pressure (Figure 9.2) and to decrease with standoff (Figure 9.3). Most of the penetration took place in the first 5 s (Figure 9.4). The rapid reduction of penetration with depth may be due to interference of water leaving the hole and breakup of the jet with travel distance.

The interference of returning water can be avoided by (1) traversing or (2) interrupting the jet. Traversing (rotating the target) was the most effective method of optimizing penetration (Figure 9.5). Brook and Summers (1969) also noted that a rotating nozzle with an angled jet was capable of drilling holes in both hard and soft rock.

## 9.2  CROW THEORY

One hypothesis and its implementing assumptions serve as the basis for the development of a theory of water jet cutting of rock by a moving jet at 90° and other angles of attack are again primarily for softer, porous rocks of fairly fine grain composition (Crow, 1973, Crow and Hurlburt, 1974), whose theory was checked with data obtained from cutting experiments on a fine-grained, moderately strong, impermeable sandstone (Wilkeson). The mathematical expression proposed for rock cutting by fine jets, assuming the pressure is high enough to remove rock particles, was

$$c = \frac{k\tau_o}{\eta f \mu_r g} \tag{9.1}$$

where $k$ is the permeability, $\tau_o$ the shear strength, $\mu_r$ the coefficient of internal friction of the rock, $g$ the grain diameter (Crow and Hurlburt, 1974), $\eta$ the viscosity of the cutting fluid, and $f$ the porosity of the rock. In the above equation, $c$ is defined as the intrinsic speed of cutting in units of length per unit time. Thus, the depth of a slot cut by a jet depends on the feed rate $v$ and the speed $c$. It appears that $c$ may also be considered as the rate of rock removal per unit cross section of jet.

Where the jet is moving in a straight line with respect to the rock, the slot deepens and represents in general a nonsteady condition but is approximated by a steady-state condition (Figure 9.6), the jet is assumed to have a diameter $d_o$, the rock is moving with a velocity $v$ with respect to the jet, and the ultimate slot has a depth $h$. The development of a theory and mathematics involves the determination of $h$ as a function of the above parameters and the appropriate properties of the rock and water jet.

Crow (1973) reports that Russian investigators found that for three types of stone—granite, limestone, and marble—with pressures $P_o$ to 2000 atm, data showed $h$ to be directly proportional to $(P_o - P_c)$, where $P_c$ is the critical cutting pressure, inversely proportional to cutting resistance or shear strength $\tau_o$. The critical pressure

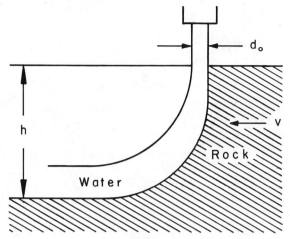

**Figure 9.6** Cutting by a steady high-speed water jet (Crow, 1973).

$P_c$ increased with feed rate $v$ and $P_o$, and cutting stopped at high values of $v$. Also, $h$ is independent of $v$ for rates up to about 10 in./s and decreases slowly with $v$ until cutting stops. Results were summarized by the equation

$$h = d_o \frac{(P_o - P_c)}{\tau_o} F(v) \tag{9.2}$$

where $P_c$ increases with $v$, and $F(v)$ is constant at low $v$ and decreases at higher $v$. Results indicate that $F(v)$ varies as $v^{-0.33}$ in the interval 20–70 in./s and that no simple power law applies for all $v$.

Here $F(v)$ is dimensionless, and $c$ is of the order 10–100 in./s. Also, for example, the speed of a water jet for 1000 atm $P_o$ is about 1500 ft/s, and the speed of shear waves in rock about 5000 ft/s, which is much higher. The theory of hydraulic rock cutting should include factors such as cavitation, brittle fracture, and permeability. Water flowing at high velocity over a granular surface exerts little shear stress, because of cavitation. The traveling jet (Figure 9.6) with the curvature causes a high surface pressure, which pressurizes cavity bubbles and exposes the grains to direct impact. The permeability creates a pore pressure beneath the cutting surface, which removes the grains.

Data from Olson and Thomas (Crow, 1973) cover a three-decade range of $v$, which confirm Crow and Hurlburt's (1974) theory and fix the value of the coefficient $\mu_w$ of Coulomb friction between water and rock for cavitational conditions.

The theory developed by Crow (1973) and modified by Crow and Hurlburt (1974) is based on an erosion rather than a fracture model (Figure 9.7). The square jet of dimension $d_o$ is assumed to make a slot of the same width, the cutting being on the forward contact or face of the jet, where friction decelerates the jet as the depth increases (Figure 9.8). Initial experiments were made with $\theta_o = 90°$, but mathematical developments indicate that deeper cuts occur for $\theta_o > 90°$.

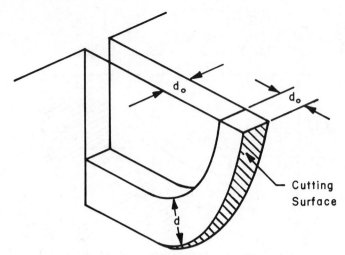

**Figure 9.7**  Geometry of the idealized cut (Crow, 1973).

Cavitation in flowing water may be due to several causes such as the flow itself near a solid surface, movement of a solid in the water, rapid flow, and so on. Bubble dynamics have been the subject of much research. Incipient cavitation is defined (Birkhoff and Sarantello, 1957) as the rapid formation and collapse of minute bubbles at locations of negative pressure in high-speed flow. Minute nuclei of gas are believed to be required to initiate the growth of the cavities. Intense sound waves may cause dissolved gas to come out of solution (acoustic cavitation), and bubbles may be caused by local vorticity.

Cavitation "pitting" is a well-known occurrence in propellers, in turbines, and in spillways. Research results indicate that it is due to the collapse of vapor-filled cavities against solid surfaces. According to the ideal theory (Rayleigh), the pres-

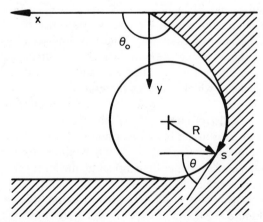

**Figure 9.8**  Geometry of the cutting surface. The jet enters with a positive rate (Crow, 1973).

sure would be infinite if the energy of collapse is concentrated at a point. Pitting has occurred in materials having yield points of $10^5$–$10^6$ psi, but this may be as low as $10^3$–$10^4$ psi, although an intermediate range of $10^4$–$10^5$ psi is usually assumed.

One suggested mechanism for cutting by a moving continuous jet is that cavity bubbles form behind grains so that the upstream faces of the grains experience a pressure $P_s$ and the downstream faces the vapor pressure of the water $P_v$ and the grain is subject to a shear stress

$$\tau = \mu w (P_s - P_v) \tag{9.3}$$

where $\mu w$ is the Coulomb coefficient of friction between the water and the rock.

Where the jet is of small diameter compared with the radius of curvature of its path, the friction law and laws of hydrodynamics can be (Crow, 1973) combined to give an equation for slot depth:

$$h = 2_{\mu_w} d_o P_o \int_0^{\theta_o} \frac{e^{\mu w}(\theta - \theta_o) \sin \theta}{\tau} \tag{9.4}$$

where $d_o$ is the jet diameter, $P_o$ the total pressure, and $\theta_o$ the angle of incidence of the jet. The failure shear stress of the rock is given by the Mohr–Coulomb failure condition

$$\tau = \tau_o + \mu_r (P_s - P) \tag{9.5}$$

where $\tau$ is the shear strength of the rocks, $\mu_r$ the internal friction, and $P$ the pore pressure.

The pressure difference $(P_s - P)$ for one grain diameter within the rock is determined by a Taylor series expansion

$$(P_s - P) = g \frac{\partial p}{\partial n} \tag{9.6}$$

where $(\partial p)/(\partial n)$ is the pressure gradient beneath the surface. Application of Darcy's law for flow in a porous medium gives

$$\frac{\partial p}{\partial n} = \frac{\eta f}{k} v \sin \theta \tag{9.7}$$

where $k$ is the permeability of the rock, $f$ the porosity, $\eta$ the viscosity of the water, and $v \sin \theta$ the rate of rock movement normal to the cutting surface of the jet.

Equations 9.3 through 9.7 may be combined to give

$$h = 2\mu_w \frac{d_o P_o}{\tau_o} \int_0^{\theta_o} \frac{e^{\mu w(\theta - \theta_o)} \sin \theta}{1 + (v/c) \sin \theta} \tag{9.8}$$

**TABLE 9.1. Properties of Wilkeson sandstone**

$$k = 0.58 \text{ millidarcys}$$
$$f = 0.096$$
$$\tau_o = 11.0 \text{ MN/m}^2$$
$$\mu_r = 1.23$$
$$g = 0.16 \text{ mm}$$
$$c = 33.8 \text{ cm/s}$$

*Source:* Crow (1974).

As $v/c \to 0$, $h \to$ constant, and as $v/c$ becomes large, $h$ decreases as $(v/c)^{-1}$, and an important rock property is the permeability, which may also determine the critical pressure.

The rock properties that were found to be related to water jet cutting were obtained by Crow and Hurlburt (1974) for Wilkeson sandstone (Table 9.1).

For an incompressible fluid the permeability of a porous medium is

$$k = \frac{nuL}{(P_o - P_a)} \tag{9.9}$$

where $u$ = volume of flow (rate per unit area), $L$ = sample length, $P_o - P_a$ = pressure drop over length $L$, and the latter is measured by means of a permeameter. Porosity may be calculated or measured by weighing before and after saturation. Shear strength and internal friction are inferred from triaxial test results (Mohr envelope). Grain size was measured by screening or differential settling methods.

For cutting tests, pressurized water was provided by an intensifier, and cutting was performed on rotating or linearly moving blocks of sandstone in front of the

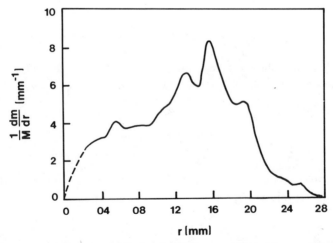

**Figure 9.9** Differential grain size distribution for Wilkeson sandstone (Crow and Hurlburt, 1974).

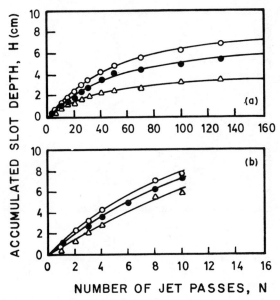

**Figure 9.10** Variation of accumulated slot depth with number of jet passes for total jet pressures of (*a*) 52 MN/m² and (*b*) 128 MN/m² (Crow and Hurlburt, 1974).

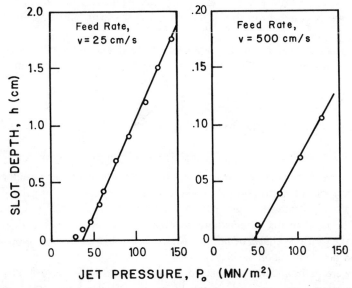

**Figure 9.11** Variation of slot depth with jet pressure for feed rates of (*a*) 25 cm/s and (*b*) 500 cm/s (Crow and Hurlburt, 1974).

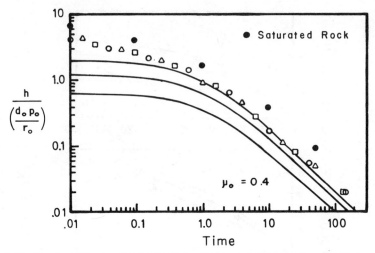

**Figure 9.12** Slot depth as a function of feed rate, in universal dimensionless coordinates. The open points represent three samples of dry Wilkeson sandstone, and the black points represent saturated sandstone (Crow and Hurlburt, 1974).

jets. It was found that (1) standoff is an important factor, (2) the effect of multiple passes decreases with depth, (3) the instability of cutting with depth is due to variations in permeability, (4) permeability affects cutting in accordance with theory, and (5) slot depth is proportional to the difference between jet pressure and critical pressure.

The cumulative effect of multiple passes decreases with depth (Figures 9.9, 9.10) and increases with pressure for a constant feed rate and standoff (1.2 cm). The slot depth is nonuniform, but it increases linearly with pressure and decreases with feed rate (Figure 9.11).

The theoretical expression for depth versus feed rate is in overall agreement with experimental results for Wilkeson sandstone (Figure 9.12), although the theory does not apply to rock saturated prior to cutting.

## 9.3   ROCK FAILURE UNDER JET IMPACT

It was proposed by Foreman and Secor (1973) that the failure of rock under jet impact may be due to (1) the impact causing stress waves and sustained stress fields in the rock medium, and (2) penetration of pores and fractures by the water and the creation of internal pressure coupling with the stress field to cause tensile fracture. The ideal theory of elasticity was applied to calculate stress distributions and to use the Griffith criteria of failure. Analytical results explained some observed phenomena but were inconclusive.

Rocks were experimentally impacted with water jets above threshold pressures with and without thin sheets of copper between. The latter prevented damage to the rock, showing that penetration into the pores is essential to jet cutting. Prop-

**TABLE 9.2. Poisson's Ratio, Tensile Strength, and Measured Threshold Pressure for Three Rocks.**

| Rock Type | Poisson's Ratio | Tensile Strength (psi) | Measured Threshold Pressure (psi) |
|---|---|---|---|
| Berea sandstone | 0.14 | 170 | 2000 |
| Indiana limestone | 0.268 | 750 | 3500 |
| Georgia granite | 0.19 | 450 | 6000 |

*Source:* Foreman and Secor (1973).

erties of rock related to jet cutting were found to be as enumerated in Tables 9.2 and 9.3.

The tensile stress level $\sigma_t$ to cause failure was found to be $2.5\sigma_t$ for Indiana limestone, $3.5\sigma_t$ for Barre granite, and $3.0\sigma_t$ for Palisades diabase.

Various mechanical processes have been proposed for explanation of the cutting capability of water jets, including abrasion, erosion, pressure in pores and micro-fractures, and shear due to water flow. Also, in the case of abrasive particles in water jets, the cutting process is characterized as erosion. As with other rock fragmentation processes, more than one mechanism appears to be operative, the degree to which it affects the rock removal or breakage depending largely on rock properties and the parameters of the fragmentation agent—in this case a water jet or a jet with abrasive particles injected into it. Thus, in the cutting of soft, porous rocks porosity is a dominating factor, whereas in hard, crystalline rocks penetration into pores has little effect. Cavitation has been found to erode hard metals and plays an important part in jet cutting with water, and water hammer may affect penetration where pressures are high enough.

Very high velocity impact (hypervelocity) is found in metallic jets from shaped explosive charges where the impact pressure exceeds the strength of the target

**TABLE 9.3. Standard Properties of Three Rocks Related to Jet Cutting.**

| Rock | Young's Modulus (psi $\times$ 10$^{-6}$) | Poisson's Ratio | Tensile Strength (psi) | Compressive Strength (psi) | Permeability Viscosity (in.$^4$/lb-s) |
|---|---|---|---|---|---|
| Indiana lime-stone | 5.0 | 0.25 | 750 | 10,000 | $5.75 \times 10^{-4}$ |
| Barre granite | 5.56 | 0.31 | 1200 | 30,000 | $5.56 \times 10^{-8}$ |
| Palisades diabase | 11.88 | 0.28 | 1650 | 35,000 | $8.9 \times 10^{-12}$ |

*Source:* Foreman and Secor (1973).

material and the depth of penetration is determined by the relative density of the jet and target material. The threshold pressure for the hydrodynamic region where true shock waves occur is higher for iron targets but is much lower for plastics such as Plexiglas. Jet pressure required for water-abrasive jet erosion of steel is in a much lower range; the pressure required for cutting porous rocks is the lowest.

## 9.4 REHBINDER THEORY

Rehbinder (1977, 1978) derived a theoretical model for cutting rock with a steady high-pressure water jet at 90° to the rock face and compared this with data for eight different rocks, including sandstones, granites, diabase, and porphyry. The .theory and experiments indicate that the depth of the slot is a function of the diameter of the jet, the ratio of jet pressure and threshold pressure of the rock, the time of exposure (i.e., number of passes times the diameter of the jet divided by the traversing velocity), and its permeability and grain size. A critical factor is that the erosion resistance of a rock is related to its permeability.

If a water jet impacts on a flat rock face at a right angle, the water penetrates the pore spaces between the grains. For a moving jet, the penetration takes place during the time of residence,

$$T = d/v \tag{9.10}$$

where $d$ is the diameter and $v$ is the traversing speed. If the stagnation pressure of the jet is greater than the threshold pressure, the grains are removed at a rate equal to the mean rate at which the water passes a grain, and the slot is deepened linearly with time. However, the stagnation pressure of the jet is not constant, and it is assumed that the stagnation pressure at the bottom of a slot decreases exponentially:

$$p_o/p_r = e^{-\beta h/D} \tag{9.11}$$

where $p_o$ is the stagnation pressure at the bottom of the slot, $p_r$ is the stagnation pressure of the jet, $h$ is the depth, $D$ is the width of the slot, and $\beta$ is an experimental constant. For fine-grained rock (i.e., $d/l > 1$), the ratio $d/D \simeq 0.25$, where $l$ is the average diameter of the grains of the rock and $\beta = 0.025$. For the average speed of water penetration in porous rock Darcy's law gives

$$u = \frac{\kappa p_o}{\mu l} \tag{9.12}$$

where $u$ is the average penetration speed of the water, $\mu$ is the dynamic viscosity of the water, $\kappa$ is the permeability, and $l$ is the average grain size of the rock. Thus the equation for the depth of the slot can be written as follows:

$$\frac{dh}{dt} = \begin{cases} \dfrac{\kappa p_o}{\mu_l} & t < T \\[2mm] 0 & t > T \end{cases} \tag{9.13}$$

$$h(0) = 0$$

This holds for values of $p_o$, which is also a function of $h$, greater than the threshold pressure $p_{th}$. The solution of equation 9.13 is

$$\frac{h}{d} = 100 \ln\left(1 + \frac{\beta\kappa p_o}{\mu l D}\right) \tag{9.14}$$

and the maximum depth is given by

$$\left(\frac{h}{d}\right)_{max} = 100 \ln\left(\frac{p_o}{p_{th}}\right) \tag{9.15}$$

For shallow slots—that is, if

$$\frac{\beta\kappa p_o}{\mu l D} T \ll 1 \tag{9.16}$$

equation 9.14 can be expanded in a series, the first term being

$$h = \frac{\kappa}{\mu l} p_o T \tag{9.17}$$

In Figure 9.13, equations 9.14 and 9.15 are shown in dimensionless form.

The jet was produced by a conventional pressure intensifier of 500 MPa and a power pack of 0.5 MW. The nozzle diameter was less than 0.9 mm and traversed at a maximum speed of 0.6 m/s. The different kinds of rocks included sedimentary rocks and crystalline rocks such as sandstones, granites, porphyries, and diabase. The standoff distance was constant and as small as possible.

For each pass of the jet the depth of the slot was measured. The time of jet residence for $n$ is

$$T = \frac{dn}{v} \tag{9.18}$$

Lemunda sandstone is the most erodable, while the crystalline rocks are more resistant. The results show that (1) the average depth increases linearly with time at first but becomes constant later and (2) the deviation of range of depth is considerable.

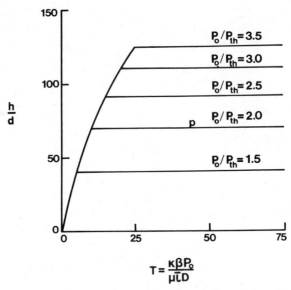

$$T = \frac{\kappa \beta P_o}{\mu \bar{L} D}$$

**Figure 9.13** Dimensionless depth of a slot as a function of dimensionless time of exposure, the dimensionless jet pressure as a parameter. Theory according to equations 9.13 and 9.14 (Rehbinder, 1977).

Three of the rocks were analyzed for physical properties that control erodability (Table 9.4), and the depth of the slot was predicted (Figures 9.14, 9.15). The specific energy (Figure 9.16) was determined for five rocks, but, whereas qualitative agreement is good, quantitative agreement is not, and the theoretical maximum depth agrees surprisingly well. In the initial stage the theory predicts depths too large for Lemunda sandstone and too small for Stockholm and Bohus granite. For granite (Figure 9.15$d$), the agreement is poor, the reason for this being that the condition $d/l > 1$ is not satisfied. That is, $d = 0.5$ mm, $l = 0.5$, and the jet does not have unlimited access to the pores. The discrepancy between theory and experiment is not considered serious since the time constant in equation 9.14 cannot be empirically determined by one experiment. For Bohus granite this may be done

**TABLE 9.4. Physical Properties of Three Rocks.**

| Rock | Permeability $(m^2)$ | Average Grain Size (mm) | Threshold Pressure (MPa) | Compressive Strength (MPa) | Tensile Strength (MPa) |
|---|---|---|---|---|---|
| Bohus granite | $2.4 \times 10^{-15}$ | 0.5 | 90 | 180 | 10 |
| Stockholm granite | $1.0 \times 10^{-15}$ | 0.6 | 100 | 210 | 12 |
| Lemunda sandstone | $2.5 \times 10^{-12}$ | 0.2 | 40 | 60 | 3 |

*Source:* Rehbinder (1977).

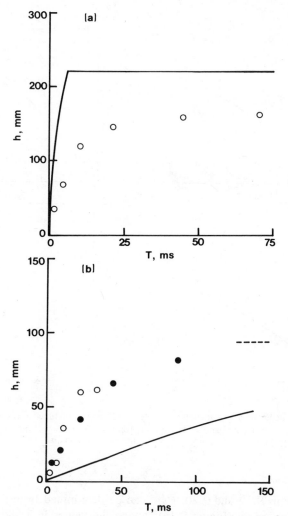

**Figure 9.14** The depth of a slot in rock as a function of time of exposure. Comparison between theory and experiment. $d = 1.3$ mm; $p_o = 220$ MPa. (*a*) Lemunda sandstone, (*b*) Stockholm granite (Rehbinder, 1977).

in three different ways. The theoretically calculated characteristic quantity $\kappa/\mu l = 4.8 \times 10^{-9}$ m$^3$/Ns is too small. Differentiation of equation 9.17 gives

$$\frac{\partial h}{\partial T} = \frac{\kappa}{\mu l} p_o \qquad (9.19)$$

and

$$\frac{\partial h}{\partial p_o} = \frac{\kappa}{\mu l} T \qquad (9.20)$$

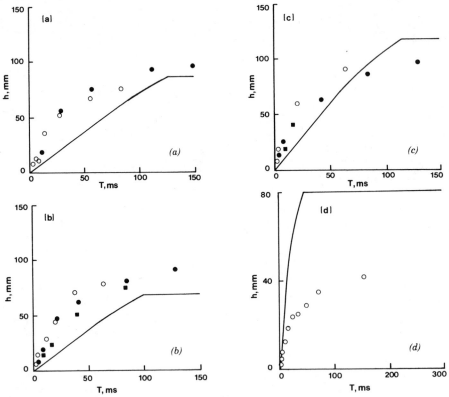

**Figure 9.15** The depth of a slot in Bohus granite as a function of time of exposure. Comparison between theory and experiment. (*a*) $d = 1.7$ mm; $p_o$ 105 MPa. (*b*) $d = 1.3$ mm; $p_o = 150$ MPa. (*c*) $d = 1.3$ mm; $p_o$ 220 MPa. (*d*) $d = 0.5$ mm; $p_o = 460$ MPa (Rehbinder, 1977).

From equations 9.19 and 9.20, $\kappa/\mu l$ can be determined from the value of the derivatives at the origin (Figure 9.15), giving the values in Table 9.5. Thus the characteristic quantity $\kappa/\mu l$ can be calculated from experiments for prediction of cutting depth as it includes the viscosity of the fluid, the permeability, and the grain size. The characteristic quantity is a measure of the erosion resistance of a rock and is a property of the rock and fluid. The specific energy as a function of the time of exposure is much greater for Bohus granite than Lemunda sandstone (Figure 9.16). The specific energy was calculated from the depth of the slot and the diameter and pressure of the jet. The volume per second of the slot, $\partial V/\partial t$, is

$$\partial V/\partial t = hDv \tag{9.21}$$

and the power of the jet is

$$p = \pi d^2 \rho U^3/8 \tag{9.22}$$

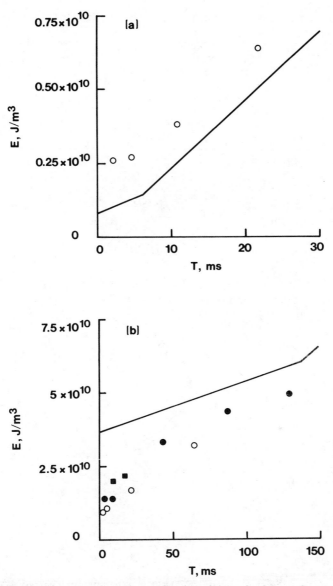

**Figure 9.16** The specific energy for rock as a function of time of exposure. Comparison between theory and experiment. $d = 1.3$ mm; $p_o = 220$ MPa for (*a*) Lemunda sandstone and (*b*) Bohus granite (Rehbinder, 1977).

**TABLE 9.5. Calculation of the Characteristic Quantity $\kappa/\mu l$ for Bohus Granite**

| Case | Fig. No. | $\left(\dfrac{\partial h}{\partial T}\right)_{T=0}$ (m/s) | $p_0$ (Pa) | $\left(\dfrac{\partial h}{\partial p_0}\right)_{T=0}$ (m³/N) | $T$ (s) | $\dfrac{\kappa}{\mu l}$ (m³/Ns) |
|---|---|---|---|---|---|---|
| 1 | 9.5 b | 2.50 | $1.5 \times 10^8$ | | | $1.67 \times 10^{-8}$ |
| 2 | 9.5 c | 3.85 | $2.2 \times 10^8$ | | | $1.75 \times 10^{-8}$ |
| 3 | 9.5 b,c | | | $3.3 \times 10^{-10}$ | $2 \times 10^{-2}$ | $1.56 \times 10^{-8}$ |

*Source:* Rehbinder (1977).

where $\rho$ is the density and $U = \sqrt{2p_o/\rho}$ is the velocity of the jet. The specific energy $E$ then becomes

$$E = \pi d^2 \rho U^3 / 8hDv \tag{9.23}$$

where $h = h(T)$ (equation 9.14). The experimental values of specific energies were obtained by integration of the depth of the slot. If $s$ is a position coordinate along the slot, then $V$ is

$$V = D \int_0^{s_o} h(s)\, ds \tag{9.24}$$

The energy consumed for $s_o$ is

$$W = P \frac{s_o}{v} \tag{9.25}$$

and

$$E = \frac{\pi d^2 \rho U^3 s_o}{8vD \int_0^{s_o} h(s)\, ds} \tag{9.26}$$

Thus, the deeper the slot, the more costly it is to cut, and five to ten times more energy is required to erode Bohus granite than Lemunda sandstone. This comparison was made for only three rocks, since the permeability was not determined for the others.

Rehbinder (1977, 1978) concluded that

1. The slot depth is not only a function of the parameters

$$h = f(n, v, d, p_o, \kappa, l, p_{th}, \mu) \tag{9.27}$$

but also of the parameters:

$$\frac{h}{d} = g\left(\frac{p_o}{p_{th}}, \frac{\kappa p_o n}{\mu l v}\right) \tag{9.28}$$

where the function $g$ is an expression derived from Darcy's law for viscous flow through porous media; and

2. Rock cuttability is characterized by

$$l/\kappa, \text{ erosion resistance, and } p_{th}, \text{ threshold pressure} \tag{9.29}$$

Parameters such as compressive strength, tensile strength, fracture toughness, and so on, do not enter the theory explicitly but may be involved in the properties determining threshold pressure.

## 9.5  HOLE DRILLING WITH WATER JETS

Water jet drilling of holes in rock is used mostly for small diameters in relatively soft pourous rocks, or as an assist to mechanical drilling for harder, more resistant rocks for larger diameters, 3 in. or more.

Rehbinder's (1977, 1978) theory of linear cutting of rocks may be applied (Vijay et al., 1982) and is based on permeability and pressure between the grains if the net hydraulic force on a grain overcomes the cohesive forces. The rate at which water penetrates the pores is found from Darcy's equations of fluid flow through porous media, from which is derived an equation for the depth of a slot cut by a water jet traversing the surface of a rock:

$$\frac{h}{W} = \frac{1}{\beta}\left(1 - e^{-\beta D\kappa P/\mu l W u}\right) \tag{9.30}$$

where (for $h/W \geq 10$), $h$ = depth of the slot, $W$ = width of the slot, $D$ = diameter of the nozzle, $\kappa$ = permeability of the rock, $p$ = nozzle pressure, $\mu$ = viscosity of water, $l$ = average grain size of the rock, $u$ = velocity (traverse speed) of the jet with respect to the rock, and $\beta$ is a constant relating the pressure to the depth of the slot:

$$P(h) = Pe^{-\beta h/W} \tag{9.31}$$

For fine-grained rocks ($l/D > 1$), with $W/D \simeq 4$ and $\beta = 0.025$, and for $\beta D\kappa P/\mu l W u \ll 1$, equation 9.30 becomes

$$\frac{h}{D} = \frac{\kappa}{\mu l} \times \frac{p}{u} \tag{9.32}$$

Differentiation of equation 9.32 gives

$$\frac{1}{D}\frac{h}{P} = \frac{\kappa}{\mu l} \times \frac{1}{u} \tag{9.33}$$

and

$$\frac{1}{D}\frac{h}{u} = \frac{\kappa}{\mu l} \times \frac{P}{u^2} \tag{9.34}$$

The parameter $\kappa/\mu l$ contains the factors for permeability and grain size and is a measure of the erosion resistance of the rock. These factors can be determined from the experimental data or by plotting $h$ against $P$ or $1/u$. The rate of penetration $(R)$ and specific energy $(E)$ obtained with a rotating nozzle can be derived from equation 9.30 or equation 9.32. From Figure 9.17, the tangential traverse speed of the jet relative to the rock is

$$u_o = \text{speed of outer jet}$$

$$= 2\pi N(a + S \tan \theta_o) \tag{9.35}$$

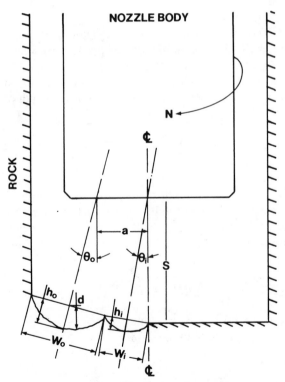

**Figure 9.17**  Penetration of a rock with twin rotating water jets (Vijay et al., 1982).

and

$$u_i = \text{speed of inner jet}$$

$$= 2\pi NS \tan \theta_i \tag{9.36}$$

where $N$ = rotational speed and $S$ = standoff distance.

Since $u_i$ is greater than $u_o$, the inner jet penetrates deeper than the outer jet, and the rate at which the nozzle advances is determined by the outer jet. If $d$ equals the penetrations per revolution, then

$$R = Nd = Nh_o \cos \theta_o \tag{9.37}$$

For equation 9.32

$$h_o = \frac{\kappa}{\mu l} \times D_o \times \frac{P}{u_o}$$

$$= \frac{\kappa}{\mu l} \times D_o \times \frac{P}{2\pi N(a + S \tan \theta_o)} \tag{9.38}$$

and

$$R = \frac{\kappa}{\mu l} \times \frac{D_o}{2\pi} \times \frac{P \cos \theta_o}{a + S \tan \theta_o} \tag{9.39}$$

The diameter of the hole ($D_H$) depends on the widths $W_o$ and $W_i$ cut by the jets (Figure 9.17), and

$$D_H = 2b(D_o \sec \theta_o + D_i \sec \theta_i) \tag{9.40}$$

where $W = bD$ and $b$ is a constant that varies from 2 to 5. The rate of volume removal is

$$V = \frac{\pi}{4} D_H^2 R \tag{9.41}$$

Substitution for $R$ and $D_H$ from equations 9.38 and 9.39 gives

$$V = \frac{\kappa}{\mu l} \times \frac{b^2}{2} \times \frac{D_o P(D_o \sec \theta_o + D_i \sec \theta_i)^2 \cos \theta_o}{(a + S \tan \theta_o)} \tag{9.42}$$

The total hydraulic power of the jets is

$$H_p = \text{total hydraulic power} = (H_p)_o + (H_p)_i$$

$$= kD_o^2 P^{3/2} + kD_i^2 P^{3/2}$$

$$= kP^{3/2}(D_o^2 + D_i^2) \tag{9.43}$$

where $k$ is a constant. Thus

$$E = \text{specific energy} = H_p/V$$

$$= k' \times \frac{\mu l}{\kappa} \times \frac{(a + S \tan \theta_o)(D_o^2 + D_i^2)}{D_o P^{1/2} \cos \theta_o (D_o \sec \theta_o + D_i \sec \theta_i)^2} \qquad (9.44)$$

where $k'$ is a constant.

This expression for $E$ shows that it is directly proportional to $l$ and inversely proportional to $\kappa$, agreeing with experimental results. Also, for small values of $\theta_o$ and $\theta_i$, $S \simeq 0$ and $D_o \simeq D_i$, equation 9.43 reduces to

$$E = \kappa'' \frac{\mu l}{\kappa} \times \frac{a}{D_o} \times \frac{1}{P^{1/2}} \qquad (9.45)$$

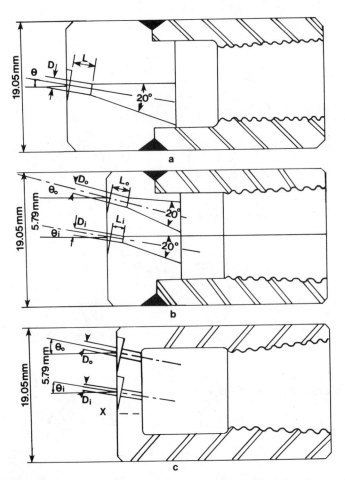

**Figure 9.18** Details of effective nozzles (see Table 9.6) (Vijay and Brierly, 1980).

Thus $E$ can be reduced by reducing $a$, the distance between orifices with the other factors being held constant, which is useful for designing nozzle bodies to drill small holes ($\approx 1$ cm) or to cut deep narrow slots in rocks.

In experimentation and analysis reported by Vijay and Brierly (1980), tests were made for the evaluation of the design of bit parameters such as nozzle size and orientation and spacing of the jet orifices (Figure 9.18; Table 9.6). The single-orifice nozzle with an inclined jet would not cut a hole large enough to accommodate the body of the bit. Most of the bits were made with tapered orifices (Figure 9.18b). The rock blocks tested in this initial program were all granite (Table 9.7), with important properties related to jet cutting being the porosity and grain size.

Test data included rpm ($N$), rate of penetration ($R$), diameter of the hole ($D_H$), the ratio of the orifice area to the nozzle body ($A_p = A_h/A_N$), and the specific energy ($E$). The specific energy was used as a basis for the evaluation of test results for best designs, and water jet drilling results were compared to those from diamond drilling (Table 9.8).

These conclusions were drawn from the results of tests in granite:

1. The optimum speed of rotation was about 200 rpm, which is equivalent to a linear traverse velocity of the jet of about 15–20 cm/s.

**TABLE 9.6. Design Specification of Nozzle Bodies (Figure 9.18); Diameter of Nozzle Body = 19.05 mm; $A_T/A_N$ = 0.008**

| Nozzle Body Identification | $D$ or $D_I$ (mm) | $D_O$ (mm) | $\theta$ or $\theta_O$ (deg.) | Remarks |
|---|---|---|---|---|
| SO1 | 1.702 | — | 10 | |
| SO2 | 1.702 | — | 20 | |
| SO3 | 1.702 | — | 10 | $\alpha = 180°$ |
| A1 | 1.180 | 1.180 | 10 | $\alpha = 20°$ for all |
| A2 | 1.090 | 1.320 | 10 | nozzles except |
| A3 | 0.914 | 1.400 | 10 | where noted |
| A4 | 0.742 | 1.510 | 10 | $\theta_I = 10°$ for all |
| B1 | 1.180 | 1.180 | 15 | dual-orifice nozzle |
| B2 | 1.090 | 1.320 | 15 | bodies |
| B3 | 0.914 | 1.400 | 15 | |
| C1 | 1.180 | 1.180 | 20 | |
| C2 | 1.090 | 1.320 | 20 | |
| C3 | 0.914 | 1.400 | 20 | |
| SB1 | 1.180 | 1.180 | 15 | $\alpha = 180°$ |

*Source:* Vijay and Brierley (1980).

**TABLE 9.7. Measured Properties of Granite**

| Property | Mean Value | Range | Coefficient of Variation (%)[a] |
|---|---|---|---|
| $\rho_R$ | 2627 | 2603–2649 | 0.5 |
| $\sigma_t$ | 8.0 | 5.3–10.0 | 16.6 |
| $\sigma_c$ | 165.5 | 139–175 | 7.2 |
| C | 4115 | 3767–4456 | 4.8 |
| $E_R$ | 44,540 | 37,230–52,400 | 9.9 |
| Grain size, mm | 1.98 | 0.2–4.4 | 14.9 |
| Porosity, % | 1.235 | 0.61–2.65 | 56.9 |

[a] Standard deviation expressed as a percentage of the mean value.
*Source:* Vijay and Brierley (1980).

2. The best value of the ratio $A_o/A_t$, the ratio of the area of the outer orifice to the total cross section of the orifices, is 0.50–0.60.
3. The optimum value of the angle of inclination of the outer orifice for this diameter of hole and type of rock is about 20°.
4. Bit C2 (Table 9.6) appeared to be the best configuration for cutting the granite tested.

The nomenclature for hole drilling includes the following:

$A_H$ = mean cross-sectional area of the drilled hole, cm$^2$

$A_I$ = cross-sectional area of the inner orifice in the nozzle body, cm$^2$

**TABLE 9.8. Drilling Data Obtained with Coring Bit**

Outer diameter of coring bit = 6.35 cm

Inner diameter of coring bit = 5.72 cm

Power of drill = 1.98 kW

$N \approx 1500$ rpm, No. of tests = 5

Mean value of E based on the volume of hole (6.35 cm) drilled
= $1.35 \times 10^3$ J/cm$^3$

Mean value of E based on the actual cutting action of the drill (material removed in the annulus) = $7.1 \times 10^3$ J/cm$^3$

Mean value of R = 2.8 cm/min

*Source:* Vijay and Brierley (1980).

$A_N$ = cross-sectional area of nozzle body, cm$^2$

$A_O$ = cross-sectional area of outer orifice in the nozzle body, cm$^2$

$A_R$ = $A_H/A_N$

$A_T$ = total cross-sectional area of orifices in the nozzle body, cm$^2$

$C$ = compression wave velocity in the rock, m/s

$D$ = diameter of orifice, mm

$D_H$ = diameter of hole, cm

$D_I$ = diameter of inner orifice, mm

$D_O$ = diameter of outer orifice, mm

$E$ = specific energy, J/cm$^2$

$E_R$ = modulus of deformation of rock, MN/m$^2$

$H_P$ = total hydraulic power of jets, kW

$L$ = length of cylindrical outlet of nozzle, mm

$N$ = rotational speed of drill, rpm

$P$ = pressure at nozzle inlet, MPa

$R$ = mean drilling rate (feed rate), cm/min

$\theta$ = angle of inclination of orifice, degrees

$\theta_I$ = angle of inclination of inner orifice, degrees

$\theta_O$ = angle of inclination of outer orifice, degrees

$\alpha$ = nozzle angle entry, degrees

$\rho_R$ = dry bulk density of rock, kg/m$^3$

$\sigma_c$ = uniaxial compressive strength of rock, MN/m$^2$

$\sigma_t$ = uniaxial tensile strength of rock, MN/m$^2$

## 9.6 COAL—JET SLOTTING

In addition to the mining of coal by large-diameter jet hydromining, similar to the monitors used in placer mining of gold, research with fine jets has been performed for possible applications as an assist in mechanical methods of mining coal. One such mining device is a type of coal plough assisted by jets designed to work on a long face, where slot dimensions are determined by the jet diameter and the pressure, traverse velocity, standoff, and applicable properties of the coal. Russian researchers (Kuzmich, 1972) found that for constant traverse velocity and nozzle diameter

$$h = C(P_o - P_k) \tag{9.46}$$

where $P_k$ is the critical pressure (Figure 9.19). Also, the relation between depth and traverse velocity (Figure 9.20) is

$$h = h_o e^{-dV} m \tag{9.47}$$

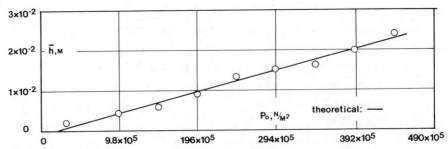

**Figure 9.19** Variation in the coal penetration depth versus jet hydrodynamic pressure (Kuzmich, 1972).

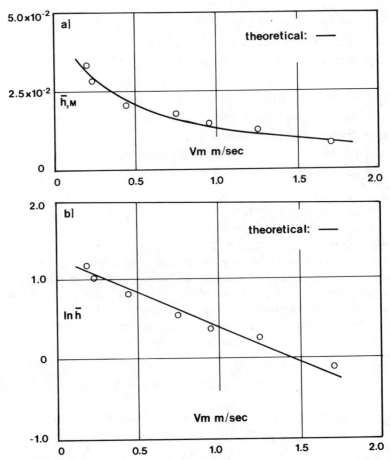

**Figure 9.20** Variation in the coal penetration depth versus jet traverse velocity (Kuzmich, 1972).

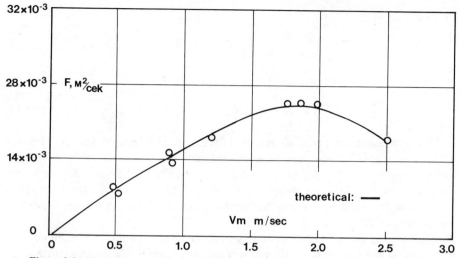

**Figure 9.21** Variation in the slot lateral area versus jet traverse velocity (Kuzmich, 1972).

where $h_o$ is the maximum penetration and $d$ is a penetration factor that is a function of coal properties. Thus, the resistance of coal to penetration, $C$, is

$$C = C_o e^{-dV} m \qquad (9.48)$$

and

$$h_o = C_o(P_o - P_k) \qquad (9.49)$$

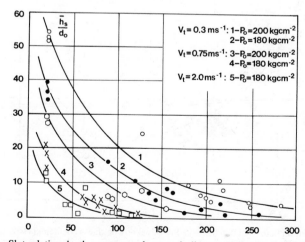

**Figure 9.22** Slot relative depth versus nozzle-to-rock distance (Nikonov and Goldin, 1972).

and, finally,

$$h = C_o(P_o - P_k)e^{-dV}m \qquad (9.50)$$

The lateral area of the slot cut per unit time is given by

$$F = h \cdot V_m \qquad (9.51)$$

This increases to an optimum and then decreases with traverse velocity (Figure 9.21). The water jet parameters varied in the tests were pressures of 50–500 atm, nozzle diameters of 0.7–3.5 mm, jet traverse velocity of 0.28 and 3.83 m/s on coals with Protodyaknov coefficient of $f = 0.37$–$2.56$. As noted in the following, the Protodyaknov coefficient does not appear to be an accurate parameter to predict coal penetrability.

Nikonov and Goldin (1972) found that for coal and rock the maximum cutting effectiveness is obtained with minimum standoff (Figure 9.22). In jet-cutting tests coal was cut with jets 2.17–2.20 mm in diameter at 150-atm pressure and with a traverse velocity of 0.3–0.4 m/s, perpendicular to the "fracture planes" of the coal. The relationship between cutting and the Protodyaknov number was established, but the scatter of data was too great to indicate the applicability of the equation derived (Figure 9.23).

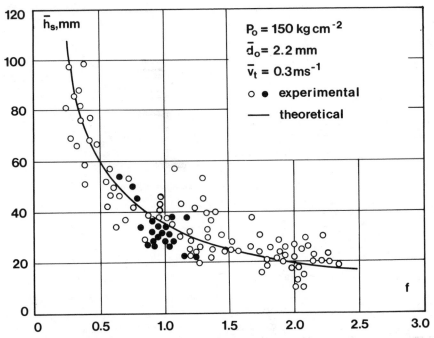

**Figure 9.23** Variation in the slot depth versus Protodyaknov hardness coefficient (Nikonov and Goldin, 1972).

## 9.7   COAL MINING AND COAL CUTTING

Investigations with larger-diameter high-pressure jets ($\frac{1}{4}$ and $\frac{3}{8}$ in.) for coal mining were made (Fowkes and Wallace, 1968) in an underground coal mine of the pressure distribution in jets, effective nozzle design, jet parameters, and coal cutting (mining) rates. Axial jet stream pressure was measured with a pitot tube, and it was found that the pressure could be represented by a Gaussian curve:

$$P = Ae^{BR^2} \tag{9.52}$$

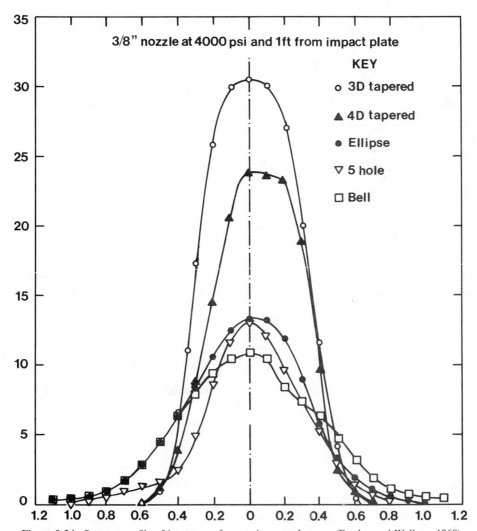

**Figure 9.24**  Pressure profile of jet streams from various nozzle types (Fowkes and Wallace, 1968).

where $A$ and $B$ are experimentally determined parameters for given test conditions and $R$ is the radial distance normal to the axis of the jet. Several types of tapered and curved internal profile nozzles were tested, giving different pressure profiles in the jets (Figure 9.24). The tapered profiles gave the highest center pressure. One-quarter-inch-diameter nozzle of 10-diameter straight section and different tapers showed that pressure for the smallest taper, 8°, was highest (Figure 9.25), and $\frac{3}{8}$-in. nozzles of 3-diameter straight section showed a rapid change in pressure and pressure distribution at 1-, 3-, and 5-ft distances from the nozzle (Figure 9.26).

For fine water-jet slotting of rock the high nozzle pressure is usually the most important parameter, but in the mining of coal with larger jets at lower pressure

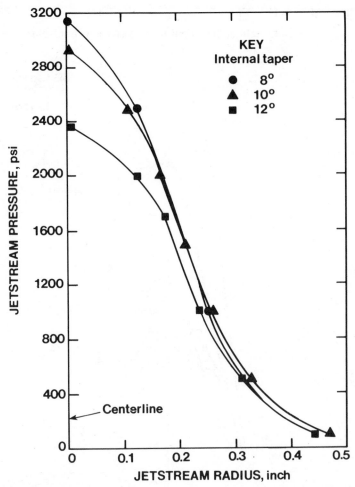

**Figure 9.25** Pressure profiles of $\frac{1}{4}$-in. 10D nozzle streams taken 1 ft from impact (Fowkes and Wallace, 1968).

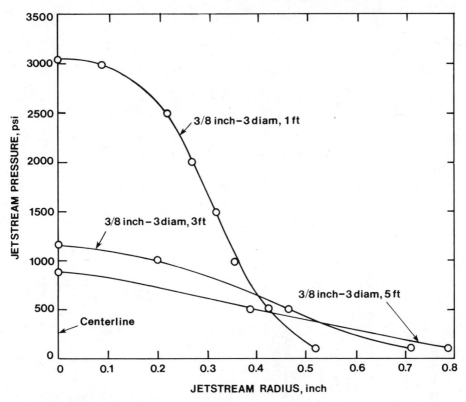

**Figure 9.26** Pressure distribution of $\frac{3}{8}$-in.-diameter 3D nozzle at 1, 3, and 5 ft from impact; 4000-psi pump pressure (Fowkes and Wallace, 1968).

the coal is not slotted or drilled but is fragmented from the face in various-size pieces.

For coal-mining tests the monitor was mounted over the conveyor of a coal-loading machine. While the maximum traverse speed was 24 in./s with 1-ft stand-off, 4 in./s was used for mining, the depth of the cut being about 2 ft. The monitor was traversed across the face (11 ft), raised 4 in., and traversed again. Measured data at the face and calculated jet characteristics were recorded (Table 9.9). The cutting rate showed considerable scatter when plotted against total force, total force to the 100-psi contour, the flow rate, and the kinetic energy at the nozzle (Figure 9.27). The appropriate jet parameters may be calculated from the appropriate formula for equation 9.52.

Correlation coefficients for the data in Table 9.10 were determined for the important parameters, that between the cutting rate and the total force being 0.91, kinetic energy 0.84, and flow rate 0.84. A change in pressure from 4000 to 3000 psi has a significant effect on cutting rate, but quantitative values for this and the effects of traverse rates were not determined in this set of experiments.

**TABLE 9.9. Summary of Measured and Calculated Jet Stream Characteristics**

| | | Nozzle | | | Face Data | | | | | | | | | Laboratory Data | | | | | |
| | | Cutting Rates, tpm | | | Pressure, psi | | | | Flow Rate, gpm | Calculated Parameters | | Force, lb | | Radius Out to 100 psi on Impact, in. | Kinetic Energy per Unit Time, ft lb/s | | Average Velocity Out to 100 psi on Impact, fps |
| Type | Remarks | Test 1 | Test 2 | Test 3 | At Pump | At Monitor | Maximum on Impact | Average Out to 100 psi on Impact | | A | B | Total | Out to 100 psi on Impact | | Average Out to 100 psi on Impact | At Nozzle | |
|---|---|---|---|---|---|---|---|---|---|---|---|---|---|---|---|---|---|
| 1 | | 0.87 | 0.84 | — | 4,000 | 3,030 | 2,980 | 1,078 | 226 | 3,259 | 9.29 | 1,102 | 1,068 | 0.617 | 65,800 | 220,000 | 31.2 |
| | | | | | 4,000 | 3,135 | 3,050 | 1,121 | 222 | 3,385 | 8.99 | 1,183 | 1,148 | 0.626 | 66,400 | 223,000 | 31.6 |
| 1 | Old | 0.73 | — | — | 4,000 | 3,115 | 2,760 | 1,033 | 220 | 3,078 | 8.59 | 1,126 | 1,090 | 0.632 | 61,300 | 220,000 | 30.5 |
| 1 | | 0.69 | 0.57 | — | 4,000 | 3,135 | 2,830 | 1,071 | 216 | 3,210 | 8.99 | 1,121 | 1,086 | 0.621 | 62,100 | 217,000 | 31.0 |
| | | | | | 4,000 | 3,135 | 2,810 | 1,044 | 218 | 3,116 | 9.65 | 1,015 | 982 | 0.597 | 61,400 | 219,000 | 30.7 |
| 1 | Polished | 0.62 | — | — | 4,000 | 3,185 | 2,880 | 1,107 | 216 | 3,335 | 8.73 | 1,200 | 1,165 | 0.634 | 63,900 | 221,000 | 31.5 |
| 1 | Do | 0.52 | — | — | 3,000 | 2,350 | 2,230 | 853 | 192 | 2,464 | 8.49 | 912 | 875 | 0.615 | 45,400 | 164,000 | 28.1 |
| 1 | | 0.54 | 0.47 | — | 3,000 | 2,375 | 2,130 | 841 | 192 | 2,423 | 8.81 | 864 | 829 | 0.602 | 44,900 | 146,000 | 28.0 |
| | | | | | 3,000 | 2,350 | 2,180 | 859 | 192 | 2,487 | 9.06 | 863 | 828 | 0.596 | 45,800 | 164,000 | 28.2 |
| 1 | | 0.44 | 0.41 | — | 3,000 | 2,305 | 2,270 | 864 | 196 | 2,500 | 10.16 | 773 | 743 | 0.563 | 46,900 | 165,000 | 28.3 |
| | | | | | 3,000 | 2,305 | 2,270 | 874 | 196 | 2,536 | 10.23 | 779 | 748 | 0.562 | 47,400 | 165,000 | 28.4 |
| 2 | | 0.52 | 0.42 | — | 4,000 | 3,660 | 3,420 | 1,256 | 132 | 3,858 | 19.94 | 608 | 592 | 0.428 | 43,600 | 155,000 | 33.2 |
| | | | | | 4,000 | 3,660 | 3,440 | 1,250 | 132 | 3,857 | 21.20 | 569 | 554 | 0.415 | 45,400 | 155,000 | 33.2 |
| 2 | | 0.29 | 0.28 | — | 3,000 | 2,700 | 2,400 | 938 | 120 | 2,750 | 19.37 | 446 | 430 | 0.413 | 30,800 | 104,000 | 29.3 |
| | | | | | 3,000 | 2,700 | 2,470 | 958 | 124 | 2,820 | 18.34 | 483 | 466 | 0.427 | 32,400 | 107,000 | 29.6 |
| 3 | | 0.33 | 0.32 | — | 4,000 | 3,635 | 2,920 | 962 | 136 | 2,835 | 19.95 | 446 | 431 | 0.410 | 35,700 | 159,000 | 29.6 |
| | | | | | 4,000 | 3,660 | 2,900 | — | 140 | — | — | — | — | — | — | 164,000 | — |

| | | | | | | | | | | | | | | | | |
|---|---|---|---|---|---|---|---|---|---|---|---|---|---|---|---|---|
| 3 | | 0.24 | — | 3,000 | 2,700 | 2,050 | 801 | 125 | 2,290 | 18.33 | 392 | 375 | 0.413 | 36,900 | 108,000 | 27.4 |
| | | 0.23 | | 3,000 | 2,700 | 2,120 | 794 | 125 | 2,268 | 19.84 | 359 | 343 | 0.397 | 25,400 | 108,000 | 27.3 |
| 4 | 12° taper | 0.40 | — | 4,000 | 3,660 | 2,420 | 907 | 125 | 2,647 | 16.96 | 490 | 472 | 0.439 | 31,200 | 147,000 | 28.9 |
| | | 0.31 | | 4,000 | 3,735 | 2,330 | 879 | 125 | 2,551 | 16.97 | 472 | 454 | 0.437 | 30,400 | 150,000 | 28.5 |
| 4 | 10° taper | 0.37 | — | 4,000 | 3,610 | 3,040 | 1,106 | 125 | 3,332 | 18.99 | 551 | 535 | 0.430 | 36,100 | 145,000 | 31.1 |
| | | 0.28 | | 4,000 | 3,760 | 3,270 | 1,189 | 125 | 3,270 | 18.66 | 610 | 593 | 0.439 | 39,300 | 151,000 | 32.4 |
| 4 | | 0.36 | — | 4,000 | 3,660 | 2,970 | 1,071 | 125 | 3,209 | 17.42 | 579 | 561 | 0.446 | 36,000 | 147,000 | 31.0 |
| | | 0.26 | | 4,000 | 3,660 | 2,910 | 1,051 | 125 | 3,140 | 16.99 | 581 | 562 | 0.450 | 35,400 | 147,000 | 30.8 |
| 4 | 10° taper | 0.34 | — | 4,000 | 3,760 | 1,120 | 456 | 125 | 1,171 | 8.72 | 422 | 386 | 0.531 | 17,500 | 151,000 | 21.6 |
| | | 0.25 | | 4,000 | 3,710 | 1,080 | 433 | 125 | 1,100 | 7.08 | 488 | 444 | 0.582 | 16,700 | 149,000 | 21.2 |
| 4 | 10° taper | 0.30 | — | 4,000 | 3,785 | 400 | 176 | 125 | 355 | 2.98 | 374 | 268 | 0.652 | 8,100 | 152,000 | 14.7 |
| | | 0.22 | | 4,000 | 3,785 | 750 | 308 | 125 | 727 | 4.91 | 465 | 401 | 0.635 | 12,700 | 152,000 | 18.4 |
| 4 | | 0.26 | — | 4,000 | 3,710 | 1,720 | 671 | 125 | 1,860 | 12.34 | 474 | 448 | 0.487 | 24,200 | 149,000 | 25.4 |
| | | 0.25 | | 4,000 | 3,845 | 1,480 | 612 | 125 | 1,560 | 10.81 | 453 | 424 | 0.472 | 24,300 | 154,000 | 25.5 |
| 4 | 10° taper | 0.26 | — | 4,000 | 3,810 | — | — | 125 | — | — | — | — | — | — | 153,000 | — |
| | | 0.24 | | 4,000 | 3,875 | 1,260 | 508 | 125 | 1,336 | 9.74 | 431 | 398 | 0.516 | 19,120 | 155,000 | 22.6 |
| 4 | Do | 0.24 | — | 3,000 | 2,745 | 2,230 | 834 | 125 | 2,398 | 16.43 | 458 | 439 | 0.440 | 29,000 | 110,000 | 27.9 |
| | | 0.23 | | 3,000 | 2,820 | 2,250 | 853 | 125 | 2,466 | 16.30 | 475 | 456 | 0.444 | 29,600 | 113,000 | 28.1 |
| 4 | 12° taper | 0.23 | — | 4,000 | 3,810 | 860 | 314 | 125 | 742 | 5.02 | 464 | 401 | 0.632 | 12,800 | 153,000 | 18.5 |
| | | 0.22 | | 4,000 | 3,870 | 1,270 | 500 | 125 | 1,309 | 9.38 | 438 | 405 | 0.524 | 18,900 | 155,000 | 22.5 |
| 4 | | 0.25 | — | 3,000 | 2,720 | — | — | 125 | — | — | — | — | — | — | 109,000 | — |
| | | 0.17 | | 3,000 | 2,725 | 2,200 | 844 | 125 | 2,434 | 19.15 | 399 | 383 | 0.408 | 29,300 | 109,000 | 28.0 |
| 4 | 12° taper | 0.21 | — | 3,000 | 2,745 | 1,730 | 658 | 125 | 1,875 | 16.57 | 355 | 337 | 0.421 | 24,300 | 110,000 | 22.3 |
| | | 0.18 | | 3,000 | 2,795 | 1,930 | 757 | 125 | 2,142 | 17.98 | 374 | 357 | 0.413 | 26,700 | 112,000 | 26.7 |
| 4 | | 0.22 | — | 3,000 | 2,795 | 1,000 | 422 | 125 | 1,069 | 10.09 | 333 | 302 | 0.485 | 16,400 | 112,000 | 21.0 |
| | | 0.17 | | 3,000 | 2,820 | 1,000 | 407 | 125 | 1,020 | 9.03 | 355 | 320 | 0.507 | 15,900 | 113,000 | 20.6 |
| 4 | 12° taper | 0.18 | — | 3,000 | 2,795 | 550 | 247 | 125 | 550 | 6.04 | 286 | 234 | 0.531 | 10,600 | 112,000 | 16.9 |
| | | 0.18 | | 3,000 | 2,890 | 1,040 | 434 | 125 | 1,105 | 10.34 | 336 | 305 | 0.482 | 16,800 | 116,000 | 21.2 |

TABLE 9.9. (Continued)

| Type | Remarks | Test 1 | Test 2 | Test 3 | At Pump | At Monitor | Maximum on Impact | Average Out to 100 psi on Impact | Flow Rate, gpm | A | B | Total | Out to 100 psi on Impact | Radius Out to 100 psi on Impact, in. | Average Out to 100 psi on Impact | At Nozzle | Average Velocity Out to 100 psi on Impact, fps |
|---|---|---|---|---|---|---|---|---|---|---|---|---|---|---|---|---|---|
| | | Cutting Rates, tpm | | | Pressure, psi | | | | | Calculated Parameters | | Force, lb | | | Kinetic Energy per Unit Time, ft lb/s | | |
| 4 | 10° taper | 0.18 | 0.15 | — | 3,000 | 2,890 | 530 | 235 | 125 | 518 | 5.94 | 274 | 221 | 0.526 | 10,100 | 116,000 | 16.4 |
| | | | | | 3,000 | 2,915 | 980 | 407 | 125 | 1,023 | 10.44 | 308 | 278 | 0.472 | 15,900 | 117,000 | 20.6 |
| 4 | Do | 0.15 | 0.14 | — | 3,000 | 2,720 | 380 | 184 | 125 | 377 | 3.91 | 302 | 174 | 0.582 | 8,400 | 109,000 | 15.0 |
| | | | | | 3,000 | 2,745 | 540 | 236 | 125 | 518 | 10.31 | 158 | 128 | 0.400 | 10,200 | 110,000 | 16.5 |
| 4 | | 0.12 | 0.11 | — | 3,000 | 2,820 | 1,330 | 548 | 125 | 1,463 | 12.95 | 355 | 331 | 0.455 | 20,400 | 113,000 | 23.4 |
| | | | | | 3,000 | 2,915 | 1,130 | 472 | 125 | 1,221 | 11.38 | 337 | 310 | 0.469 | 18,000 | 117,000 | 21.9 |
| 5 | Orifice only | 0.66 | 0.53 | — | 4,000 | 3,185 | 730 | 326 | 215 | 778 | 3.10 | 788 | 687 | 0.813 | 22,800 | 220,000 | 18.8 |
| | | | | | 4,000 | 3,185 | 770 | 329 | 215 | 787 | 2.91 | 849 | 741 | 0.842 | 23,000 | 220,000 | 18.9 |
| 5 | Orifice +25/64-in. insert | 0.48 | 0.45 | — | 4,000 | 3,135 | 980 | 391 | 210 | 974 | 3.76 | 814 | 731 | 0.778 | 25,900 | 211,000 | 20.3 |
| | | | | | 4,000 | 3,160 | 1,010 | 415 | 210 | 1,047 | 3.89 | 846 | 765 | 0.777 | 27,200 | 213,000 | 20.8 |
| 5 | Orifice +27/64-in. insert | 0.56 | 0.46 | — | 4,000 | 3,115 | 730 | 326 | 212 | 778 | 3.00 | 814 | 710 | 0.827 | 25,500 | 212,000 | 18.8 |
| | | | | | 4,000 | 3,190 | 980 | 427 | 212 | 1,084 | 3.87 | 879 | 798 | 0.784 | 28,100 | 217,000 | 21.0 |
| 5 | | 0.46 | — | | 4,000 | 3,060 | 1,070 | 450 | 214 | 1,154 | 4.05 | 896 | 818 | 0.777 | 29,600 | 210,000 | 21.5 |
| 5 | Orifice +27/64-in. insert | 0.44 | 0.44 | — | 3,000 | 2,280 | 540 | 239 | 190 | 527 | 2.70 | 612 | 496 | 0.784 | 15,700 | 139,000 | 16.6 |
| | | | | | 3,000 | 2,330 | 710 | 327 | 190 | 782 | 3.58 | 687 | 599 | 0.758 | 20,200 | 142,000 | 18.9 |
| 5 | Orifice +25/64-in. insert | 0.38 | 0.34 | — | 3,000 | 2,350 | 730 | 320 | 188 | 760 | 3.75 | 637 | 553 | 0.736 | 19,600 | 161,000 | 18.7 |
| | | | | | 3,000 | 2,375 | 760 | 336 | 188 | 808 | 4.02 | 632 | 553 | 0.721 | 20,400 | 162,000 | 19.1 |

| | | | | | | | | | | | | | | | |
|---|---|---|---|---|---|---|---|---|---|---|---|---|---|---|---|
| **5** Orifice only | 0.36 | 0.34 | — | 3,000 | 2,295 | 390 | 181 | 190 | 369 | 2.22 | 521 | 379 | 0.766 | 12,600 | 140,000 | 14.9 |
| | | | | 3,000 | 2,305 | 440 | 197 | 190 | 413 | 2.48 | 523 | 396 | 0.756 | 13,400 | 160,000 | 15.4 |
| **6** Enlarged inlet | 0.52 | 0.42 | — | 4,000 | 3,485 | 1,320 | — | 182 | — | — | — | — | — | — | 204,000 | — |
| | | | | 4,000 | 3,445 | 1,380 | 399 | 183 | 997 | 4.52 | 692 | 623 | 0.713 | 22,900 | 202,000 | 20.5 |
| **6** | 0.33 | 0.29 | — | 4,000 | 3,485 | 1,320 | 502 | 182 | 1,317 | 6.80 | 609 | 562 | 0.616 | 27,600 | 204,000 | 22.5 |
| | | | | 4,000 | 3,445 | 1,380 | — | 183 | — | — | — | — | — | — | 202,000 | — |
| **6** | 0.34 | 0.26 | — | 3,000 | 2,550 | 1,060 | 355 | 158 | 865 | 7.40 | 367 | 325 | 0.540 | 18,000 | 129,000 | 19.5 |
| | | | | 3,000 | 2,575 | 1,030 | 373 | 156 | 918 | 8.25 | 350 | 311 | 0.518 | 18,500 | 129,000 | 19.9 |
| **7** | 0.63 | 0.58 | 0.52 | 4,000 | 3,135 | 2,380 | 895 | 216 | 2,609 | 7.64 | 1,072 | 1,031 | 0.653 | 53,300 | 217,000 | 28.7 |
| | | | | 4,000 | 3,185 | 2,470 | 933 | 215 | 2,736 | 8.09 | 1,063 | 1,024 | 0.640 | 55,000 | 220,000 | 29.2 |
| **7** | 0.44 | 0.43 | 0.42 | 3,000 | 2,305 | 1,820 | 714 | 194 | 2,000 | 7.82 | 803 | 763 | 0.619 | 39,500 | 163,000 | 26.1 |
| | | | | 3,000 | 2,305 | 1,730 | 687 | 194 | 1,912 | 7.50 | 801 | 759 | 0.627 | 38,200 | 163,000 | 25.7 |
| **8** | 0.62 | 0.61 | — | 4,000 | 3,125 | 1,080 | 424 | 220 | 1,072 | 3.67 | 917 | 831 | 0.804 | 28,900 | 221,000 | 21.0 |
| | | | | 4,000 | 3,085 | 1,020 | 410 | 220 | 1,031 | 3.96 | 818 | 739 | 0.768 | 28,200 | 218,000 | 20.7 |
| **8** | 0.66 | 0.57 | — | 4,000 | 3,135 | 830 | 341 | 216 | 823 | 3.14 | 823 | 723 | 0.819 | 23,800 | 217,000 | 19.2 |
| | | | | 4,000 | 3,135 | 770 | 324 | 216 | 772 | 3.30 | 735 | 640 | 0.787 | 22,800 | 217,000 | 18.8 |
| **8** | 0.43 | 0.38 | — | 3,000 | 2,305 | 730 | 319 | 196 | 758 | 3.90 | 610 | 530 | 0.722 | 20,400 | 165,000 | 18.7 |
| | | | | 3,000 | 2,305 | 720 | 319 | 200 | 757 | 3.95 | 602 | 522 | 0.716 | 20,800 | 168,000 | 18.7 |
| **8** | 0.36 | 0.34 | — | 3,000 | 2,305 | 580 | 255 | 192 | 573 | 3.23 | 557 | 459 | 0.735 | 16,700 | 161,000 | 17.1 |
| | | | | 3,000 | 2,305 | 530 | 240 | 194 | 532 | 3.14 | 532 | 432 | 0.730 | 16,100 | 163,000 | 16.6 |
| **9** | 0.61 | 0.57 | — | 4,000 | 3,110 | 1,350 | 553 | 228 | 1,414 | 4.64 | 957 | 899 | 0.755 | 36,300 | 228,000 | 23.1 |
| | | | | 4,000 | 3,110 | 1,330 | 518 | 224 | 1,366 | 4.55 | 943 | 874 | 0.758 | 34,800 | 224,000 | 22.8 |
| **9** | 0.36 | 0.35 | — | 3,000 | 2,305 | 1,050 | 427 | 198 | 1,082 | 4.73 | 719 | 652 | 0.710 | 26,200 | 166,000 | 21.0 |
| | | | | 3,000 | 2,350 | 1,000 | 384 | 200 | 952 | 4.26 | 702 | 629 | 0.727 | 24,300 | 171,000 | 21.0 |

*Source*: Fowkes and Wallace (1968).

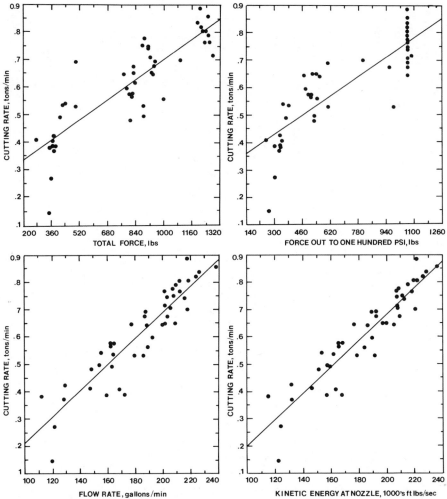

**Figure 9.27**   Cutting rate as a function of measured and calculated quantities (Fowkes and Wallace, 1968).

The depth of the jet penetration or removal of coal by mining was found to obey the law (Palovitch and Malenka, 1964)

$$y = kt^n \tag{9.53}$$

where $y$ = depth in inches, $t$ = time in seconds, and $k$ and $n$ are parameters depending on nozzle, pressure, and so on, but these two factors were usually held constant for a test. Traverse speeds were varied from 1.7 to 24.0 in./s with a standoff of 1 ft, the coal cutting (volume mined) increased linearly with traverse speed, but the depth of cut decreased exponentially. The cutting rate also decreased with standoff.

Several field tests were made of the hydraulic jet method of mining both bituminous and anthracite coal. While the method has a high rate of production when the jet is operating but a low percentage time of jetting, it may find its greatest applicability in the mining of pitching seams.

## 9.8  WATER JET APPLICATIONS

While water jets are effective in cutting slots and small-diameter holes in porous rocks, they do not generate high stresses at sufficient depth in the rock and at effective distances from the water–rock interface to cause fracture and breakage. This is also true for softer materials such as coal. Hence, they are most effective in volume breakage of softer porous rocks and coal only when they are used in combination with a mechanical method of excavation such as roller cutters, drag bits, or other types of mechanical rock excavators. There has perhaps been more development of water-jet-assisted rock excavation equipment in Europe than in the United States. The primary reasons why equipment development, usage, and sales have been slow may be due to the following or related reasons:

1. High-pressure water pumps have a high capital cost.
2. Pump maintenance problems are frequent and costly.
3. Water jet nozzles wear rapidly and are costly.
4. High-pressure tubing and plumbing require special care in installation and maintenance.
5. Water jet equipment interfacing with other mechanical equipment also involves special engineering and design problems.
6. With some types of excavating machines, the use of water at the face may be detrimental to the muck-handling process or clog up the cutters and muck chutes.
7. The increase in the rate of penetration is not necessarily a measure of the increase in overall rate of advance or of the cost per foot of advance.

For the above and other related reasons the adaptation of high-pressure water jets to excavation processes in the United States appears to be limited to drilling of small-diameter holes in a restricted class of drillable rocks.

# 10

# EXPLOSIVES THEORY AND CALCULATIONS

In the fragmentation of earth materials by chemical high explosives, it is advantageous for the excavation engineer to understand the source of energy. High-explosive energy is developed by exothermic chemical reactions that are initiated by a shock wave in the explosive.

An understanding of the detonation of high explosives requires a knowledge of both the shock waves and the thermochemistry involved. The first high explosives used were composed primarily of the combined atoms of carbon (C), hydrogen (H), oxygen (O), and nitrogen (N), some with small additions of compounds containing metallic elements. It is this type of explosive with which we are primarily concerned here.

Many modern explosives may contain fairly large amounts of sodium, aluminum, or other elements. These types of explosives must be analyzed by special methods.

**334**

This chapter is primarily for engineers who have a knowledge of elementary physical chemistry and physics, to serve as a basis for calculation of the properties of near oxygen-balanced explosives with only minor amounts of metallic elements in them.

## 10.1  HISTORY OF EXPLOSIVES

The history of industrial explosives began with the use of black powder for blasting in the 1600s. Black powder could not be used with safety until the invention of safety fuse in 1831. It is a low-power explosive because it does not *detonate* but rather *deflagrates*. A portion of the products of reaction is in the form of solids, which can do no useful work in blasting, and the energy per unit time produced is small relative to that generated by high explosives.

An explosive is usually composed of a *fuel* and an *oxidizer*, the reaction of which gives off a large amount of energy in a rapid manner. In the case of black powder, the sodium (or potassium) nitrate furnishes oxygen for the reaction, and this oxygen then combines with the sulfur and carbon. This same principle holds for dynamites and other types of high explosives. Single compound explosives are substances such as nitroglycerin, TNT, PETN, and others. Some of these materials may be unstable when heat or impact energy is applied, and when they break down and recombine chemically the reaction is violently exothermic.

Nitroglycerin (NG) and nitrocellulose (NC) were first discovered by Sobrero in Italy in the 1840s, but NG could not be used for blasting because of its sensitivity to shock and heat. Alfred Nobel "discovered" one basic principle of dynamite compositions when some nitroglycerin accidentally spilled into the kieselguhr (powdered $SiO_2$) in which containers were packed. Kieselguhr will absorb about three times its weight of nitroglycerin, and the resulting mixture can be packed in cartridges that are safe to handle. This mixture was the first dynamite, but, because kieselguhr is inert and cannot react with the explosive ingredients, it limits the amount of energy given off per unit weight of explosive. The kieselguhr was replaced with active absorbents such as wood pulp and sodium nitrate to give an *oxygen-balanced* mixture (see the following discussion). Wood pulp will absorb a maximum of 60% NG, and, hence, this is the maximum strength of *straight dynamite*. Various grades of straight dynamite are manufactured, with part of the NG being replaced by other substances such as ammonium nitrate (AN).

Nobel later discovered that when NG is combined with nitrocellulose (NC) in a 92/8 ratio, a stiff gel is formed that is called blasting gelatin (BG) and is one of the strongest commercial explosives. It also can be mixed with other active agents such as AN to give various strengths of gelatin dynamites. Many other types of dynamite were designed, including semigelatins that contain nitrocotton as a gelling agent.

During most of the period of development and increased use of dynamite, blasting caps containing mercury fulminate were employed for initiation. Fulminate caps were later replaced by safer, less sensitive caps containing both lead azide (a primary explosive) and a secondary explosive, such as PETN.

Both the fulminate blasting caps and the NG-based dynamites were sensitive to initiation by shock and heat, and frozen-thawed NG dynamites were often highly sensitive. Because of the greater safety in using AN plus its low cost, it was employed whenever possible as an oxidizer in dynamites. However, AN is very hygroscopic, and it could not be used under wet blasting conditions. One of the more effective methods of waterproofing AN explosives was to package them in cans—that is, AN (oxidizer) plus a fuel such as wax or oil; they were sold only in large diameters, 4–6 in.

Although it had long been known that crystalline AN with some type of fuel could be detonated, this would only occur when the mixture was packaged in a large diameter—that is, larger than its *critical diameter* of 4–6 in.

The successful development and use of ANFO (a mixture of ammonium nitrate

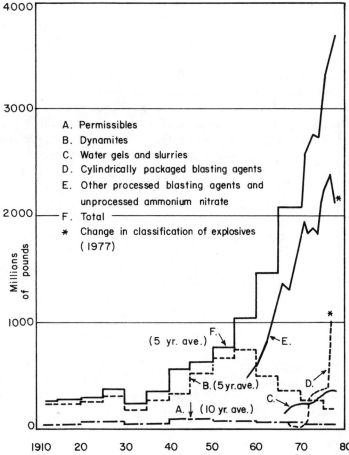

**Figure 10.1**  Industrial explosives and blasting agents sold for consumption in the United States (Clark, 1959).

prills and fuel oil) came about by coincidence. Before 1945, most of the manufacturers of fertilizer-grade AN manufactured it in prilling towers in which the molten AN was sprayed down in a cooling tower, with almost all of the contained moisture removed before prilling. Beginning in 1945, so as to reduce manufacturing costs and to improve the free running quality of the prills, usually −8 + 20 mesh, the moisture content before prilling was increased. The small amount of moisture left after prilling was evaporated following the prill formation, and, as a result, porous prills were formed. This porosity permitted the fuel oil to become more intimately mixed with the AN and exposed more of its surface area to chemical reaction, in turn reducing the critical diameter at which the ANFO would sustain a detonation and increasing its sensitivity to initiation by an explosive booster. If the particle size of AN is reduced sufficiently, mixtures of AN plus fuel oil may become cap-sensitive or require only small boosters to initiate them.

Because of the great affinity of AN for water (hygroscopicity), efforts were made to develop AN explosives that could be used in wet drill holes. This was one of the factors that led to development of the explosive slurries. A slurry usually consists of a saturated aqueous solution of AN used to disperse the other constituents, including solid oxidizers and sensitizing fuels, the latter being combustible materials such as TNT, smokeless powder, hydrocarbons, or heat-producing particulate metals such as aluminum. Because slurries contain water, they are among the safest explosives to handle, being insensitive to shock, bullet impact, and friction. They are made resistant to water infiltration by means of guar gum, which prevents flow of water in and out of the explosive. Most water-based explosives are used in large-diameter holes and require a powerful booster to initiate them. However, they can be sensitized by the addition of certain agents that make them cap-sensitive to a number 6 blasting cap, and they will propagate a detonation in diameters as small as 1 in.

ANFO is classed as a "blasting" agent, as are those slurries that contain a fuel that is not cap-sensitive. Some of these can be mixed at the site where they are to be used and can be placed in the hole in bulk, rather than being packaged. These factors, plus safety in handling and the relatively low cost of their ingredients, have resulted in ANFO-type explosives replacing dynamites in many blasting operations (Figure 10.1).

## 10.2 CLASSIFICATION OF EXPLOSIVES

Explosives may be classified according to their explosion characteristics or by their field of use. Considering the *usage* classification first, one designates explosives as those that are employed for (1) commercial or (2) military purposes. Dynamites, ammonium nitrate–fuel oil (ANFO) explosives, and water-based explosives are examples of the first type, and TNT, PETN, and RDX are examples of the latter. Explosives such as ANFO and slurries, which require a large booster—that is, they are not cap-sensitive—are further classed as "blasting agents" as indicated above.

They may also be classified as high explosives (HE) or low explosives (LE). A high explosive *detonates* when it is properly initiated, and a low explosive *deflagrates*. The velocity and rate of reaction of the former are measured in microseconds, and those of the latter in milliseconds. Corresponding pressures developed by HE are $3.4 \times 10^3$ to $2.7 \times 10^5$ atm ($5 \times 10^4$ to $4 \times 10^6$ psi), whereas LE develop only up to $3.4 \times 10^3$ atm ($5 \times 10^4$ psi). The commercial and military explosives listed above detonate, and hence are classed as HE. Black powder and some propellents are examples of LE.

Most military explosives are not used for blasting, because they are usually more costly, give off noxious fumes, and their strength cannot be easily varied for different rock conditions.

High explosives may further be classified as *primary* or *secondary*. Primary high explosives can usually be detonated in small amounts by thermal ignition, spark, or impact, whereas secondary explosives require a detonator or a booster that usually contains a small amount of primary explosive as the activating element.

Nuclear explosives are in a separate classification because the release of energy is by fusion or fission. They have had little commercial application because of the amount of energy released by optimum size charges, radiation produced, and related factors.

The principal chemical characteristic of the oxidizers in high explosives is that they contain oxygen which may be freed from its original molecule and thus be made available to recombine with other molecules to form more stable compounds. Some compounds carry their own oxygen (NG, for example). An exception to this is lead azide $Pb(N_3)_2$ which contains no oxygen but is very sensitive and will detonate with a high, rapid release of energy. A typical explosive reaction is usually one in which the original molecules of the explosive are broken down to form gaseous molecules, such as $CO$, $CO_2$, $H_2O$, $NO$, $CH_4$, $H_2$, and solids, such as C, CaO, and $Al_2O_3$. When these reactions take place, there is also a rapid evolution of a large amount of heat of the order of a kilocalory per gram of explosive material. The rate of the reaction is favored by such physical characteristics as small grain size, the intimacy of mixture, and other factors.

As a practical consideration, a *high-explosive* material must be stable enough to be handled and transported under ordinary conditions and, at the same time, must be susceptible to initiation by some external energy source that is practical to apply in blasting operations. An explosion is initiated by the application of a sufficient amount of energy in a proper manner, and the subsequent course of events depends on a large number of factors, both physical and chemical. If the reaction is not inhibited by some chemical or physical factor, its final rate of reaction and velocity will ultimately be limited by laws of propagation of shock wave motion, and a characteristic steep-fronted detonation wave will be formed. Lower rates can be obtained by a suitable decrease of the concentration of the reacting molecules or by increasing the grain size of the components. The shattering effect of a high-explosive process, which is sometimes called "brisance," would be destructive rather than useful for a propellant but is very useful where large quantities of energy must be made available in microseconds so as to generate the high-order stress waves required in hard rock blasting.

## 10.3 DEFLAGRATION AND DETONATION

Slower exothermic chemical processes than detonations may be described as burning or deflagration. A detonation, however, represents the shock wave propagation for the maximum rate at which explosive ingredients can react, and the velocity of detonation and the final state of the products are determined by the composition of the explosive material, the initial density, and the detonation temperature. The initial pressure for *solid explosives* before the explosion takes place is usually negligible (atmospheric), but it may be taken into consideration in computing detonation velocities for *explosive gaseous mixtures*. Maximum or ideal detonation velocities are not always attained for a number of reasons: lack of confinement by surroundings, which permits lateral expansion of the explosion products before the reaction is complete and causes a falling off of pressure behind the detonation front; insufficient supply or duration of initiating energy during the period of initiation; limiting effects of the physical state of the explosive, such as homogeneity, composition, density, grain size, and the like.

In the more sensitive explosives, a detonation wave is initiated almost instantaneously by the shock wave generated by the detonation of a small amount of explosive in a blasting cap. In others, such as ANFO and most slurries, a large booster is required to initiate a detonation, and the resulting initial velocity of the detonation wave in the ANFO may depend on the ''strength'' of the booster. In certain other types of explosive, the initiator may induce a high-order deflagration in the explosive, which rapidly develops into a stable detonation. If an explosive mixture is not capable of sustaining a stable detonation, a strong booster may induce an unstable detonation in it, which deteriorates into a deflagration and then dies out entirely.

For chemical explosives, the explosion may be either high- or low-order, depending on the rate of release of energy and other factors involved in the explosion. Both detonations and deflagrations are rapid exothermic chemical reactions.

A distinguishing characteristic of a useful high explosive is that it detonates when an explosion is initiated in it. *Detonation* is the process of propagation of a shock wave through an explosive, which is accompanied by a chemical reaction that furnishes energy to maintain the shock wave propagation in a stable manner. Some typical high explosives are those used for military purposes, such as TNT, PETN, RDX, and those used commercially, such as dynamites, blasting gelatins, water-based explosives, and ammonium nitrate blasting agents. They may be either single compounds or mixtures, usually containing combinations of the elements hydrogen, nitrogen, oxygen, and carbon. As noted earlier, many military explosives are seldom used as such in underground blasting because their composition cannot be varied and they yield noxious gases on detonation. Commercial explosives are usually mixtures whose strength may be varied by changing their composition, and they are designed so that they have a zero oxygen balance (see below).

Low explosives, on the other hand, do not exhibit a shock wave in the explosive but burn with an extremely rapid rate that is characterized by the term *deflagration*. As discussed, a typical low explosive is black powder.

Some of the properties of high explosives that are important in blasting are density, velocity of detonation, temperature and pressure of the ''detonation state,'' temperature and pressure of the ''explosion state,'' heat of explosion, and available energy. The density of the explosive is determined largely by its composition, grain size, and the density of the constituents. The velocity of detonation is the only property that can be readily measured. Other properties may be calculated from thermochemical and hydrodynamic relationships involving liberated heat, temperature, and shock wave parameters.

It is notable that certain gaseous mixtures of fuels and oxidizers will detonate, as will liquid and solid or condensed explosives. Gaseous detonations served as a means of studying the detonation phenomena and as a basis for deducing the hydrodynamic theory.

Nobel's discoveries of the safe means of handling and initiating nitroglycerin explosives revolutionized blasting practices. Whereas black powder had to be carefully stemmed and confined to produce effective blasting because it deflagrated slowly, the new dynamites detonated with very high velocities, even when unconfined, and were much more effective in breaking rock.

Taylor (1952) states that one fundamental difference between detonation and deflagration is in the physical–chemical processes taking place, because some explosives will both deflagrate and detonate. It has been found, for example, that if the mass of a given explosive is small, *thermal ignition* usually results in burning. However, if the mass exceeds a certain critical value, it is possible for burning to become so rapid that it results in the generation of a shock wave and detonation takes place. The critical mass is characteristic of the given explosive. For lead azide, the critical mass is too small to measure, but that for trinitrotoluene (TNT) is about 1 ton, and that of ammonium nitrate is even larger. On the other hand,

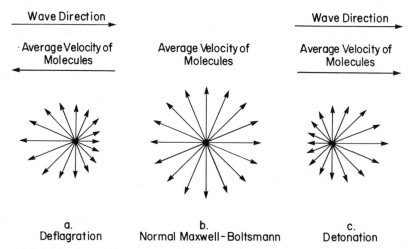

**Figure 10.2** Molecular velocity and direction distribution for deflagration, still air, and detonation (after Cook, 1954) (Clark, 1959).

small charges of TNT can be made to detonate by initiating a high-order explosion with a blasting cap containing a primary explosive such as lead azide.

The important ability of such primary explosives as lead azide and mercury fulminate to detonate in small quantities by means of thermal ignition makes them especially valuable for initiation of detonations in explosives which themselves must be stable enough to handle safely in bulk.

As stated previously, the primary characteristic of a detonation is that the mechanism of propagation is by means of a shock wave that is accompanied by an exothermic chemical reaction, the energy from which supports the propagation of the shock wave. The mechanism of the movement of the materials in front of and behind the shock front is different for a deflagration and a detonation (Figure 10.2). In a deflagration, the hot molecules and atoms are moving away from the burning zone. By comparison, the average movement of molecules in air is the same in all directions; that is, the pressure is the same in all directions. In a detonation, however, the average movement of the flow of high-temperature particles is into the wave front.

## 10.4 COMMERCIAL EXPLOSIVES

The oxygen for combustion or explosion is supplied by substances other than air called "oxidizers." Typical materials widely used for this purpose are sodium nitrate and, more commonly, ammonium nitrate. These are mixed with fuels and other substances to vary their detonation characteristics.

*Black powder* is an explosive that was invented before the 1200s in China, was brought to Europe in the 1300s, and was subsequently used for blasting from about 1500 until the 1860s. It is still used today as an igniter for fuses. It is composed of a grained mixture of finely ground sodium nitrate or potassium nitrate, sulfur, and carbon. The equation for the reaction of the constituents, when it is oxygen balanced, is (Cook 1974):

a. $2NaNO_3 + S + C = Na_2SO_4 + CO_2 \uparrow 195$ kcal
$Q = 920$ kcal/kg   No. of moles of gas $= n = 4.7$ mole/kg

Other reactions are possible, but they do not produce as much heat:

b. $2NaNO_3 + S + \frac{3}{2}C = Na_2SO_3 + \frac{3}{2}CO_2$
$Q = 620$ kcal/kg   $n = 6.9$

c. $2NaNO_3 + S + 3C = NaS_2 + 3CO_2$
$Q = 620$ kcal/kg   $n = 12.6$

d. $2NaNO_3 + S + 3C = Na_2SO_4 + 2CO$
$Q = 680$   $n = 8.9$

Although the energy of the reaction is somewhat dependent upon the oxygen balance, the *strength* is likewise dependent upon both the energy and the number of moles of gas produced. Hence, Cook (1974) points out that the optimum mixture is somewhere between b and d.

Other important properties of black powder are that (1) it will not detonate, (2) because of its high explosion temperature and solid products of reaction, it is an excellent ignition agent for military uses, and (3) because of its low explosion pressure, it has been used for breaking lump coal and large blocks of dimension stone.

If cellulose (cotton), starch, or glycerin is treated with nitric acid, the resulting product is called *nitrocellulose* (guncotton or nitrocotton), nitrostarch, and nitro-glycerin, respectively.

*Nitroglycerin* is produced by mixing below 22°C (72°F) sulfuric acid (to absorb water) and nitric acid in a steel tank and then adding glycerin. The nitroglycerin is washed several times with cold water and once with caustic soda to remove any remaining trace of acid. Nitroglycerin is insoluble in water and is poisonous either when in contact with the skin or when it is breathed as a vapor. It usually produces a violent headache. Nitroglycerin freezes or becomes crystalline at from 4 to 8°C (40 to 46°F). It is shock-sensitive and especially dangerous to handle. It explodes when heated to about 200°C (392°F), and it can also be exploded by a violent blow or shock. Its properties made it especially suitable for ''shooting'' oil wells in the past.

The chemical formula for NG is $C_2H_5(NO_3)_3$, and it has a natural density of 1.60 g/cc. It is oxygen-positive, and the equation for its reaction in the detonation process may be written, assuming that the reaction goes to completion, as shown:

$$C_3H_5O_9N_3 = 3CO_2 + \tfrac{1}{2}H_2O + \tfrac{1}{4}O_2 + 1\tfrac{1}{2}N_2$$

The heat of reaction for the above is 1503 cal/g referred to a base temperature of 298°K. The comparable heat of reaction for the detonation of ammonium nitrate is only 378 cal/g, which indicates one of the primary reasons why NG is such a powerful explosive. It has a detonation velocity of approximately 7926 m/s (26,000 ft/s), and the pressure developed in the detonation wave is almost 250,000 atm.

In the early 1950s, it was found that commercial-grade fertilizer, ammonium nitrate of 8/20 mesh particle size, could be successfully detonated in +76.2 mm (+3 in.) diameter holes when it was intimately mixed either with the proper amount of finely divided carbon or with No. 2 fuel oil. Enough carbon or fuel oil is added to give an oxygen-balanced explosive:

$$NH_4NO_3 + \tfrac{1}{2}C = N_2 + 2H_2O + \tfrac{1}{2}CO_2 + 929.0 \text{ cal/g}$$

$$NH_4NO_3 + \tfrac{1}{3}CH_2 = N_2 + \tfrac{7}{3}H_2O + \tfrac{1}{3}CO_2 + 927.0 \text{ cal/g}$$

The heats of explosion are approximately the same for mixtures of fuel oil and carbon, whereas that for pure AN is only 343.1 cal/g versus 927 cal/g for either mixture.

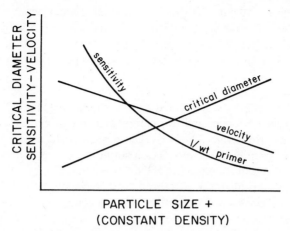

**Figure 10.3**  Change of sensitivity, velocity, and critical diameter with particle size (Clark, 1959).

As described earlier, fertilizer-grade AN is made up of prills from which the last 5% of water is evaporated after the prills are solidified. This produces a porous prill which has much more surface area per unit weight than crystalline AN or solid prills. It is the porosity of prills that makes them detonable in charges of as small a diameter as 50.8 mm (2 in.). That is, the surface area of the porous prills exposed to chemical reaction in the detonation wave increases the rate of mass reactivity sufficiently to sustain a detonation, whereas pure crystalline AN requires a much larger diameter charge to propagate a detonation, depending on the particle size.

The factors that determine the sensitivity of such "blasting agents" to initiation and the propagation velocity are "effective" particle size, charge diameter, density, and homogeneity of mixing. These factors are shown in Figures 10.3 and 10.4. Thus, if a charge is made more dense by tamping, its velocity increases, but its

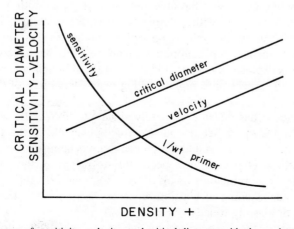

**Figure 10.4**  Change of sensitivity, velocity, and critical diameter with charge density (Clark, 1959).

sensitivity decreases. If particle size is decreased, both velocity and sensitivity increase, and the critical diameter decreases. Producers of AN–fuel oil blasting agents for large hole blasting may adjust the particle size so that the blasting agent is not sensitive to a No. 6 blasting cap to keep it classified as a blasting agent rather than a high explosive.

A mixture of ammonium nitrate and TNT was used in World War I as a military explosive (amatol), but it is not adapted to commercial blasting. TNT has a markedly negative oxygen balance, and free carbon and carbon monoxide are produced when it detonates. However, AN is oxygen-positive, and a mixture of the two which gives an oxygen balance, produces a strong explosive. Amatol was usually cast with grains of AN in a matrix of TNT for maximum density. A simple mixture of granular TNT and prilled AN contains a high percentage of void space, and the maximum pouring density is about 0.95 g/cc. If enough water is added to fill the voids, a slurry of density of 1.40 g/cc may be obtained. Even though part of the heat given off by the explosive reaction of such slurries is absorbed to change the water to steam and heat it to the resulting detonation temperature, the thermohydrodynamics and chemical equilibria involved are such that the increase in density markedly increases the detonation velocity and the amount of effective work of the explosive.

## 10.5  NEAR OXYGEN-BALANCED EXPLOSIVES

The manufacture and use of commercial explosives has changed drastically in the past two decades owing largely to the discovery that commercial-grade ammonium nitrate prills will detonate when mixed with fuel oil. As noted earlier, this development came about because of the method of manufacture of prills, which leaves them porous so that fuel oil can easily be absorbed, thus creating a more homogeneous mixture and exposing greater surface area of the prills to chemical reaction. Another contributing factor was the invention of slurries, permitting the design of a wide variety of explosives with different sensitivities, velocities, strengths, and other variable properties, as well as providing explosives that are very safe to handle and have excellent fume properties. Thus, ANFO has taken over a large portion of the bulk explosives market, and the slurries, although not quite as flexible as dynamites in all respects, are rapidly taking over a large part of the market for dynamite.

Explosive science has developed rapidly since World War II, when the thermohydrodynamic theory of detonation was made available in the literature. Although the theory of detonation and the related thermochemistry have been widely known by scientists, engineers have been slow to recognize the usefulness of a knowledge of the basic principles involved and their vital importance in furnishing a base for the understanding of blasting processes which is essential for the most efficient design of blasting rounds.

The hydrodynamic theory of detonation, which was initially developed in studies of the detonation of gaseous mixtures, deals with the application of plane shock wave phenomena and associated physics, chemistry, and mathematics to the process of propagation of a detonation in an explosive. The physics of plane shock waves deals with the application of the laws of conservation of mass, momentum, and energy of the material in front of and behind a shock front; that is, it relies on the inertial response of matter to very rapid acceleration, rather than the laws of motion used in conventional wave analysis.

Shock waves are of two general types: *nonreactive*, in which the wave receives its energy at the source of the shock and is propagated without any energy being added as the wave travels its course; and *reactive*, in which the front of the shock wave also is the front of a chemical reaction zone from which reaction energy is constantly added to the material in the wave front. For a reactive shock (detonation), the chemical energy released in the shock wave must be considered, as well as the equations that govern the behavior of the gaseous products, the thermodynamic and thermochemical equations, and related material balance equations. Although for some explosives the applicable equations and required calculations may be complex, for most (oxygen-balanced) explosives of interest, the calculations are quite tractable.

The use of the basic equations for the detonation process consists of combining the parameters of plane shock waves with those in an appropriate equation of state for gases at high pressure and temperatures, together with thermodynamic parameters, equilibrium constants, and material balance requirements. The wave velocity, temperature, pressure, density, and heat evolved in the shock wave define the "detonation state." After the detonation wave has passed through a contained explosive and the products of the explosion occupy the volume originally occupied by the explosive, the temperature, pressure, density (the latter being the same as that of the explosive), and heat are the parameters that define the "explosion state." The explosion state pressure and heat determine the maximum available energy that the explosive yields to perform work on the rock surrounding a borehole.

## 10.5.1 Explosive Property Calculations

Properties of the explosive that affect its detonation behavior, or its *detonation state parameters*, are composition, grain size, rate of reaction, grain density, loading (bulk) density, heat of formation, and the thermochemical properties of the products of the detonation reaction, such as the heat of formation and the heat capacity of gases and solids formed. These properties determine the detonation parameters, such as (1) detonation velocity, (2) pressure, (3) heat of detonation, and (4) detonation temperature. In the explosion state, the products of the explosion are assumed to occupy for a few microseconds the volume previously occupied by the unreacted explosive, and their characteristics under these conditions constitute the *explosion state parameters*, which are (1) borehole pressure, (2) temperature, and (3) available work.

The problem of solving for the performance parameters of high explosives is fairly detailed, the complexity of the solution depending largely upon the composition of the explosive. AN explosives of interest, including fuel oil mixtures, dynamites, some military explosives, and other CHNO explosives, fall within the category of *high explosives* to which the thermohydrodynamic method of analysis is readily applicable. More specifically, the appropriate laws of physics and chemistry and mathematical procedures provide a means of solving for the important *ideal* explosive performance parameters. The calculation of these parameters does not take into account such factors as particle-size effects, charge diameter, reaction rate, and similar properties and processes, but it assumes that they are such that the detonation may be accurately described by the derived equations or that the detonation is "ideal."

The evaluation of explosive parameters (for a given density and composition) for a non-oxygen-balanced explosive requires the solution of a number of nonlinear simultaneous algebraic equations by iterative procedures. This type of problem can be readily programmed and solved on a digital computer. The solution for oxygen-balanced explosives can usually be obtained more easily because the explosion products can be calculated directly and an iteration is required only for a detonation factor, $\beta$. See equation 10.61.

One of the detonation parameters that can be measured directly and accurately is the detonation velocity; the detonation pressure may be measured by indirect experimental methods. It must be kept in mind that any one of the detonation state or the explosion state parameters by itself may not be a good indicator of the "strength" of the explosive. Usually more than one parameter must be considered in relation to the characteristics of the rock being blasted. These may include the explosion state pressure, duration of the pressure pulse, available work, and the amount of gas produced per unit weight of explosive.

## 10.5.2 Physics of Shock Waves

In ordinary sound waves in solids, liquids, and gases, the velocity of the wave is practically independent of the pressure, and the change in pressure at the wave front is small and relatively gradual. On the other hand, a shock wave in explosives is characterized by three properties: (1) a very rapid pressure rise is almost instantaneous at the wave front; (2) the magnitude of pressure may be small or as high as $7.5 \times 10^5$ atm ($10^6$ psi) in the detonation of solid (condensed) explosives; and (3) its velocity is greater than that of a sound wave in the medium. Shock waves may be *nonreactive* or *reactive*.

An example of the first is a sonic boom, a low-pressure shock wave in air whose energy is supplied initially by the kinetic energy of an airplane, but no energy is added to it after it leaves the source. When an explosive mixture of gases, liquids, or solids detonates, energy is continuously supplied by an exothermic chemical reaction that accompanies the wave and whose products of combustion make up the body of the wave immediately behind the wave front.

The phenomenon of detonation was first observed by early investigators (Jost, 1946) in their study of the propagation of flames in tubes containing explosive gaseous mixtures. It was found that these mixtures would propagate flames at velocities from 1000 to 3000 m/s (3280 to 9840 ft/s), which is several times the velocity of sound at ordinary temperatures and pressures. The formation of detonation waves in solid explosives is somewhat analogous to that in gases. The following model was derived for gases, but the concepts can also be used for describing the processes involved in liquid or solid explosives.

If flammable gas mixtures capable of supporting a detonation are placed in a long tube that is open or closed at the ends and the mixture is ignited at one end, the flame rapidly accelerates until it reaches a usually high stable (characteristic) velocity. This velocity is maintained regardless of the length of the tube. If the burned gases flow out of the ignition end of the tube at a high velocity, a back pressure will be built up in the fresh gas at the front of the flame. The compression of the unburned gases produces a shock wave that travels faster than the velocity of sound in the explosive media, and the reaction zone is coupled with the shock wave. The detonation velocity is a characteristic constant for a given mixture of gases—that is, dependent upon their chemical nature, percentage composition, and density. Also, the speed of the shock wave is influenced very little by moderate changes in initial temperature and pressure of the unburned gases.

Early observers also noted another phenomenon connected with the detonation process. As a flame accelerated and reached its detonation velocity, it was found that a similar pressure wave (nonreactive) of high velocity was formed, traveling in the opposite direction. That such a wave should be formed is reasonable because of the conservation of momentum; that is, the forward transmitted impulse must be compensated for by that of a "retonation" wave traveling in the opposite direction. This results in a rarefaction wave that follows well behind and in the wake of the detonation wave, caused by the flow of gases in both directions away from the rear of the reaction zone.

The origin of a detonation wave in a tube has been compared to the processes accompanying the acceleration of a piston. The initial movement of the piston starts a wave traveling down the tube. If the velocity of the piston is increased in increments, successive waves will be formed, each with a greater velocity than the former. The later waves will overtake the ones in front until the piston reaches some high terminal velocity. Thus, a wave with a very high-pressure gradient is created; that is, a shock or detonation wave is produced which moves with a velocity greater than that of sound.

Basic relationships among velocity, temperature, pressure, and specific volume in the shock wave may be derived by means of application of the laws of conservation of mass, momentum, and energy of the material ahead of, and immediately behind, the detonation wave front, together with appropriate thermodynamic laws. The detonation wave is a reactive shock wave; that is, it is characterized by a very sharp rise to a high pressure and supersonic velocity coupled with the chemical reaction of the explosive. A simplified profile for a detonating column of explosive is shown in Figure 10.5.

The explosive represents ideally a column of reactive material which is a section of an infinite mass of explosive in which a plane detonation wave is traveling from left to right; that is, no molecules are escaping laterally from the column, or the number escaping is balanced by the entering molecules.

The method indicated in Figure 10.5b,d uses a coordinate system moving to the

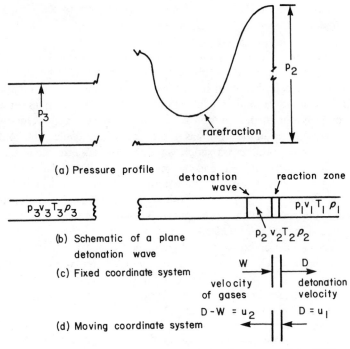

(a) Pressure profile

(b) Schematic of a plane detonation wave

(c) Fixed coordinate system

(d) Moving coordinate system

**Figure 10.5** Detonation system schematic for plane wave (Clark, 1959).

right with a velocity $D$. The mass of material flowing across a unit cross section per unit time must be the same in front of and behind the detonation wave front. Hence,

$$u_1\rho_1 = u_2\rho_2 \quad \text{or} \quad \frac{u_1}{v_1} = \frac{u_2}{v_2} \tag{10.1}$$

where

$$v = \frac{1}{\rho}$$

Also, the material transmits an impulse which is the sum of that due to (1) its mass flow and that due to (2) its hydrostatic pressure, the latter being due to the velocity of the molecules within the flowing mass. Thus, the impulse per unit mass is given by

$$\frac{u_1^2}{v_1} + p_1 = \frac{u_2^2}{v_2} + p_2 \tag{10.2}$$

Likewise, the energy per unit mass of explosive and explosion products is composed of the internal energy $E$, the volume energy $pv$, and the kinetic energy $(u^2)/2$, yielding

$$E_1 + \frac{u_1^2}{2} + p_1 v_1 = E_2 + \frac{u_2^2}{2} + p_2 v_2 \qquad (10.3)$$

From equations 10.1 to 10.3, the following relations are obtained:

$$u_1 = D = v_1 \sqrt{\frac{p_2 - p_1}{v_1 - v_2}} \qquad (10.4)$$

$$u_2 = v_2 \sqrt{\frac{p_2 - p_1}{v_1 - v_2}} \qquad (10.5)$$

$$E_2 - E_1 = \tfrac{1}{2}(p_1 + p_2)(v_1 - v_2) \qquad (10.6)$$

$$W = u_1 - u_2 = \left(v_1 - v_2 \sqrt{\frac{p_2 - p_1}{v_1 - v_2}}\right) \qquad (10.7)$$

where $u_1$ = rate of flow of material into the detonation wave which is numerically equal to the detonation velocity $D$, $u_2$ = rate of flow of material away from the detonation front, and $W$ = particle velocity into the wave front.

The above equations are all for a nonreactive shock wave. However, for a reactive shock or a detonation, the amount of heat $Q_2$ is released in the reaction zone, and equation 10.6 becomes:

$$E_2 - E_1 = \tfrac{1}{2}(p_1 + p_2)(v_1 - v_2) + Q_2 \qquad (10.8)$$

### 10.5.3  Thermochemistry of Explosives

The shock wave or hydrodynamic equations provide relationships that exist in the shock wave in terms of specific volumes, energies, pressures, particle velocity, and wave velocity. The only factors that are "known" are the ambient pressure, composition, heat of formation, and specific volume ($v_1$) or density ($\rho_1$) of the explosive. Energy determination involves a knowledge of the *material balance equation*, the *amount of heat* given off by the chemical reactions (heat of reaction), and the laws of thermodynamics. These, in turn, are related to the temperature and pressure by means of *heat capacities* and an *equation of state* for the gaseous products. When the reactions do not all go to completion, appropriate chemical *equilibria* must be included by means of equilibrium constants. Inasmuch as temperature and pressure in a detonation wave are very high, corrections are made to the equilibrium constants using a fugacity factor and to the equation of state in terms of a co-volume factor to account for deviation from ideal behavior.

The first law of thermodynamics is essentially a statement of the law of conservation of energy. Inasmuch as the chemical reactions involved in calculating the detonation parameters and the available work from explosives are concerned primarily with the energy of the gaseous products of detonation, it is necessary to define a practical concept of *internal energy*. Within the conditions under which chemical explosions take place, internal energy may be considered in terms of *changes* in energy content. These include potential energy and the energy associated with the addition or subtraction of heat.

In an exothermic chemical process such as that occurring in an explosion, the change in internal energy $\Delta E$ as the system changes from state A to B is equal to the sum of the heat added $q$ and the work done on the system $-w$. The convention used is $+w$ for work done by the system. Therefore,

$$\Delta E = E_B - E_A = q - w \tag{10.9}$$

While both $q$ and $w$ may depend on the path the system takes from condition $A$ to condition $B$, their difference is independent of the path.

Equation 10.9 may also be written for a differential change, in the form

$$dE = dq - dw \tag{10.10}$$

For a reversible process,

$$\int_A^B dw = \int_A^B p\,dv \tag{10.11}$$

This differential expression cannot usually be integrated to find the total work, because $p$ is dependent upon the temperature and the path along which the work is done. However, $\int_A^B dE$ can always be performed, because the internal energy of a system is a function only of its initial and final states. Mathematically, then, the first law of thermodynamics states that $dE$ is an exact differential and $dq$ and $dw$ are inexact.

Most of the processes that take place in an explosion are considered to be *adiabatic* in character; that is, heat is neither taken from nor added to the system during the detonation or the transition to the explosion state. That is, $\Delta q = 0$. This is the condition assumed for a detonation wave where there is no lateral loss of heat from the reaction zone.

The *heat capacity* of a substance is the amount of heat required to raise its temperature a unit amount; *specific heat* is the amount per g, and molar heat capacity the amount per mole. In terms of differentials, where the heat capacity is not constant over a given range of temperature, the heat capacity may be defined by

$$C = \frac{dq}{dT} \tag{10.12}$$

Heat capacities may be defined either at constant volume or constant pressure:

$$C_v = \frac{dq_v}{dT} = \left(\frac{\partial E}{\partial T}\right)_v \tag{10.13}$$

$$C_p = \frac{dq_p}{dT} = \left(\frac{\partial H}{\partial T}\right)_p \tag{10.14}$$

where $H$ is the enthalpy and is defined by

$$H = E + pv \tag{10.15}$$

It can be shown that the relation between the two types of heat capacities is

$$C_p - C_v = \left[p - \left(\frac{\partial E}{\partial v}\right)_T\right]\left(\frac{\partial v}{\partial T}\right)_p \tag{10.16}$$

The term $p((\partial v)/(\partial T))_p$ represents the contribution to the specific heat $C_p$ caused by expansion of the system against an external pressure $p$. The term $((\partial E)/(\partial v))_T \cdot ((\partial v)/(\partial T))_p$ is the work done in expansion against the internal cohesive or repulsive forces of the substance, represented by a change in energy with volume at constant temperature. The term $((\partial E)/(\partial v))_T$ is known as the internal pressure, which in the case of gases is small compared with the pressure $p$. Hence, in most computations for explosives, it is considered to be zero.

Evaluation of the *heat of reaction* of the exothermic reactions that take place in chemical explosions deals with the conversion of chemical energy into heat energy, which in turn heats the products of the explosion, and the heat energy of the gaseous products is used as mechanical energy in blasting. Analysis is based on the first law of thermodynamics which states essentially that *the expenditure of a certain amount of work always produces the same amount of heat.*

When applied to an isolated chemical system, such as an exploding stick of dynamite enclosed in rock, the law requires that the *total* energy of the system must remain constant, although the energy may change from one form to another. The process of reaction of liquids and gases in the laboratory are usually carried out at *constant pressure*, whereas reactions of explosives are assumed to be carried out at *constant volume*—that is, the volume of the borehole. The heat absorbed or released at constant pressure by a chemical process is denoted as $Q_p$, and this is equal to the difference between the total heat content of the products of the reaction and that of the reactants at constant pressure or volume. That is, Heat of reaction = Heat content of products − Heat content of reactants, or

$$Q_p = \Delta H_p = \Sigma H_{(\text{products})} - \Sigma H_{(\text{reactants})} \tag{10.17}$$

Tabulated values for the heat content $H$ of a substance are all based on a specified reference temperature, usually 25°C (298°K) and a pressure of 1 atm. The summation signs below designate the sum of the heat contents of all of the molecular species in the products of reaction and the reactants, respectively. The assumption is made that all of the reactants are converted to products—that is, that the reaction goes to completion. In some coarse-grained explosives, the reaction is not completed within a critical time, and the effective heat of reaction and the detonation parameters are reduced accordingly.

Two important applicable thermochemical laws are that (1) the quantity of heat required to decompose a chemical compound into its elements is equal to the heat evolved when the compound is formed from its elements, and (2) the heat change of a reaction, at constant pressure or constant volume, is the same whether it takes place in one or several stages. Thus, according to the first law, thermochemical equations can be reversed if the sign of $\Delta H$ is changed, that is,

$$CH_4(g) + 2O_2(g) = CO_2(g) + 2H_2O(\ell) \quad \Delta H = +212.80 \text{ kcal}$$

and                                                                                                  (10.18)

$$CO_2(g) + 2H_2O(\ell) = CH_4(g) + 2O_2(g) \quad \Delta H = -212.80 \text{ kcal}$$

The second law of thermodynamics states that the amount of heat given off or added depends only on the initial and final states of the chemical system. This law also provides, therefore, that thermochemical equations can be added and subtracted in an algebraic manner. The heat contents of the elements and compounds are definite quantities, and the addition or removal of a substance from the system increases or decreases the heat content accordingly. Thus, heat changes can be accurately calculated without being determined experimentally. This can be shown algebraically, given the following reactions occurring at 25°C:

a. $C_2H_4(g) + 3O_2(g) + 2CO_2(g) + 2H_2O(\ell) \quad \Delta H = +337.3 \text{ kcal}$

b. $H_2(g) + \frac{1}{2}O_2(g) = H_2O(\ell)$ $\qquad\qquad\qquad \Delta H = + 68.3 \text{ kcal}$   (10.19)

c. $C_2H_6 + 3\frac{1}{2}O_2(g) = 2CO_2(g) + 3H_2O(\ell) \quad \Delta H = +372.8 \text{ kcal}$

To determine the heat of reaction of

$$C_2H_4(g) + H_2(g) = C_2H_6(g) \text{ at } 25°C \tag{10.20}$$

add $a$ and $b$ and subtract $c$ to give

$$C_2H_4(g) + H_2(g) = C_2H_6(g) \quad \Delta H = +32.8 \text{ kcal} \tag{10.21}$$

The *heat of formation* of a compound is the change in heat content when 1 mole is formed from its elements at a specified temperature and pressure, usually 25°C and 1 atm. As indicated by the physical state, gaseous, liquid, or solid, the heat content also depends upon physical conditions. For example, the heat of formation of water vapor at 25°C is less than that of liquid water at the same temperature, the difference being the heat of vaporization of water at 25°C. The *standard state* is the stable form of the substance at 1 atm and 25°C (Table 10.1). The heat of formation of elements is taken as zero.

In Table 10.1 and the preceding equations, a minus ($-$) sign denotes that heat is absorbed and a plus ($+$) sign denotes that heat is given off in the indicated reaction (this is the opposite of nomenclature often used in texts on physical chemistry). Usually, if the heat of formation is negative, the compound is unstable at standard conditions. Also, the values of the heat of formation of a compound may indicate the affinity of the constituent elements for each other. In the determination of the affinity of oxygen, which is the critical element in most explosives, the "pecking" order in which the metallic elements absorb oxygen is approximated by the value of the heat of formation of the compound formed.

The heat of reaction of methane is calculated as follows:

$$CH_4(g) + 2\ O_2(g) = CO_2(g) + 2H_2O(\ell)$$

$$+\ 17.89 \qquad 0 \qquad 94.05 + 2 \times 68.32 \qquad\qquad (10.22)$$

$$\Delta H = (94.05 + 2 \times 68.32) - (17.89 + 0) = +212.08 \text{ kcal}$$

## TABLE 10.1. Standard Heats of Formation (kcal/mole)

### *Inorganic Compounds*

| Substance | ΔH | Substance | ΔH |
|---|---|---|---|
| $H_2O(\ell)$ (g) | 68.32, 57.80 | $Na_2O$(s) | 99.4 |
| $H_2S$(g) | 4.8 | $K_2O$(s) | 86.4 |
| $NH_3$(g) | 11.04 | NaOH(s) | 101.99 |
| $SO_2$(g) | 70.96 | KOH(s) | 101.78 |
| $SO_3$(g) | 94.45 | $Al_2O_3$(s) | 384.84 |
| CO(g) | 26.42 | NaC(s) | 98.23 |
| $CO_2$(g) | 94.05 | KC(s) | 104.18 |
| $N_2O$(g) | $-19.49$ | $Na_2SO_4$(s) | 330.9 |
| NO(g) | $-21.6$ | $K_2SO_4$(s) | 342.66 |
| $NO_2$(s) | $-\ 8.0$ | | |

### *Organic Compounds*

| | | | |
|---|---|---|---|
| Methane, (g) $CH_4$ | 17.89 | Methanol, (g) $CH_3OH$ | 48.1 |
| Ethane, (g) $C_2H_6$ | 20.24 | Methanol, ($\ell$) $CH_3OH$ | 57.0 |
| Propane, (g) $C_3H_8$ | 24.82 | Ethanol, (g) $C_2H_5OH$ | 56.3 |
| Ethylene, (g) $C_2H_4$ | $-12.56$ | Ethanol, ($\ell$) $C_2H_5OH$ | 66.4 |
| Propylene, (g) $C_3H_6$ | $-\ 4.96$ | Phenol, (s) $C_6H_5OH$ | 38.4 |
| Acetylene, (g) $C_2H_2$ | $-54.23$ | | |

Notes: $\ell$ = liquid
      g = gas
      s = solid

*Source:* Glasstone (1946).

The above calculation can be reversed; that is, if the heat of reaction is known and the heats of formation of all but one substance are known, the unknown heat of formation can be calculated. The heat of reaction of methane and oxygen is also known as the *heat of combustion* of these two substances. The amount of heat required to change a substance from solid to liquid or from liquid to vapor (change of state) is also of interest in the study of chemical reactions in explosives, particularly for water, which is one of the major products of both combustion and most chemical explosive reactions. It is also an important ingredient in water-based explosives such as slurries. That is, the standard heat of formation of water in the liquid state at 25°C is

$$H_2(g) + \tfrac{1}{2} O_2 = H_2O(\ell) \quad \Delta H = +68.32 \text{ kcal} \qquad (10.23)$$

The heat of vaporization of water is given by

$$H_2O(\ell) = H_2O(g) \quad \Delta H = -10.52 \qquad (10.24)$$

and, hence,

$$H_2(g) + \tfrac{1}{2} O_2(g) = H_2O(g) \quad \Delta H = +57.80 \text{ kcal} \qquad (10.25)$$

This value, +57.80, is used in explosives calculations where water is produced as a reaction product in the form of vapor. However, where water exists in the explosive, as is the case for slurries, the heat of vaporization must be used in the calculation of the heat of formation of the explosive.

In the following paragraphs, the flame temperature is calculated for methane, a hydrocarbon gas that occurs in bituminous coal mines. Its high explosion temperature when mixed with air illustrates one of the reasons why it will initiate coal dust explosions, and the calculation of the flame temperature illustrates the chemical thermodynamic principles involved in such high temperature reactions.

Some of the processes of methane combustion are similar to those involved in the detonation of high explosives. The heat produced is considered to be used to heat the products of reaction, and the heat capacities of gases at high temperatures must be known. In detonations, high pressures are also involved, and heat capacities at constant volume $C_v$ are used. Other parameters also must be considered and are described later.

Most heats of reaction for purposes of standardization are given under the assumption that the reaction takes place at 25°C and that the heat liberated (if the reaction is exothermic) is removed by the surroundings as it is produced. Thus, the chemical equation for the combustion of methane for an oxygen-balanced mixture is

$$CH_4(g) + 2O_2(g) = CO_2(g) + 2H_2O(\ell) + 212,800 \text{ cal/mol} \quad (10.26)$$

Equation 10.26 shows that the water is assumed to be formed in the liquid state and that the products of the reaction are at the same temperature as the reactants. Under these circumstances, 212,800 calories or 212.8 kcal are given off when the combustion of 1 mol of methane is carried to completion. If it is assumed that such a reaction takes place under an adiabatic condition (no heat loss to surroundings) and that the reaction takes place as a constant pressure process, then the products of the reaction will be heated to a much higher temperature. That is, all of the heat of reaction is used in heating the products of the reaction. It is then possible to calculate the final temperature of such a chemical system, and this final temperature is known as the *maximum flame* or *explosion temperature*.

One of the simplest procedures is to assume that the reaction takes place at 25°C—that is, that this is the initial temperature of reactants—and then determine to what temperature the products can be raised by the heat of reaction. For the combustion of methane in air, a ratio of 1 $O_2$ to $4N_2$ is taken as the composition of the air. Since the water will be in the form of vapor in the actual combustion, the heat of reaction yielding vapor instead of liquid $H_2O$ is required:

$$CH_4(g) + 2O_2(g) = CO(g) + 2H_2O(\ell) + 212.8 \text{ kcal} \tag{10.27}$$
$$2H_2O(\ell) = 2H_2O(g) - 2 \times 10.52 \text{ kcal}$$

adding, and including the $N_2$ in air,

$$8N_2(g) + CH_4(g) + 2O_2(g) = CO_2(g) + 2H_2O(g) + 8N_2(g) \tag{10.28}$$
$$+ 191.76 \text{ kcal}$$

That is, the nitrogen does not take part in the reaction, and the heat content of liquid water is 10.52 kcal/mol less than the heat content of the same amount of vapor. The calculated quantity of heat is then taken as the amount that will heat the products $1CO_2$, $2H_2O$, and $8N_2$ to the flame temperature. If $T_2$ is this temperature and if the molar heat capacities of the products are constants, then $T_2$ could be calculated from the following equation:

$$191,760 = \Sigma\, nC_p(T_2 - 298°K) \tag{10.29}$$

However, the heat capacities of each of the combustion products are functions of temperature as shown below; consequently, the temperature $T_2$ must be determined by means of integration:

$$191,760 = \int_{298}^{T_2} \Sigma\, nC_p\, dt \tag{10.30}$$

In this case,

$$\Sigma\, nC_p = C_p(CO_2) + 2C_p(H_2O, g) + 8C_p(N_2) \tag{10.31}$$

where

$$C_p(CO_2) = 6.396 + 10.100 \times 10^{-3}T - 3.405 \times 10^{-6}T^2$$

$$C_p(H_2O, g) = 7.219 + 2.374 \times 10^{-3}T + 0.267 \times 10^{-6}T^2 \qquad (10.32)$$

$$C_p(N_2) = 6.449 + 1.413 \times 10^{-3}T - 0.0807 \times 10^{-6}T^2$$

The number of mols in the products of reaction are

$$C_p(CO_2) = 6.396 + 10.100 \times 10^{-3}T - 3.405 \times 10^{-6}T^2$$

$$2 \times C_p(H_2O) = 14.438 + 4.748 \times 10^{-3}T + 0.534 \times 10^{-6}T^2 \qquad (10.33)$$

$$8 \times C_p(N_2) = 51.592 + 11.304 \times 10^{-3}T - 0.646 \times 10^6 T^2$$

$$\Sigma \, nC_p = 72.43 + 26.15 \times 10^{-3}T - 3.517 \times 10^{-6}T^2 \text{ cal/deg} \qquad (10.34)$$

or

$$191,760 = \int_{298}^{T_2} [72.43 + 26.15 \times 10^{-3}T - 3.517 \times 10^{-6}T^2] \, dT \qquad (10.35)$$

Integration and substitution of limits yields

$$214,475 = 72.43 \, T_2 + 13.08 \times 10^{-3}T_2^2 - 1.172 \times 10^{-6}T_2^3 \qquad (10.36)$$

This cubic equation may be solved by successive approximations to yield

$$T_2 = 2250°K \text{ or } 1977°C \qquad (10.37)$$

Measured flame temperatures are usually about 100° below the calculated maximum value. There are several reasons why calculated values are higher than those determined experimentally. First, it is difficult to carry out an experiment under adiabatic conditions, that is, where there is no heat loss to the surroundings. Also, it is probable that the theoretical amount of air required for complete combustion is insufficient under actual conditions and that excess air must be added. The effect of excess oxygen that does not actually take part in the reaction and of the additional nitrogen or other inert gas will be to lower the flame temperature. The maximum temperature for combustion of methane in pure oxygen has been found to be near 4000°K, showing that nitrogen and other inert gases in air lower the temperature by nearly 2000°.

Another reason that theoretical temperatures are not reached is that the combustion reaction may not go to completion. Also, the dissociation of water into hydrogen and oxygen and carbon dioxide into carbon monoxide and oxygen is appreciable. These dissociation reactions absorb considerable amounts of heat, which should be included in the calculations for greater accuracy.

**Problem 1.** Assuming methane to be burned in air atmosphere of $1 O_2$, $2N_2$, and $2CO_2$, calculate the flame temperature.

**Problem 2.** What is the flame temperature for the combustion of ethane $(C_2H_6)$ in air if its heat of combustion is 372.8 kcal/mol at 25°C?

For the calculation of the detonation wave energy parameters, it is necessary to relate the variables of the shock wave equations to equations of chemical thermodynamics. From the laws of chemical thermodynamics, the difference in internal energy as a function of temperature is given as

$$E_2 - E_1 = \int_{T_1}^{T_2} C_v dT = \overline{C}_v(T_2 - T_1) \qquad (10.38)$$

or

$$E_2 - E_1 = \overline{C}_v(T_2 - T_1) \qquad (10.39)$$

where $\overline{C}_v$ is the average heat capacity between $T_1$ and $T_2$.

Equations 10.1 through 10.7 apply to a nonreactive shock wave—that is, one in which no chemical reaction takes place. Where the energy to maintain a stable shock wave is furnished by a chemical reaction, the quantity of heat $Q_2$ liberated by the reaction must be taken into account. From equations 10.6 and 10.39, the energy for a detonation wave in an explosive becomes

$$\overline{C}_v(T_2 - T_1) = \tfrac{1}{2}(p_1 + p_2)(v_1 - v_2) + Q_2 \qquad (10.40)$$

As described previously, the equations of the hydrodynamic theory do not in themselves lead to the calculation of a given set of detonation conditions. Additional required relations include a properly chosen equation of state and laws of thermodynamics in conjunction with the hydrodynamic equations for a solution of the parameters that characterize the detonation state. That is, in the equations discussed to this point, only $p_1$, $T_1$, and $v_1$ are known.

## 10.5.4   Equation of State

For solid explosives and slurries, the primary interest is in an *equation of state* for the gases that are produced by the detonation, and this equation, if it defines the pressure–volume–temperature relations at high temperatures and pressures with sufficient accuracy, may then be used for calculation of both the detonation and explosion state parameters.

The equation of state for ideal gases

$$pv = nRT \qquad (10.41)$$

is applicable to gases at only moderate temperatures and pressures. In a detonation wave and in the gases that occupy a borehole after the detonation has passed through the explosive column, the pressures are of the order of $10^4$ atm or more. In these cases, the volume of the molecules must be accounted for, which the ideal equation of state does not take into consideration. One equation that has been used to correct for the fact that molecules have finite size is Van der Waal's:

$$p + \frac{a}{v^2} (v - b) = nRT \tag{10.42}$$

The factor $b$ is a constant covolume factor representing a correction which is about four times the actual volume of the molecules. The second correction $a/v^2$ was proposed to allow for the attraction between the molecules. While this equation is useful to account for the behavior of gases at intermediate pressures, it is not applicable to gases at the pressures developed by high explosives.

A virial equation of state is used by Taylor (1952) and is written in the form

$$\frac{pv}{nRT} = 1 + \frac{b}{v} + 0.625 \frac{b_0^2}{v^2} + 0.2869 \frac{b_0^3}{v^3} + 0.1928 \frac{b_0^4}{v^4} \tag{10.43}$$

where $b$ and $b_0$ are constants depending on the composition of the gases. A more generalized form of an equation of state may be written as follows, or expanded as a series:

$$pv = nRT\phi \tag{10.44}$$

where

$$\phi = c^x \tag{10.45}$$

$$x = K(v) \frac{T^c}{v}$$

and the parameter $c$ is assumed to take on certain values within a limited range, depending on the explosive. Other equations are used by particular authors.

One of the equations most tractable to systematized calculations is a modified form of Abel's equation of state, which has been used by both Brown (1941) and Cook (1958):

$$p(v - \alpha) = nRT \tag{10.46}$$

The term $\alpha$ is a covolume factor (the effect of the actual volume of the molecules and atoms) and is used by Cook (1958) as a function of the specific volume of the gases $v$. It was found, by means of computing the values of covolume for a large number of explosives from measured detonation velocities, that there is a consistent

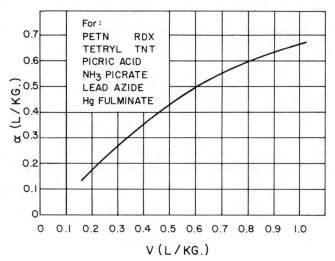

For:
PETN     RDX
TETRYL   TNT
PICRIC ACID
$NH_3$ PICRATE
LEAD AZIDE
Hg FULMINATE

**Figure 10.6**  Variation of $\alpha$ with $v$ (Cook, 1958).

relationship between $\alpha$ and $v_2$ (Figure 10.7). These experimentally determined values can be used to calculate the ideal detonation velocities of other explosives by a procedure that is the inverse of that used to determine the relationship between specific volume and covolume.

### 10.5.5  Material Balance

The chemical reactions accompanying and supporting a detonation wave are always exothermic, and the principles of thermochemistry state that the heat of any given reaction is equal to the heat of formation of the products of the reaction minus the heat of formation of the reactants. Thus, before the heat of reaction can be calculated, the composition of the explosive must be known, as well as the composition of the products of the reaction. For a known explosive composition, the first explosive calculations are concerned with the determination of the products of reaction. For an oxygen-balanced explosive or an explosive that is nearly oxygen-balanced, the products can be easily calculated, but for those with a negative or positive oxygen balance, the calculations are more complex, because chemical equilibria must be considered.

For those explosives that are composed only of the elements H, N, C, and O with minor amounts of other elements, a generalized form of the material balance equation may be written in the following form:

$$H_a N_b C_c O_d \rightarrow xN_2 + yH_2O + zCO_2 + \cdots + Q \qquad (10.47)$$

That is, in 100 g of explosive, there are $a$, $b$, $c$, and $d$ gram atoms of the four elements. The total number of gram atoms of the elements must be the same on both sides of the equation. For purposes of calculating a material balance equation, it is convenient to write the equation in tabular form.

The number of gram atoms per 100 g of a substance is determined as follows:

$$\text{gram-at.}/100\text{ g} = \frac{100 \times n}{\text{mol wt}} \tag{10.48}$$

where $n$ = the number of gram atoms per mol.

Example: How many gram atoms of $H_0$, $N_0$, and $O_0$ are there in ammonium nitrate?

$$\text{mol wt} = 80.05$$

$$\text{Formula:}\quad NH_4NO_3$$

$$H_0 = \frac{100 \times 4}{80.05} = 4.997$$

$$N_0 = \frac{100 \times 2}{80.05} = 2.498$$

$$O_0 = \frac{100 \times 3}{80.05} = 3.748 \tag{10.49}$$

For a mixture of fuels, oxidizers, and other ingredients, the number of gram atoms per 100 g of each element in each substance is multiplied by the percentage present. Then the amounts of each element per 100 g of explosive are totaled to give the gram atoms of each element in 100 g of explosive. We chose 100 g as the base amount of explosive for calculations as a matter of convenience. The composition and heats of formation of several common explosive ingredients are given in Tables 10.2 and 10.3.

Most commercial explosives are designed with near zero oxygen balance. That is, the usual constituent elements hydrogen, nitrogen, carbon, and oxygen are so proportioned that in the gases resulting from the detonation all of the hydrogen combines to form water (vapor), the combined nitrogen is released to form molecular nitrogen, and the carbon reacts to form carbon dioxide. If there is just enough oxygen present to form water and carbon dioxide, then the explosive is said to be oxygen-balanced. If it has a deficiency, it has a negative oxygen balance, and if it has an excess, it has positive oxygen balance. A deficiency will result in the formation of carbon monoxide and other compounds, the amounts present being dependent on the temperature and chemical equilibrium constants. Either an excess or a deficiency of oxygen will usually yield a lower heat of reaction than a balanced composition. Where an explosive does not contain other elements such as Na, Ca,

**TABLE 10.2. Heats of Formation of Explosives and Explosive Ingredients**

| Name | M.Wt. | Formula | Composition (gram atoms/100 gm) | | | | | Heat of Formation | |
|---|---|---|---|---|---|---|---|---|---|
| | | | $C_o$ | $H_o$ | $N_o$ | $O_o$ | Other | (kg cal/mole) | (kg cal/kg) |
| Guar | 110 | $C_6H_{10}O_5$ | 5.4545 | 9.0909 | — | 4.5454 | — | 228 | 2072 |
| Gilsonite | 13.4 | $CH_1$ | 7.4627 | 10.4477 | — | — | — | 10 | 750 |
| Nitroglycerin | 227.09 | $C_3H_5(ONO_2)_3$ | 1.321 | 2.202 | 1.321 | 3.963 | — | 90.83 | 400 |
| Ethylene glycol dinitrate | 152.97 | $C_2H_4(NO_3)_2$ | 1.315 | 2.630 | 1.315 | 3.946 | — | 56.00 | 366.0 |
| Nitrocellulose | | | | | | | | | |
| 11.05% $N_2$ | | | 2.390 | 3.190 | 3.570 | 0.790 | — | — | 754.0 |
| 11.64% $N_2$ | | | 2.320 | 3.030 | 3.570 | 0.830 | — | — | 699.0 |
| 12.20% $N_2$ | | | 2.250 | 2.870 | 3.620 | 0.870 | — | — | 664.0 |
| 12.81% $N_2$ | | | 2.180 | 2.720 | 3.650 | 0.910 | — | — | 605.0 |
| 13.45% $N_2$ | | | 2.100 | 2.540 | 3.670 | 0.960 | — | — | 558.0 |
| 14.12% $N_2$ | | | 2.020 | 2.360 | 3.700 | 1.010 | — | — | 500.0 |
| Trinitrotoluene (2-4-6) | 227.13 | $C_6H_2CH_3(NO_2)_3$ | 3.082 | 2.201 | 1.321 | 2.642 | — | 13.40 | 78.5 |
| Dinitrotoluene | 182.13 | $C_7N_2O_4H_6$ | 3.843 | 3.294 | 1.098 | 2.196 | — | 6.900 | 38.00 |
| Lead Azide | 291.3 | $Pb(N_3)_2$ | — | — | 2.060 | — | $Pb_o = 0.340$ | 104.3 | −364.0 to −386.0 |
| Mercury fulminate | 284.65 | $Hg(CNO)_2$ | 0.7026 | — | 0.7026 | 0.7026 | $Hg_o = 0.3513$ | −64.0 | −226 |
| S:G Pulp | | | 4.170 | 6.300 | — | 2.740 | — | — | 1050 |
| X Pulp | | | 4.050 | 5.850 | — | 2.800 | — | — | 1000 |
| Paraffin | | | 7.100 | 14.800 | — | — | — | — | 500 |
| Cellulose | | | 3.710 | 6.180 | — | 3.090 | — | 2270 | 1400 |
| Ammonium nitrate | 80.05 | $NH_4NO_3$ | — | 4.997 | 2.498 | 3.748 | | 87.27 | 1090 |
| Sodium nitrate | 85.01 | $NaNO_3$ | — | — | 1.176 | 3.529 | $Na_o = 1.176$ | 114.179 | 1309 |
| Calcium carbonate | 100.09 | $CaCO_3$ | 0.9999 | — | — | 3.000 | $Ca_o = 0.9999$ | 298.5 | 2859 |
| Fuel Oil | 14.3 | $(CH_2)$ | 7.1276 | 14.2552 | — | — | 2.4± | | 17 |
| Water | 180.16 | $H_2O$ | — | 11.1012 | — | 5.5506 | — | 57.8 | 3208 |
| Sulfur | 32.06 | $S$ | — | — | — | — | $S_o = 3.1186$ | 0 | 0 |
| Aluminum | 26.08 | $Al$ | — | — | — | — | $Al_o = 2.7042$ | 0 | 0 |

*Source:* Cook (1958).

**TABLE 10.3. Atomic Composition and Heat of Formation per 100 g**

| Constituent | % Comp | $C_o$ | $H_o$ | $N_o$ | $O_o$ | Ca | Na | $H_f$ (kg-cal/kg) |
|---|---|---|---|---|---|---|---|---|
| NG | 18 | 0.2378 | 0.3964 | 0.2378 | 0.7133 | —— | —— | 72 |
| TNT | 3 | 0.0925 | 0.0660 | 0.0396 | 0.0793 | —— | —— | 2 |
| $NH_4NO_3$ | 57 | —— | 2.8483 | 1.4239 | 2.1364 | —— | —— | 621 |
| $NaNO_3$ | 10 | —— | —— | 0.1176 | 0.3529 | —— | 0.1176 | 131 |
| S pulp | 10 | 0.4170 | 0.6300 | —— | 0.2740 | —— | —— | 105 |
| $CaCO_3$ | 2 | 0.0200 | —— | —— | 0.0600 | 0.0200 | —— | 57 |
| | 100 | 0.7673 | 3.9407 | 1.8189 | 3.6159 | 0.0200 | 0.1176 | 988 kg-cal/kg |

$$OB = O_o - 2C_o - 1/2H_o - Ca_o - 1/2Na_o$$

$$OB = 3.6159 - (2 \times 0.7673 + 1/2 \times 3.9407 + 0.0200 + 1/2 \times 0.1176)$$

$$= 3.6159 - (1.5346 + 1.9704 + 0.0200 + 0.0588)$$

$$= 3.6159 - 3.5838 = +0.0321 \text{ gm atoms/100 gm}$$

Thus, this mixture of explosive ingredients has a 0.0321 gm atoms/100 gm positive oxygen balance and the heat of formation of the explosive reactants is 988 kg-cal/kg.

*Source:* Clark (1959).

or Al which have a high affinity for oxygen, the oxygen balance equation may be written as

$$OB = O_0 - 2CO_2 - H_2O \qquad (10.50)$$

or it may be expressed with Ca, Na, and Al present and assuming that CaO, $Na_2O$, and $Al_2O_3$ are formed

$$OB = O_0 - 2C_0 - \tfrac{1}{2} H_0 - Ca_0 - \tfrac{1}{2} Na_0 - \tfrac{2}{3} Al_0 \qquad (10.51)$$

where $O_0$, $C_0$, and $H_0$ represent the number of gram atoms of these elements per 100 g of the explosive. The value of $O_0$ is corrected for the amount combined with the metallic elements to form solid products such as CaO and $Na_2O$ where such elements are present. That is, it is assumed that Ca and Na have a greater affinity for oxygen than hydrogen or carbon do.

**Problem.** Calculate the oxygen balance of an explosive of the following composition:

| | |
|---|---|
| Nitroglycerin | 18% |
| TNT | 3 |
| Ammonium nitrate | 57 |
| Sodium nitrate | 10 |
| S pulp | 10 |
| Calcium carbonate | 2 |
| | 100% |

**Problems.** (1) Ammonium nitrate dynamite contains 40% AN. What percentages of wood pulp and NG should it contain to be oxygen-balanced? Small amounts of $CaCO_3$ and $NaNO_3$ may be added also. What is the heat of formation of the mixture? (2) Calculate the percentages by weight for an oxygen-balanced mixture of AN and TNT. What is the heat of formation? What is the heat of reaction, assuming only $H_2O$, $N_2$, and $CO_2$ are formed?

## 10.5.6  Rates of Reaction

The effects of chemical *rates of reaction* upon the detonation process are complex and are the subject of some differences in interpretation. In general, the mass rate of reaction depends on the activity of the reactants, their concentration, how well they are mixed, temperature, and other factors.

Thus, for some substances that react at ambient temperatures, an increase in temperature of 10°C will double or treble the rate of a given chemical reaction (Glasstone, 1946). It is suggested that in every chemical system, an equilibrium exists between normal and "active" molecules, the latter being the only ones that can take part in a chemical reaction. The activation energy $E_a$ represents the difference between an active and a normal molecule, and it is acquired as a result of collision with other molecules. All molecules of a gas do not possess the same energy, and it is only those possessing an energy above the critical amount $E_a$ that are capable of taking part in a reaction.

The simple molecular collision theory is not adequate to explain chemical reaction rates, and the theory of *absolute reaction rates* has been developed. This states that only those molecules that reach the top of the "energy barrier" react. Hence, the rate of reaction is equal to the concentration of the activated molecules multiplied by the frequency of crossing the energy barrier. The reaction rate equation can be written

$$\text{Rate of reaction} = C*(kT/h) \qquad (10.52)$$

where $C*$ = concentration of activated molecules, $k$ = Boltzmann constant, $T$ = temperature (K), and $h$ = Planck's constant.

The rate of reaction is a function of the temperature, and for very high pressures, the rate of reaction must be corrected by a *fugacity factor* to account for the nonideal behavior. These theories apply to single compounds that are decomposed, or to mixtures of compounds; both types of materials are used as chemical explosives. For heterogeneous mixtures such as dynamites, the same principles apply approximately to the chemical reactions in the reaction zone in a detonation wave, provided the ingredients are reasonably well mixed and all of the constituent solid particles are small enough.

For explosives containing coarse grains, the grains may not be consumed rapidly enough for the reaction to take place within a critical distance of the front of the detonation wave. Thus, all of the heat produced by the reaction is not released in the detonation wave, and only part of the energy supports the velocity of atoms, molecules, and ions in the wave that contribute to its velocity. Consequently, the velocity of the wave is less than the ideal or theoretical value.

Coarse grains of any constituent are assumed to burn inward from their outer surface at a constant radial rate (Eyring et al., 1949), depending on the temperature. If the grains are large enough that they are not consumed within a distance $a_0$ from the wave front, then the velocity will be nonideal (Cook, 1958). The value of $a_0$ varies with the effective diameter $d$ of the explosive charge. That is, the detonation wave is considered to be triangular in shape unless the explosive is completely confined laterally (Figure 10.7). If the reaction of all materials takes place within a value of $a_0/d$ less than 1, then the detonation will be ideal.

This "geometrical model" provides one explanation of critical diameter effects

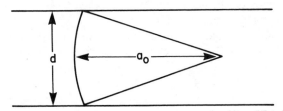

**Figure 10.7**   Geometry of detonation wave used for calculation of $a_o/d$ ratio (Clark, 1959).

in nonideal explosives such as coarse TNT and ANFO. If $a_0/d$ is too large, it is not possible to propagate a stable detonation. This also offers one explanation of why coarse-grained material requires large-diameter charges to maintain a stable detonation.

This analysis has dealt primarily with absolute reaction rates and indirectly with mass reaction rates. If one considers the relationship of surface area to particle size, it is found that the surface area of particles per unit weight (specific area) varies as the inverse of the particle diameter. Hence, for small particles, the surface area per unit weight exposed to reaction is greater, which facilitates both the initiation and the propagation of a detonation, depending on the packing density.

Other factors that affect the velocity of detonation and critical diameter are the *bulk density* of the particulate mass and the degree of *confinement*. The effects of density are considered in conjunction with the solution of the chemical thermodynamics of detonation waves. Theoretically, the detonation velocity will increase indefinitely with the increase in density of the explosive. However, if many solid and liquid explosives are compressed too much, the density of the mass inhibits the attack of the explosive by the hot gaseous molecules, and initiation and propagation of detonation will not take place; that is, the explosive is "dead pressed." Similar phenomena have been observed in slurries where they were compressed by the weight of a long column of explosives.

As stated earlier, commercial ammonium nitrate prills are porous owing to their method of manufacture. Consequently, the fuel oil can penetrate the prills, and more surface area is thus exposed to reaction with high-velocity molecules, atoms, and ions. Thus, the mass reaction rate is greater than for solid prills of the same diameter, the rate of release of energy is more rapid, adequate energy is released to permit initiation of an explosion, and enough of the mass reaction occurs close enough to the wave front to support a stable detonation wave if the charge diameter is larger than the critical value $d_c$.

In order to calculate (1) the detonation state and (2) the explosion state parameters, it is necessary to develop appropriate mathematical expressions relating the shock wave equations to the *thermochemical equations* and an equation of state for high temperatures and pressures.

### 10.5.7  Heat Capacities and Temperatures

Inasmuch as the heat capacities of the products of the reaction in a detonation wave determine the temperature of the gases, the pressure, and the wave velocity, it is necessary to calculate the amount of each chemical product, its heat capacity, and the heat of reaction, and to use these quantities to determine the detonation parameters. However, to calculate the temperature, pressure, and heat of reaction for the explosion state, it is necessary to use the thermochemical equations only; that is, the shock wave parameters do not determine explosion state parameters.

The heat capacities $C_v$ of many of the products of reaction have been determined as functions of temperature and may be calculated from appropriate equations or be taken from tables. For computer calculations, it is more convenient to use the equations. To facilitate some of the calculations that do not require the value of the ideal heat capacity of $C_v$ at a specific temperature, an average heat capacity $\overline{C}_v$ is defined from equation 10.39 as follows:

$$\overline{C}_v(T_2 - T_1) = \int_{T_1}^{T_2} C_v \, dT \tag{10.53}$$

If $T_1$ is taken as 298°K, then for each gas product $\overline{C}_v$ may likewise be calculated or tabulated as a function of $T_2$.

The detonation process may be considered to be adiabatic with no loss in heat to the surroundings. Then equation 10.10 becomes, with $q = 0$ for an adiabatic process,

$$dE = -dw \tag{10.54}$$

and the work is defined for an infinitesimal change in energy as

$$dE = -p \, dv \tag{10.55}$$

Also, $dE$ is a total differential and is a function of $p$, $v$, and $T$. Therefore,

$$dE = \left(\frac{\partial E}{\partial v}\right)_T dv + \left(\frac{\partial E}{\partial T}\right)_v dT \tag{10.56}$$

By definition, the heat capacity at constant volume is

$$C_v = \left(\frac{\partial E}{\partial T}\right)_v \tag{10.57}$$

Substituting 10.56 and 10.57 into 10.58 and rearranging terms, one obtains

$$-C_v \, dT = \left[\left(\frac{\partial E}{\partial v}\right)_T + p\right] dv \tag{10.58}$$

The equation of state $p(v - \alpha) = nRT$ may be introduced by differentiating it to yield

$$dT = \frac{p \, dv + v \, dp - \alpha \, dp - p \, d\alpha}{nR} \tag{10.59}$$

Substitution in equation 10.58 yields

$$-\left(\frac{dp}{dv}\right) = \left\{\frac{p\beta}{(v - \alpha)}\right\} \tag{10.60}$$

where

$$\beta = \frac{nR + C_v}{C_v} - \left(\frac{d\alpha}{dv}\right) + \left[\left(\frac{\partial E}{\partial v}\right)_T \frac{nR}{C_v p}\right] \qquad (10.61)$$

The last term on the right in equation 10.61 is very small and may be neglected, hence,

$$\beta = \frac{nR + C_v}{C_v} - \frac{d\alpha}{dv} \qquad (10.62)$$

Equation 10.4 may be written in the form

$$D^2 = v_1^2\left(\frac{p_2 - p_1}{v_1 - v_2}\right) = v_1^2\left(-\frac{dp}{dv}\right) \qquad (10.63)$$

Substitution of 10.60 into 10.63 gives

$$D^2 = \frac{v_1^2 p_2 \beta}{v_2 - \alpha} \qquad (10.64)$$

where in 10.63 the subscript 2 has been applied to indicate detonation conditions. If the value of $p$ from the equation of state is substituted in 10.64, one obtains

$$D^2 = \frac{v_1^2 nRT_2 \beta}{(v_2 - \alpha)^2} \qquad (10.65)$$

Also, since $p_2 > p_1$ for condensed explosives, one obtains from equation 10.60

$$-\left(\frac{dp_2}{dv_2}\right) = \frac{p_2 - p_1}{v_1 - v_2} = \frac{p_2}{v_1 - v_2} \qquad (10.66)$$

hence,

$$\beta = \frac{v_2 - \alpha}{v_1 - v_2} \qquad (10.67)$$

From 10.65 and 10.68

$$\alpha = v_1 - v_2 \frac{(\beta + 1)(nRT_2)^{1/2}}{D\beta^{1/2}} \qquad (10.68)$$

Also, from equation 10.7

$$W = \left(\frac{nRT_2}{\beta}\right)^{1/2} \tag{10.69}$$

and

$$E_2 - E_1 = \frac{nRT_2}{2\beta} \tag{10.70}$$

From 10.70 and 10.8 the value of the detonation temperature is for a single gas

$$T_2 = \frac{\left(Q_2 + T_1 \sum n\overline{C}_v\right)\beta}{\beta \sum n\overline{C}_v - \frac{1}{2}R \sum n} \tag{10.71}$$

where $\sum n\overline{C}_v$ = the sum of the heat capacities of the solids and gases produced, $T_1 = 298°\text{K}$, and $\sum n$ = sum of moles of gases.

The explosion state temperature does not involve detonation state parameters and can be written

$$Q_3 = \sum n\overline{C}_v(T_3 - T_1) \tag{10.72}$$

or

$$T_3 = T_1 + \frac{Q_3}{\sum_n \overline{C}_v} \tag{10.73}$$

where $T_3$ = explosion state temperature, $Q_3$ = heat released at $T_3$, and $\sum_n \overline{C}_v$ = sum of average heat capacities of gases and solids.

### 10.5.8   Calculations—Oxygen-Balanced Explosives

The solution of the preceding equations requires that the products of the reaction be evaluated. If the explosive is close to oxygen balance, and contains only small amounts of metallic elements, then the following procedure may be used: (see also Tables A-1 to A-5, Appendix A)

1. Given: explosive composition and density
2. Calculate gram atoms per 100 g of C, H, N, and O in the explosive
3. Subtract the amount of O required to form oxides of Na, Al, or Ca ($Na_2O$, $Al_2O_3$, or CaO) if they are present

4. Assume a detonation temperature $T_2$ in K
5. Assume a value of $v_2$, approximated from the empirical relation:

$$v_2 = v_1[0.72 + 0.1(p_1 - 0.9)] \qquad (10.74)$$

6. Calculate the oxygen balance
7. Calculate $\beta$ from equation 10.61 for gases only. Adjust $\rho_2(v_2)$ until 10.59 and 10.64 are satisfied
8. Check the temperature $T_2$ using equation 10.71 with both gaseous and solid products
9. Adjust $\beta$, $v_2$, and $T_2$ by iteration until equations 10.62, 10.67, and 10.71 are all satisfied
10. Calculate detonation pressure from equation 10.46, using $R = 0.08207$ liter atm/mol K
11. Calculate the detonation velocity from 10.65:

$$D^2 = \frac{v_1^2(nRT_2\beta)}{(v_2 - \alpha)^2} \qquad (10.75)$$

For the calculations of the explosion state parameters:

1. Repeat steps 1–3 for the detonation state
2. Assume a temperature $T_3$
3. Calculate average heat capacity of products
4. Iterate temperature with equation 10.72
5. Calculate explosion pressure with equation 10.46

## 10.6  NON-OXYGEN-BALANCED EXPLOSIVES

Although there is some dissociation taking place when oxygen-balanced explosives react, the assumption that the reactions go to completion gives reasonably accurate values of detonation and explosion state parameters. However, for non-oxygen-balanced explosives, there are many possible equilibrium reactions that can take place, and these must be used in the calculations of both detonation state and explosion state parameters.

Most of the important chemical reactions taking place in an explosion are exothermic, and with explosives that contain a large number of elements a variety of products may be formed. Usually with a limited amount of oxygen available, which is one of the controlling factors in the effectiveness of explosives, it is desired that the oxygen and other elements combine in such a way that a maximum amount of heat is produced. For example, for each gram-atom of oxygen in CO, $CO_2$, and $Al_2O_3$, the relative amounts of heat, using CO as a base, are 1, 1.78, and 4.95, respectively. Hence, it could appear that a maximum amount of Al should be included in explosive compositions, if costs permit, or such other elements whose heats of reaction with oxygen are the greatest. However, there are other factors that affect the performance of the explosive as well, and these must be considered in addition to the heat of reaction.

For explosives containing metallic elements, the solids formed in the explosion products do not affect the pressure or the detonation parameters involving $\beta$. Hence, the quantity $C_v$ in equation 10.62 and the $n$ in 10.65 and 10.71 are for gases only. Wherever $\overline{C}_v$ occurs, it should include the solids and gases—that is, in equations 10.71 and 10.72.

If there are relatively large amounts of solids in the explosion products, the available work is not approximated by the heat of explosion (of the explosion state) but must be reduced by a factor that depends upon the value of $\Sigma n \overline{C}_v$ (solids). That is, as the gases expand to do work, the (finely disseminated) solids can perform no work and cannot give up all of the heat they contain to the surrounding gases rapidly enough to make all of their heat (energy) available. However, some of the heat is transferred to the gases, and, as an approximation, a factor of $0.5 \Sigma n \overline{C}_v$ (solids) is sometimes used as part of the available energy.

In earlier studies of thermochemistry (Glasstone, 1946), it was suggested that the heat involved in a chemical reaction was a measure of the affinity of the reacting substances. Although this assumption is not completely correct, it is a close approximation, especially for exothermic reactions at ordinary temperatures. That is, while exothermic reactions predominate at ordinary temperatures, endothermic reactions may take place at high temperatures.

In determining which molecules have the greatest affinity for oxygen, the heat of reaction is often used to establish a "pecking order." That is, the metallic atoms combine with oxygen to give off more heat than carbon and hydrogen. The metallic atoms are therefore assumed to combine with oxygen first, and only the remaining amount of oxygen participates in the equilibrium reaction with the hydrogen and carbon.

### 10.6.1  Chemical Equilibria

The *law of mass action* and the development of *chemical equilibrium* concepts are directly related to chemical *affinity*. The law of mass action states that the rate of a chemical reaction is proportional to the active masses of the reacting substances and that the molecular or atomic concentration in solution or in the gas phase is taken as the active mass. The concepts of affinity gave way to "coefficients of velocity" of reactions which in turn led to the use of *equilibrium constants*.

Thus, for a reversible homogeneous gas reaction,

$$aA + bB \rightleftharpoons cC + dD \tag{10.76}$$

the rate of reaction from right to left in terms of the concentrations $c_A$ and $c_B$ is $K c_A c_B$ and that from left to right $K' c_C c_D$. At equilibrium, these two rates are equal; therefore,

$$Kc_A c_B = K' c_C c_D \tag{10.77}$$

or

$$\frac{c_C c_D}{c_A c_B} = \frac{K}{K'} = K \tag{10.78}$$

where $K$ is defined as the equilibrium constant. The *law of equilibrium* states that, no matter what the initial concentrations of the substances $A$, $B$, $C$, and $D$ are, under equilibrium conditions they are related to each other by an equation of the form of equation 10.78.

Where the coefficients of the substances in a reaction are expressed in a more general form

$$aA + bB + cC + \cdots \rightleftharpoons lL + mM + nN + \cdots \tag{10.79}$$

the equilibrium constant is given by

$$\frac{(c_L)^l (c_M)^m (c_N)^n}{(c_A)^a (c_B)^b (c_C)^c} = K \tag{10.80}$$

This equation may also be expressed, for an ideal gas, in terms of the partial pressures $p_A$, $p_B$, and so on, or in terms of the chemical potential of the substances.

For gases, whether ideal or nonideal, it can be shown that the equilibrium constant can be written in terms of the activity $a$ of each substance:

$$\frac{(a_L)^l (a_M)^m (a_N)^n}{(a_A)^a (a_B)^b (a_C)^c} = K \tag{10.81}$$

The free energy for a nonideal reaction is expressed as

$$-\Delta F = nRT \left[ \ln K - \Sigma \nu \ln a \right] \tag{10.82}$$

where $\Delta F$ = change in free energy and $\Sigma \nu$ = exponents $(a + b + c) - (l + m + n)$. That is, equation 10.82 gives the decrease in free energy, resulting in the transfer of reactants into products at given activities.

Also of importance in explosive reactions is the principle of mobile equilibrium, which is stated by Glasstone (1946) as follows:

If a change occurs in one of the factors, such as temperature or pressure, under which a system is in equilibrium, the system will tend to adjust itself to annul, as far as possible, the effect of that change. For example, if the pressure of a gas mixture at equilibrium is increased—the volume occupied by the number of molecules is diminished, the reaction will tend to occur involving a decrease in the number of molecules, and the equilibrium will be shifted in that direction; similarly, a decrease in pressure will result in a movement in the direction of an increased number of molecules. If the temperature is raised, that reaction will occur in which heat is absorbed, whereas a lowering of the temperature will move the equilibrium in the direction of an exothermic reaction.

At high pressures, the activities of the various gas species are not proportional to the partial pressure, and the fugacities must be used in their place. That is,

$$K_p(T) = \frac{f_A^a \times f_B^b \times \cdots}{f_M^m \times f_N^n \times \cdots} \qquad (10.83)$$

where $f_i$ is the partial fugacity (analogous to partial pressure) of the $i$th component of the gaseous mixture in the detonation wave. The total fugacity (analogous to total pressure) is related to the pressure and specific volume by the equation

$$nRT \left( \frac{d \ln f}{dp} \right)_T = v \qquad (10.84)$$

which, when combined with the equation of state $p(v - \alpha) = nRT$, yields

$$nRT \left( \frac{d \ln f}{dp} \right)_T = \frac{nRT}{p} + \alpha \qquad (10.85)$$

Integration of equation 10.85 between pressure limits of zero and $p$ gives the following:

$$\frac{f}{p} = \exp \left[ \frac{1}{nRT} \int_o^p \alpha \, dp \right] \qquad (10.86)$$

Just as the partial pressure of a component of a gaseous mixture is related to the total pressure, the partial fugacity of any given component is related to the total fugacity by

$$f_i = fN_i \qquad (10.87)$$

where $f_i$ is the partial fugacity of a given component and $N_i$ is the mole fraction of that component in the mixture. Substituting this value of $f$ in equation 10.86 and taking $v_i$th power of both sides, this expression becomes

$$\left( \frac{f_i}{pN_i} \right)^{v_i} = \exp \frac{v_i}{nRT} \int_o^p \alpha \, dp \qquad (10.88)$$

where $v_i$ is the coefficient of the particular gas in equation 10.81.

Substitution of this value for $(f_i)\nu_i$ in equation 10.83 gives

$$K_p(T) = \frac{(pN_A)^a \times (pN_B)^b \cdots}{(pN_M)^m \times (pN_N)^n \cdots} \exp \frac{\Delta\nu}{nRT} \int_o^p \alpha\, dp \qquad (10.89)$$

where $\Delta\nu = (a + b + \cdots) - (m + n + \cdots)$ or

$$K_p(T) = \frac{(N_A)^a \times (N_B)^b \times \cdots}{(N_M)^m \times (N_N)^n \times \cdots} p^{\Delta\nu} \exp \frac{\Delta\nu}{nRT} \int_o^p \alpha\, dp \qquad (10.90)$$

To adapt the equilibrium constant for the chosen equation of state and 100 g of explosive, a new equilibrium constant may be written

$$K_i = K_p \left(\frac{1.2181}{T}\right)^{\Delta\nu} \qquad (10.91)$$

For any gaseous component, the chosen equation of state holds, and

$$pN_i = \frac{N_i nRT}{v - \alpha} = \frac{A_i T}{a 1.2181} \qquad (10.92)$$

in which $a = (v - \alpha)$ and $A_i$ represent any one of the components on either side of equation 10.77 which is taking part in the equilibrium reaction in moles per 100 g of explosive. The quantities $v$ and $\alpha$ are in liters per kilogram, and $p$ is in atmospheres.

By means of equations 10.90, 10.91, and 10.92, a new equilibrium constant may be written in terms of moles of gases present:

$$K_i = \frac{(A)^a \times (B)^b \times \cdots}{(M)^m \times (N)^n \times \cdots} \frac{\exp \Delta\nu z}{a^{\Delta\nu}} \qquad (10.93)$$

where

$$z = \frac{1}{nRT} \int_o^p \alpha\, dp \qquad (10.94)$$

For purposes of calculation, the correction factor is denoted as $F$, called a fugacity factor:

$$\frac{e^z}{a} = F \qquad (10.95)$$

The known chemical equilibria that may be involved in explosive reactions in detonating high explosives are given in Table 10.4.

## TABLE 10.4. Chemical Reactions and Definitions of Equilibrium Constants

3. $H \rightleftharpoons \dfrac{1}{2} H_2$  $\qquad K_{p3} = \dfrac{f(H)}{f(H_2)^{1/2}}$  $\qquad K_3 = \dfrac{(H)}{(H_2)^{1/2}} F^{1/2}$

5. $CO + H_2O \rightleftharpoons CO_2 + H_2$  $\qquad K_{p5} = \dfrac{f(CO)f(H_2O)}{f(CO_2)f(H_2)}$  $\qquad K_5 = \dfrac{(CO)(H_2O)}{(CO_2)(H_2)}$

6. $N \rightleftharpoons \dfrac{1}{2} N_2$  $\qquad K_{p6} = \dfrac{f(N)}{f(N_2)^{1/2}}$  $\qquad K_6 = \dfrac{(N)}{(N_2)^{1/2}} F^{1/2}$

8.† $2CO \rightleftharpoons CO_2 + C$  $\qquad K_{p8} = \dfrac{f(CO)^2}{f(CO_2)}$  $\qquad K_8 = \dfrac{(CO)^2}{(CO_2)} F$

9. $O_2 + 2CO \rightleftharpoons 2\,CO_2$  $\qquad K_{p9} = \dfrac{f(O_2)f(CO)}{f(CO_2)^2}$  $\qquad K_9 = \dfrac{(O_2)(CO)^2}{(CO_2)^2} F$

12. $CH_4 + CO_2 \rightleftharpoons 2CO + 2H_2$  $\qquad K_{p12} = \dfrac{f(CH_4)f(CO_2)}{f(CO)^2 f(H_2)^2}$  $\qquad K_{12} = \dfrac{(CH_4)(CO_2)}{(CO)^2(H_2)^2} F^{-2}$

13. $CO + OH \rightleftharpoons CO_2 + \dfrac{1}{2} H_2$  $\qquad K_{p13} = \dfrac{f(CO)f(OH)}{f(CO_2)f(H_2)^{1/2}}$  $\qquad K_{13} = \dfrac{(CO)(OH)}{(CO_2)(H_2)^{1/2}} F^{1/2}$

14. $NO + CO \rightleftharpoons \dfrac{1}{2} N_2 + CO_2$  $\qquad K_{p14} = \dfrac{f(NO)f(CO)}{f(N_2)^{1/2} f(CO_2)}$  $\qquad K_{14} = \dfrac{(NO)(CO)}{(N_2)^{1/2}(CO_2)} F^{1/2}$

15. $\dfrac{1}{2} N_2 + \dfrac{3}{2} H_2 \rightleftharpoons NH_3$  $\qquad K_{p15} = \dfrac{f(N_2)^{1/2} f(H_2)^{3/2}}{f(NH_3)}$  $\qquad K_{15} = \dfrac{(N_2)^{1/2}(H_2)^{3/2}}{(NH_3)} F$

16.† $HCN \rightleftharpoons C + \dfrac{1}{2} N_2 + \dfrac{1}{2} H_2$  $\qquad K_{p16} = \dfrac{f(HCN)}{f(N_2)^{1/2} f(H_2)^{1/2}}$  $\qquad K_{16} = \dfrac{(HCN)}{(N_2)^{1/2}(H_2)^{1/2}}$

21. $CO + 2H_2 \rightleftharpoons CH_3OH$  $\qquad K_{p21} = \dfrac{f(CO)f(H_2)^2}{f(CH_3OH)}$  $\qquad K_{21} = \dfrac{(CO)(H_2)^2}{(CH_3OH)} F^2$

22. $CH_2O_2 \rightleftharpoons H_2 + CO_2$  $\qquad K_{p22} = \dfrac{f(CH_2O_2)}{f(H_2)f(CO_2)}$  $\qquad K_{22} = \dfrac{(CH_2O_2)}{(H_2)(CO_2)} \dfrac{1}{F}$

23.† $CH_2O \rightleftharpoons CO + H_2$  $\qquad K_{p23} = \dfrac{f(CH_2O)}{f(CO)f(H_2)}$  $\qquad K_{23} = \dfrac{(CH_2O)}{(CO)(H_2)} \dfrac{1}{F}$

24. $O + CO \rightleftharpoons CO_2$  $\qquad K_{p24} = \dfrac{f(O)f(CO)}{f(CO_2)}$  $\qquad K_{24} = \dfrac{(O)(CO)}{(CO_2)} F$

25.† $C_2H_4 \rightleftharpoons 2C + 2H_2$  $\qquad K_{p25} = \dfrac{f(C_2H_4)}{f(H_2)^2}$  $\qquad K_{25} = \dfrac{(C_2H_4)}{(H_2)^2} \dfrac{1}{F}$

26.† $C_2H_6 \rightleftharpoons 2C + 3H_2$  $\qquad K_{p26} = \dfrac{f(C_2H_6)}{f(H_2)^3}$  $\qquad K_{26} = \dfrac{(C_2H_6)}{(H_2)^3} \dfrac{1}{F^2}$

27.† $C_2H_5OH \rightleftharpoons C + CO + 3H_2$  $\qquad K_{p27} = \dfrac{f(C_2H_5OH)}{f(CO)f(H_2)}$  $\qquad K_{27} = \dfrac{(C_2H_5OH)}{(CO)(H_2)^3} \dfrac{1}{F^3}$

28. $Al_2O + 2CO_2 \rightleftharpoons Al_2O_3 + 2CO$  $\qquad K_{p28} = \dfrac{f(Al_2O)f(CO_2)^2}{f(CO)^2}$  $\qquad K_{28} = \dfrac{(Al_2O)(CO_2)^2}{f(CO)^2} F$

29. $AlO + \dfrac{1}{2} CO_2 \rightleftharpoons$

$\qquad \dfrac{1}{2} Al_2O_3 + \dfrac{1}{2} CO$  $\qquad K_{p29} = \dfrac{f(AlO)f(CO_2)^{1/2}}{f(CO)^{1/2}}$  $\qquad K_{29} = \dfrac{(AlO)(CO_2)^{1/2}}{(CO)^{1/2}} F$

---

† These reactions apply only in the presence of free carbon.
*Source:* Cook (1958).

It must be remembered that the equilibria are functions of temperature and change continuously from $T_2$ to $T_3$. They also change in the process of doing work from $T_3$ and $p_3$ to the final temperature and pressure, near atmospheric.

## 10.6.2  Calculations—Non-Oxygen-Balanced Explosives

In accordance with the "pecking order" described earlier, only the four elements H, N, O, and C are used in equilibrium calculations outlined below. Where other elements are present, it is usually possible to determine the products formed from them without considering their thermodynamic equilibria. The composition in terms of elements includes the important ones that are found in many commercial high explosives. Some oxygen-poor reactions result in the generation of free carbon, but, inasmuch as most commercial explosives are nearly oxygen-balanced, there is no free carbon formed, and these types of reactions need not be considered. The missing equilibrium constants in Table 10.4 are for equations that are concerned with elements other than H, N, O, and C or which are not required in this method of calculation.

The number of reactions that must be included for detonation calculations is somewhat arbitrary. Thus, only applicable ionization equations are included, because, although high temperatures favor ionization, high pressures inhibit its occurrence.

The occurrence of only four basic elements that participate significantly in equilibria reactions makes it possible to write all of the reactions so that they include one or more of four control molecules, which in turn provides a more tractable means of solution. The control molecules must, however, (1) contain among them all four basic elements and (2) be independent of each other. It is desirable to choose molecules as simple as possible, and those that best serve this purpose are $H_2$, $N_2$, $CO$, $H_2O$, and $CO_2$, inasmuch as carbon is not present as a gas. The molecule $O_2$ is also always a minor constituent and is not used as a control molecule, although it may be present in sufficient quantity to be included in the calculation of the total number of moles of gas produced. Thus, $CO$ and $CO_2$ are used in place of C and $O_2$, respectively. All reactions involving the four basic elements can be written in terms of one or more of the four molecules, although the reaction denoting the formation of a molecule does not necessarily involve one of the four.

Each additional equilibrium reaction introduces a new molecule, and hence one new unknown concentration and one equilibrium equation. The use of four control molecules also introduces four more unknowns, for which four additional equations must be found. These are furnished by the material balance equations, giving as many equations as there are unknowns.

The reactions listed in Table 10.4 may be solved for other molecules in terms of $H_2$, $N_2$, $CO$, and $CO_2$, resulting in the following:

$$(H_2O) = K_5(H_2)(CO_2)/(CO) \qquad (10.96)$$

$$(NH_3) = \frac{F}{K_{15}} (N_2)^{1/2} (H_2)^{1/2} \tag{10.97}$$

$$(CH_4) = K_{12} F^2 (H_2)^2 (CO)^2 / (CO_2) \tag{10.98}$$

$$(CH_3OH) = \frac{F^2}{K_{21}} (H_2)^2 (CO) \tag{10.99}$$

$$(CH_2O_2) = K_{22} F (H_2) (CO_2) \tag{10.100}$$

$$(O_2) = \frac{K_9}{F} (CO_2)^2 / (CO)^2 \tag{10.101}$$

$$(OH) = \frac{K_{13}}{F^{1/2}} (H_2) (CO_2) / (CO) \tag{10.102}$$

$$(NO) = \frac{K_{14}}{F^{1/2}} (N_2)^{1/2} (CO_2) / (CO) \tag{10.103}$$

$$(H) = \frac{K_3}{F^{1/2}} (H_2)^{1/2} \tag{10.104}$$

$$(N) = \frac{K_6}{F^{1/2}} (N_2)^{1/2} \tag{10.105}$$

$$(O) = \frac{K_{24}}{F} (CO_2) / (CO) \tag{10.106}$$

For material balance, the total number of atoms of each element in all of the compounds or elements listed in equations 10.96 through 10.106 in the reaction products must be equal to the number of atoms of the same element in the explosive. Thus, the total number of gram-atoms of H in the explosion products must equal the number of gram-atoms in the unreacted explosive, denoted by $H_0$. The same holds true for $N_0$, $O_0$, and $C_0$. The material balance equations may then be written as follows:

$$H_0 - 2(H_2) - 2(H_2O) - 3(NH_3) - 4(CH_4) - 4(CH_3OH)$$
$$- 2(CH_2O_2) - (OH) - (H) = 0 \tag{10.107}$$

$$N_0 - 2(N_2) - (NH_3) - (NO) - (N) = 0 \tag{10.108}$$

$$O_0 - 2(CO_2) - (CO) - (H_2O) - (CH_3OH)$$
$$- 2(CH_2O_2) - 2(O_2) - (OH) - (NO) - (O) = 0 \tag{10.109}$$

$$C_0 - (CO_2) - (CO) - (CH_4) - (CH_3OH) - (CH_2O_2) = 0 \tag{10.110}$$

For use of these and related equations for computer calculations, it has been found convenient to retain five significant figures for numerical values. For mathematical reasons, the equations involving the chosen independent variables and the dependent variables are placed in a different form from that given previously.

It has been found that the magnitude of two of the independent variables, $N_2$ and $CO_2$, does not vary greatly with temperature or fugacity, and hence, they do not require modification. In equations 10.96 through 10.106, $N_2$ appears only to the one-half power, and so the variable $(N_2)^{1/2}$ is used to facilitate calculations. Also, with the exception (equations 10.99 and 10.100), $CO_2$ always appears in a ratio with CO, and the number of arithmetic operations is reduced by using $CO_2/CO$ as a variable with a modification. The amount of CO varies widely with fugacity, as does $H_2$, both decreasing as F increases. If both CO and $H_2$ are multiplied by $F^{1/2}$, the variation is considerably reduced.

The independent variables are therefore chosen as combinations of the four control molecules as follows:

$$X_1 = (H_2)F^{1/2} \tag{10.111}$$

$$X_2 = (N_2)^{1/2} \tag{10.112}$$

$$X_3 = (CO_2)/(CO)F^{1/2} \tag{10.113}$$

$$X_4 = (CO)F^{1/2} \tag{10.114}$$

The independent variables are as follows, using $y_i$'s to denote the individual components:

$$y_1 = (H_2) \qquad = \gamma_1 x_1 \qquad \gamma_1 = 1/F^{1/2} \tag{10.115}$$

$$y_1 = (N_2) \qquad = \gamma_2 x_2^2 \qquad \gamma_2 = 1 \tag{10.116}$$

$$y_3 = (CO_2) \qquad = \gamma_3 x_3 x_4 \qquad \gamma_3 = 1 \tag{10.117}$$

$$y_4 = (CO) \qquad = \gamma_4 x_4 \qquad \gamma_4 = 1/F^{1/2} \tag{10.118}$$

$$y_5 = (H_2O) \qquad = \gamma_5 x_1 x_3 \qquad \gamma_5 = K_5 \tag{10.119}$$

$$y_6 = (NH_3) \qquad = \gamma_6 x_1^{3/2} x_2 \qquad \gamma_6 = F^{1/4}/K_{15} \tag{10.120}$$

$$y_7 = (CH_4) \qquad = \gamma_7 x_1^2 x_4/x_3 \quad \gamma_7 = K_{12} \tag{10.121}$$

$$y_8 = (CH_3OH) \quad = \gamma_8 x_1^2 x_4 \qquad \gamma_8 = F^{1/2}K_{21} \tag{10.122}$$

$$y_9 = (CH_2O_2) \qquad = \gamma_9 x_1 x_3 x_4 \qquad \gamma_9 = K_{22}F^{1/2} \tag{10.123}$$

$$y_{10} = (O_2) \qquad = \gamma_{10} x_3^2 \qquad \gamma_{10} = K_9 \tag{10.124}$$

$$y_{11} = (OH) \qquad = \gamma_{11} x_1^{1/2} x_3 \quad \gamma_{11} = K_{13}/F^{1/4} \tag{10.125}$$

$$y_{12} = (NO) \qquad = \gamma_{12} x_2 x_3 \qquad \gamma_{12} = K_{14} \tag{10.126}$$

$$y_{13} = (H) \qquad = \gamma_{13} x_1^{1/2} \qquad \gamma_{13} = K_3/F^{3/4} \tag{10.127}$$

$$y_{14} = (\text{N}) \qquad = \gamma_{14}x_2 \qquad \gamma_{14} = K_6/F^{1/2} \qquad (10.128)$$

$$y_{15} = (\text{O}) \qquad = \gamma_{15}x_3 \qquad \gamma_{15} = K_{24}/^{1/2} \qquad (10.129)$$

The material balance equations in terms of the $y_i$'s are then:

$$H_0 - 2y_1 - 2y_5 - 3y_6 - 4y_7 - 4y_8 - 2y_9 - y_{11} - y_{13} = R_1 \qquad (10.130)$$

$$N_0 - 2y_2 - y_6 - y_{12} - y_{11} = R_2 \qquad (10.131)$$

$$O_0 - 2y_3 - y_4 - y_5 - y_8 - 2y_9 - 2y_{10} - y_{11} - y_{12} - y_{15} = R_3 \qquad (10.132)$$

$$C_0 - y_3 - y_4 - y_7 - y_8 - y_9 = R_4 \qquad (10.133)$$

Inasmuch as the equilibrium constants are functions of temperature and the fugacity a function of specific volume of the explosion products, neither of which is known at the outset of a problem, the preceding equations (many of which are nonlinear) must be solved by iteration. Also, the values of $R_1$, $R_2$, $R_3$, and $R_4$ in equations 10.126–10.129 will only be zero (the material balance equations satisfied) when the proper values of the $y_i$'s are calculated for a given value of $T_2$ and $v_2$.

Temperature plays a dominant role in explosive calculations, and it must be chosen as one of the iteration control variables. The other two variables that might be called subiteration controls are $v_2$, the specific volume, and $\beta$, a thermodynamic factor. The last two can usually be approximated quite closely at the beginning of a problem and are relatively easily adjusted once the correct temperature and composition of the detonation products are calculated. In a calculation of explosion state parameters, $v_3 = v_1$, $\beta$ does not enter the problem, and two subiterations are eliminated.

The step-by-step calculation of detonation state parameters for an HNOC non-oxygen-balanced explosive with small amounts of metallic elements is as follows:

1. Given the explosive composition, calculate the number of gram atoms per 100 g of each element in the explosive
2. Subtract the amount of O required to form oxides of base metals or other elements ($CaO$, $Na_2O$, $Al_2O$, etc.)
3. Assume a detonation temperature $T_2$
4. Approximate the empirical equation value of specific volume $v_2$:

$$v_2 = v_1[0.72 + 0.1(\rho_1 - 0.9)]$$

5. Calculate the fugacity factor $F$ and the required equilibrium constants for the above values of $v_2$ and $T_2$, respectively
6. Solve equations 10.107–10.133 to obtain moles of products for chosen temperatures
7. Calculate a value of $\beta$ from equation 10.27 and iterate value with equation 10.32

8. Check assumed detonation temperature with equation 10.36
9. Adjust $T_2$, $v_2$, and $\beta$ by iteration until all equations are satisfied
10. Calculate detonation velocity and pressure
11. Calculate explosion state conditions with the same procedure, assuming $v_1$ = $v_3$, neglecting the effect of $\beta$

Of the preceding calculations, the iterations of the material balance equations are the most tedious. They involve the solution of four simultaneous nonlinear algebraic equations with the controlling factor of temperature $T_2$. Cook (1958) has used a method of iteration that is similar to the Newton–Raphson method. The latter is used herein because of its wide use in digital machine programming.

A computer program for calculating the detonation and explosion state parameters follows essentially the steps listed earlier. The input to the computer consists of the explosive composition in gram atoms per 100 g of explosive, the heat of formation of the reactants, and the density of the explosive. Also, starting values are required for the detonation temperature $T_2$, the explosion state temperature $T_3$, and the specific volume of the detonation product $v_2$ (Appendix A).

The equilibrium constants (Table 10.5), fugacity factors (Table 10.6), heat capacities (Table 10.7), and thermohydrodynamic factors (Table 10.A1) can be calculated from information stored in the computer. The iteration procedures to obtain correct values of $v_2$ and $\beta$ are programmed, and those for the correction values for $X_i$ in the material balance equations use standard programs for the solution of simultaneous linear equations. Most computer programs for simultaneous equations

## TABLE 10.5. Equilibrium Constant Approximations

Range 1500 °K ≤ T ≤ 6000 °K

$\ln K = A + T'[B + T'(C + DT')]$, $T' = 1000/T$

| K | A | B | C | D | $\Delta_R max^*$ |
|---|---|---|---|---|---|
| 3 | 2.471053 | −22.736291 | −6.220603 | 3.551435 | 0.0037 |
| 5 | 3.495409 | −4.001861 | 0.002038 | −0.001249 | 0.0003 |
| 6 | 3.499324 | −39.744317 | −6.211124 | 3.536575 | 0.0033 |
| 9 | 10.731676 | −58.535992 | −12.429842 | 7.086523 | 0.0068 |
| 12 | −16.280249 | 17.426424 | 16.214966 | −6.962426 | 0.0324 |
| 13 | 7.059063 | −32.088330 | −6.213579 | 3.538873 | 0.0034 |
| 14 | 6.896320 | −40.140647 | −6.233845 | 3.557053 | 0.0036 |
| 15 | 4.969711 | 0.761493 | −8.133611 | 3.500181 | 0.0188 |
| 21 | 9.182171 | 3.181536 | −16.055044 | 6.851758 | 0.0339 |
| 22 | −4.161277 | −10.876618 | 8.341794 | −3.623585 | 0.0156 |
| 24 | 8.564920 | −55.070173 | −12.459500 | 7.108882 | 0.0066 |

$*$ $\Delta_R max = \dfrac{\text{(computed value} - \text{tabular value)}}{\text{tabular value}}$

*Source:* Cook (1958).

**TABLE 10.6. Fugacity Factor, Alpha $(\partial\alpha/\partial v)$ Approximations**

$$\ln F, \quad \alpha = A + \rho[B + \rho(C + D\rho)], \quad (\partial\alpha/\partial v) = A + B\rho$$

| Function | A | B | C | D | $\Delta_R$max* | Range |
|---|---|---|---|---|---|---|
| ln F | -2.39474 | 7.984487 | -4.0581941 | 2.2507612 | 0.014 | $0.5 < \rho \leq 1.1$ |
| ln F | 4.77218 | -9.810743 | 11.339655 | -2.4017373 | 0.064 | $1.1 < \rho \leq 2.2$ |
| $\alpha$ | 1.2506041 | -0.85411083 | 0.3175400 | -0.05957424 | 0.012 | $0.5 < \rho \leq 2.2$ |
| $(\partial\alpha/\partial v)$** | -0.10 | 0.5 | — | — | 0 | $0.5 < \rho < 0.9$ |
| $(\partial\alpha/\partial v)$ | -0.01 | 0.4 | — | — | 0 | $0.9 < \rho \leq 1.2$ |
| $(\partial\alpha/\partial v)$ | 0.11 | 0.3 | — | — | 0 | $1.2 < \rho \leq 2.2$ |

\* $\Delta_R$max $= \dfrac{\text{computed value} - \text{tabular value}}{\text{tabular value}}$

\*\* (Cook 1958)

*Source:* Cook (1958).

**TABLE 10.7. Quadratic Approximations in $T$ for Heat Capacities for the Range $2000 < T < 6000°K$**

$$C_v^* = A + \frac{B \times 10^3}{T}\left[1 + \frac{C \times 10^3}{T}\right] \text{(cal/mol/K°)}$$

$$\bar{C}_v^* = \bar{A} + \frac{\bar{B} \times 10^3}{T}\left[1 + \frac{\bar{C} \times 10^3}{T}\right] \text{(cal/mol/K°)}$$

| Substance | A | B | C | Δmax | $\bar{A}$ | $\bar{B}$ | $\bar{C}$ | Δmax |
|---|---|---|---|---|---|---|---|---|
| $H_2$ | 8.487 | −5.962 | −0.4582 | 0.01 | 7.638 | −7.429 | −0.8281 | 0.02 |
| $N_2$ | 7.478 | −2.060 | −0.3408 | 0.01 | 7.193 | −3.634 | −0.5823 | 0.01 |
| $CO_2$ | 14.616 | −5.692 | −0.5344 | 0.02 | 13.851 | −8.569 | −0.6243 | 0.02 |
| $CO$ | 7.478 | −1.994 | −0.4157 | 0.01 | 7.204 | −3.379 | −0.5623 | 0.01 |
| $H_2O$ | 12.221 | −2.758 | +1.225 | 0.02 | 11.887 | −10.751 | −0.6200 | 0.01 |
| $NH_2$ | 18.113 | −2.341 | +2.793 | 0.02 | 17.854 | −15.564 | −0.5278 | 0.01 |
| $CH_4$ | 24.089 | −2.434 | +3.782 | 0.02 | 23.830 | −21.671 | −0.4431 | 0.01 |
| $CH_3OH$ | 29.025 | −2.212 | +4.892 | 0.04 | 28.794 | −24.003 | −0.4282 | 0.01 |
| $CH_2O_2$ | 24.034 | −1.898 | +3.089 | 0.02 | 23.829 | −14.948 | −0.4356 | 0.01 |
| $O_2$ | 9.235 | −5.995 | −0.5193 | 0.03 | 8.470 | −7.664 | −0.8948 | 0.02 |
| $OH$ | 8.642 | −7.084 | −0.7045 | 0.03 | 7.679 | −7.274 | −0.8506 | 0.02 |
| $NO$ | 7.759 | −3.157 | −0.7906 | 0.02 | 7.321 | −3.437 | −0.6433 | 0.01 |
| $H$ | 2.981 | 0.000 | 0.0000 | 0.00 | 2.981 | 0.000 | 0.0000 | 0.00 |
| $N$ | 6.476 | −18.840 | −1.2880 | 0.15 | 3.744 | −4.210 | −1.4980 | 0.05 |
| $O$ | 4.029 | −5.326 | −1.2560 | 0.05 | 3.279 | −1.498 | −1.3720 | 0.02 |
| $HCN$ | 13.002 | −0.866 | +3.7410 | 0.01 | 12.910 | −7.465 | −0.4634 | 0.01 |
| $CH_2O$ | 18.005 | −1.127 | +4.8630 | 0.01 | 17.804 | −12.209 | −0.3430 | 0.01 |
| $C_2H_4$ | 30.026 | −2.287 | +4.4800 | 0.03 | 29.786 | −23.412 | −0.4254 | 0.01 |
| $C_2H_6$ | 41.010 | −2.817 | +5.4320 | 0.05 | 40.716 | −34.096 | −0.3993 | 0.01 |
| $C_2H_5OH$ | 46.101 | −3.832 | +3.8130 | 0.02 | 45.677 | −35.188 | −0.4123 | 0.01 |
| $C(s)$ | 8.494 | −9.090 | +0.9468 | 0.09 | 7.257 | −7.444 | −0.8137 | 0.04 |

*Source:* Cook (1959).

use a Gaussian reduction that requires a reasonably stable matrix with the larger values of the coefficients on the diagonal.

When computed values of $T_2$ and $T_3$ are obtained, they are compared with the assumed values. The next assumed $T_2$ and $T_3$ are chosen by adjusting to values halfway between the assumed and calculated values, unless the original assumed values are too much in error.

An investigation of the values of the control variable (Cook, 1958) indicated that for HNOC explosives that produce no free carbon in the detonation products, the following starting values of the control variables gave good results:

$$X_1 = 1 \qquad X_2 = 1 \qquad X_3 = 0.1 \qquad X_4 = 10 \qquad (10.135)$$

The assumed value of a trial detonation temperature $T_2$ determines the value of the equilibrium constants $K_i$ (Table 10.5), and the approximated value of $v_2(\rho_2)$ fixes the value of the fugacity factor $F$ (Table 10.6).

Starting values (equations 10.111 through 10.114) for the control variables $X_1$, $X_2$, $X_3$, and $X_4$ are then inserted, and the values of $y_i$'s are calculated. The four material balance equations 10.130–10.133 are for the correct values of the $X_i$'s if

$$
\begin{aligned}
H_0 - \Sigma\, y_i(H) &= 0 \\
N_0 - \Sigma\, y_i(N) &= 0 \\
O_0 - \Sigma\, y_i(O) &= 0 \\
C_0 - \Sigma\, y_i(C) &= 0
\end{aligned}
\qquad (10.136)
$$

where all species containing the hydrogen element are listed in the first equation, and so on. These are used for iteration to determine the correct values of $X_i$ and consequently of $y_i$ to satisfy the material balance equations for the assumed and corrected values of $T_2$ and $v_2$. These values of the gaseous constituents, together with iterated values of $v_2$ and $\beta$, are used to calculate the temperature (equation 10.71). The values of the temperature are compared with the assumed value, and if the difference is more than $20°$, a new value of the temperature is selected, and the whole process of iteration is repeated.

Considering the material balance equations in which the four control variables $X_i$ may occur to odd powers, these constitute four nonlinear equations that must be solved by iterative procedures. The iteration requires that values for the correction of the $X_i$ variables be determined from the results of each iterative step. That is, differences between terms in equations 10.136 will usually not be equal to zero, and the remainder $R_i$ is then used to determine four corrective factors for the $X_i$'s.

Cook (1958) used a method for correcting values of the variables based on a similar method by Brinkley and Wilson (Brinkley and Smith, 1949). The Newton–Raphson method may also be employed, but the method used by Cook (1958) may converge more rapidly.

To apply this latter method, the equations for the gaseous species may be written in the form

$$y_i = \gamma_i X_1^a X_2^b X_3^c X_4^d \qquad (10.137)$$

where $a$, $b$, $c$, and $d$ may take on the values $-1$, $0$, $1/2$, $1$, or $3/2$, and the $\gamma_i$'s are functions of $K_i$'s and $F$. The relation between the values of the $X_i$'s which result in $R_i$'s instead of zero and the correct values $X_i$ are determined as follows: Assume that the correct value $X^*$ is to be found in terms of a correction value:

$$X^* = (1 + h_i) X_i \qquad (10.138)$$

where the $h_i$'s are the required correction factors for each of the four control variables to give the correct values. The approximate value of $X_i$ is given by:

$$X_i = \frac{X_i^*}{1 + h_i} = X_i^*(1 - h_i \cdots) \qquad (10.139)$$

or if $X_i$ occurs to a power $\nu$

$$X_i^\nu = (X^*)^\nu (1 - \nu_i h_i + \cdots) \qquad (10.140)$$

where the higher powers of $h$ are neglected.

The material balance equations may be written for the approximated $X_i$:

$$H_0 - \Sigma\, y_i(H) = R_1$$
$$N_0 - \Sigma\, y_i(N) = R_2$$
$$O_0 - \Sigma\, y_i(O) = R_3$$
$$C_0 - \Sigma\, y_i(C) = R_4 \qquad (10.141)$$

The quantities $X_i^\nu = (X_c^*)^\nu (1 - \nu_i h_i)$ are then substituted into the material balance equations, resulting in

$$H_0 - \Sigma\, y_i(H) + \Sigma\, \nu_i h_i\, y_i(H) = R_1$$
$$N_0 - \Sigma\, y_i(N) + \Sigma\, \nu_i h_i\, y_i(N) = R_2$$
$$O_0 - \Sigma\, y_i(O) + \Sigma\, \nu_i h_i\, y_i(O) = R_3$$
$$C_0 - \Sigma\, y_i(C) + \Sigma\, \nu_i h_i\, y_i(C) = R_4 \qquad (10.142)$$

If the adjusted values of $X_i^*$ are correct, the sum of the two first terms on the left side of equation 10.142 is equal to zero. This gives the four expressions in terms of $h_i$ and $R_i$ for determining the values of $h_i$, the correction factors. That is, when the expressions under the summation signs are written out in complete form, a set of four linear simultaneous equations in $h_i$ is obtained

$$A_{11}h_1 + A_{12}h_2 + A_{13}h_3 + A_{14}h_4 = R_1$$

$$A_{21}h_1 + A_{22}h_2 + A_{23}h_3 + A_{24}h_4 = R_2$$

$$A_{31}h_1 + A_{32}h_2 + A_{33}h_3 + A_{34}h_4 = R_3$$

$$A_{41}h_1 + A_{42}h_2 + A_{43}h_3 + A_{44}h_4 = R_4 \tag{10.143}$$

where the values of the $A$'s and $R$'s are as follows:

$$A_{11} = 2y_1 + 2y_5 + 4.5y_6 + 8y_7 + 8y_8 + 2y_9 + 0.5y_{11} + 0.5y_{13}$$
$$A_{12} = 3y_6$$
$$A_{13} = 2y_5 - 4y_7 + 2y_9 + y_{11}$$
$$A_{14} = 4y_7 + 4y_8 + 2y_9$$
$$R_1 = H_0 - 2y_1 - 2y_5 - 3y_6 - 4y_7 - 4y_8 - 2y_9 - y_{11} - y_{13}$$
$$A_{21} = 1.5y_6 = 1/2A_{12}$$
$$A_{22} = 4y_2 + y_6 + y_{12} + y_{14}$$
$$A_{23} = y_{12}$$
$$A_{24} = 0$$
$$R_2 = N_0 - 2y_2 - y_6 - y_{12} - y_{14}$$
$$A_{31} = y_5 + 2y_8 + 2y_9 + 0.5y_{11} \tag{10.144}$$
$$A_{32} = y_{12} = A_{23}$$
$$A_{33} = 2y_3 + y_5 + 2y_9 + 4y_{10} + y_{11} + y_{12} + y_{15}$$
$$A_{34} = 2y_3 + y_4 + y_8 + 2y_9$$
$$R_3 = O_0 - 2y_3 - y_4 - y_5 - y_8 - 2y_9 - 2y_{10} - y_{11} - y_{12} - y_{15}$$
$$A_{41} = 2y_7 + 2y_8 + y_9 = 1/2A_{14}$$
$$A_{42} = 0$$
$$A_{43} = y_3 - y_7 + y_9$$
$$A_{44} = y_3 + y_4 + y_7 + y_8 + y_9$$
$$R_4 = C_0 - y_3 - y_4 - y_7 - y_8 - y_9$$

An alternate method for determining the correction factors, that is, solving the nonlinear equations for $X_i$ values, is by means of the Newton–Raphson procedure. In this method, the functions of the corrections $(X_i + h_i)$ are expanded in a Taylor's series, and four simultaneous equations are determined for the $h_i$'s. The resulting values of the $A_{ij}$ coefficients are similar in form to those in equation 10.143, but the $A_{ij}$'s obtained by the Newton–Raphson methods differ by a factor of $1/X_1$, $1/X_2$, $1/X_3$, and $1/X_4$ for each of the equations, respectively. The method of Brinkley and Smith (1949) apparently gives a more stable matrix—that is, with the higher values on the diagonal.

# 11

# ANFO AND SLURRIES

For several reasons, primarily cost, ammonium nitrate–based explosives have taken over a large part of the explosives market in the United States in the past three decades. ANFO does not meet all of the needs for explosives, especially in very hard rock and in wet conditions, and slurries have proved to be an effective substitute for dynamites for these and other uses.

## 11.1   AMMONIUM NITRATE

Because of its oxygen content, favorable heat of formation, and low cost, ammonium nitrate (AN) has been used as an ingredient of high explosives and as a substitute for nitroglycerin. Mixtures of AN and carbonaceous materials had been patented as explosives as early as 1885, but it has only been a few years since mixtures with hydrocarbons (ANFO) have become usable and, because of their low cost, have captured a large portion of the explosives trade in the United States. This type of blasting agent is inexpensive, is safe to use, and will perform well under a wide variety of conditions. Its limitations are due largely to its relatively low velocity of detonation, low energy yield, and affinity for water.

Also, AN is corrosive, because it acts as a weak acid when it is wet and will react with most metals, that is,

$$NH_4NO_3 + H_2O \leftrightarrows HNO_3 + NH_4OH \qquad (11.1)$$

A second property that affects its use is that it is very hygroscopic. When AN is in an atmosphere of 60% relative humidity in an open container, it will dissolve itself. In practice, exposure of prilled AN to moisture causes it to cake, which reduces its sensitivity to detonation. AN also recrystallizes at 23.1°C (89.8°F), and heating and cooling above and below this temperature cause the prills (or crystals) to disintegrate, which also increases its sensitivity to initiation and accelerates caking if moisture is present.

Almost all of the physical and chemical properties of AN (Table 11.1) have an effect on its use as an ingredient in explosives, particularly for field mixing. Most AN for use as fertilizers is made by a prilling process in which a molten 95% solution is sprayed in a prilling tower against a countercurrent of air. The droplets of AN solidify as they fall through a prilling tower. They are then collected at the bottom and are conveyed through a series of rotary dryers to remove the remaining 5% of water. This produces a relatively porous, low-density prill. It is this porosity that makes fertilizer-grade ammonium nitrate (FGAN) usable as an oxidizer in ammonium nitrate–fuel oil (ANFO) blasting agents.

Prior to the Texas City disaster, where two ships loaded with FGAN caught fire and exploded, prills were coated with 2% wax to prevent caking. However, the wax is a hydrocarbon and acts as both a fuel and a sensitizer. Since that time, prills have been coated with finely ground inert materials such as diatomaceous earth.

**TABLE 11.1. Physical and Chemical Properties of Ammonium Nitrate**

| Property | Value |
|---|---|
| Formula | $NH_4NO_3$ |
| Formula Weight | 80.05 |
| Heat of Formation | 87.27 kcal/mol (2070 Btu/lb) |
| Heat of Fusion | 16.2 kcal/gm (29.2 Btu/lb) |
| Heat of Solution | 6.3 kcal/mol (142 Btu/lb) |
| Melting Point | 169.6° C (337.3° F) |
| Decomposition Point | Not defined (between 200-260° C) |
| Physical Form | White crystalline solid |

CRYSTAL FORMS:

| Process | State Initial → | Final | Temperature |
|---|---|---|---|
| Transition | V | IV | 18° C ( 0.4° F) |
| Transition | IV | III | 32.1° C ( 89.8° F) |
| Transition | III | II | 84.2° C (183.6° F) |
| Transition | II | I | 125.2° C (257.4° F) |
| Transition | I | liq. | 169.6° C (337.3° F) |

DENSITY:

| | |
|---|---|
| Form V | 1.66 g/cc (103.6 lb/ft³) |
| Form IV | 1.73 g/cc (107.5 lb/ft³) |
| Form III | 1.65 g/cc (103.0 lb/ft³) |
| Form II-Liquid | 1.40 g/cc |

Solubility of ammonium nitrate in water and density of aqueous solutions:

| Temperature °C | Wt. % AN in Water | Density of Solution |
|---|---|---|
| 0 | 54.5 | |
| 20 | 66.0 | 1.310 |
| 40 | 74.0 | 1.345 |
| 60 | 80.5 | 1.370 |
| 100 | 91.0 | 1.425 |

The rapid increase in use of FGAN as an explosive agent since the introduction in 1955 of AN–fuel mixtures (e.g., Akremite) led to more extensive investigations of the effects on their explosive properties of various fuels, sensitizers, density, and other explosibility factors. Earlier, AN had found wide usage as an explosive ingredient in dynamites and some military explosives, its function being to furnish both oxygen and heat. The explosive characteristics of AN mixed with substances other than fuel oil have been investigated extensively to define limits of applicability and comparative safety.

A large amount of laboratory and field research has been performed to evaluate the pertinent physical and chemical properties of AN–fuel blasting agents, their influence on detonation and explosion parameters, and, ultimately, their influence on field performance. Initiation sensitivity tests usually require only a "yes" or "no" answer, whereas other properties are compared by means of measurement of detonation pressure and velocity or field performance.

The value of AN as an ingredient of explosives is its relatively low cost and ability to furnish oxygen and heat when it decomposes at high temperatures—that is, as an oxidizer. In large charge diameters, pure AN of proper particle size can sustain a stable detonation. The equation for its reaction, assuming no oxides of nitrogen are formed, and its heat of reaction are given by

$$NH_4NO_3 = N_2 + 2H_2O + \tfrac{1}{2}O_2 + 343.1 \text{ kcal/kg} \qquad (11.2)$$

When AN is properly mixed with a fuel such as fine carbon or fuel oil, the oxygen reacts with it to increase the released energy by a factor of almost 3. For "do-it-yourself" AN blasting agents using fuel oil, the reaction equation for a stoichiometric or oxygen-balanced mixture (i.e., where there is just enough oxygen in the explosive to yield only $CO_2$, $H_2O$, and $N_2$) becomes

$$NH_4NO_3 + \tfrac{1}{3}CH_2 = N_2 + \tfrac{1}{3}CO_2 + \tfrac{7}{3}H_2O + 927.0 \text{ kcal/kg} \qquad (11.3)$$

producing $\tfrac{11}{3}$ moles of gas per mole of AN.

For a stoichiometric mixture of ammonium nitrate and carbon, the reaction is

$$NH_4NO_3 + \tfrac{1}{2}C_2 = N_2 + \tfrac{1}{2}CO_2 + 2H_2O + 929.0 \text{ kcal/kg} \qquad (11.4)$$

which produces $\tfrac{7}{2}$ moles of gas per mole of AN.

Thus, the explosive energy obtained by mixing either the proper amount of carbon or fuel oil is approximately the same, but fuel oil produces about 5% more gas in the explosion products. Also, the fuel oil is much more readily handled in preparation procedures to obtain a uniform mixture. The chemical energy yield drops off rapidly when either too much or too little oil is added. Calculated (ideal) energies (heats of explosion) for various percentages of oil show that the maximum occurs for an oxygen-balanced condition.

## 11.2   PERFORMANCE PARAMETERS—ANFO

The pertinent explosive properties and performance parameters of ANFO and the factors that affect them are given (Table 11.2) in a classification under four general headings: (1) chemical properties, (2) physical properties, (3) detonation state parameters, and (4) explosion state parameters. Most of these are interrelated or dependent on the others. For example, detonation velocity is dependent on loading density, particle size, composition, and other physical and chemical factors. The

**TABLE 11.2. Physical and Chemical Explosion Parameters**

| Parameters of Explosive | | Parameters of Explosion | |
|---|---|---|---|
| Chemical | Physical | Detonation State | Explosion State |
| Composition | Particle density | Temperature | Temperature |
| Absolute reaction rate | Loading density | Pressure | Pressure |
| Mass reaction rate | Charge diameter | Velocity | Available work |
| Available work | Charge confinement | Heat of detonation | Heat of explosion |
| Prill coating | Particle size | Density of products | |
| Moisture content | Particle shape | Sensitivity | |
| Explosion products | Moisture | | |
| Heat of formation | | | |

calculation of detonation parameters by means of the hydrodynamic theory and related thermochemical equations, however, gives the ideal values but does not take confinement, particle size, or charge diameter into account.

## 11.2.1 Confinement

The equations for an ideal detonation wave assume that mass and energy are conserved across the detonation front, as well as the average momentum of particles in the direction of propagation. This requires (1) that the explosive be a semi-infinite mass or (2) that the length of the reaction zone be very small relative to the diameter. One might assume that if the explosive column is sufficiently confined and thus most of the atoms of molecules that have lateral motion are reflected back in the reaction zone within a critical distance of the detonation front, thus contributing to the energy of the shock wave, an ideal velocity would be attained. However, for small-diameter ANFO charges with ''infinite'' confinement, the maximum velocity is a function of the diameter.

For primary explosives, such as lead azide, the reaction zone is very small, as it is for secondary explosives such as PETN, and their detonation velocities are equal to ideal values even when they are of small diameter or unconfined. Explosives that are composed of large grains and that have relatively slow absolute reaction rates will exhibit detonation velocities less than ideal unless they are adequately confined and of large diameter.

To evaluate effects of confinement, experimental values of detonation velocity of ANFO were determined (Figure 11.1) as a function of W/C, the weight of explosive charge per foot divided by the weight per foot of the casing, the ratio, W/C, being used as a measure of confinement. The tests were made with three different charge diameters—3, 4, and 6 in.—and in casings of stove pipe, light-weight steel pipe, standard black iron pipe, and concrete-jacketed pipes with varying thickness of concrete (see also Figure 11.14).

The effect of increased confinement is similar in some respects to that of an increase of charge diameter. The curves rise rapidly to a critical value beyond which a further increase in the W/C ratio no longer has an appreciable effect on the detonation velocity. For each test diameter, the W/C ratio for black iron pipe was slightly beyond the inflection of the curve or on the horizontal section. The W/C ratios corresponding to black iron pipe are as follows:

|  | 3-in. diameter | 4-in. diameter | 6-in. diameter |
|---|---|---|---|
| W/C | 2.52 | 2.35 | 1.72 |

It therefore appears that detonation velocity test conditions in standard iron pipe closely approximate the confinement of field shots in rock, and experimental results from tests in iron pipe can be used to predict field performance. As might be expected, smaller charge diameters require a larger W/C ratio for a maximum stable detonation velocity to be established.

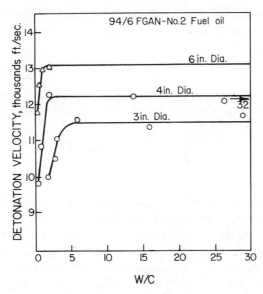

**Figure 11.1** Influence of steel confinement on detonation velocity (Clark et al., 1961).

## TABLE 11.3. Borehole Versus Iron Pipe Test Results

| Explosive Mixture | Density | 9-1/2-in. Strip-Mine Avg. Det. Vel (ft/sec) | Test Shots in 3-in. Iron Pipe Avg. Det. Vel. (ft/sec) |
|---|---|---|---|
| 95% 50% Reg. AN (50% Fines) 5% Fuel Oil No. 2 | 0.86 | 13,690 | 12,280 |
| 95% 50% Reg. AN Unc. (50% Fines) 5% Fuel Oil No. 2 | 0.86 | 12,370 | 11,820 |
| 95% AN Reg. Unc. 5% Fuel Oil No. 2 | 0.83 | 12,970 | 11,760 |
| 95% AN Reg. 5% Fuel Oil | 0.83 | 12,970 | 11,930 |

*Source:* Clark et al. (1961).

An equation for the curve through the inflection points on the curves can be written in the following form:

$$D = D^* \exp\left[-K(\rho, R)\text{W/C}\right], \tag{11.5}$$

where $K(\rho, R)$ = constant for given values of $\rho$ and $R$, $D$ = optimum detonation velocity for a given $\rho$ and $R$, and $D^*$ = ideal detonation velocity for a given $\rho$. Hence,

$$\text{W/C} = \frac{1}{K} \ln \frac{D^*}{D} \tag{11.6}$$

which gives the relationship between the optimum confinement and the maximum velocity for a given explosive density and charge diameter.

As a further study of the effects of confinement and a check on the results obtained in controlled experimental work, detonations in four $8\frac{1}{2}$-in. diameter shot holes were instrumented at a strip coal mine. The results of the tests checked reasonably well with measurements made in steel pipe when an adjustment is made for the larger diameter of the boreholes (Table 11.3). The effect of confinement was also evaluated further in determining the combined effects of particle size and confinement (see particle size discussions in next section).

## 11.2.2   Critical Diameter

Every explosive of a given composition, density, and grain size distribution has a critical charge diameter below which it will not propagate a detonation. For primary explosives such as lead azide, it is too small to be measured. The critical diameter of pure crystalline AN unconfined is 6–9 in., that of fertilizer-grade ANFO is in the neighborhood of 4 in., and confinement decreases it to about $1\frac{1}{2}$ in. Eyring et al. (1949) have shown that the relationship between reaction zone length and the detonation velocity is controlled by a heat of activation parameter $\Delta H^*/R_o T_i$ where $R_o$ = gas constant, $T_i$ = temperature of ideal detonation, and $\Delta H^*$ = heat of activation.

According to Eyring, there exists a minimum radius of an unconfined charge of a given explosive with other properties held constant, below which a stable detonation will not propagate. The critical radius (defined in units of the ideal reaction zone length) and the corresponding critical velocity (defined in units of the ideal velocity) depend only on the heat of activation parameter $\Delta H^*/R_o T_i$. If the reaction zone length approaches a value that too greatly exceeds that of the radius of the charge, then detonation will fail.

However, the porosity of AN prills increases the effective specific surface area exposed to reaction and thus decreases the effective particle size. Hence, the reaction zone of ANFO is shorter than the prill diameter otherwise indicates.

The rate of release of mass energy in relation to a given charge diameter is as critical as the activation energy criterion. These factors in conjunction with density, particle size, confinement, and other parameters govern detonation failure of ANFO at critical diameters. Thus, a certain portion of the energy required to produce ideal detonation must be released within the length of the propagation detonation head, which is itself a function of charge diameter. In any event, a maximum detonation velocity is usually achieved for an unconfined charge when the $a/d$ ratio (reaction zone length divided by charge diameter) becomes less than 1. In other words, when the reaction zone becomes shorter than the detonation head, all of the explosive heat (energy) is released within the detonation wave, and the ideal temperature, pressure, and velocity are achieved. For ANFO, the reaction zone is longer than the detonation head in blast hole diameters less than 8–10 in., giving less than ideal velocity.

### 11.2.3  Charge Diameter

The influence of charge diameter on detonation velocity is pronounced (Figure 11.2). The two top curves for regular fertilizer-grade AN and the recirculate product (see particle size discussion) have the same general shape, but the curve for fertilizer-grade AN at the pouring density and essentially no confinement (in stovepipe) has a different rate of change of slope. Therefore, in an analysis of charge diameter effect, it is important to include the influence of confinement. For example, at no confinement, the critical diameter for fertilizer grade ANFO at 0.80 density is 4 in., the optimum diameter (to achieve ideal detonation) is much greater than for confined charges, and the critical diameter for confinement in iron pipe is 1 ½ in. Critical diameter, as defined earlier, is that minimum diameter at which a stable detonation will propagate, and optimum diameter, the minimum diameter at which an increase in diameter no longer has an appreciable effect on the detonation velocity.

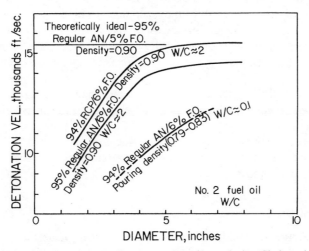

**Figure 11.2**   Influence of charge diameter on detonation velocity (Clark et al., 1961).

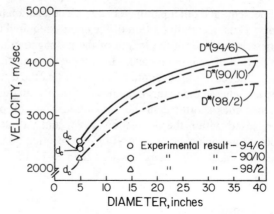

**Figure 11.3**   Charge diameter velocity curves for prilled AN–fuel explosives (Cook, 1958).

The velocity of detonation is dependent, for small-diameter charges, on the mass reaction rate, which is in turn a function of particle size and absolute reaction rate. That is, for a slow-reacting material, the chemical reactions may not all take place within the detonation wave or ''head'' and will therefore contribute only part of the total available heat of reaction to support the detonation wave. The approximate fraction of the reaction $N$ completed in the detonation wave is expressed by Cook (1958) as

$$N = \left(\frac{D}{D*}\right)^2 \tag{11.7}$$

where $N$ = fraction of reaction, $D$ = actual detonation velocity, and $D*$ = ideal detonation velocity. Thus, if the measured detonation velocity is about 0.7 that of the ideal, approximately one-half of the reaction takes place in the detonation wave zone. As the charge diameter is increased, the length of the detonation wave becomes longer, and the percentage of the reaction taking place in the detonation wave increases to an optimum where the reaction is completed within the detonation wave of $D = D*$.

The calculated OB velocity–diameter curves and experimental points determined by Cook (1958) (Figure 11.3) correspond roughly to the lower experimental curve (Figure 11.2) for charges with very little confinement. Therefore, a confinement of W/C of 2 has the effect of reducing the critical diameter by a factor of about 3 and the optimum diameter by a factor of approximately 10 according to both calculated and measured results.

### 11.2.4   Fuel Oil Content

The heat of reaction of commercial prills and fuel oil is greatest at oxygen balance (Figure 11.4), whereas the detonation velocity is a maximum with a fuel oil content

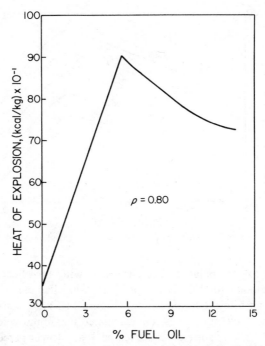

**Figure 11.4** Calculated heat of explosion of ammonium nitrate–fuel oil mixtures at 0.80 density (Clark et al., 1961).

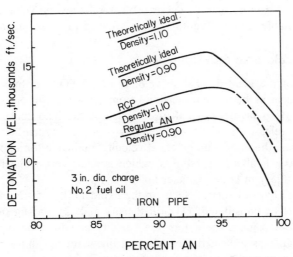

**Figure 11.5** Influence of oil content on detonation velocity (iron pipe) (Clark et al., 1961).

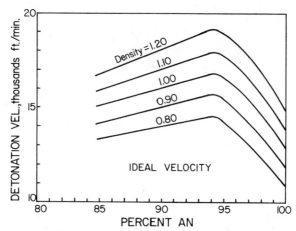

**Figure 11.6**  Influence of oil content on detonation velocity (ideal velocity) (Clark et al., 1961).

slightly less than oxygen balance. The curves showing the effect of oil content on the detonation velocity for FGAN and RCP in 3-in.-diameter steel pipe (the undersize minus 8-mesh reject from prilling) (Figure 11.5) are similar and are nearly parallel to the ideal velocity curves, the calculated maximum velocity occurring at OB (Figure 11.6). The experimental curves peak at a lower percentage of oil than the theoretical curves, probably owing to an excess of oil on the surface of the prills for OB. That is, for low percentages of oil at the near surface portion of the prills that react first, a stoichiometric condition exists that causes a higher detonation velocity because there is near oxygen balance for that portion of the prill that reacts within the detonation head.

However, a large excess of oil is less detrimental to the detonation velocity than a shortage. As the oil content decreases and the oxygen balance becomes more positive, there is a more marked decrease in the detonation velocity, because the maximum available energy (Figure 11.2) is affected in a like manner.

### 11.2.5   Particle Size and Density

The increase of the ideal detonation velocity of ANFO with bulk density is approximately linear in the range of densities from 0.80–1.20 for oil contents ranging from 0–10%. If all factors other than composition are held constant, it is found that maximum release of energy and velocity occur at oxygen balance. For AN–fuel oil mixtures, the maximum practical velocity attainable for a borehole pouring density of 0.80± is about 12,000 ft/s, which is effective for blasting many types of rock, particularly in large-diameter holes.

In the evaluation of density effects on the explosibility of AN prills, two types of density must be considered: (1) prill density and (2) loading or bulk density. For a given composition, the detonation velocity is a function of the loading or bulk density, and, as indicated previously, a linear relationship between velocity and density is commonly used for prediction purposes.

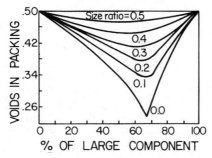

**Figure 11.7** Voids for two components with 0.5 voids for single component (Dallavalle, 1948).

### 11.2.5.1 Bulk Density of Spherical Packings

The bulk density of particulate matter is a function of the density of the particles and the percentage of void space. Theoretically, for uniform-size spheres, the void space varies from about 27–48% for different types of packing from 100% rhombohedral to 100% cubic. The percentage void space for prills, which are fairly uniform in size and shape, falls within these two values, usually toward the higher value because of random arrangement of the prills. The bulk density depends on the shape of the particles, their density, and their geometrical arrangement.

According to the grain burning theory proposed by Eyring et al. (1949), in which the grains of explosives are assumed to burn from the surface inward, the efficiency of burning and detonation depends on the number of contact points between grains (see below). In the closest (rhombohedral) packing arrangement, every sphere touches 12 other spheres, whereas in the most uniform packing (cubic) arrangement, contact exists with only six adjoining spheres. Theoretically, rhombohedral packing should yield higher velocities and more efficient combustion than cubic, both because of higher bulk density (as predicted by the hydrodynamic theory) and because of a higher mass reaction rate due to a larger number of contact points between grains.

Systems of one size component are not amenable to a high bulk density as are multiple-component systems, and binary systems (or two sizes of spherical particles) do not reduce the void space as much as three or more component systems (Dallavalle, 1948) (Figures 11.7, 11.8).

It can be shown (Dallavalle, 1948) that the absolute volume $\phi$ of the solids in a system with $n$ components is given by

$$\phi = \frac{1}{1 + v} (1 + v + v^2 \cdots v^{n-1}) \tag{11.8}$$

where $v$ is the initial void space for each component and is identical for all components. The minimum voids for two-, three-, and four-component systems of known initial voids of 40–60% vary with particle size (Figure 11.8), whereas the composition of packings for minimum voids for two components (Table 11.4)

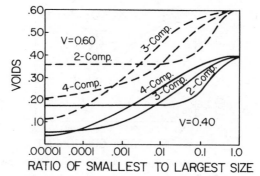

**Figure 11.8** Minimum voids for various component systems for initial voids of 0.40 and 0.60 (Dallavalle, 1948).

shows that only small amounts of small-size material are required for minimum void space.

For two or more components, values of ratios of diameters greater than 0.003 required two components. For spherical packings of 40% initial void space, three-component systems are almost as effective as four-component systems. For a three-component system of sizes 1, 0.003, and 0.001 in., the critical ratio is 0.001, and the percentages of each size for maximum density are 64.2, 25.6, and 10.2%, respectively. Under ideal packing conditions, this would yield a void space of somewhat less than 10%. In practice, however, these low values (ideal packings) are not attainable.

For a binary system where the smaller component is just large enough to fit in the interstices of the large component, the number of contact points between spheres would be increased by at least 4 for each included small particle. On the basis of the mechanics of the contact-point theory of combustion, the mass reaction rate for this type of binary mixture should be high compared with a single-component system. Also, where the packing is not ideal and the smaller spheres are not located

**TABLE 11.4. Packing of Minimum Void Space**

| Initial Voids in Packing of Uniformly Sized Particles | Number of Components | Volume (Percent of Each Component) | | | |
|---|---|---|---|---|---|
| (Percent) | | $d_1$ | $d_2$ | $d_3$ | $d_4$ |
| 40 | 2 | 71.5 | 28.5 | -- | -- |
| | 3 | 64.2 | 25.6 | 10.2 | -- |
| | 4 | 61.7 | 24.6 | 9.8 | 3.9 |
| 60 | 2 | 62.5 | 37.5 | -- | -- |
| | 3 | 51.0 | 30.6 | 18.4 | -- |
| | 4 | 46.0 | 27.6 | 16.5 | 9.9 |

*Source:* Dallavalle (1948).

in the interstices between the larger particles, they will hold the larger particles apart. That is, the introduction of small grains, according to Eyring et al. (1949), decreases the number of contact points between grains and thus inhibits the propagation of intergrain reactions and has been demonstrated experimentally for ammonium picrate at densities of 0.95 and 1.00. It also appears to be true for *dense-particle* prilled AN.

Presumably greater particle density should enhance the detonation velocity, but for small-diameter charges of 3 in. it has been found that a dense crystalline AN–fuel mixture cannot be detonated with a nominal booster but that less dense fertilizer prill–fuel oil mixtures can be detonated consistently. Hence, the porous structure of fertilizer-grade prills makes them more amenable to initiation and enhances their capability for sustaining a detonation. Inasmuch as the mass reaction rate and, hence, the portion of the explosive reacting within the detonation wave are among the critical factors in determining the stability of detonation, particle-size effects are of first-order importance in the explosibility of AN. The surface area available for reaction, inasmuch as the particles, for the most part, will react from the surface inward, is dependent on the particle size of the explosive. The specific surface area for spherically shaped particles is inversely proportional to the particle diameters, and the surface area per unit volume increases very rapidly with a decrease in particle size, especially at smaller diameters. This results in a much larger surface area per unit volume of material accessible to chemical reactions—that is, to attack by gaseous molecules and atoms moving into the front of the reaction zone.

A simplified scheme of the progress of a detonation wave through a granular material, such as prilled AN, involves the rate at which the explosive mass completely reacts, which in turn depends on the absolute reaction rate of the explosive material and the particle size. The relationship between particle size and absolute mass reaction rates for spherical particles was analyzed by Eyring et al (1949). It is postulated that heat conduction in the solid AN is of minor importance and therefore that the chemical reaction must proceed preferentially at the surface of the grains. If $\tau$ is the time required for a grain of a given radius to burn at a linearly constant rate along the radius (Figure 11.9), the percentage of material $N$ consumed varies for three different types of burning: (1) radially inward, (2) radially outward, and (3) from one point ignited on the surface. For a given size grain and a given reaction rate, the time for complete burning is the same for a sphere ignited over its complete surface and one ignited at the center, whereas the time of burning for a sphere ignited at one point on its surface is three times as great. For a sphere ignited over its whole surface, 90% of the material is consumed in 60% of the total time, whereas for a similar sphere ignited at its center, only 20% of the material is consumed in 60% of the time. For one-point ignition on the surface, the total time is twice as great, and at an equivalent time only approximately 10% of the mass is consumed. If a spherical particle is ignited at two points, the rate of burning would be doubled over the first portion of the burning period. It is this latter type of burning (through contact points) that affects the continuity of reaction in coarse-grained material such as AN prills.

For smaller particle sizes, then, the reaction time is reduced in proportion to

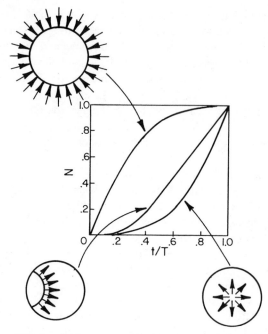

**Figure 11.9** Rates of burning for three types of ignition for spherical grains (Eyring et al., 1949).

the radius of the particle, and the mass reaction rate per unit volume of explosive is thus increased depending on the geometry of burning. Hence, assuming any one of the three types of burning takes place, the grain reaction time is proportional to the grain radius. But if more energy is released in the detonation zone, the temperature is higher, and the efficiency of the detonation is enhanced by the temperature increase.

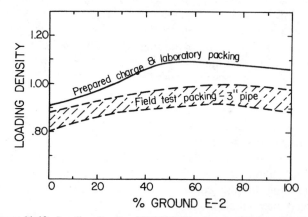

**Figure 11.10** Loading density of E-2/FGAN mixtures (Clark et al., 1962).

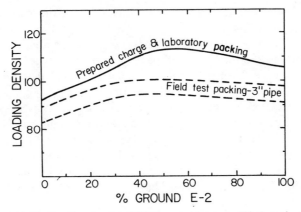

**Figure 11.11** Loading density of E-2/uncoated mixtures (Clark et al., 1962).

Obviously, surface burning is the most efficient of the three types, but, in all probability, a prill begins to burn at one or more points on its surface and in many cases may approach the surface burning conditions, which may be expressed mathematically in different form (Eyring et al., 1949) as

$$\frac{dN}{dt} = \frac{3k_r(1 - N)^{2/3}}{R_g} \tag{11.9}$$

or, in an integrated form,

$$N = 1 - \left(1 - \frac{t}{\tau}\right)^3, \tag{11.10}$$

where $t$ = reaction time and $\tau$ = time required for consumption of grain.

Equation 11.9 is known as the two-thirds power law which states that the rate of mass consumption varies as the two-thirds power of the amount of material left in the grain. Equation 11.10 shows that the amount of material consumed varies

**TABLE 11.5. Particle Size E-2 Prills**

| Mesh | Percent |
|---|---|
| +20 | 10.0 |
| 20/40 | 26.9 |
| 40/60 | 12.0 |
| 80/100 | 8.8 |
| -100 | 10.5 |

*Source:* Clark et al. (1962).

as a cube of a function of the ratio of the reaction time $t$ to the total time $\tau$ required for the consumption of the grain.

Mixtures of FGAN and ground E-2 (dense) prills were tested for bulk densities obtained by pouring and by vibrational packing in pipes. Higher packing densities were obtained by careful vibration packing in the laboratory than for field packing (Figures 11.10, 11.11). The particle size distribution of ground E-2 varies somewhat, but a representative distribution is given in Table 11.5. This product, when mixed with standard prills, would approximate very roughly a four-component mixture whose ratio of smallest to largest size is about $1/100$. Ideally, it should be possible to reduce the void space by 10% with the proper percentage of components, or an increase of 30% in bulk density. The results indicate that this order of increase in bulk density was obtained but was partially due to the greater density of ground E-2.

### 11.2.6 Detonation Velocity

Particle size and particle size distribution as well as bulk density are vital factors in determining the sensitivity, stability, and rate of detonation of condensed explosives, particularly those that have a slow absolute reaction rate and in which the grain size is normally $> 20$ or 30 mesh. Several AN prilled products were detonated in 3-in.-diameter steel pipes to check the effects of particle size, both for discrete screen fractions and for given particle size distributions. The particle size distributions of regular AN prills and the recirculation product (RCP), which is the undersize reject in the production of standard plus 20-mesh prills (Figure 11.12), indicate that about 60% of RCP is plus 20 mesh after the size separation process.

Experimental velocity–density curves (Figure 11.13) for regular fertilizer-grade ANFO are almost linear in the range tested. As the pouring density approaches the

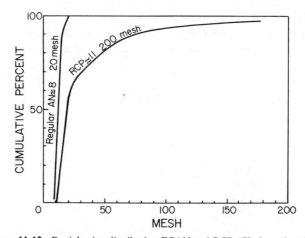

**Figure 11.12** Particle size distribution FGAN and RCP (Clark et al., 1962).

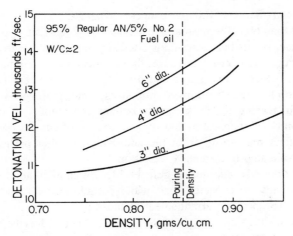

**Figure 11.13**  Influence of charge density on detonation velocity (Clark et al., 1962).

maximum (near 0.90), the curves show a slight increase in slope. It is difficult to increase the loading density of ANFO beyond 0.90 without physically tamping the explosive column.

Measured detonation velocities over a range of loading densities from 0.75–1.10 (Figure 11.14) are lower than the ideal velocity. For a loading density of about 0.90, the detonation velocity increases with decrease in prill size. The apparently anomalous behavior of the 60/100-mesh material beyond this density is probably due to experimental error.

The effect of packing, proposed by Eyring et al. (1949), would offer a possible explanation of the apparent anomalous detonation behavior of some AN particle sizes. Two of the most obvious of these are (1) that regular AN fuel has a velocity

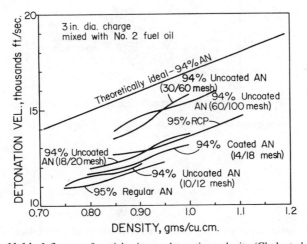

**Figure 11.14**  Influence of particle size on detonation velocity (Clark et al., 1962).

consistently lower than 60/100 mesh uncoated and (2) that 30/60 mesh has a higher velocity than 60/100 mesh at densities above 0.95. One hypothesis offered is that the contact points between grains constitute hot spots due to compressional heating and friction, and at such hot spots the chemical reaction in individual grains begins. Also, a reduction in loading density reduces the number of contact points for uniform particle sizes. For mixed grain sizes, small grains may fit in between larger grains, increasing the bulk density but not the number of contact points. Thus, while combined increase in density and number of contact points is desired, the opposite effect may result from the addition of smaller particles unless the material is consolidated by extensive vibration.

Also, both of the AN mixtures (Figure 11.14), which exhibited lower velocities than might normally be expected, had wider particle size distributions than the other mixtures. The packing arrangement of uniform-size particles versus nonuniform obtained without effective vibration could account for the differences in detonation velocity.

In each of the previous tests, a standard primer of four half-size sticks of 60% ammonia dynamite was used. The primer is known to exceed the minimum booster required to detonate an OB mixture of prilled AN–fuel oil with a density of 0.90 but may be less effective than boosters with a higher detonation pressure. No difficulty was experienced in detonating any of the particle size mixtures at the densities in the particle size series of tests (Figure 11.15). In all mixtures the

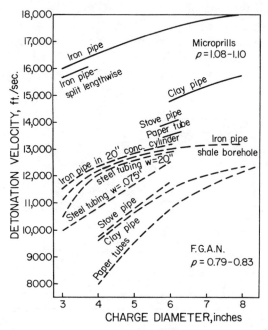

**Figure 11.15**  Detonation velocity for various densities and confinement—FGAN and micro-prill (Clark et al., 1962).

maximum density tested was that obtained by vibration without physical tamping, 0.79–0.83 for FGAN and 1.08–1.10 for small E-2 prills. The detonation velocity of stoichiometric fuel mixtures depends on prill density, bulk density, charge diameter, particle size, and confinement. This is shown by the results of a large number of tests of FGAN; fertilizer-grade AN, which is 8/20-mesh; and microprills, which are −20-mesh dense stabilized prills (Figure 11.15). The detonation velocity consistently increases with increase in bulk density, confinement, and charge diameter. The velocity for one shot in a shale borehole was slightly less than that for iron pipe of the same diameter, probably owing to lower (pouring) density.

Although E-2 microprills, which are stabilized, dense prills of −20-mesh size, showed superior performance (18,000 ft/s), the cost of production for the commercial market does not justify their production, and they were available only on an experimental basis.

### 11.2.7  Sensitivity Tests

For practical applications, such as measurement of the sensitivity of AN–fuel oil and similar type blasting agents, two useful sensitivity tests have been employed: (1) the determination of the confined critical diameter and the minimum primer at the unconfined critical diameter, and (2) the minimum primer for the 3-in.-diameter confined charges. The first method is suited for comparison of the sensitivity of AN blasting agents, and the latter may be used for measuring relatively small changes in the sensitivity of ANFO with such variables as size, shape, structure of AN particles, oil content, particle and bulk density, and percentage of inert coatings.

The sensitivity of an explosive can be described in a general way as the energy required to initiate a detonation in it. An exact definition has not been established in a chemical or thermodynamic sense. Therefore, most sensitivity ratings for explosives are based on tests designed to measure sensitivity to detonation by friction, shock, heat, or sensitivity which results from chemical instability. Explosivity is determined not only by the chemical nature of the explosive but also by its physical state. Size, shape, and density of particles; packing density; homogeneity; charge mass; and confinement are important physical properties. The sensitivity tests usually employed for commercial explosives, such as air-gap, impact, and friction, are not adaptable for field testing of ANFO. For insensitive blasting agents, minimum primer tests are commonly used.

Because of the range of sensitivity of the kinds of available ANFO and the factors that affect their sensitivity, several types of measurements have been devised. Three methods were employed by Yancik et al. (1959):

**1.** Related to critical diameter. a. Confined critical diameter. b. Unconfined critical diameter. c. Minimum primer at unconfined critical diameter. For small primers, bundles of No. 6 caps were employed, and for large primers, a plastic explosive. Charge lengths were 24 in. to determine stable or unstable detonation.

Diameters were varied from 2- to 8-in.; steel pipe was used for confined charges, and cardboard tubing for unconfined charges.

**2.** Minimum primer for 3-in. confined charge. Various types of ANFO were tested, a 3-in. pipe being assumed to represent the confinement of a borehole in rock. Primers were chosen as in paragraph 1 above.

**3.** Minimum Primacord sensitivity. Tests may be conducted with bundles of Primacord of different lengths, although this test is not extensively used.

For blasting operations, sensitivity tests serve two primary functions: to assess the reliability of field performance of blasting agents and to evaluate safety hazards of the mixtures. In addition to the first two methods described above, Cook (1958) recommends a sympathetic detonation test using either air-gap, cards, metals, or glass plates between a donor and an acceptor charge. For determining critical diameters, Cook used 50/50 cast pentolite boosters initiated with a No. 6 blasting cap; for large diameters (9 or more in.), 320-g boosters; and for diameters less than 2 in., a 40-g booster.

The products tested by Yancik et al. (1959) were the following:

FGAN, Fertilizer-grade AN: 8/20-mesh, porous prills, coated and uncoated.

E-2 prills. Solid prills stabilized against recrystallization: 8/20-mesh or screen fractions.

E-3 prills. Solid prills: 95% −60 mesh.

E-4 prills. Solid prills: 95% −80 mesh.

E-2 ground prills. E-2 prills ground to varying particle size distributions.

AN RCP. Recirculate prills—that is, −20-mesh reject prills which are usually reprocessed.

A typical particle size distribution of ground E-2 (Table 11.5) indicates a large percentage of fine material.

The sensitivity values obtained for unconfined charges (Table 11.6) show that the minimum primer decreases as the charge diameter increases. For confined shots, however, the opposite is true. The minimum primer values (Table 11.6) are for 15 shots fired at each diameter, and the values can therefore be considered to be representative. Two possible factors that might account for required minimum primer are that (1) the ratio of primer diameter to charge diameter increases with charge diameter, and (2) the distance the shock wave from the primer must travel to encounter the steel wall increases as the charge diameter increases. Hence, the MP for confined ANFO is less than that for unconfined ANFO.

Since variation of physical properties of AN–fuel oil mixtures affects the sensitivity, two standard procedures were adopted to reduce effects of preparation procedures: (1) charge density was the pouring density in a 6-in. pipe, and (2) test charges were thoroughly mixed. Thorough mixing resulted in near maximum velocities within a short time after mixing, while velocities for charges prepared by pouring the oil into an open bag were less than those for well-mixed prills.

**TABLE 11.6. Minimum Primer Versus Charge Diameter and Confinement**

| 94% AN Uncoated Prills | 6% Fuel Oil | Density = 0.83 |
|---|---|---|
| Steel Pipe Diameter Inches | No. 6 Caps Required for Detonation | Primacord Full Length of Charge |
| 1-1/2 | 3 | -- |
| 2 | 3 | -- |
| 3 | 4 | 150 grain |
| 4 | 6 | -- |
| 6 | 15 | 200 grain |
| 8 | 15 | -- |

*Source:* Clark et al. (1962).

Another vital factor is the change in sensitivity caused by the crystal transitions of AN at 89.8°F, which are accompanied by a 3.6% volume expansion. These may occur prior to usage; therefore, the temperature history of the prills affects both sensitivity and velocity. For controlled tests of recrystallization effects, the transition was carried out in an automatic cycling oven. The sensitivity (Figure

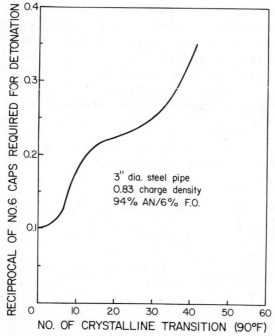

**Figure 11.16**  Sensitivity of clay-coated prilled AN versus number of crystalline transitions (Clark et al., 1961).

11.16) caused by the degradation of the prills, which is visible under low magnification, is increased almost to single-cap initiation.

### 11.2.7.1  Oil Content and Sensitivity

The most sensitive ANFO mixture does not necessarily have the highest detonation velocity and may not be the most desirable for blasting purposes. With other properties affecting sensitivity held constant, the influence of oil content on the sensitivity of uncoated FGAN prills is essentially constant from 1–7% percent fuel oil (Figure 11.17) and decreases rapidly for higher oil content.

The sensitivity of ANFO is important in both boostering and safety; hence, it should be emphasized that the values given in Figure 11.17 are for uncoated, porous, low-density, fertilizer-grade AN prills confined in 3-in. pipe. Sensitivity for other types of prilled AN or AN products may not be the same. For example, dense AN prills have a maximum sensitivity with approximately 2% fuel oil, because the oil does not penetrate the surface as with porous prills. Cook (1958) also found that the maximum sensitivity for wax-coated prills, where the wax did not penetrate the prills, was approximately 1.5% wax content. Thus, for a good blasting agent design, the percentage of fuel oil should not be dictated by sensitivity but by the chemical energy released. On the other hand, one of the advantages of ANFO is that it requires a booster for initiation.

### 11.2.7.2  Sensitivity Versus Bulk Density

Evaluation of the relationship between density and sensitivity is concerned primarily with bulk density and not the prill density. For an OB mixture, the normal pouring density in a 3-in.-diameter pipe is approximately 0.80 g/cc. For experimental charges, it can be increased to 0.90 g/cc by mechanical vibration of the pipe. To obtain a charge density above 0.90, it was necessary to tamp the charge at frequent intervals as it was poured into the pipe. The sensitivity is almost constant for usual loading densities, but it decreases rapidly above a value of 0.94 (Figure

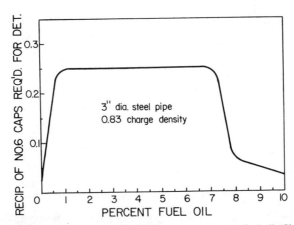

**Figure 11.17**  Sensitivity of uncoated prilled AN versus percentage fuel oil (Clark et al., 1961).

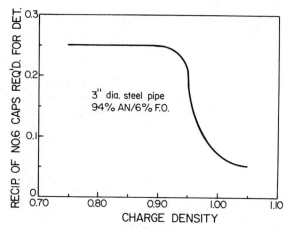

**Figure 11.18** Sensitivity of uncoated prilled AN versus charge density (Clark et al., 1962).

11.18). Detonation velocity may also decrease at high density if the condition approaches a "dead pressed" density.

### 11.2.7.3 Particle Size and Sensitivity

In the manufacture of many explosives, the control of grain size is of prime importance, because it has marked influence on blasting properties. Eyring et al.'s (1949) grain-burning theory explains on a fundamental basis its importance. Even though ANFO is a polymodular mixture, it approaches a state of a homogeneous granular composition because of the oil penetration into the prill. The porous prill structure reduces the effective grain size compared to that normally determined by screen analysis. Prilled AN is $-8-+20$ mesh, and to obtain small screen-size fractions the prills must be ground, which produces particles with angular, irregular

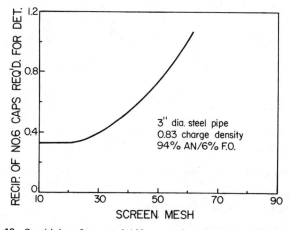

**Figure 11.19** Sensitivity of uncoated AN versus size of particles (Clark et al., 1961).

shapes. Hence, for ground prills, the ratio of surface area to particle diameters differs from that of prills. The sensitivities of mixes of different particle sizes— that is, of (1) prilled AN 8/20 mesh and (2) five separate screen size fractions, 10/12, 14/16, 20/60, 60/100, and −100 mesh—show that cap-sensitive AN can be obtained by grinding the AN to a −60 mesh size (Figure 11.19).

A factor that may alter sensitivity of ANFO is the growth of small crystals on the prills and the conglomeration (caking) of ground AN. These effects can be minimized by prompt use of fresh lots of AN and blasting with the ground AN as soon as possible after grinding and mixing.

A summary of the effects on sensitivity of particle size, shape, and structure (Table 11.7) shows that shape as well as size is critical. The influence of shape on sensitivity is illustrated by a comparison of dense flaked AN with dense AN prills,

### TABLE 11.7. Influence of Particle Size, Shape, and Structure on Sensitivity

| Mixture | Min. No. of No. 6 Caps Reqd. for Complete Detonation | Remarks |
|---|---|---|
| 94% AN Unc. Prills 6% Fuel Oil | 4 | Density 0.83 +20 Mesh |
| 94% AN RCP 6% Fuel Oil | 2 | Density 0.88 -20 Mesh |
| 94% Flaked AN 6% Fuel Oil | 10 | Density 0.88 Avg. Size 2 × 2 × 0.08 mm Thickness |
| 94% (E-2)* Prills 6% Fuel Oil | Failed 35 | Density 1.00 +20 Mesh |
| 94% (E-2)* RCP 6% Fuel Oil | Failed | Density 1.05 -20 Mesh |
| 94% (E-2)* Flaked 6% Fuel Oil | 10 | Density 0.90 Avg. Size 2 × 2 × 0.08 mm Thickness |
| 94% (E-2)* Ground 6% Fuel Oil | Unconfined $d_c$-1 in. | Density 0.80 65% -100 Mesh |
| 85% (E-2)* Prills 15% Aluminum | Unconfined $d_c$-1 in. | Density 0.86 +20 Mesh |

All charges were shot in 3-inch diameter steel pipes, 24 inches in length except as noted. Mixtures were aged a minimum of 2 hours.

*Trademark name for dense, stabilized ammonium nitrate.

*Source:* Clark et al. (1962).

the two of which have the same particle density but not the same bulk density. The reasons for the higher sensitivity of flaked material (10 caps), which has roughly one half the specific surface area of the dense AN prills (35 caps failed), are not apparent. The influence the structure of the prills has on sensitivity becomes apparent when regular AN prills (4 caps) are compared with dense E-2 prills (35 caps failed). Dense AN has approximately the same particle size and particle size distribution as regular AN, but the dense AN is nonporous and capable of holding only about 3% fuel oil, which is all on the surface of the prills. An illustration of the influence of these three factors can be seen in the comparison between the dense AN prills and the dense ground AN material. The high-density prills with fuel are insensitive, but when ground they become sensitive. However, there is a marked difference in bulk density between the dense prills (1.00–1.10) and the ground prills (0.80). An E-2/aluminum mixture was found to be sufficiently sensitive that it could be initiated with one cap.

The addition of dense prills to regular fertilizer-grade prills resulted in a rapid desensitization of the mixture as the percentage of high-density prills was increased (Table 11.8). To keep the bulk density constant, because each mixture with a

## TABLE 11.8. Mixture of Dense and Porous Prills

| Mixture | Min. No. of No. 6 Caps Reqd. for Complete Detonation | Detonation Velocity (ft per sec) | Density (g/cc) |
|---|---|---|---|
| 94% AN Unc. Prills<br>  6% Fuel Oil | 4 | 11,200 | 0.83 |
| 94% 90% AN Prills<br>      10% (E-2)* Prills<br>  6% Fuel Oil | 7 | 9,760 | 0.90 |
| 94% 85% AN Prills<br>      15% (E-2)* Prills<br>  6% Fuel Oil | 8 | 9,600 | 0.90 |
| 94% 80% AN Prills<br>      20% (E-2)* Prills<br>  6% Fuel Oil | 8 | 9,600 | 0.90 |
| 94% 70% AN Prills<br>      30% (E-2)* Prills<br>  6% Fuel Oil | 20 | -- | 0.90 |

All sensitivity measurements were made in 3-inch steel pipes, 24 inches in length. Detonation velocity charges were 48 inches in length.

*Trademark name for dense, stabilized ammonium nitrate.

*Source:* Clark et al. (1962).

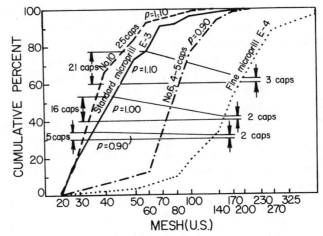

**Figure 11.20**  Cumulative particle size distribution, density, and sensitivity (Clark et al., 1962).

higher percentage of dense AN would have a slightly higher pouring density, all mixtures were packed at density of 0.90. As little as 10% of dense prills caused a marked decrease in the detonation velocity and a somewhat smaller decrease in the sensitivity. The 70/30 mixture exhibited only a low-order explosion.

Although a few tests were made of the sensitivity of discrete particle size fractions (Figure 11.19), the result of reducing the effective average particle size in the overall cumulative distribution upon sensitivity to initiation is of even greater interest. As the percentage of fine prills is increased, the number of No. 6 caps required for initiation decreases rapidly (Figure 11.20). The number of caps is also affected by bulk density. The finer particle sizes correspond to those produced by the crushing of the prills and are readily initiated by two caps or a small booster.

## 11.2.8  Density and Field Performance

The interrelated criteria used in selection of the best explosive for a given set of blasting conditions are (1) the detonation velocity, (2) available energy per unit weight, (3) density of the explosive, and (4) pressure–time curve. For a very hard rock that is difficult to break, one would select an explosive with a high detonation velocity and available energy, and, conversely, for a soft, easily broken rock, one would select a low-velocity, low-energy explosive. These factors may be used as a guideline for initial selection. Generally, blasting experience coupled with tests will ensure a good choice of explosive. The pressure–time curve of an explosive is of major importance. From reaction rate data, the maximum available work, and the properties of the explosive, it is possible to approximate the $p$–$t$ curves. However, this approximation involves lengthy computations.

An approximation may be made by calculating the explosion state and estimating the total reaction time. These two quantities determine to some extent the shape

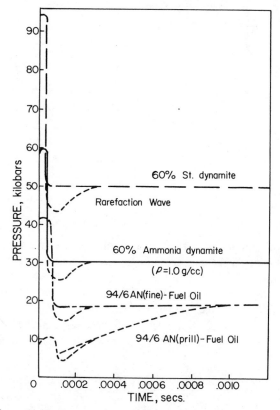

**Figure 11.21**  Pressure–time curves for explosives (Cook, 1958).

of the $p$–$t$ curves for two dynamites and ANFO (Figure 11.21), which show the influence of mass reaction rates and explosion state pressures upon the pressure–time history (Cook, 1958). The more nonideal the explosive, the greater the deviation is from the ideal $p$–$t$ curve. The detonation pressure $p_2$ can be approximated by the equation

$$p_2 = 0.0009\rho_1 \frac{D^2}{4} \text{ atm} \qquad (11.11)$$

The sensitivity of the mixtures of E-2/FGAN and E-2/uncoated FGAN (Figure 11.22), which contain about 2.5% of diatomaceous earth, vary considerably. The charges fired as part of the packed E-2/uncoated test series, which contained 100% of E-2/ground, were consistently detonated with three caps, whereas the charges containing 100% E-2 (un-ground) required as many as seven caps to initiate.

With about 1 to 2% content of diatomaceous earth, the sensitivity of FGAN decreases very rapidly (Figure 11.23), requiring four No. 6 caps at 1% and increasing rapidly to 50 caps at 5% DE. A possible reason for the desensitizing effect is found

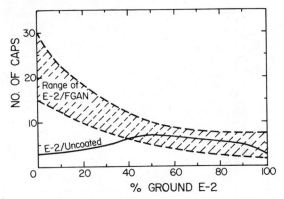

**Figure 11.22**  Sensitivity of FGAN/ground E-2 mixtures (Clark et al., 1962).

in the "hot spot" theory of initiation of explosion (Bowden and Yoffe, 1952). According to Bowden's theory, detonations result from the existence and growth of a critical number of hot spots in the explosive. Thus, while the presence of an optimum amount (approximately 2.5%) of DE results in good adsorption conditions for oil, all of the DE is present on the surface of the prills. The coating of inert material reduces the number of contact points between grains of AN, which offers a physical barrier to high-temperature gaseous molecules and thus reduces the

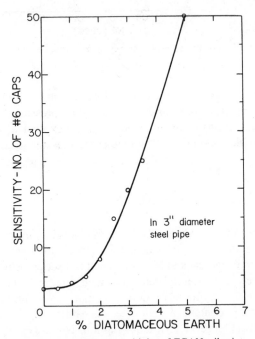

**Figure 11.23**  Effect of diatomaceous earth on sensitivity of FGAN–oil mixtures (Clark et al., 1962).

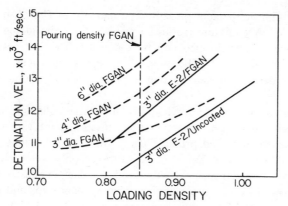

**Figure 11.24**  Detonation velocity versus density for constant and variable particle size (Clark et al., 1962).

probability of the formation of the necessary number of hot spots to initiate and maintain a stable detonation. The DE also absorbs heat and thus reduces the temperature of the reaction products.

## 11.2.9  Particle Size and Detonation Velocity

A comparison of the detonation velocities of FGAN and E-2 FGAN versus density (Figure 11.24) and percent E-2 fines (Figure 11.25) indicates that the percentage of fines, as well as the bulk density, affects the detonation velocity. Within experimental error, the E-2 addition of fines initially enhances the detonation velocity

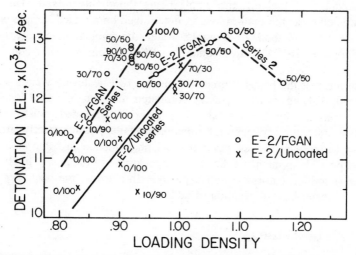

**Figure 11.25**  Combined effects of particle size and loading density on detonation velocity (Clark et al., 1962).

almost linearly with the percent present. The increase of $D$ with bulk density for the 50/50 E-2/FGAN mixture (series 2) was less than the increase caused by a combined change in particle size and bulk density. The last data point for E-2/FGAN series 2 (all of 50/50 composition) represents a loss in velocity with increase in bulk density, indicating that a "dead pressed" condition was being approached.

The polymodular blends, the applicability of packing behavior, and its effect on detonation velocity and sensitivity were further investigated for mixes of four components of screened stabilized prills. Their particle size distribution, sensitivity, and velocity are recorded in Table 11.9. Although only a very few shots using such spherical polymodular material were fired, the data emphasize that the largest particle size, as well as the proper choice of component sizes and perce tages, is important in determining detonation properties. Mix No. 1, whose largest size component is 40/60 mesh, has approximately the same sensitivity as ground E-2. It has a considerably higher pouring density and detonation velocity: 1.16 pouring density compared to approximately 1.00 for E-2 ground, and 15,700 ft/s detonation velocity compared to 13,000 ft/s for ground E-2. These brief tests indicate that spherical particles will pack more readily than angular particles and that their overall physical characteristics, which control chemical reactions in the detonation wave, are more favorable to faster mass reaction rates than those of angular particles.

### 11.2.10    Moisture Content

Since AN is highly hygroscopic and soluble in water, the amount of moisture contained in an AN explosive mixture has a marked effect on its performance. Varying percentages of water were added to fertilizer-grade prills and then the desired percentge of fuel oil, the whole being thoroughly mixed. The percent moisture was taken as the total moisture, determined on a mixed sample using the Karl Fisher titration method. At about 4% water, the detonation velocity decreases sharply (Figure 11.26). Mixtures containing 9% or more of water could not be detonated.

Considering that the heat required to vaporize 10% water by weight would be well over 50 kcal/kg moist explosive compared to about 850 kcal available to support detonation, plus the necessary heat to bring the added moisture to detonation temperature, the explanation for detonation failure seems readily attributable to heat lost to these processes. As a parallel phenomenon in some cases, it has been found possible to quench a detonation by adding as little as 5% finely divided (nonreacting) sodium chloride to an explosive. In slurries, the heat loss is compensated by greater heat of reaction of additives.

### 11.2.11    AN Sensitizers

The sensitivity of AN without fuels or sensitizers is low, particularly for large grain sizes or when the grains have maximum density. Standard prill porosity and pen-

**TABLE 11.9. Particle Size Distribution, Sensitivity, and Velocity of Polymodular Blends**

| Polymodular Mix No. | Particle Size Distribution | | Bulk Density | Sensitivity | |
|---|---|---|---|---|---|
| | Mesh | Percent | | No. of No. 6 Caps in 3-in. Pipe | Velocity in 3-in. Pipe |
| 1 | 40/60 | 87.8 | 1.10 | 6 | 15,730 |
| | 100/140 | 6.3 | 1.13 | | |
| | 170/230 | 2.0 | | | |
| | 230/270 | 3.9 | | | |
| 2 | 12/20 | 87.8 | 1.09 | 4-1/2 Stick Primer 60% Dynamite | Partial Detonation |
| | 40/50 | 6.3 | 1.10 | | |
| | 70/100 | 2.0 | | | |
| | 100/140 | 3.9 | | | |
| 3 | 10/12 | 87.8 | 1.10 | 4-1/2 Stick Primer 60% Dynamite | Partial Detonation |
| | 18/20 | 6.3 | 1.16 | | |
| | 30/40 | 2.0 | | | |
| | 40/50 | 3.9 | | | |

*Source:* Clark et al. (1962).

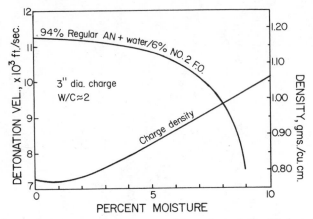

**Figure 11.26** Influence of moisture content on detonation velocity (Clark et al., 1962).

etration of the oil effectively decrease the grain size, which enhances the sensitivity and increases the detonation velocity. However, sensitivity and detonation velocity do not always vary in the same manner.

Studies by numerous investigators indicate that sensitivity of explosives is governed almost wholly by the heat balance involved in initiation processes. The central idea is that "hot spots" are created in the explosive by impact, thermal ignition, spark, or other means, and, if the reaction can be accelerated sufficiently owing to production of heat, a shock wave will be formed. The present concept of ignition involves both (1) the heat of reaction and (2) the heat of activation of the explosive. Only those shock waves in which the temperature is high enough to cause an appreciable fraction of the explosive to react within a short critical distance of the shock front become stable detonation waves. Most users of ANFO prefer nonsensitized material for safety reasons.

## 11.3 SAFETY AND ANFO

### 11.3.1 Fume Qualities of ANFO

Because of its relatively large critical diameter, lower than ideal detonation velocity, and so on, there was concern during the early development and use of ANFO in small-diameter holes in underground mines, where the possible generation of toxic fumes might occur. Of primary interest are carbon monoxide, oxides of nitrogen, and oxides of sulfur if this element is present in fuel oil or is otherwise added.

For complete combustion, optimum detonation velocity for a given diameter and bulk density, and best blasting performance, mixtures of commercial-size prills and fuel oil are satisfactory for large-hole diameters usually used in open-pit blasting. The low cost of AN prills and fuel oil made them attractive for all blasting uses. Means were sought to make ANFO usable in small-diameter holes, such as reducing the particle size by grinding or adding of sensitizers.

The noxious gases that may be generated by detonating explosives include carbon monoxide, oxides of nitrogen, and oxides of sulfur.

*Carbon monoxide* (CO) is colorless, odorless, tasteless, and lighter than air, but it is both toxic and explosive when mixed with air. It has a high affinity for elements in the bloodstream, and its effects are cumulative. Hence, prolonged exposure to concentrations as low as 0.01% are dangerous, and above 0.04% may be lethal.

*Nitrogen oxides* ($NO_2$, $N_2O$, NO) have an irritating odor, are red-brown in color, and are bitter to the taste. They are very toxic, and the usual allowable concentration in air is 0.0005%, the lethal percentage being 0.005%.

*Sulfur dioxide* ($SO_2$) may result from a detonation if there is sulfur (or a sulfur compound) in an explosive. It has an irritating odor and an acid taste. The allowable concentration is 0.005%, and the fatal amount is 0.1%.

The Bichel and Crenshaw-Jones methods were used by Tournay et al. (1959) in their investigation of AN–fuel oil mixtures, the latter consisting of seven compositions of commercial-prilled AN with keiselguhr coating, six without, with a semi-gelatin dynamite as a standard. The No. 2 diesel oil contained 13.5% sulfur, and inert coating of prills was 3%. Oxygen balance of the ANFO varied from plus 20 to minus 13.8 g/100 g.

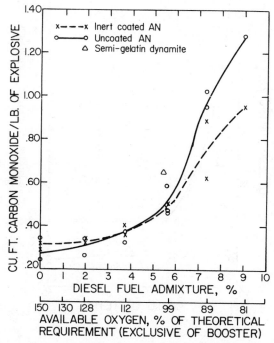

**Figure 11.27** Effect of fuel content of AN–DF compositions on production of carbon monoxide in the Crenshaw–Jones apparatus. A reference value for a semi-gelatin dynamite is shown (Cook, 1974).

The Crenshaw-Jones apparatus consists of a steel cannon $2\frac{1}{4}$-in.-diameter bore in which the charge, 300 g plus 35 g of booster, is loaded and stemmed. The products are discharged into an air-evacuated tube and sampled for analysis by Orsat apparatus while the oxides of nitrogen were determined by colorimetric method.

The carbon monoxide increases slowing from about 0.3 ft$^3$/lb of explosive to oxygen balance (Figure 11.27), probably owing to the booster and more rapidly with increase in fuel oil content, the increase being somewhat greater for uncoated AN prills. The amount of CO was somewhat greater for the semi-gelatin dynamite employed as a standard.

The gases from the tetryl booster markedly affected the formation of CO and nitrous oxides (Figure 11.28), the latter decreasing with the weight of the booster.

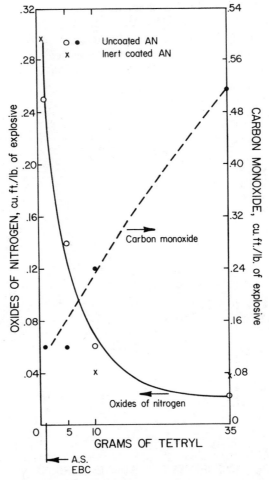

**Figure 11.28**  Effect of varying the booster–initiator on production of carbon monoxide and oxides of nitrogen by an AN–DF composition containing 5.7% oil (Cook, 1974).

The tetryl booster has a large oxygen deficiency, a 35-g booster for most tests representing 10% of the charge. The extent to which the detonation products of the booster and explosive are mixed is not known. The available oxygen in the booster was not considered in evaluating the test results. Thus, the data indicate too much CO and too little $NO_2$.

The oxides of nitrogen decrease rapidly from zero fuel content to small values near oxygen balance (Figure 11.29), being nearly the same here as for the reference explosive. The presence of water also increases the amount of nitrogen oxides for oxygen balance ANFO and mixtures with fuel oil percentages to 7.4 (Figure 11.30).

Van Dolah et al. (1959) found that because of the tetryl booster, results from tests in the C–J apparatus could not be correlated with results from field studies.

In further laboratory tests in the C–J apparatus, PETN and dynamite boosters were used. As before, the CO increased with fuel oil percent, and nitrous oxides decreased (Figure 11.31). The addition of water to 4, 6, and 8% ANFO had little effect on the oxides of nitrogen to 4% water, but more water increased the amount of oxides.

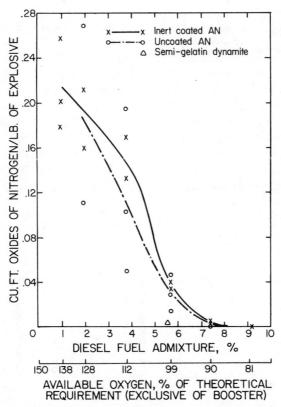

**Figure 11.29** Effect of fuel content of AN–DF compositions on production of oxides of nitrogen in the Crenshaw–Jones apparatus. A reference value for a semi-gelatin dynamite is shown (Cook, 1974).

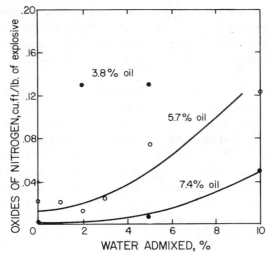

**Figure 11.30** Effect of water admixed to AN–DF compositions on production of oxides of nitrogen (Cook, 1974).

Experimental field results (Table 11.10) obtained at Carlsbad are compared to those from the Crenshaw–Jones apparatus with 94/6 ANFO mixtures primed with the same high explosives. The volume of the CO and NO$_2$ per pound of explosive was calculated from the average gas concentration in the samples that were collected 15 min after the round was fired. The results are given as a CO/NO$_2$ ratio, because of sampling difficulties with fumes in the mine where gases escape through the curtain closure and are trapped in a muck. The CO/NO$_2$ ratios were expected to be about the same for the mine and laboratory tests if the fume-producing reactions are similar. The CO/NO$_2$ ratios for the mine shots varied from 0.80 3.7, which agrees reasonably with the data for ANFO initiated by dynamites. Other than dynamite B, which has a small amount of explosive oil, the CO/NO$_2$ ratios

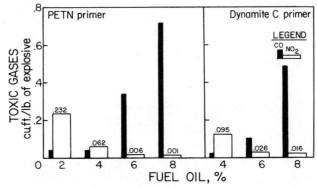

**Figure 11.31** Effect of fuel oil content on production of toxic gases by ANFO compositions initiated with different-strength primers C–J (Van Dolah et al., 1959).

**TABLE 11.10. Toxic Fumes Produced by Ammonium Nitrate–Fuel Oil Mixtures in Crenshaw–Jones and Field Tests**

| Explosive | Primer | Carlsbad Tests cu ft/lb | | | Crenshaw-Jones Tests cu ft/lb | | |
|---|---|---|---|---|---|---|---|
| | | CO | $NO_2$ | $CO/NO_2$ | CO | $NO_2$ | $CO/NO_2$ |
| ANFO | Dyn. A[1] | .081 | 0.47 | 1.72 | .427 | .229 | 1.88 |
| ANFO | Dyn. B[1] | .057 | .071 | .80 | .399 | .202 | 1.98 |
| ANFO | Dyn. C[1] | .058 | .039 | 1.49 | .346 | .157 | 2.20 |
| ANFO | PTEN | — | — | — | .341 | .006 | 57.00 |
| Dyn. B | EBC[3] | .097 | .114 | .85 | .104 | .184 | .56 |
| Dyn. C | EBC[3] | .108 | .041 | 2.63 | .314 | .012 | 26.00 |
| Dyn. D[2] | EBC[3] | .114 | .031 | .368 | .255 | .008 | 32.00 |

[1] Economy type dynamite.
[2] 60 percent semigelatinous dynamite.
[3] Electric blasting caps used were: Carlsbad tests, No. 6; Crenshaw-Jones tests, No. 8.

*Source:* Van Dolah et al. (1959).

for the dynamites themselves are higher for the test apparatus than for the mine by a factor of nearly 10. For ANFO initiated by PETN (C–J apparatus), addition of water increased the amount of nitrous oxides (Figures 11.32, 11.33) and gave a higher ratio than dynamite. These data indicate a difference in the character of the detonation reactions for variations in confinement and initiation.

Primer strength influences the production of $NO_2$ in the Crenshaw–Jones apparatus (Figure 11.34). Strong primers, like PETN, cause low amounts of $NO_2$ and high $CO/NO_2$ ratios similar to those from a high-strength dynamite. Weak primers, economy dynamites, and Army special electric blasting caps give high yield of $NO_2$ and low $CO/NO_2$ ratios.

As shown earlier in the description of ANFO test results, the detonation velocity is a good indication of the completeness of the chemical reaction in the detonation zone for a given density and charge diameter. Confinement also has an effect on detonation velocity. The factors that control the sensitivity and velocity thus also affect the chemical reaction and the fumes produced.

Van Dolah et al. (1959) concluded that dry, well-mixed, oxygen-balanced ANFO mixtures, initiated with a strong primer, produce oxides of nitrogen comparable to dynamites. The U.S. Bureau of Mines did not at that time recommend the use of ANFO underground. However, proper mixing and usage have led to the adoption of ANFO underground, where its blasting properties are suited for breaking the rock (USB of M, 1960, 1963, 1972).

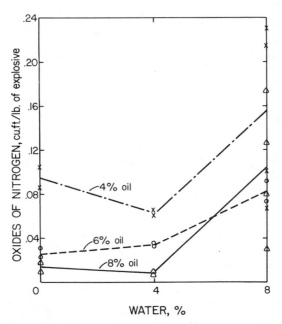

**Figure 11.32** Effect of water on production of oxides of nitrogen by ANFO compositions initiated with dynamite C primer, C–J apparatus (Van Dolah et al., 1959).

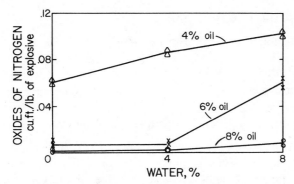

**Figure 11.33**  Effect of water on production of oxides of nitrogen by ANFO compositions initiated with PETN primers, C–J apparatus (Van Dolah et al., 1959).

## 11.3.2  Fire and Detonation—AN

Under certain conditions, burning of AN may lead to a detonation, particularly in large masses or when it is confined (Van Dolah et al., 1966ac). To give an indication of this tendency, the sensitivity of AN was investigated by card-gap techniques. It was found that critical diameter decreases with increase in temperature and that water in small amounts increases shock sensitivity. Burning to detonation is difficult to obtain in free-flowing beds of AN and ANFO, and AN burning in a vented vertical vessel did not produce burning to detonation. There are cases on record, however, in which large masses of AN have detonated when storage bins have burned.

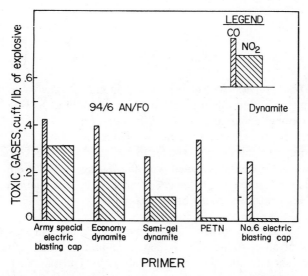

**Figure 11.34**  Effect of primer strength on the production of toxic gases from 94/6 ANFO compositions, C–J apparatus (Van Dolah et al., 1959).

## 11.3.3 Sympathetic Detonation

It has also been found that sympathetic detonation can be initiated in AN and ANFO at unexpectedly large distances (Van Dolah et al., 1966a,b). The safe distance for AN is considered to be twice that of Nitromon, which for ANFO lies between that for dynamite and smaller quantities of explosives.

## 11.3.4 Sensitivity—ANFO and Similar Explosives

A Bureau of Mines sensitivity survey (Watson, 1973) included drop-weight and friction testing and sensitivity to initiation by projectile impact and to a No. 8 blasting cap. Some AN prills produce cap-sensitive mixtures when combined with No. 2 diesel fuel (FO); nitromethane can also result in cap-sensitive mixtures. Mixtures with 1- and 2-nitropropane were not cap-sensitive, and powdered or grained aluminum in ANFO possibly increased its sensitivity to friction.

ANFO has several disadvantages. It has a relatively large critical diameter, requires a primer, and may not provide adequate energy for good fragmentation. Some other ammonium nitrate–blasting agents are designed for improved energy release and sensitivity characteristics, such as additional powdered or flaked aluminum for improved energy release, and other liquid fuels are used to augment the sensitivity and energy release. New prilling methods produce more sensitive prills. Watson (1973) has summarized results on the sensitivity of several experimental explosive systems with an ammonium nitrate prill base.

### 11.3.4.1 Experimental Procedures

Testing includes drop-weight and friction testing and determining sensitivity to initiation by projectile impact and a No. 8 blasting cap. Special tests use card-gap testing at ambient and elevated temperatures, thermal stability tests, and others. However, the projectile impact test gives excellent correlation between test results and the card-gap test (Watson, 1973). All of these tests are completely described by Mason and Aiken (1972) and are briefly described below.

*Drop-Weight Tests.* Drop-weight tests use a modified version of the Bureau's drop-weight apparatus that employs a 5.0-kg weight and a 0.73-kg intermediate weight. The modified apparatus uses a 2.0-kg drop weight and a 2.0-kg intermediate weight tightly coupled to an expendable plunger of about 35 g. The changes were made to improve the energy transfer to the specimen (Becker et al., 1972). For ANFO, five trials were conducted with the 2.0-kg weight dropped from 250 cm upon samples of volumes of 0.4 cm$^3$, confined in $\frac{1}{2}$-in.-diameter steel cups, with the number of positive events being recorded.

*Friction Tests with Bureau Tester.* The apparatus described in Mason and Aiken (1972) consists of an A-frame, a weighted pendulum to which either a steel of hard fiber-faced shoe is attached, and a steel anvil. The sample is spread over parallel grooves cut in the anvil across its surface. The pendulum, which is weighted with 20 kg and released from a height of 1.5 m, is adjusted in height so that it swings $18 \pm 1$ times before coming to rest; $7 \pm 0.1$ g of material is then spread in an even layer in and about the grooves on the anvil, and the pendulum is released. If there is no explosion, burning, or local crackling in a fixed number of trials, testing is discontinued. With the AN materials reported here, 10 trials were conducted with the steel shoe.

*German BAM Friction Test.*    The U.S. Bureau of Mines, using a friction tester developed by the German Federal Institute for Materials Testing (Bundesanstalt fur Material-prufungen; BAM), tested small samples (approximately 50 mg) placed between a stationary porcelain pin and a moving porcelain plate, both having a standard roughness. The pin (approximately $\frac{13}{32}$ in. diameter by $\frac{5}{8}$ in. long) is rounded (approximately $\frac{1}{2}$ in. radius) at each end and is mounted on a level arm loaded with one of nine weights which may be placed in one of six positions, giving a total of 454 load increments ranging from 0.5–36.0 kg. For testing, a switch is turned on, and the anvil, upon which the porcelain sample plate is mounted, reciprocates once and automatically shuts off. The relative sensitivities indicated by the BAM tester are ranked in terms of a threshold initiation limit (TIL), the maximum load in kilograms giving no reactions in 10 consecutive trials. The BAM test with the 36-kg weight appears to be more severe than the Bureau's friction test.

*Projectile Impact Tests.*    Here, the sample is confined in a 1.5-in.-diameter schedule 40 steel pipe nipple of length of 3.0 in. and a nominal wall thickness of 0.145 in. The ends of the container are sealed with a polyethylene film. The charges are then impacted by brass projectiles fired from a smooth-bore, 0.50-caliber gun along the axis of the charge to impact the unconfined surface of the explosive. The Bruceton up-and-down technique was used to determine the projectile velocity corresponding to a 50% initiation probability ($V_{50}$), rough estimates of $V_{50}$ being made by averaging the highest projectile velocity for which detonation did not occur and the lowest velocity resulting in detonation. Ten firings were considered adequate for estimating $V_{50}$ in this fashion. Mason et al. 1963.

*Cap Sensitivity Test.*    This evaluation provides a simple means for distinguishing an explosive from a blasting agent. A No. 8 detonator is inserted into material contained in a 1-qt paperboard cylinder with a cover. If initiation to detonation occurs, the material is classified as an explosive; if not, as a blasting agent.

### 11.3.4.2  *Materials Tested*

*Ammonium Nitrate Fuel Oil.*    Five different types of ammonium nitrate prills were mixed with 5% No. 2 diesel fuel as follows:

*Key No. 940:*    Prilled ammonium nitrate with a small amount of surfactant and about 0.5% clay or diatomaceous earth coating. The prills have a bulk density of approximately 0.85 g/cm$^3$ and are commonly used for preparing commercial ANFO in the United States.

*Key No. 1490:* New lot of Key No. 940.

*Key No. 1485:* Prilled ammonium nitrate with a bulk density of approximately 0.80 g/cm$^3$ and a clay content of about 0.8%, used for preparing ANFO in Canada.

*Key No. 1486:* Essentially pure ammonium nitrate prills with superfine porosity to enhance sensitivity; bulk density of about 0.78 g/cm$^3$.

*Key No. 1487:* Same as Key No. 1486 except with 0.1% organic additive serving as an anti-caking agent; the bulk density approximately 0.78 g/cm$^3$. In addition, prills of Key No. 1486 were mixed in the proportions 97.7 AN/2.3 FO and 90.7 AN/9.3 FO to determine the effect of oxygen balance.

*Key No. 1488:* Same as Key No. 1486 except with 0.3% organic additive; the bulk density approximately 0.77 g/cm$^3$.

***Ammonium Nitrate–Nitromethane.*** Ammonium nitrate Key No. 940 was mixed with commercial nitromethane (NM) in the proportions 86.1 AN/13.9 NM, 84 AN/16 NM, and 81.0 AN/19.0 NM, and tested.

***Ammonium Nitrate–Nitropropane–Methanol.*** Ammonium nitrate Key No. 940 was tested mixed with 1-nitropropane, 2-nitropropane, both neat and mixed with methanol. Tests were to determine the effect of oxygen balance and methanol content on sensitivity.

***Ammonium Nitrate–Fuel Oil–Aluminum.*** Ammonium nitrate Key No. 940 with 5% fuel oil was tested with 15% aluminum added to increase energy. Two types of aluminum were used: Al-Meg Exxo 9030 aluminum granular and Alcoa 120 atomized.

### 11.3.4.3 Experimental Results and Discussion

In the experimental results (Table 11.11), the AN–fuel oil mixtures ranged in sensitivity depending on fuel oil content and the type of prills. For prills manufactured in the United States, $V_{50}$'s were 960 m/s or higher depending on oxygen balance, the mixture with OB = −13.76 being the least sensitive. No significant difference in the projectile impact sensitivities was found for mixtures of 2.3 and 5.0% fuel oil; that with 2.3% fuel oil had an oxygen balance of +11.65, and the mixture with 5.0% fuel oil had an OB of +1.85. U.S. prills were not responsive to the Bureau friction test, the BAM friction test, the drop-weight test, or a No. 8 blasting cap.

All of the 95 AN/5 FO of Canadian prills were more sensitive than those of U.S. prills. Prills with superfine porosity and containing no additives (Key No. 1486) were the most sensitive to projectile impact and could also be initiated with a No. 8 blasting cap. All four mixtures were also sensitive to the BAM friction test, having TIL levels from 16.0 to 24 kg, but not to the Bureau friction and drop-weight tests.

Prills and nitromethane constituted the most sensitive explosive composition tested; 85 AN/16 NM (OB = + 10.56) was the most sensitive to projectile impact

**TABLE 11.11. Sensitivity Measurements of Prilled Ammonium Nitrate with Various Fuels**

| Composition (wt-%) | AN Key No. | Loading density | V50, m/sec | Bureau friction (steel shoe) | BAM friction (TIL level in kg) | Drop weight (2 kg at 250 cm) | No. 8 cap |
|---|---|---|---|---|---|---|---|
| AMMONIUM NITRATE/FUEL OIL | | | | | | | |
| 97.7 AN/2.3 FO | 1490 | 0.90 | 960 | 0/10 | >36 | 0/5 | No |
| 95 AN/5 FO | 1490 | 0.94 | 990 | 0/10 | >36 | 0/5 | No |
| 90.7 AN/9.3 FO | 1490 | 0.97 | 1425 | 0/10 | >36 | 0.5 | No |
| 95 AN/5 FO | 1485 | 0.89 | 849 | 0/10 | 24 | 0/5 | No |
| 95 AN/5 FO | 1486 | 0.84 | 638 | 0/10 | 16 | 0/5 | Yes |
| 95 AN/5 FO | 1487 | 0.85 | 771 | 0/10 | 24 | 0/5 | No |
| 95 AN/5 FO | 1488 | 0.83 | 765 | 0/10 | 24 | 0/5 | No |
| AMMONIUM NITRATE/NITROMETHANE | | | | | | | |
| 86.1 AN/13.9 NM | 940 | 0.99 | 558 | 0/10 | 28 | 0/5 | Yes |
| 84.0 AN/16.0 NM | 940 | 1.00 | 421 | 0/10 | 24 | 1/5 | Yes |
| 81.0 AN/19.0 NM | 940 | 0.97 | 513 | 0/10 | 12 | 0/5 | Yes |
| AMMONIUM NITRATE/1-NITROPROPANE/METHANOL | | | | | | | |
| 87 AN/13 1-NP/O M | 940 | 1.00 | 826 | — | >36 | 0/5 | No |
| 84 AN/8 1-NP/8 M | 940 | 1.00 | 1062 | — | >36 | 0/5 | No |
| AMMONIUM NITRATE/2-NITROPROPANE/METHANOL | | | | | | | |
| 94.7 AN/5.3 2-NP/O M | 940 | 0.91 | 785 | — | >36 | 0/5 | No |
| 87.0 AN/13 2-NP/O M | 940 | 1.00 | 718 | — | >36 | 0/5 | No |
| 77.5 AN/22.5 2-NP/O M | 940 | 0.94 | 813 | — | >36 | 0/5 | No |
| 95 AN/2.5 2-NP.2.5 M | 940 | 0.92 | 715 | — | >36 | 0/5 | No |
| 84 AN/8 2-NP/8 M | 940 | 1.00 | 992 | — | >36 | 0/5 | No |
| 79 AN/10.5 2-NP/10.5 M | 940 | 0.92 | 996 | — | >36 | 0/5 | No |
| PRILLED AMMONIUM NITRATE/FUEL OIL/ALUMINUM | | | | | | | |
| 85 AN-FO/15 Al Meg 9030 | 940 | 0.90 | 840 | 0/10 | 19.2 | 0/5 | No |
| 85 AN-FO/15 Atomized Al | 940 | 0.90 | 950 | 0/10 | 19.2 | 0/5 | No |

*Source:* Watson (1973).

with a $V_{50}$ of 421 m/s. Compositions with 13.9% NM (OB = +11.80) and 19.0% NM (OB = +8.79) were equally sensitive having $V_{50}$'s of 558 and 513 m/s, respectively. More than 19% NM in AN prills produced a slurrylike mixture, excluding tests with systems of OBs lower than +8.79. The NM mixtures showed sensitivity in the BAM friction test but not in the Bureau friction test. The 16.0% NM mixtures gave one positive test with the drop weight. All three could be detonated with a No. 8 blasting cap. This agrees with experience showing that explosive compositions having $V_{50}$'s below approximately 650 m/s are cap-sensitive. This velocity also serves as a quantitative criterion for defining blasting agents as compared with high explosives according to one definition that uses sensitivity to a No. 8 blasting cap as one criterion (Institute of Makers of Explosives, 1970).

Ammonium nitrate with nitropropane/methanol mixtures was found to be of intermediate sensitivity to projectile impact test with $V_{50}$ ranging from 715–1,062 m/s with no positive responses in the other sensitivity tests. Aluminum added to ANFO (prills) did not affect the projectile impact test sensitivity or that in the Bureau friction apparatus, the drop-weight test, or a No. 8 blasting cap. There appears to have been an increase in the BAM friction sensitivity.

Watson (1973) concluded or recommended the following: (1) Certain types of AN prills produce cap-sensitive mixtures with No. 2 diesel fuel which do not meet the requirements for classification as blasting agents. (2) AN prills mixed with nitromethane produce cap-sensitive mixtures. (3) Prills and neat 1- and 2-nitropropane are more sensitive than ANFO made with U.S. prills but are not cap-sensitive. (4) Powdered or grained aluminum does not sensitize ANFO except to increase its sensitivity to friction. (5) The safety recommendations for sensitized ammonium nitrate blasting agents are outlined in U.S. Bureau of Mines Information Circular 8179 (U.S. Bureau of Mines, 1963) which should be consulted by those interested in these materials.

## 11.4  SLURRIES

The chemical products and the detonation and explosion properties of slurries must be adjusted to account for the water and the possible metallic elements in them. This can be done by means of the computer program described herein (Appendix A) utilizing the Abel equation of state, the thermodynamic equilibria, and other related factors. First, adjustments must be made for minor amounts of metallic elements for either slurries or dry explosives. These are assumed to have a very high affinity for oxygen, and their stable oxides will be formed first. The atoms necessary to form these are subtracted from the elements available to otherwise react, and adjustments are made for the heat of reaction and the oxygen consumed. The heat capacities of the metallic oxides may be approximated as having constant values with temperature, and they are assumed to be in the solid state at temperatures $T_2$ and $T_3$. The slurry water is added to the reactants, and its effects are primarily a reduction of the reaction temperatures and an increase of the pressures $P_2$ and $P_3$. Because of its particular oxides and because their heats of reaction are different at different temperatures, aluminum must be given special mathematical treatment.

# 12

## CHARACTERIZATION OF FRAGMENTED ROCK

Several mathematical models for different types of fragmentation and comminution have been proposed to represent the relationship of the particle size distributions that result from drilling and blasting processes, from mechanical comminution and from other fragmentation operations. However, though some of these models apply to the data and describe the products for given blasting operations done under experimental or other specific conditions, they may not apply to the results of tests performed under different conditions. No model appears to fit data from both explosive and mechanical processes or from both small-scale and large-scale operations (i.e., blasting in conventional mining) and a large scale for nuclear charges in the kiloton and megaton range. Some of the models include factors to describe the explosive quantity and configuration but appear to have application only where local conditions can be accounted for in fitting the data to a given model. However, such analyses have shed considerable light on the basic physical process involved, those factors that are apparently common to each and those factors that differ. Although the total amount of rock material broken in the mining industry by mechanical means is relatively small (except coal), drilling and boring of rock are vital processes.

The size of the fragments and their distribution in the broken material produced by conventional blasting operations and the action of nuclear explosives on geologic materials (rock) can be classified by the size of the blast, the explosive distribution, the *in situ* structure of the rock, and the structure of the blast, including geometry, explosive distribution, and confinement. The size (energy) of the fragmentation event may vary from large nuclear blasts to small shots involving just a few grams of chemical explosive, such as the charges used in laboratory experimentation. Confinement of charges may vary from above-surface bursts for nuclear weapons to surface contact as in mud capping of boulders, to a high degree of confinement where explosive is distributed in holes or nuclear charges are confined at depth in the earth. Blast geometries vary from those involving a semi-infinite medium to benches, tunnel headings, underground working faces, and so on, to small blocks having a maximum of free faces. All of these factors, with appropriate emphasis on the structure of the rock and its physical properties, affect the character of the fragmentation.

There has been no attempt to standardize the reporting of results of tests, or experimentation on rock fragmentation, or of testing techniques, either by blasting or by mechanical means such as drilling or boring. Consequently it is difficult to compare data and analyses. However, some basic trends and effects can be identified, and these serve both as a basis for further understanding of mechanisms of breakage and for recommendations for standardization in future research.

## 12.1 FRAGMENT PROPERTIES

Some physical characteristics of fragmented rock materials, such as specific energy, internal friction, and size distribution, have been studied for a wide range of sizes, properties, and for several engineering purposes:

1. Very fine: −1 in.—crushing and grinding.
2. Medium fine: −1 in. to 1 ft—crushing, drilling, and conventional blasting.
3. Coarse: −1 in. to 10 ft—commercial blasting and mining operations.
4. Very coarse: −1 in. to 100 ft—large-scale explosive and nuclear cratering.

While the emphasis of fine-grinding operations is on the production of powder and finer particles less than crystal grain size (to separate valuable minerals from gangue materials) for subsequent treatment, fine material is the major product of these operations. Here, the major concern with material of the sizes produced by conventional mining operations, such as drilling and cratering, is of interest because of their possible effects on subsequent particle size reduction.

Quantitative studies of fragmented rock have been mostly concerned with smaller sizes and factors such as (1) the energy of fine grinding as well as (2) the mechanics of production blasting, (3) the stability of rock slopes, (4) loading and haulage, or (5) the stability and flow of fragmented rock and ore in mining operations. Recent research has been performed on the comminution of mining processes as related to that later required in ore beneficiation.

Particle size distribution may be represented graphically either by the percentage distribution of particles by weight or volume in the range of sizes included in the mass, or by the cumulative distribution additive from the small to large or vice versa, with specified increments of particle size diameter. For finer sizes these fractions are the amounts passing a given screen size and retained in the next-smaller screen. For larger than screen sizes, these are represented in the same graphic manner. For large blocks and fragments, the amounts are usually approximated by methods such as in-place drilling, photographic techniques, or similar means.

## 12.1.1  Fragmentation Energy

While the amount of energy required to fragment rock in standard excavation processes as a cost factor is usually not critical, other factors being more important, it is critical in determining the feasibility of a given process. Drilling and blasting are the most energy-efficient of the conventional processes (Figure 12.1). Melting or fracturing of rock by lasers, electrical heat, electron beams, or similar means is the most inefficient.

## 12.1.2  Broken Rock—Very Large Scale

An example crater (Figure 12.2) from large surface explosion has a rubble zone below which the fracture zone may be thicker. The fracture density of *in situ* rock

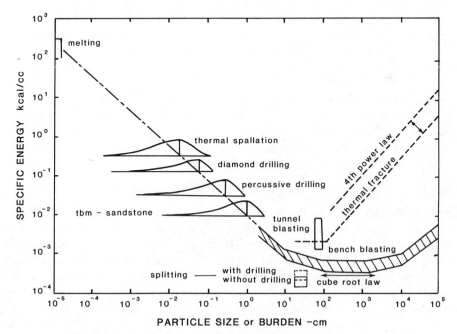

**Figure 12.1**  Specific energy consumption—rock fragmentation processes.

CROSS SECTION OF A CRATER 60MT SURFACE BURST, SANDSTONE/ SHALE

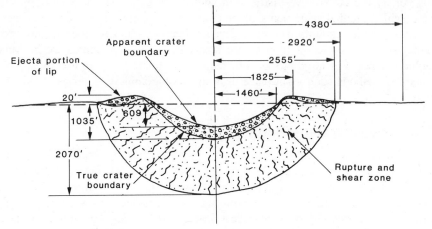

**Figure 12.2** 60MT crater (private communication).

varies with distance from the explosion at its upper boundary from very intense, closely spaced fracture to that of the natural rock.

The physical and engineering properties and the behavioral (flow) characteristics of explosively produced rubble of the size limit and distributions have been studied for materials produced in commercial mining operations and in slopes of explosively produced craters for construction of waterways (Banks, 1968; Banks and MacIver, 1969). In the latter, slope stability is of primary concern, including the angle of internal friction, angle of repose, and circle of slope failure. The angle of internal friction may affect the apparent strength of a rubble mass. Studies have been made of some of the basic mechanics of stability of flow of explosively produced masses of ore and rock in the sublevel caving method of mining. Also, the flow of broken rock in chimneys created by underground nuclear explosions such as Hardhat, Piledriver, and so on, are of indirect interest, as is the flow of rubble that is explosively produced or degraded by flow while under the pressure of the overlying burden, the latter type occurring in block caving. These studies have been largely qualitative in nature, however.

The properties of rubble in large craters have been studied to determine its stability in open cuts. The factors of importance to crater stability are (1) geometry and size of zones, slope angle, and shape; (2) size distribution of rubble; (3) bulk density, bulking factor, and porosity; (4) degree and concentration of blast fracturing and permanent displacement of rock in the fracture zone; (5) mass strength of the rubble and the rupture-zone materials; (6) permeability and seepage character of two zones; and (7) compressibility of the zones of materials. Data for the size distribution of crater rubble have been compiled for small craters in basalt, rhyolite, and saturated clay shale and proposed for a crater of nuclear surface bursts in sandstone (Fig. 12.3). These are compared with the *in situ* spacing of fractures

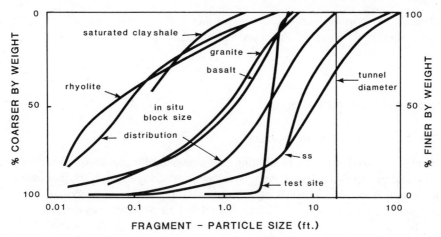

**Figure 12.3** Cumulative fragment size distribution explosively produced craters.

(block size) of basalt and rhyolite, the *in situ* size of blocks having been determined by borehole photography and core length measurement. Comparison of fragment size with the *in situ* block size distribution indicates that the basalt is more brittle.

Data on the bulk density, bulking factor, and porosity (Table 12.1) of crater material show that bulking factors are greatest for dry, hard, and friable rock in the range of 1.4–1.6. Permeability and seepage were found to be $10^3$ cm/s and 1–10 cm/s, respectively. Compressibility has not been measured but appears to occur mostly at the time of fallback and increases with rock softening, is greater for angular fragments, and increases with nonuniformity of particle size.

The existing fractures in the rock are widened by explosions and new ones are formed, which causes bulking of the rupture zone material. Bulking is determined in terms of effective porosity from the equation

$$\gamma_p = (1 - n_e)\,\gamma_i \qquad (12.1)$$

**TABLE 12.1. Bulk Densities, Bulking Factors, and Increased Porosities for Fallback and Ejecta**

| Medium | Preshot *In Situ* Bulk Density $\gamma_i$ (pcf) | Postshot Bulk Density $\gamma_p$ (pcf) | Bulking Factor Increased $BF = \dfrac{i}{\gamma_p}$ | Porosity (%) |
|---|---|---|---|---|
| Hard dry rock (basalt) | 165 | 100–112 | 1.45–1.6 | 31–37 |
| Dry friable rock (rhyolite) | 143 | 105 | 1.4 | 29 |
| Weak saturated rock (clay shale) | 126 | 104–112 | 1.15–1.23 | 14–19 |

*Source:* Lawrence Livermore Lab. NC (1970)

where $\gamma_p$ = postshot density, $\gamma_i$ = preshot bulk density, and $n_e$ = predicted postshot effective porosity. From pre- and postshot densities the bulking factor = $\gamma_i / \gamma_p$.

## 12.2   SIZE DISTRIBUTION

### 12.2.1   Particle Sizing Models

There are several methods of plotting and analyzing the results of particle sizing tests, and two of them, which were developed for sizing analysis of finer materials in mineral dressing (Hassialis and Behre, 1945), have been proposed for basic analysis of fragmentation in blasting and related excavation processes.

The data from crushing and grinding processes are usually plotted as size–frequency curves on Cartesian coordinates, cumulative distributions on Cartesian or semi-log, or log probability paper, if the form of the distribution is normal or skewed probability. This curve fitting is done in an attempt to obtain usable closed mathematical expressions that will represent the distribution curve with reasonable accuracy. The commonly used size distribution equations for commercial comminution processes were reviewed by Harris (1968), and these have some applicability to blasted material for both conventional and nuclear sizes. One equation, variously termed the Gates–Gaudin–Schuhmann equation, is suitable for describing some cumulative particle size distributions:

$$y = (x/k_s)^n \tag{12.2}$$

where $x$ = particle size; $n$ = a distribution parameter, constant, and the slope of the distribution straight line on a log–log plot; and $k_s$ = maximum particle size. This equation offers a direct correlation for cratering parameters and some crushing and grinding. Another equation that has applicability to certain particulate aggregates produced in mineral processing is the Rosin–Rammler equation, which may be expressed in the following form:

$$y = 1 - \exp(-bx^n) \tag{12.3}$$

where $b$ = constant.

Both of these equations have the effect of expanding the fine-size region, while the upper limit of the R–R equation is infinite. Faddeenkov (1975) found that this equation (R–R) would fit data for blasted materials from small-scale experiments when plotted on double-log versus log coordinates.

Data for large craters, with the exception of one shot in rhyolite, can be expressed by the Schuhmann equation (Figure 12.4). The maximum size of fragment is given by the value $k_s$. Except for the smaller craters, the slope of the lines is approximately 0.5.

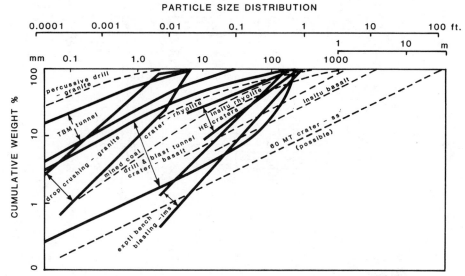

**Figure 12.4** Particle size distribution for rock fragmentation processes.

Thus, the parameter $k_s$ appears to be related to the yield of the explosion and the *in situ* block size of the rock material, and the slope of the line is related to *in situ* block size distribution and the properties of the rock.

The Schuhmann plot may be either in te⌐..s of percentage passing or cumulative percentage retained (Figure 12.5a; Table 12.2). If the latter plot is a straight line, the Schuhmann equation is assumed to apply.

The Rosin–Rammler function (Rosin and Rammler, 1933) has been used recently (Cunningham, 1983) for blasting analysis and has a scientific basis (Bennett, 1936):

$$y = 100 \exp\left(-x/x_c\right)^n \tag{12.4}$$

where $x$ = mean particle diameter, $y$ = cumulative percent, and $x_c$ and $n$ = constants. The constants $x_c$ and $n$ are best determined from the linear plot of data (Figure 12.5b; Tables 12.3, 12.4). If the equation is rewritten

$$\ln\left(100/y\right) = \left(x/x_c\right)^n \tag{12.5}$$

or

$$\ln\left[\ln\left(100/y\right)\right] = n\left[\ln\left(x\right) - \ln\left(x_c\right)\right] \tag{12.6}$$

$n$ is the slope of the line, and $n \ln x_c$ is the intercept on the vertical axis. If $x = x_c$, then $y = 36.8\%$, termed the *absolute size constant* by Hassialis and Behre (1945). The slope $n$ is also called the *dispersion constant* and has values rarely less than 0.6 for standard types of comminution products. For these materials, if it exceeds 3.0, a Gaussian probability function may be a best fit to use.

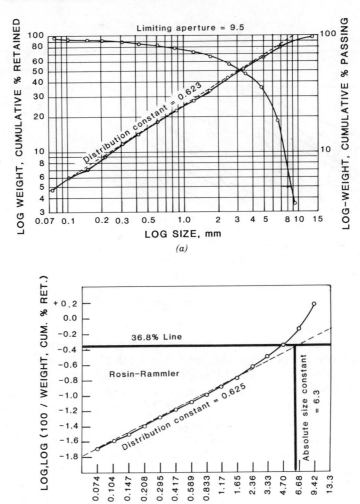

**Figure 12.5** Methods of plotting sizing tests. (*a*) Log–log method and (*b*) log versus log–log. (Hassialis and Behre, 1945).

## 12.2.2 Particle Size—Explosive Parameter Model

The development of the Kuznetsov–Rammler (K–R) model follows. Kuznetsov's (1973) equation is

$$\bar{x} = A\left(\frac{V_o}{Q}\right)^{0.8} Q^{1/6} \tag{12.7}$$

where $\bar{x}$ = mean fragment size in cm; $A$ = a rock factor (7 for medium rocks, 10 for hard, highly fissured rocks, and 13 for hard, weakly fissured rocks; limits appear

**TABLE 12.2. Sizing Test Plotted on Figure 12.5**

| Mesh | Aperture, mm | Weight, %[a] | Weight, Cum. % Through | Weight, Cum. % Retained | Reciprocal of Aperture | % Limiting Aperture | Log Log (100/ωr) | Ordinal No. | Ordinal No. × wt., % |
|---|---|---|---|---|---|---|---|---|---|
| 0.52-in. | 13.3 | 0 | 100.1 | 0 | 0.079 | 100.0 | | 0 | 0 |
| 0.37-in. | 9.4 | 3.5 | 96.5 | 3.5 | 0.105 | 70.7 | +0.167 | 1 | 3.5 |
| 3 | 6.7 | 15.5 | 81.1 | 19.0 | 0.148 | 50.3 | −0.142 | 2 | 31.0 |
| 4 | 4.7 | 17.4 | 63.7 | 36.4 | 0.213 | 35.3 | −0.358 | 3 | 52.2 |
| 6 | 3.3 | 12.0 | 51.7 | 48.4 | 0.303 | 24.8 | −0.502 | 4 | 48.0 |
| 8 | 2.4 | 9.1 | 42.6 | 57.5 | 0.417 | 17.9 | −0.619 | 5 | 45.5 |
| 10 | 1.7 | 9.2 | 33.4 | 66.7 | 0.588 | 12.8 | −0.755 | 6 | 55.2 |
| 14 | 1.2 | 6.3 | 27.1 | 73.0 | 0.834 | 9.0 | −0.865 | 7 | 44.1 |
| 20 | 0.83 | 5.4 | 21.7 | 78.4 | 1.29 | 6.2 | −0.976 | 8 | 43.2 |
| 28 | 0.59 | 4.3 | 17.4 | 82.7 | 1.69 | 4.4 | −1.084 | 9 | 38.7 |
| 35 | 0.42 | 3.1 | 14.3 | 85.8 | 2.38 | 3.2 | −1.177 | 10 | 31.0 |
| 48 | 0.30 | 2.9 | 11.4 | 88.7 | 3.33 | 2.3 | −1.284 | 11 | 31.9 |
| 65 | 0.21 | 2.4 | 9.0 | 91.1 | 4.76 | 1.6 | −1.392 | 12 | 28.8 |
| 100 | 0.15 | 1.9 | 7.1 | 93.0 | 6.67 | 1.1 | −1.503 | 13 | 24.7 |
| 150 | 0.10 | 1.2 | 5.9 | 94.2 | 10.0 | 0.8 | −1.586 | 14 | 16.8 |
| 200 | 0.074 | 1.1 | 4.8 | 95.3 | 13.5 | 0.6 | −1.679 | 15 | 16.5 |
| <200 | | 4.8 | | | | | | 16 | 76.8 |
| Total | | 100.1 | | | | | | | 587.9 |

[a]Data (112 A 154).
Source: Hassialis and Behre (1945).

**TABLE 12.3. Comparison of Rosin–Rammler and Gaudin Functions**

| Aperture, mm | Rosin–Rammler Function: D/6.3 | Log (D/6.3) | 0.625 log (D/6.3) | Log⁻¹ [0.625 log (D/6.3)] = (D/6.3)^b | e^{-(D/a)^b} | Calc. w_r | Obs. w_r | Calc. -Obs. | D/A | Derived Gaudin Function: Log (D/A) | 0.623 log (D/A) | Log⁻¹ [0.623 log (D/A)] | Calc. wt. | Obs. wt. | Calc. -Obs. | Gaudin Function: Log d | 0.632 log D | Log⁻¹ (0.632 log D) | Calc. wt. | Obs. wt. | Calc. -Obs. |
|---|---|---|---|---|---|---|---|---|---|---|---|---|---|---|---|---|---|---|---|---|---|
| 13.3 | 2.11 | 0.327 | 0.203 | 1.60 | 0.202 | 20.2 | 0 | +20.2 | 1.40 | 0.146 | 0.0909 | 1.23 | 123.0 | 100.1 | +22.9 | 1.124 | 0.710 | 5.13 | 31.6 | 0 | +31.6 |
| 9.4 | 1.49 | 0.173 | 0.108 | 1.28 | 0.278 | 27.8 | 3.5 | +24.3 | 0.99 | $\bar{1}.996$ | $\bar{1}.997$ | 0.993 | 99.3 | 96.6 | +2.7 | 0.973 | 0.615 | 4.12 | 24.7 | 3.5 | +21.2 |
| 6.7 | 1.06 | 0.025 | 0.0158 | 1.04 | 0.353 | 35.3 | 19.0 | +16.3 | 0.706 | $\bar{1}.849$ | $\bar{1}.906$ | 0.805 | 80.5 | 81.1 | -0.6 | 0.826 | 0.522 | 3.33 | 20.0 | 15.5 | +4.5 |
| 4.70 | 0.746 | $\bar{1}.873$ | $\bar{1}.921$ | 0.834 | 0.434 | 43.4 | 36.4 | +7.0 | 0.495 | $\bar{1}.695$ | $\bar{1}.810$ | 0.646 | 64.6 | 63.7 | +0.9 | 0.672 | 0.424 | 2.66 | 15.9 | 17.4 | -1.5 |
| 3.30 | 0.524 | $\bar{1}.719$ | $\bar{1}.824$ | 0.667 | 0.512 | 51.2 | 48.4 | +2.8 | 0.347 | $\bar{1}.540$ | $\bar{1}.713$ | 0.516 | 51.6 | 51.7 | -0.1 | 0.519 | 0.328 | 2.13 | 12.8 | 12.0 | +0.8 |
| 2.40 | 0.381 | $\bar{1}.581$ | $\bar{1}.738$ | 0.547 | 0.578 | 57.8 | 57.5 | +0.3 | 0.253 | $\bar{1}.403$ | $\bar{1}.628$ | 0.425 | 42.5 | 42.6 | -0.1 | 0.380 | 0.240 | 1.74 | 10.4 | 9.1 | +1.3 |
| 1.70 | 0.270 | $\bar{1}.431$ | $\bar{1}.644$ | 0.441 | 0.644 | 64.4 | 66.7 | -2.3 | 0.179 | $\bar{1}.253$ | $\bar{1}.535$ | 0.343 | 34.3 | 33.4 | +0.9 | 0.230 | 0.145 | 1.40 | 8.4 | 9.2 | -0.8 |
| 1.20 | 0.190 | $\bar{1}.279$ | $\bar{1}.549$ | 0.354 | 0.701 | 70.1 | 73.0 | -2.9 | 0.126 | $\bar{1}.100$ | $\bar{1}.439$ | 0.275 | 27.5 | 27.1 | +0.4 | 0.079 | 0.0499 | 1.12 | 6.7 | 6.3 | +0.4 |
| 0.83 | 0.132 | $\bar{1}.121$ | $\bar{1}.451$ | 0.283 | 0.753 | 75.3 | 78.4 | -3.1 | 0.0874 | $\bar{2}.942$ | $\bar{1}.341$ | 0.219 | 21.9 | 21.7 | +0.2 | $\bar{1}.919$ | $\bar{1}.9480$ | 0.887 | 5.3 | 5.4 | -0.1 |
| 0.59 | 0.0936 | $\bar{2}.971$ | $\bar{1}.357$ | 0.228 | 0.795 | 79.5 | 82.7 | -3.2 | 0.0621 | $\bar{2}.793$ | $\bar{1}.248$ | 0.177 | 17.7 | 17.4 | +0.3 | $\bar{1}.771$ | $\bar{1}.855$ | 0.716 | 4.3 | 4.3 | 0.0 |
| 0.42 | 0.0666 | $\bar{2}.824$ | $\bar{1}.265$ | 0.184 | 0.830 | 83.0 | 85.8 | -2.8 | 0.0442 | $\bar{2}.645$ | $\bar{1}.156$ | 0.143 | 14.3 | 14.3 | 0.0 | $\bar{1}.623$ | $\bar{1}.762$ | 0.578 | 3.5 | 3.1 | +0.4 |
| 0.30 | 0.0476 | $\bar{2}.678$ | $\bar{1}.174$ | 0.149 | 0.861 | 86.1 | 88.7 | -2.6 | 0.0316 | $\bar{2}.500$ | $\bar{1}.066$ | 0.116 | 11.6 | 11.4 | +0.2 | $\bar{1}.477$ | $\bar{1}.669$ | 0.467 | 2.8 | 2.9 | -0.1 |
| 0.21 | 0.0333 | $\bar{2}.522$ | $\bar{1}.076$ | 0.119 | 0.886 | 88.6 | 91.1 | -2.5 | 0.0221 | $\bar{2}.344$ | $\bar{2}.968$ | 0.0929 | 9.3 | 9.0 | +0.3 | $\bar{1}.322$ | $\bar{1}.571$ | 0.372 | 2.2 | 2.4 | -0.2 |
| 0.15 | 0.0238 | $\bar{2}.377$ | $\bar{2}.986$ | 0.0968 | 0.906 | 90.6 | 93.0 | -2.4 | 0.0158 | $\bar{2}.199$ | $\bar{2}.878$ | 0.0755 | 7.6 | 7.1 | +0.5 | $\bar{1}.176$ | $\bar{1}.479$ | 0.301 | 1.8 | 1.9 | -0.1 |
| 0.10 | 0.0159 | $\bar{2}.201$ | $\bar{2}.876$ | 0.0752 | 0.928 | 92.8 | 94.2 | -1.4 | 0.0105 | $\bar{2}.021$ | $\bar{2}.767$ | 0.0585 | 5.9 | 5.9 | 0.0 | $\bar{1}.000$ | $\bar{1}.368$ | 0.233 | 1.4 | 1.2 | +0.2 |
| 0.074 | 0.0117 | $\bar{2}.068$ | $\bar{2}.793$ | 0.0621 | 0.940 | 94.0 | 95.3 | -1.4 | 0.00779 | $\bar{3}.892$ | $\bar{2}.687$ | 0.0486 | 4.9 | 4.8 | +0.1 | $\bar{2}.869$ | $\bar{1}.285$ | 0.193 | 1.2 | 1.1 | +0.1 |

Rosin–Rammler Function: $a = 6.3$ mm, $b = 0.625$

Derived Gaudin Function: $A = 9.5$ mm., $D = 0.623$

Gaudin Function: $a = 6.0\%$, $b = 0.632$*

*Source:* Hassialis and Behre (1945)

*See source.

**TABLE 12.4. Values of Log [Log (100/$w_r$)] for Various Values of $w_r$ (after Geer and Hancey)**

| $w_r$ | Log [log (100/$w_r$)] | Difference | $w_r$ | Log [log (100/$w_r$)] | Difference | $w_r$ | Log [log (100/$w_r$)] | Difference |
|---|---|---|---|---|---|---|---|---|
| 2 | +0.2302 | 0.0000 | 42 | −0.4239 | 0.6541 | 81 | −1.0395 | 1.2697 |
| 4 | +0.1455 | 0.0847 | 44 | −0.4478 | 0.6780 | 82 | −1.0635 | 1.2937 |
| 6 | +0.0871 | 0.1431 | 46 | −0.4720 | 0.7022 | 83 | −1.0915 | 1.3217 |
| 8 | +0.0402 | 0.1900 | 48 | −0.4966 | 0.7268 | 84 | −1.1221 | 1.3523 |
| 10 | +0.0000 | 0.2302 | 50 | −0.5214 | 0.7516 | 85 | −1.1524 | 1.3826 |
| 12 | −0.0358 | 0.2660 | 52 | −0.5467 | 0.7769 | 86 | −1.1832 | 1.4134 |
| 14 | −0.0686 | 0.2988 | 54 | −0.5725 | 0.8027 | 87 | −1.2195 | 1.4497 |
| 16 | −0.0991 | 0.3293 | 56 | −0.5988 | 0.8290 | 88 | −1.2597 | 1.4899 |
| 18 | −0.1280 | 0.3582 | 58 | −0.6262 | 0.8564 | 89 | −1.2944 | 1.5246 |
| 20 | −0.1555 | 0.3857 | 60 | −0.6538 | 0.8840 | 90 | −1.3400 | 1.5702 |
| 22 | −0.1821 | 0.4123 | 62 | −0.6828 | 0.9130 | 91 | −1.3872 | 1.6174 |
| 24 | −0.2077 | 0.4379 | 64 | −0.7122 | 0.9424 | 92 | −1.4409 | 1.6711 |
| 26 | −0.2328 | 0.4630 | 66 | −0.7438 | 0.9740 | 93 | −1.5029 | 1.7331 |
| 28 | −0.2574 | 0.4876 | 68 | −0.7757 | 1.0059 | 94 | −1.5696 | 1.7908 |
| 30 | −0.2817 | 0.5119 | 70 | −0.8097 | 1.0399 | 95 | −1.6492 | 1.8794 |
| 32 | −0.3055 | 0.5357 | 72 | −0.8456 | 1.0758 | 96 | −1.7481 | 1.9783 |
| 34 | −0.3293 | 0.5595 | 74 | −0.8837 | 1.1139 | 97 | −1.8775 | 2.1077 |
| 36 | −0.3529 | 0.5831 | 76 | −0.9234 | 1.1536 | 98 | −2.0655 | 2.2957 |
| 38 | −0.3764 | 0.6066 | 78 | −0.9670 | 1.1972 | 99 | −2.3644 | 2.5946 |
| 40 | −0.4002 | 0.6304 | 80 | −1.0137 | 1.2439 | | | |

*Source:* Hassialis and Behre (1945).

to be between 8 and 12); $V_o$ = rock volume (m$^3$) broken per blasthole, equal to burden × spacing × bench height; and $Q$ = mass (kg) of TNT equivalent in energy of explosive charge in each blasthole. If $Q_e$ = mass (of explosive) per blasthole and $E$ = relative weight strength of explosions, ANFO = 100, TNT = 115, then $Q_e × E = Q × 115$ and $Q = Q_e E/115$, and equation 12.7 becomes

$$\bar{x} = A\left(\frac{V_o}{Q_e}\right)^{0.8} Q_e^{1/6}\left(\frac{E}{115}\right)^{1/6}\left(\frac{E}{115}\right)^{-0.8} \qquad (12.8)$$

and

$$\bar{x} = A\left(\frac{V_o}{Q_e}\right) Q_e^{1/6}\left(\frac{E}{115}\right)^{-19/30} \qquad (12.9)$$

The quantity $((V_o)/(Q_e))$ is the inverse of the specific charge or powder factor, $K$, in kg/m$^3$. Thus, equation 12.9 is

$$\bar{x} = A(K)^{-0.8} Q_e^{1/6}\left(\frac{E}{115}\right)^{-19/30} \qquad (12.10)$$

The Rosin–Rammler equation

$$y = e^{-(\bar{x}/x_c)^n} \qquad (12.11)$$

then gives

$$\bar{x} \text{ for } y = 0.5$$
$$0.5 = e^{-(\bar{x}/x_c)^n}$$
$$x_c = \frac{x}{(0.693)^{1/n}}$$

The value of $n$ determines the curve shape and may vary between 0.8 and 2.2, with high values for uniform sizes and low values for larger amounts of both fines and oversize.

For uniform fragmentation, high values of $n$ are preferred which vary in the model as

| Parameter | $n$ increases as parameter |
|---|---|
| burden/hole diameter | decreases |
| drilling accuracy | increases |
| charge length/bench height | increases |
| spacing/burden | increases |

For staggered patterns $n$ also increases.

Cunningham (1983) and Lounds (1986) experimentally obtained values of $n$ for the following parameters: drilling accuracy, ratio of burden to blasthole diameter, whether drilling pattern was in line or staggered, spacing/burden ratio, and ratio of charge length to bench height.

The results developed were combined with the Kuznetsov (1973) equation and was called the "Kuz–Ram" model. A routine was developed that matched the volume of a given fragment size with remaining volume of ground broken per blasthole, to predict the size of the largest boulders.

The value of $n$ is determined by

$$n = (2.2 - 14B/d)(1 - W/B)((A - 1)/2)L/H \qquad (12.12)$$

where $B$ = burden, m; $d$ = hole diameter, mm; $W$ = standard deviation of drilling accuracy, m; $A$ = spacing/burden ratio; $L$ = charge length above grade level, m; and $H$ = bench height, m. For staggered drilling, $n$ is decreased 10%.

Kuznetsov (1973) studied models of different materials with application to open-pit mines and to a nuclear blast. The degree of scatter was as expected, considering the variability of rock. Correlation was best in the model-controlled research, with greater scatter in natural full-scale rock tests. In model work reported by Cunningham (1983), the results were obtained with "permitted" explosives in 12-mm blastholes in concrete blocks. The mean fragment size was compared with the Kuznetsov equation prediction with a rock factor of 8. For F and G (Table 12.5), the mean fragment sizes are 10 and 20% larger than predicted by the Kuznetsov equation. These two explosives were slurries that gave lower calculated energy yields in small diameters. The NG explosives show unusual agreement even with variation in density and strength.

In small-scale basting experiments, Dick and Gletcher (1973) measured the fragmentation in blasting limestone with a 22-mm blasthole with results in good agreement with the K–R model (Figure 12.6) for delay blasting. Data also gave a good fit for the Schuhmann model for all shots (Dick and Gletcher, 1973; Figure

**TABLE 12.5. Explosive Properties and Fragment Size**

| Explosive | Relative Weight Strength[a] | Specific Gravity | Relative Bulk Strength[a] | Mean Fragment Size Measured (cm) | Predicted (cm) |
|---|---|---|---|---|---|
| A | 66.0 | 1.22 | 100 | 13.7 | 14.2 |
| B | 57.0 | 1.21 | 86 | 15.2 | 15.7 |
| C | 42.3 | 1.24 | 66 | 18.2 | 18.9 |
| D | 70.7 | 1.50 | 132 | 12.5 | 11.9 |
| E | 53.5 | 1.60 | 107 | 13.7 | 13.8 |
| F | 73 | 1.25 | 114 | 15.4 | 14.0 |
| G | 70 | 1.25 | 109 | 17.4 | 14.5 |

[a]Based on ANFO at S.G. 0.8 = 100.
*Source:* Cummingham (1983).

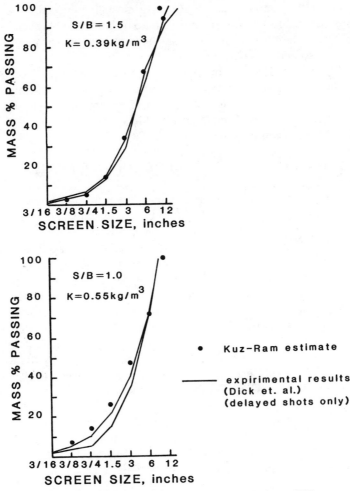

**Figure 12.6**  Fragmentation in limestone (Cunningham, 1983).

12.11). For other fragmentation studies (Winter and Ritter, 1980) in large blocks of jointed limestone, with delays between rows of 16-mm blastholes on an $0.3 \times 0.45$ m pattern (Figure 12.7), the agreement is not as close. However, variation for the measured values is less than 10%.

The large-scale blasting fragmentation was photographically obtained of broken rock from an overburden blast in sandstone at a coal mine, plotted with the K–R calculations (Figure 12.8), which give good correlation for coarser material but indicates more fines than recorded. However, in blasting, the accuracy in size prediction is more important for coarser material, the number of fines may be underestimated by the photographic method, and the fragmentation varies at different parts of the blast.

Cautions in use of the K–R model are (1) the S/B ratio should not exceed 2; (2) factors such as initiation and time should enhance fragmentation and misfires

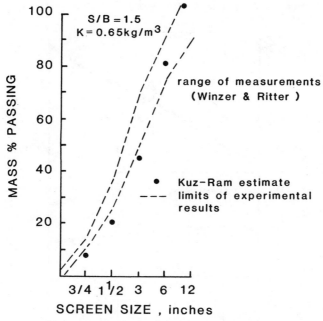

**Figure 12.7** Fragmentation in jointed limestone (Cunningham, 1983).

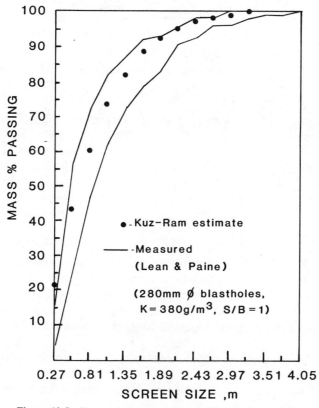

**Figure 12.8** Fragmentation in sandstones (Cunningham, 1983).

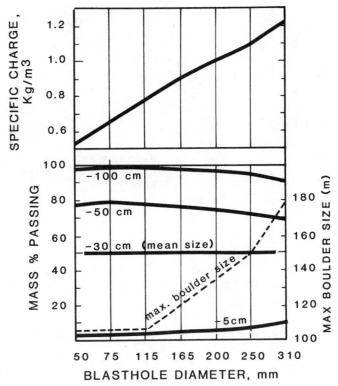

**Figure 12.9** K–R model estimates for constant mean size (Cunningham, 1983).

or cutoffs avoided; (3) the explosive energy should be near optimum; and (4) jointing and homogeneity of the rock require careful assessment, as fragmentation is affected by the rock structure, with jointing more closely spaced than the drill pattern. It is proposed by Cunningham (1983) that the K–R model can be used to predict operating parameters (Figure 12.9, 12.10); for example, where blasthole diameter is increased between 50 and 310 mm for a 12-m bench height, for ANFO, with stemming kept at 20 hole diameters and drilling accuracy at 0.45 SD at the bottoms of the holes, a rock factor of 10 applies.

Constant mean fragmentation (Figure 12.9) may be required for ore to pass through a small crusher. Note in this that, as the blasthole diameter increases, the amount of explosive increases rapidly, and boulder size increases above 115-mm spacing. (This may be because of relative drilling accuracy and distribution of explosives: the first improves and the latter deteriorates as hole size increases.) Also, for constant mean fragmentation, the amounts of both fine and coarse material increase.

The model further predicts that, for constant specific charge in overburden where fragmentation is not critical, blast design may be based on a constant powder factor (Figure 12.10) or, as diameter increases, (1) mean fragment size increases by about

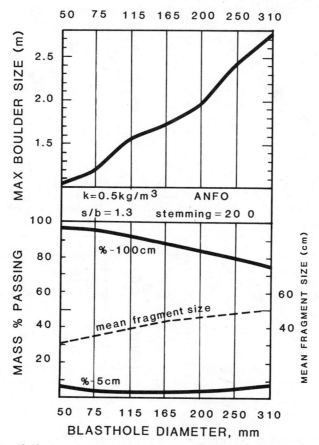

**Figure 12.10**  K–R estimates for constant specific charge (Cunningham, 1983).

60%; (2) the +100-cm fraction increases from 5 to 25%; (3) fines do not vary much but are minimal for the medium diameters (more fines are generated with small-diameter holes because of the proximity of holes and the effect of drilling inaccuracy, in large diameters by more crushing around the blasthole wall); (4) maximum boulder size increases from about 1 m to almost 2.8 m.

For a fixed drilling pattern, improvements can be obtained by increasing explosive strength (Figure 12.10) with predicted fragmentation from explosives of different strengths in a drill pattern of 3–4 m in hard rock (A-12) with 115-mm blastholes. The explosive density is 1.2 (i.e., constant specific charge), but weight strengths vary between 90 and 124. For a critical fragment size of 50 cm, a RWS 124 gives 77% passing as opposed to 56% for TWS 90, or a reduction of 33% in oversize.

The expected fragmentation of ANFO at density 0.8 and for a slurry at strength 124 at density 1.4 shows that ANFO yields 40% oversize and that the stronger explosive yields only 15% in oversize at 50cm—that is, a reduction of 62%.

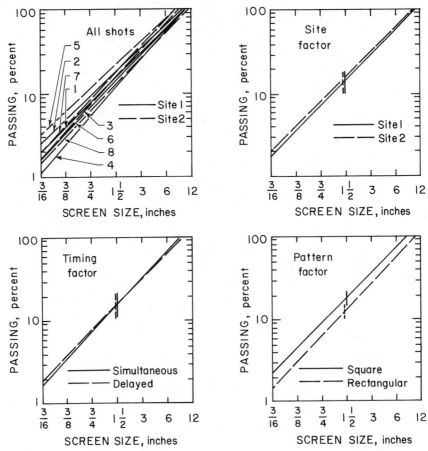

**Figure 12.11** Fragment distribution from small-scale bench blasting (Dick et al., 1973).

Cunningham (1983) concluded that a problem may exist in the assessment of fines, which can be generated both by the equipment loading of rock and by weak binding material between mineral grains in the ore. Factors other than the explosive play a major role. There are a large number of variables in blasting including geological conditions, and each operation requires individual analysis. The above calculations show illustrative relationships between blasting parameters and fragmentation. The model has been applied in South African mining for realistic designs and analysis, but more research with the model is planned.

### 12.2.3 Particle Size Model—Small-Scale Blasting

For small-scale blasting in a limestone bench, Dick and Gletcher (1973) used the cumulative percentage distribution for fragments and found that the data (Figure 12.11) were accurately described by the equation

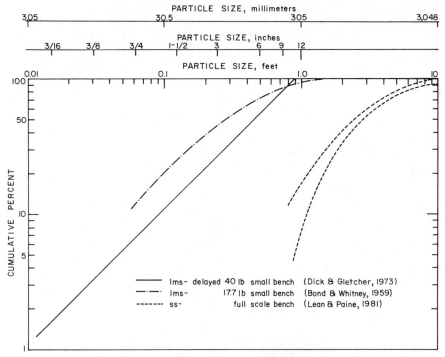

**Figure 12.12** Particle size, model, and full-scale bench blasting.

$$\log y = \log k + b \log x \qquad (12.13)$$

where $y$ = percentage of material passing, $k$ = intercept of line ($k = y$ when $s = 100\%$), $b$ = slope of line, and $x$ = screen size.

This equation is a form of the Schuhmann equation and may be written

$$y = kx^b \qquad (12.14)$$

or

$$y = 100(x/k_s)^n \qquad (12.15)$$

where $k_s$ = the 100% intercept at maximum particle size and $n$ = slope of line. Other experiments yielded curvelinear plots (Figure 12.12).

## 12.3 FLOW OF EXPLOSIVELY FRAGMENTED MATERIAL UNDERGROUND

Mining operations that involve the stability and flow of rock material broken by explosives include that in the chimneys formed by the underground nuclear shots plus mining experience gained in sublevel and block caving. There appears to be no published information on the theory of the behavior of rubble, although much experimental work on waste piles of rock in a geotechnical centrifuge is reported by Malushitsky (1981).

The significant amount of research has been done on the behavior of the stability and flow of explosively fragmented rock of about 2–3 ft in diameter. This is somewhat defined by the behavior of relatively fine material in storage bins and the coarser material handled in sublevel caving. A majority of the semi-quantitative investigations of caving material have been made in conjunction with the method of sublevel caving. The mining method consists essentially of blasting by ring drilling or similar-method horizontal sections of an ore deposit and then drawing off the broken ore through chutes or raises, the drawing process being done under the force of gravity. The overlying waste material is allowed to cave (Figure 12.13).

The following factors involved in caving are of interest in broken ore flow in commercial mining: (1) stability of broken rock mass before movement; (2) initiation of caving, conditions for; (3) first flow, its effects on pressure and structure of mass; (4) intermediate flow, its effect on pressure and structure of mass; and (5) final flow, its effects on pressure and structure of mass. That is, the ore must be well broken and fine enough so that it will flow, and the behaviorisms, including the percentage of total material recovery, are of primary interest.

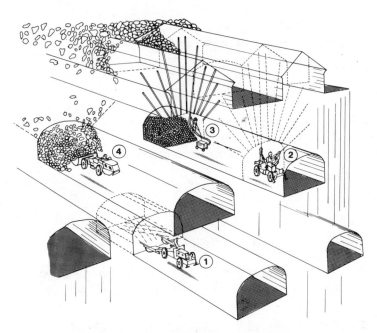

**SUBLEVEL CAVING IN KIRUNA**

1.) Drifting: 2.) Ring drilling (roof of slice). 3.) Charging of explosives
4.) Ore Loading (Kvapil, R., 1973)

**Figure 12.13**  Sublevel caving in Kiruna (Kvapil 1973).

Under certain ideal conditions the laws of gravity flow are independent of fragment size—that is, where the fragments are much smaller than the opening into which the material passes and the flow channel size and factors such as interlocking may affect the flow-mass stability. Flow will take place for small-size material, in general about one-fourth the size of the opening. The conditions for gravity flow that have been discovered in sublevel caving may occur in similar conditions where extensively fragmented material occurs (Yenge, 1981).

The factors that affect the structural stability of a rubble mass immediately above an opening are several in number: (1) the *in situ* stability of the rubble structure, its strength, and particulate composition; (2) the degree of loosening caused by the excavation; (3) the amount of flow allowed into the opening, the type and effectiveness of support; (4) the fragment size distribution and fragment shape, particularly near and above the opening; (5) the depth of the opening beneath the surface; (6) water present, amount and distribution; and (7) angle of friction and interlocking factors.

Although the angle of friction of loose materials determines the failure characteristics of open slopes, it plays a lesser role in the stability of underground openings in such material, particularly where the opening is circular in cross section. Fragmented and alluvial geologic materials have the same general range of values of angle of friction, and it decreases for both with increase of confining pressure (Figure 12.14).

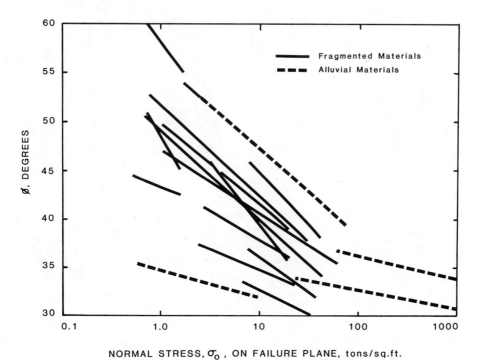

NORMAL STRESS, $\sigma_o$ , ON FAILURE PLANE, tons/sq.ft.

**Figure 12.14** Decrease in friction angle with increasing normal stress (Banks and McIver 1969).

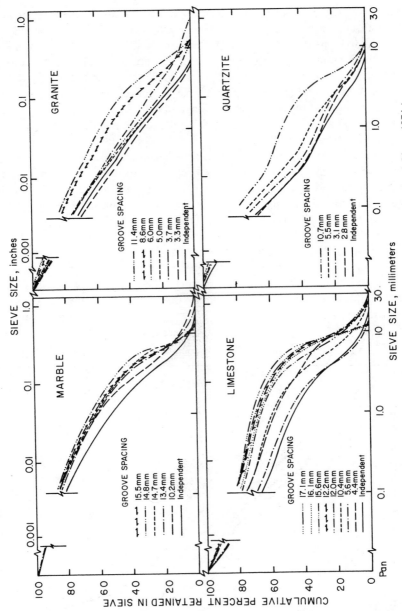

**Figure 12.15** Size distribution for grooves at pre-optimum spacing (Rad and Olson, 1974a).

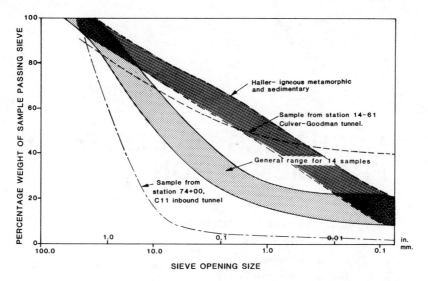

**Figure 12.16**  Muck gradations for samples collected in tunnels in limestone and shale (Nelson et al. 1984); and in igneous, metamorphic and sedimentary rocks (Haller, 1973).

## 12.4  TMB MUCK PARTICLE SIZE

The particle size distribution of muck material cut from smooth sawn surfaces in the laboratory to determine groove spacing effects does not follow the Schumann exponential law and gives consistently smaller particle sizes for smaller spacings for four types of rock (Figure 12.15; Rad and Olson, 1974a). The regions for the best effect of machine parameters according to lab tests indicate that there are optimum combinations for spacing, thrust, and cutter diameter.

For 14 muck samples from field tunneling in sandstones, limestones, and shales (Nelson et al., 1984), the particle size distribution was somewhat similar (Figure 12.16), although the cumulative plot is from large to small sieve size left to right. (Data were too numerous to replot.)

In an informative study of TBM excavation by Haller et al. (1973), a linear correlation was attempted between compressive strength, rock quality designation, dry unit weight, hardness, and ground water to give muck designation numbers (MDNs). However, those that were intended to be used for engineering design purposes have not found wide application. On the other hand, the particle size distribution studies indicate smaller particle sizes for all types of rock than for the rocks studied by Nelson et al. (1984) (Figure 12.16). One of the main influences on particle size is the *in situ* fracture distribution in addition to the physical properties.

# 13

# MECHANICS OF EXPLOSIVE FRAGMENTATION

Many qualitative but fewer meaningful investigations have been made of the breakage of rock by contained explosives, and the analyses of data and observed phenomena have given rise to different explanations of the mechanisms that occur in the stressed rock and result in rock breakage, including the effects of the different geometries that are employed in a wide variety of blasting operations. The questions arise as to whether the rock breakage mechanisms are the same for one, two, or three faces, for spherical and cylindrical geometries, for rocks of various strength and geologic structure, and for explosives of different properties. More definitive research is needed to give quantitative, complete answers to these questions.

## 13.1  MECHANISMS OF ROCK BREAKAGE

The breakage mechanisms occurring in rock fragmentation by explosives loaded in drill holes depend on the number of free faces, the burden, the hole placement and rock geometry, the physical properties and loading density of the explosive, the quality of the stemming, the rock properties, and other factors. The breakage of rock near a free face may involve the following processes, their relative influence depending on the above and related parameters:

1. Crushing of the rock immediately around the explosive cavity;
2. Initial close-in radial fracturing due to the tangential stress in the outgoing stress wave;
3. Propagation of existing fractures;
4. Secondary radial fractures formed at the surface, traveling inward, due to enhanced tangential stress accompanying free surface displacement;
5. Joining of inward traveling fractures with initially created outward radial fractures;
6. Extension of the initial radial fractures by reflected radial tensile strain at oblique angles to the surface;
7. Tensile separation and shear of rock at planes of weakness in the rock;
8. Separation of the rock due to reflected radial tensile strain (slabbing);
9. Fracture and acceleration of fragments by strain energy release;
10. Further fracture and acceleration of broken rock by late expanding gases.

The mechanical processes that take place in the rock after a stress wave is generated by a confined charge detonated within a critical distance of and reflected by a free face in natural rock are complex and cannot be accurately described by a simple model. The detonation of the explosive generates a stress wave beyond the walls of the cavity. In most rocks there is no true shock wave generated (see below), and stress waves impinging on the free surface are reflected as radial, tangential, tensile, and shear waves. If an outgoing radial compressive stress is reflected as a tensile stress and is of great enough magnitude, slabbing occurs.

With regard to the importance of multiple directional displacement relief with more than one free face, it has been found that cratering to one free surface requires more explosive per unit weight of rock broken than bench blasting where two or more free surfaces are present with the same burden (or depth of burial) for each charge. However, the same mechanisms of breakage are present in each case, and the slabbing mechanism plays an important role in both bench blasting and cratering, depending on the magnitude, shape, and duration of the reflected stress pulse. Cratering to a single free face is often used as a guide to round design.

Thus, in the design of explosive rounds it is desirable to provide as many free faces as practical, either in the round before breakage starts or by the breakage of the rock as it is blasted, to obtain assurance that the subsequently blasted holes will break the rock remaining around them.

High (chemical) explosives are designed with differences in chemical composition and density to blast rock of a wide variety of physical and structural properties, geometries, and degrees of explosive confinement. The secondary breakage of large boulders illustrates minimum confinement and is easily acccomplished by either a ''mudcapped'' charge or by a charge detonated in a shallow drill hole, where the rock surface is free to expand in all directions. In contrast, the rock in the cut portion of ''no cut'' or a ''burn cut'' round has a minimum of such freedom to fracture and to move so that rock breakage and removal can occur. Also, more explosive is required per unit volume broken because of resistance to breakage, and more energy is transmitted to the surrounding rock. The ultimate confinement is represented by an explosive charge within an infinite solid where only crushing immediately around the explosive may occur. The degree and type of breakage that result from the detonation of a confined explosive depend on the properties of the explosive and the rock, the rock structure, the shape and location of the charge with respect to free faces, the confinement, the manner of loading, the degree of existing fractures, and other variable factors. Hence, the same breakage mechanisms do not dominate in the same degree for different blasting conditions.

### 13.1.1.  Free Faces

In the effective blasting of rock with confined explosives, the stress wave generated by the explosive must encounter adequate free faces to allow the rock to break. The free surface within a critical distance of a contained detonating explosive serves at least two primary related functions: to provide displacement so that the segments of a rock can move enough that effective fracturing can take place, and to reflect stress waves and their energy back into the rock to provide adequate stress or strain levels for different possible types of fracture initiation and propagation. An adequate energy density is required to further propagate the fractures and move the rock fragments.

Experimentation with very small charges of explosive encased in Plexiglas models, with the burden designed so that fracture to the nearest surface will not occur, shows that the plastic flow occurs first, followed by multiple small radial fractures further from the blast hole, all created by the stress by the first outgoing wave. As the distance to the nearest surface is decreased, more fracturing occurs in the solid between the charge and the nearest free face until the distance is reduced sufficiently that a crater is broken out. Also, as the burden is decreased, the total percentage of energy transmitted from the explosive to the solid decreases. However, Plexiglas has many different properties from rock and does not simulate rock in all respects.

It is usually assumed that the suddenly applied pressure from a detonating explosive in a hole generates an outgoing elastic stress wave in most solid rock beyond crushed and plastic zones. If a spherical charge is enclosed near one face of a homogeneous, semi-infinite solid, the wave has the same character in all directions until that portion of the wave traveling in the direction of the free face reaches this surface. The diverging stress wave moving on into an infinite solid with no free face will continue until it dies out with distance, resulting in no effective breakage. Hence, it becomes apparent that the intensity of the stress wave and the parameters that result from reflection and the other events that take place at the free surface, together with the action of the remainder of the outgoing wave through which it passes upon reflection, are determining factors in the amount and type of breakage that will take place. Similar principles apply to both spherical and cylindrical geometries.

### 13.1.2  Crushing

When a high explosive is detonated in direct contact with rock or is confined when it detonates, the high pressure of the gases may exceed the strength of the rock and result in some crushing. The amount of crushing is a function of the magnitude and duration of the pressure at the rock–gas interface and the rock properties. If a cartridge of high explosive is simply lying on the flat surface of a rock, the rock will experience the side-on pressure of the detonation wave, which is of only a few microseconds' duration, plus the pressure of the gases as they push away from the rock surface. This latter pressure, which is about one-half of the magnitude of the detonation pressure, is also of very short duration and is quickly dissipated because the gases are not confined. Thus, with no confinement, only minor crushing and chipping of the rock surface will usually result, with a minimum of fracturing into the rock.

If a cartridge is placed vertically on the rock and detonated from the far end, the head-on pressure of the detonation wave will cause somewhat greater crushing. Further, if the explosive is covered with mud or sand (mudcapped), the mass of the covering material will increase the magnitude of the pressure against the rock because of the conservation of momentum—that is, the gases pushing simultaneously against both the rock and the mudcap. The increase in momentum is accompanied by an increase in stress and energy transferred to the rock, and, if the rock mass is a boulder of proper size, it will be fractured by the mudcapped charge, where a bare charge may only cause local damage on the rock surface.

### 13.1.3  Initial Fracturing

Likewise, for a confined explosive at a loading density of 1 (the explosive fills the hole), the peak detonation pressure $p_2$, which varies from 60–220 kbar for commercial explosives, exists for only a few microseconds at any position in the borehole and rapidly drops to the explosion pressure $p_3$. Although $p_3$ has only about one-half of the value of $p_2$, it is of longer duration, usually causes major crushing and radial fracturing in the rock immediately around the cavity, and generates a strong but rapidly diminishing outward-traveling stress wave beyond the fracture zone. The gases do little or no useful blasting work at the detonation pressure $p_2$.

In practice, one often sees sections of the periphery of drill holes in a drift or tunnel round which remain apparently unfractured after the rock is blasted. The surfaces of these holes still show the drill marks, and, although crushing or fracturing cannot be readily observed, small fractures do exist. One reason may be that the peripheral holes do not hold the gas pressure long enough for intense crushing to take place.

Langefors and Kihlstrom (1979) found that, if small charges are detonated in holes in plates of Plexiglas, radial cracks are formed by the first outgoing stress wave. No radial cracks are formed immediately around the hole, but there is an annular plastic zone instead of the crushed region found in rock. Here there are no cracks formed, but cracking takes place beyond this zone, which is arrested at a distance of 6–12 radii from the hole. However, the threshold pressure for the formation of a shock wave in Plexiglas at which plastic flow occurs is only 8 kbar. Hence, the unfractured annular section around the blasthole in Plexiglass may result from flow caused by shock pressure. The theshold shock pressures for most rocks are much above even the detonation pressure of the strongest chemical explosives, and one would expect to find crushing only at the hole, then fractures emanating from the surfaces of blastholes or the crushed zone in most brittle rocks.

Most fractures are initiated at small flaws in the stressed materials, but Plexiglas does not have as many natural flaws as rock and so does not exhibit the same processes or fracture mechanisms when subjected to transient stresses of high intensity. It is different from rock in that it has the same strength in tension as in compression and a higher Poisson's ratio, 0.5. These factors must be considered in stress and fracture analysis, as well as in comparisons with rock behavior.

Other significant laboratory and field experiments have been performed demonstrating that radial fractures are generated by outgoing stress waves from an explosive cavity (Field and Ladegaard-Pederson, 1971). Studies of waves in Perspex models showed that, when a block of the material was immersed in a liquid of matching acoustic impedance, there were no reflected waves, and the pattern of radial cracks emanating from the blasthole was fairly uniform. The pressure required to initiate a true shock wave in Plexiglas is only 8 kbar, compared to about 200 kbar for limestone and 300 kbar for granite (Cook, 1958). Because in Plexiglas models the initial radial cracking occurs only beyond a plastic flow zone, it is of limited value to simulate some of the important detailed mechanisms of blasting in rock. In addition, it has been found by the author that the reflection of light from fracture surfaces within a Plexiglas model is not an accurate measure of the extent of the fracturing. That is, where Plexigas has been fractured, a small amount of pressure normal to the cracks will bring the surfaces together so that no light is reflected. Thus, where fractures radiate from a central hole and the model is still intact, there may be sufficient stress and consequent closure pressure on some of the cracks that only a portion of their true area is visible.

Obert and Duvall (1950) have suggested that a *radius of equivalent cavity*, $r_c$ in rock, to the limit of the fractured zone, may be calculated for a spherical charge where

$$r_c = k\sqrt[3]{w} \qquad (13.1)$$

where $r_c$ = radius of equivalent cavity, $k$ = material constant, and $w$ = weight of explosives (lb).

The radius of equivalent cavity is defined as the distance from the explosive cavity where the pressure causes a tangential stress equal to the tensile strength of the rock and beyond which no fracturing due to the outgoing wave takes place. This value has not been determined in practice.

In rock failure due to radial cracking, certain factors limit the extent of the initial radial fratures. The first is that in order for tensile or shear fractures to be initiated and extended, the rock must be displaced relative to fracture surfaces. The compressibility of the surrounding rock will determine the magnitude of small initial displacements; large displacements can take place only if appropriately large sections of the rock can move relative to each other. That is, the initial radial fracturing is largely controlled by the compressibility, confinement, and transient stress field in the rock. Later fracturing is controlled by block movement which can take place only if there is a free face within a critical distance. Also, it must be considered that the dynamic breaking strength of many rocks, and rocklike materials such as concrete, is greater than the static strength.

Where the reflected wave travels back through the outgoing wave, the reflected radial stress and strain being tensile, the outward particle velocity is increased, and some of the effective confinement that inhibited the initial outward radial fracturing is removed. This release of confinement plus the reflected stress causes the initial radial fractures to progress outward. The front of the reflected triangular pulse has greater kinetic energy than the pulse tail near the free surface. Hence, fragments of material close to the blasthole may have greater velocity when broken from the parent mass and will collide with the fragments in front of them, increasing the velocity of the latter. This phenomenon has been observed in quarry blasting.

## 13.2  CRATERING*

The detailed cratering experimentation by Duvall and Atchison (1957) with point charges with one free face has yielded the most quantitative data available and indicates that the slabbing mechanism plays a dominant role in this type of breakage. The results of their work are given here in detail to illustrate small-scale cratering phenomena in rock and associated mechanisms of rock breakage.

### 13.2.1  Experimental Procedure

Crater tests that break rock to one free face (Duvall and Atchison, 1957) were made in a comprehensive project with a range of sizes of drill holes and charges to excavate craters for various depths of explosive burial (Figure 13.1). Small cylindrical charges ($<50$ lb and a length diameter ratio $<8$ stemmed with sand) were detonated in vertical drill holes with electric primers, the crater cleared of broken rock, and the dimensions of the true crater recorded.

---

*This section on cratering was rewritten from Duvall and Atchison, 1957 with modifications and additions.

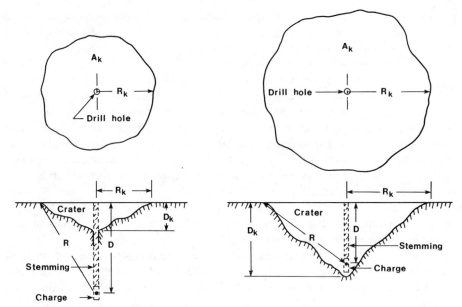

**Figure 13.1** Plan and section drawings illustrating crater tests variables (Duvall and Atchison, 1957).

The crater area, $A_k$, at the surface was measured, and the effective crater radius, $R_k$, was computed from the area. Two vertical sections of the crater were mapped, and the crater volume $V_k$ was computed.

The crater angle $\phi$ is

$$\tan \frac{\phi}{2} = \frac{R_k}{D} \qquad (13.2)$$

and the radius of rupture is

$$R = \sqrt{D^2 + R_k^2} \qquad (13.3)$$

The craters were blasted in exposed outcrops of granite, marlstone, sandstone, and chalk. Four shots were fired at greater than critical depth in chalk in horizontal holes drilled into a vertical face, with the length of the hole much greater than the depth below the horizontal surface. After the shots, a 30-in.-wide vertical saw cut was made through the center of the crater and the charge location. The breakage from the shot point and the crater surface was observed and mapped (see Figure 13.6).

Two types of stress wave measurements were made—the radial strain versus time at several distances in a linear array of gages, and the measurement of arrival time of the stress wave. In selected crater tests in sandstone, continuous strips of conducting foil were cemented to the rock surface in concentric circles around the charge hole to determine time of fracture of the rock at the surface. Flyrock velocity

**TABLE 13.1. Physical Properties of Rock Types**

| Rock type | Lithonia granite | Green River marlstone | Kanawha sandstone | Niobrara chalk |
|---|---|---|---|---|
| Formation member | Lithonia belt | Parachute creek | Homewood | Smoky Hill-Fort Hays |
| Description | Gneissic | Kerogenaceous, dolomitic limestone (oil shale) | Coarse-grained | Chalky limestone |
| Apparent specific gravity | 2.6 | 2.1 | 2.2 | 2.0 |
| Compressive strength (lb/in.²) | 30,000 | 10,000 | 10,000 | 2,000 |
| Tensile strength (lb/in.²) | 450 | Not measured | 70 | Not measured |
| Tensile breaking strain ($\mu$in./in.)[a] | 280 | do. | 500 | Do. |
| Modulus of rupture (lb/in.²) | 2,000 | 400 | 400 | 300 |
| Scleroscope hardness (scleroscope units) | 85 | 45 | 30 | 10 |
| Dynamic Young's modulus (lb/in.²) | $3.0 \times 10^6$ | $1.2 \times 10^6$ | $1.0 \times 10^6$ | $0.75 \times 10^6$ |
| Modulus of rigidity (lb/in.²) | $1.5 \times 10^6$ | $0.5 \times 10^6$ | $0.5 \times 10^6$ | $0.5 \times 10^6$ |
| Longitudinal bar velocity (ft/s) | 9,000 | 6,000 | 5,000 | 5,000 |
| Longitudinal field velocity (ft/s) | 18,500 | 13,000 | 5,000 | 7,500 |

[a] $\mu$in./in. = microinch per inch = $10^{-6}$ inch per inch.
*Source:* Duvall and Atchison (1957).

was measured with high-speed motion pictures of the blasting experiments in sandstone. These data on initial surface fracture are unique.

The crater and the strain data, which were produced by different charge sizes, and the measurements were all reduced by cube-root scaling for comparison. That is, the scaling factor $\bar{r}$ for small linear, spherical, or concentrated charges is the cube root of the charge weight, which is arbitrarily defined as a length in feet numerically equal to the cube root of the charge weight in pounds. For scaling, linear dimensions are thus scaled by dividing by $\bar{r}$, areas by $\bar{r}^2$, and volumes by $\bar{r}^3$. Strain, a dimensionless ratio, is a scaled quantity. Time is scaled by dividing by $\bar{r}$, and scaled time multiplied by velocity gives scaled distance.

## 13.2.2  Experimental Data

The physical properties of the four rock types were obtained by standard methods (Table 13.1). The average tensile breaking strain was taken as the strain at failure in simple tension. Six commercial explosives were employed (Table 13.2): Type A semi-gelatin explosive was used as a standard, and others were selected for different rates of detonation, weight, strength, and density. However, no adjustments were made in the calculations of the differences in explosive properties.

The crater data include (Tables 13.3–13.6) (1) charge data—the weight, position, and dimensions of the charge; (2) crater data—the dimensions and geometry of the crater; and (3) computed and scaled data. These data are uniquely valuable for comparative analysis.

The crater cross sections have different shapes (Figures 13.2–13.5) for granite, marlstone, sandstone, and chalk, which have different properties and structure. The breakage created in the four horizontal-hole crater tests in chalk was (Figures 13.6, 13.7) a crushed zone around the explosive cavity and a shallow crater at the

## TABLE 13.2.  Properties of Explosives

| Type of Explosives | Weight Strength[a] (%) | Bulk Strength[b] (%) | Density[c] (lb/ft³) | Velocity of Detonation[d] (ft/s) |
|---|---|---|---|---|
| Semi-gelatin, type A | 65 | 45 | 72 | 12,000 |
| Semi-gelatin, type B | 65 | 60 | 80 | 13,000 |
| Gelatin, type A | 60 | 35 | 85 | 20,000 |
| Gelatin, type B | 80 | 70 | 80 | 21,000 |
| Ammonia gelatin | 60 | 60 | 88 | 8,500 |
| Ammonia dynamite | 65 | 45 | 70 | 6,500 |

[a] Strength equivalent to same weight of straight dynamite of percentage indicated.
[b] Strength equivalent to same volume of straight dynamite of percentage indicated.
[c] Average cartridge density; actual charge density varied somewhat with loading conditions.
[d] For $1\frac{1}{4} \times 8$ in. diameter, unconfined; field rates were sometimes higher because of larger charge diameter and good confinement.
*Source:* Duvall and Atchison (1957).

**TABLE 13.3. Granite Crater Test Data**

| Charge Data | | | | Crater Data | | | | Computed and Scaled Data | | | | | | |
|---|---|---|---|---|---|---|---|---|---|---|---|---|---|---|
| Weight (lb) | Depth (ft) | Length (ft) | Diameter (ft) | Radius (ft) | Depth (ft) | Angle (degree) | Volume (ft³) | Scale Length (ft) | Scaled Charge Depth | Charge Length Diameter Ratio | Scaled Crater Radius | Scaled Crater Depth | Scaled Crater Radius of Rupture | Scaled Crater Volume |
| Semi-gelatin, type A | | | | | | | | | | | | | | |
| 6.2 | 0.7 | 0.9 | 0.33 | 4.2 | 1.1 | 150 | 16.7 | 1.84 | 0.4 | 2.7 | 2.3 | 0.60 | 2.3 | 2.7 |
| 6.2 | 1.8 | 1.0 | 0.33 | 5.2 | 1.4 | 150 | 40.0 | 1.84 | 1.0 | 3.0 | 2.8 | 0.76 | 3.0 | 6.5 |
| 6.2 | 3.0 | 0.9 | 0.33 | 5.8 | 1.1 | 158 | 33.0 | 1.84 | 1.6 | 2.7 | 3.1 | 0.60 | 3.5 | 5.3 |
| 6.2 | 4.2 | 1.0 | 0.33 | 3.4 | 0.6 | 160 | 6.2 | 1.84 | 2.3 | 3.0 | 1.8 | 0.33 | 2.9 | 1.0 |
| 6.2 | 5.3 | 1.0 | 0.33 | 2.3 | 0.5 | 154 | 2.7 | 1.84 | 2.9 | 3.0 | 1.3 | 0.27 | 3.2 | 0.4 |
| 32.0 | 5.8 | 2.2 | 0.50 | 5.7 | 1.5 | 105 | 46.0 | 3.17 | 1.8 | 4.4 | 1.8 | 0.47 | 2.5 | 1.5 |
| Semi-gelatin, type B | | | | | | | | | | | | | | |
| 4.5 | 0.9 | 0.6 | 0.33 | 4.1 | 1.1 | 152 | 14.0 | 1.65 | 0.5 | 1.8 | 2.5 | 0.67 | 2.5 | 3.1 |
| 4.5 | 2.0 | 0.7 | 0.33 | 3.5 | 0.9 | 152 | 11.0 | 1.65 | 1.2 | 2.1 | 2.1 | 0.55 | 2.4 | 2.4 |
| 4.3 | 3.2 | 0.7 | 0.33 | 2.6 | 0.5 | 156 | 3.8 | 1.63 | 2.0 | 2.1 | 1.6 | 0.30 | 2.6 | 0.9 |
| 4.3 | 4.5 | 0.6 | 0.33 | Just broke surface | | | | 1.63 | 2.8 | 1.8 | | Just broke surface | | |
| 4.3 | 5.5 | 0.7 | 0.33 | No crater | | | | 1.63 | 3.4 | 2.1 | | No crater | | |
| 1.2 | 1.9 | 1.0 | 0.17 | 3.6 | 0.9 | 152 | 9.4 | 1.06 | 1.8 | 5.9 | 3.4 | 0.85 | 3.9 | 7.8 |
| 4.1 | 2.8 | 1.5 | 0.25 | 5.8 | 1.1 | 158 | 38.1 | 1.60 | 1.8 | 6.0 | 3.6 | 0.69 | 4.0 | 9.3 |
| 9.6 | 3.9 | 1.8 | 0.33 | 4.8 | 1.6 | 144 | 34.6 | 2.12 | 1.8 | 5.3 | 2.3 | 0.76 | 2.9 | 3.6 |
| 19.0 | 4.8 | 2.1 | 0.42 | 6.2 | 1.6 | 152 | 64.6 | 2.67 | 1.8 | 5.0 | 2.3 | 0.60 | 2.9 | 3.4 |
| 32.0 | 5.9 | 2.2 | 0.50 | 9.0 | 2.0 | 154 | 125.0 | 3.17 | 1.9 | 4.4 | 2.8 | 0.63 | 3.4 | 3.9 |
| Gelatin, Type B | | | | | | | | | | | | | | |
| 1.2 | 1.9 | 0.9 | 0.17 | 4.4 | 0.8 | 160 | 13.2 | 1.06 | 1.8 | 5.3 | 4.2 | 0.75 | 4.6 | 11.0 |
| 4.1 | 2.9 | 1.2 | 0.25 | 4.3 | 1.2 | 148 | 20.1 | 1.60 | 1.8 | 4.8 | 2.7 | 0.75 | 3.2 | 4.9 |
| 9.6 | 4.1 | 1.2 | 0.33 | 7.0 | 1.1 | 162 | 53.8 | 2.12 | 1.9 | 3.6 | 3.3 | 0.52 | 3.8 | 5.6 |
| 19.0 | 5.1 | 1.4 | 0.42 | 11.6 | 1.6 | 164 | 186.0 | 2.67 | 1.9 | 3.3 | 4.3 | 0.60 | 4.7 | 9.8 |

*Source:* Duvall and Atchison (1957).

## TABLE 13.4. Marlstone Crater Test Data

| | Charge Data | | | | Crater Data | | | | Computed and Scaled Data | | | | | | |
|---|---|---|---|---|---|---|---|---|---|---|---|---|---|---|---|
| Weight (lb) | Depth (ft) | Length (ft) | Diameter (ft) | Radius (ft) | Depth (ft) | Angle (degree) | Volume (ft³) | Scale Length (ft) | Scaled Charge Depth | Charge Length Diameter Ratio | Scaled Crater Radius | Scaled Crater Depth | Scaled Crater Radius of Rupture | Scaled Crater Volume |
| Semi-gelatin, type A | | | | | | | | | | | | | | |
| 3.4 | 1.2 | 1.1 | 0.25 | 3.1 | 1.2 | 138 | 15.0 | 1.50 | 0.8 | 4.4 | 2.1 | 0.8 | 2.2 | 4.4 |
| 3.4 | 2.4 | 1.1 | 0.25 | 3.3 | 2.6 | 104 | 32.0 | 1.50 | 1.6 | 4.4 | 2.2 | 1.7 | 2.7 | 9.4 |
| 3.4 | 4.2 | 1.1 | 0.25 | 4.2 | 2.5 | 118 | 54.0 | 1.50 | 2.8 | 4.4 | 2.8 | 1.7 | 4.0 | 15.9 |
| 3.4 | 6.0 | 1.1 | 0.25 | 0.6 | 0.4 | 112 | 0.2 | 1.50 | 4.0 | 4.4 | 0.4 | 0.3 | 4.0 | 0.1 |
| 1.5 | 4.7 | 1.2 | 0.17 | No crater | | | | 1.14 | 4.1 | 7.1 | No crater | | | |
| 3.4 | 4.5 | 1.0 | 0.25 | 0.3 | 0.3 | 90 | 0.2 | 1.50 | 3.0 | 4.0 | 0.2 | 0.2 | 3.0 | 0.1 |
| 7.6 | 4.5 | 1.0 | 0.38 | 6.0 | 4.4 | 108 | 182.0 | 1.97 | 2.3 | 2.6 | 3.0 | 2.2 | 3.8 | 23.9 |
| 13.5 | 4.1 | 1.4 | 0.46 | 6.2 | 4.0 | 114 | 182.0 | 2.38 | 1.7 | 3.0 | 2.6 | 1.7 | 3.1 | 13.5 |

*Source:* Duvall and Atchison (1957).

# TABLE 13.5. Sandstone Crater Test Data

| Charge Data | | | | Crater Data | | | | | Computed and Scaled Data | | | | | |
|---|---|---|---|---|---|---|---|---|---|---|---|---|---|---|
| Weight (lb) | Depth (ft) | Length (ft) | Diameter (ft) | Radius (ft) | Depth (ft) | Angle (degree) | Volume (ft³) | Scale Length (ft) | Scaled Charge Depth | Charge Length Diameter Ratio | Scaled Crater Radius | Scaled Crater Depth | Scaled Crater Radius of Rupture | Scaled Crater Volume |
| Semi-gelatin, type A | | | | | | | | | | | | | | |
| 8.0 | 0.4 | 0.7 | 0.46 | 3.0 | 1.3 | 134 | 16 | 2.00 | 0.2 | 1.5 | 1.5 | 0.7 | 1.5 | 2.0 |
| 8.0 | 0.4 | 0.7 | 0.46 | 2.7 | 1.1 | 136 | 9 | 2.00 | 0.2 | 1.5 | 1.4 | 0.6 | 1.4 | 1.1 |
| 8.0 | 0.9 | 0.7 | 0.46 | 3.5 | 1.6 | 130 | 36 | 2.00 | 0.5 | 1.5 | 1.8 | 0.8 | 1.9 | 4.5 |
| 8.0 | 0.9 | 0.7 | 0.46 | 3.3 | 1.7 | 126 | 31 | 2.00 | 0.5 | 1.5 | 1.7 | 0.9 | 1.8 | 3.9 |
| 8.0 | 1.8 | 0.8 | 0.46 | 4.0 | 2.5 | 116 | 63 | 2.00 | 0.9 | 1.7 | 2.0 | 1.3 | 2.2 | 7.9 |
| 8.0 | 2.4 | 0.9 | 0.46 | 5.8 | 3.0 | 126 | 130 | 2.00 | 1.2 | 2.0 | 2.9 | 1.5 | 3.1 | 16.3 |
| 8.0 | 2.4 | 0.6 | 0.46 | 5.5 | 3.5 | 116 | 100 | 2.00 | 1.2 | 1.3 | 2.8 | 1.8 | 3.0 | 12.5 |
| 8.0 | 2.9 | 0.6 | 0.46 | 5.9 | 3.4 | 120 | 150 | 2.00 | 1.5 | 1.3 | 3.0 | 1.7 | 3.4 | 18.8 |
| 8.0 | 3.9 | 0.6 | 0.46 | 7.4 | .9 | 166 | 100 | 2.00 | 2.0 | 1.3 | 3.7 | 0.5 | 4.2 | 12.5 |
| 8.0 | 4.7 | 0.8 | 0.46 | 8.0 | 2.3 | 148 | 150 | 2.00 | 2.4 | 1.7 | 4.0 | 1.2 | 4.7 | 18.8 |
| 8.0 | 6.8 | 0.8 | 0.46 | | Just broke surface | Just broke surface | | 2.00 | 3.4 | 1.7 | | Just broke surface | Just broke surface | |
| 8.0 | 6.9 | 0.7 | 0.46 | | Just broke surface | Just broke surface | | 2.00 | 3.5 | 1.5 | | Just broke surface | Just broke surface | |
| 8.0 | 9.8 | 0.8 | 0.46 | | No crater | No crater | | 2.00 | 4.9 | 1.7 | | No crater | No crater | |
| 8.0 | 9.9 | 0.8 | 0.46 | | No crater | No crater | | 2.00 | 5.0 | 1.7 | | No crater | No crater | |
| Gelatin, type A | | | | | | | | | | | | | | |
| 8.0 | 1.8 | 0.6 | 0.46 | 6.1 | 2.7 | 132 | 120 | 2.00 | 0.9 | 1.3 | 3.1 | 1.4 | 3.2 | 15.0 |
| 8.0 | 2.5 | 0.6 | 0.46 | 5.5 | 3.2 | 120 | 140 | 2.00 | 1.3 | 1.3 | 2.8 | 1.6 | 3.1 | 17.5 |
| 8.0 | 2.7 | 0.5 | 0.46 | 5.7 | 3.2 | 122 | 140 | 2.00 | 1.4 | 1.1 | 2.9 | 1.6 | 3.2 | 17.5 |
| 8.0 | 2.9 | 0.6 | 0.46 | 5.4 | 3.3 | 118 | 130 | 2.00 | 1.5 | 1.3 | 3.7 | 1.7 | 3.1 | 16.3 |
| 8.0 | 3.9 | 0.6 | 0.46 | 6.0 | 0.7 | 166 | 100 | 2.00 | 2.0 | 1.3 | 3.0 | 0.4 | 3.6 | 12.5 |
| 8.0 | 4.9 | 0.6 | 0.46 | | Just broke surface | Just broke surface | | 2.00 | 2.5 | 1.3 | | Just broke surface | Just broke surface | |
| 8.0 | 6.8 | 0.7 | 0.46 | | Just broke surface | Just broke surface | | 2.00 | 3.4 | 1.5 | | Just broke surface | Just broke surface | |
| 8.0 | 6.9 | 0.6 | 0.46 | | No crater | No crater | | 2.00 | 3.5 | 1.3 | | No crater | | |
| 8.0 | 9.9 | 0.6 | 0.46 | | No crater | No crater | | 2.00 | 5.0 | 1.3 | | No crater | | |
| 8.0 | 9.9 | 0.5 | 0.46 | | No crater | No crater | | 2.00 | 5.0 | 1.1 | | No crater | | |
| Ammonia gelatin | | | | | | | | | | | | | | |
| 8.0 | 0.5 | 0.4 | 0.50 | 2.4 | 1.1 | 130 | 6 | 2.00 | 0.3 | 0.8 | 1.2 | 0.6 | 1.2 | 0.8 |
| 8.0 | 1.0 | 0.4 | 0.50 | 3.5 | 1.3 | 140 | 17 | 2.00 | 0.5 | 0.8 | 1.8 | 0.7 | 1.9 | 2.1 |
| 8.0 | 2.5 | 0.5 | 0.50 | 6.2 | 2.9 | 130 | 116 | 2.00 | 1.3 | 1.0 | 3.1 | 1.5 | 3.4 | 14.5 |
| 8.0 | 6.6 | 0.4 | 0.50 | 2.0 | 0.8 | 136 | 4 | 2.00 | 3.3 | 0.8 | 1.0 | 0.4 | 3.4 | 0.5 |
| 8.0 | 10.5 | 0.4 | 0.50 | | No crater | | | 2.00 | 5.3 | 0.8 | | No crater | | |
| Ammonia dynamite | | | | | | | | | | | | | | |
| 8.0 | 0.4 | 0.7 | 0.46 | 3.4 | 1.2 | 142 | 33 | 2.00 | 0.2 | 1.5 | 1.7 | 0.6 | 1.7 | 4.1 |
| 8.0 | 0.4 | 0.7 | 0.46 | 2.5 | 1.1 | 132 | 10 | 2.00 | 0.2 | 1.5 | 1.3 | 0.6 | 1.3 | 1.3 |
| 8.0 | 0.9 | 0.7 | 0.46 | 3.0 | 1.6 | 124 | 18 | 2.00 | 0.5 | 1.5 | 1.5 | 0.8 | 1.6 | 2.3 |
| 8.0 | 0.9 | 0.8 | 0.46 | 2.6 | 1.2 | 130 | 16 | 2.00 | 0.5 | 1.7 | 1.3 | 0.6 | 1.3 | 2.0 |

*Source:* Duvall and Atchison (1957).

**TABLE 13.6. Chalk Crater Test Data**

| | Charge Data | | | | Crater Data | | | | Computed and Scaled Data | | | | | | |
| | Weight (lb) | Depth (ft) | Length (ft) | Diameter (ft) | Radius (ft) | Depth (ft) | Angle (degree) | Volume (ft³) | Scale Length (ft) | Scaled Charge Depth | Charge Length Diameter Ratio | Scaled Crater Radius | Scaled Crater Depth | Scaled Crater Radius of Rupture | Scaled Crater Volume |
|---|---|---|---|---|---|---|---|---|---|---|---|---|---|---|---|
| Semi-gelatin, type A | | | | | | | | | | | | | | | |
| | 2.0 | 1.4 | 0.7 | 0.25 | 3.4 | 1.3 | 138 | 18 | 1.26 | 1.1 | 2.8 | 2.7 | 1.0 | 2.9 | 9.0 |
| | 2.0 | 2.5 | 0.6 | 0.25 | 4.2 | 2.7 | 114 | 70 | 1.26 | 2.0 | 2.4 | 3.3 | 2.1 | 3.9 | 35.0 |
| | 2.0 | 4.4 | 0.7 | 0.25 | 4.0 | 0.9 | 154 | 20 | 1.26 | 3.5 | 2.8 | 3.2 | 0.7 | 4.7 | 10.0 |
| | 2.0 | 6.2 | 0.6 | 0.25 | No crater | | | | 1.26 | 4.9 | 2.4 | | No crater | | |
| | 0.9 | 4.7 | 0.7 | 0.17 | No crater | | | | 0.97 | 4.8 | 4.1 | | No crater | | |
| | 2.0 | 4.7 | 0.7 | 0.25 | Just broke surface | | | | 1.26 | 3.7 | 2.8 | | Just broke surface | | |
| | 4.5 | 4.7 | 0.7 | 0.38 | 6.6 | 5.0 | 106 | 261 | 1.65 | 2.8 | 1.8 | 4.0 | 3.0 | 4.9 | 58.0 |
| | 8.0 | 4.7 | 0.6 | 0.50 | 7.6 | 5.1 | 112 | 400 | 2.00 | 2.4 | 1.2 | 3.8 | 2.6 | 4.5 | 50.0 |
| Ammonia gelatin | | | | | | | | | | | | | | | |
| | 8.0 | 0.8 | 0.4 | 0.50 | 5.4 | 2.5 | 130 | 96 | 2.00 | 0.4 | 0.8 | 2.7 | 1.3 | 2.7 | 12.0 |
| | 8.0 | 1.9 | 0.3 | 0.50 | 5.5 | 3.3 | 118 | 272 | 2.00 | 1.0 | 0.6 | 2.8 | 1.7 | 3.0 | 34.0 |
| | 8.0 | 3.7 | 0.5 | 0.50 | 6.5 | 4.9 | 106 | 312 | 2.00 | 1.9 | 1.0 | 3.3 | 2.5 | 3.8 | 39.0 |
| | 8.0 | 6.2 | 0.4 | 0.50 | 7.5 | 7.5 | 90 | 448 | 2.00 | 3.1 | 0.8 | 3.8 | 3.8 | 4.9 | 56.0 |
| | 8.0 | 10.7 | 0.4 | 0.50 | No crater | | | | 2.00 | 5.4 | 0.8 | | No crater | | |
| | 0.4 | 5.2 | +[a] | + | No crater | | | | 0.74 | 7.0 | + | | No crater | | |
| | 0.8 | 5.0 | + | + | 2.4 | 0.9 | 138 | 9 | 0.93 | 5.4 | + | 2.6 | 1.0 | 6.0 | 1.1 |
| | 1.5 | 4.0 | + | + | 3.1 | 1.6 | 126 | 37 | 1.14 | 3.5 | + | 2.7 | 1.4 | 4.4 | 24.7 |
| | 3.0 | 5.8 | + | + | 4.8 | 2.3 | 128 | 99 | 1.44 | 4.0 | + | 3.3 | 1.6 | 5.2 | 33.0 |

[a] + = Horizontal shot holes.

Source: Duvall and Atchison (1957).

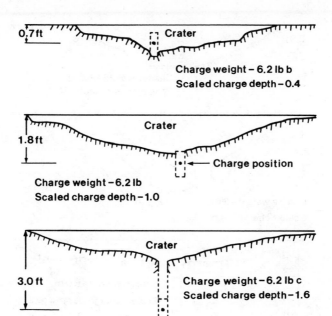

**Figure 13.2** Crater sections—ganite (Duvall and Atchison, 1957).

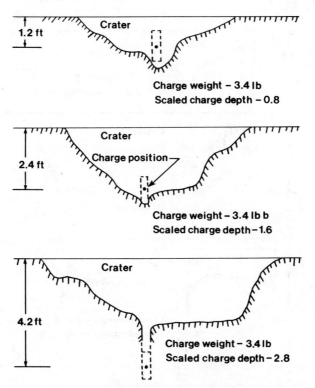

**Figure 13.3** Crater sections—marlstone (Duvall and Atchison, 1957).

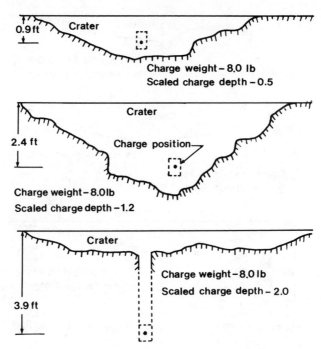

**Figure 13.4** Crater sections—sandstone (Duvall and Atchison, 1957).

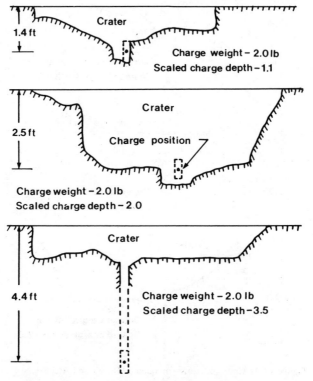

**Figure 13.5** Crater sections—chalk (Duvall and Atchison, 1957).

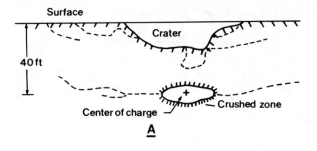

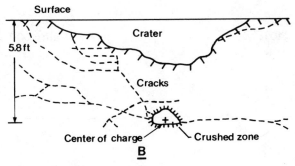

**Figure 13.6** Sections—horizontal-hole crater tests in chalk: (A) 3.5 scaled charge depth, 1.5-lb charge weight; (B) 4.0 scaled charge depth, 3.0-lb charge weight (Duvall and Atchison, 1957).

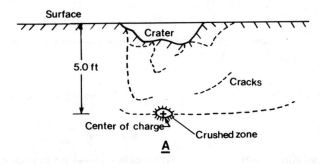

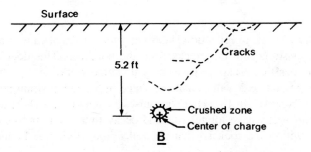

**Figure 13.7** Sections—horizontal-hole crater tests in chalk: (a) 5.4 scaled charge depth, 0.8-lb charge weight; (b) 7.0 scaled charge depth, 0.38-lb charge weight (Duvall and Atchison, 1957).

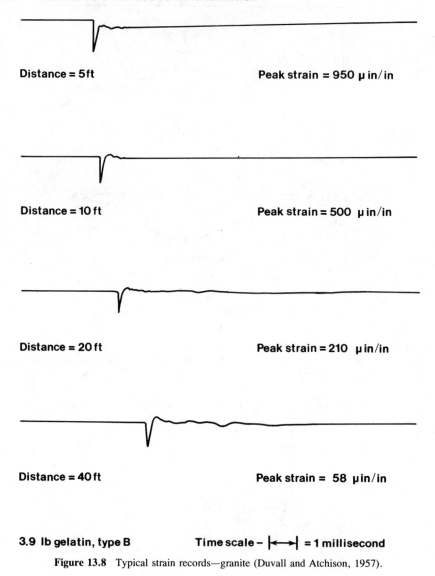

Distance = 5 ft          Peak strain = 950 μ in/in

Distance = 10 ft          Peak strain = 500 μ in/in

Distance = 20 ft          Peak strain = 210 μ in/in

Distance = 40 ft          Peak strain = 58 μ in/in

**3.9 lb gelatin, type B**          Time scale – |←→| = 1 millisecond

**Figure 13.8**  Typical strain records—granite (Duvall and Atchison, 1957).

horizontal surface due to reflection slabbing, which was not caused by the pressure of the gases pushing into the cracks from the borehole. This does not, however, preclude the idea that escaping gases may do some work.

The average and peak values and the shape of the radial strain pulses vary with rock type (Figures 13.8–13.11). (The pulses in granite are short and sharp, and those in softer rocks have longer rise and decay time.) The pulses were recorded at positions within the rock so that the reflections from free boundaries did not

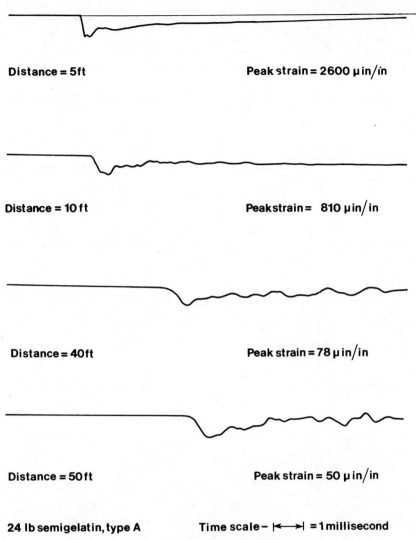

Distance = 5ft · · · · · · · · · · · Peak strain = 2600 μin/in

Distance = 10 ft · · · · · · · · · · Peak strain = 810 μin/in

Distance = 40ft · · · · · · · · · · Peak strain = 78 μin/in

Distance = 50ft · · · · · · · · · · Peak strain = 50 μin/in

**24 lb semigelatin, type A**   Time scale— |←——→| = 1 millisecond

**Figure 13.9**  Typical strain records–marlstone (Duvall and Atchison, 1957).

interfere. In the traces for marlstone and chalk, reflections from bedding planes appear as minor disturbances. Time is measured from left to right, the baseline is zero strain, compressive strain below, and tensile strain above (Figure 13.12). Where there is no tensile strain, the decrease in compressive strain is insignificant beyond $E$. Point $B$ is the time of detonation, $C$ the start of the pulse, and $BC$ the travel time.

Important parameters are the peak compressive strain at $D$ and the peak tensile strain at $E$. Other definitive parameters are the arrival time, rise time, fall time,

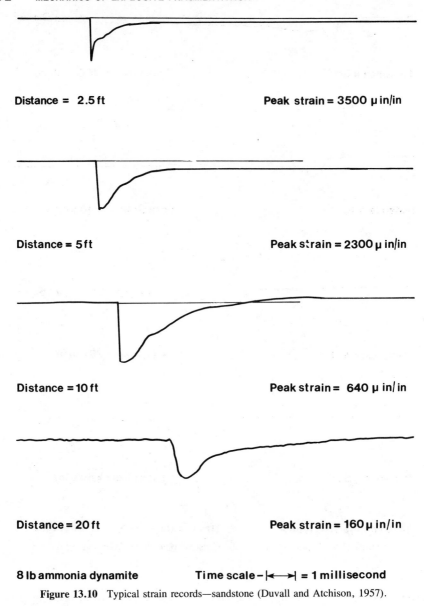

**Figure 13.10** Typical strain records—sandstone (Duvall and Atchison, 1957).

peak compressive strain, and fall strain (Figures 13.13–13.16). For granite, sandstone, and chalk, each curve represents approximately 100 data points, and for marlstone about 40 data points.

The wave velocity was determined from the slope of the plot of travel distance against arrival time. The wave velocity was constant for granite, marlstone, and

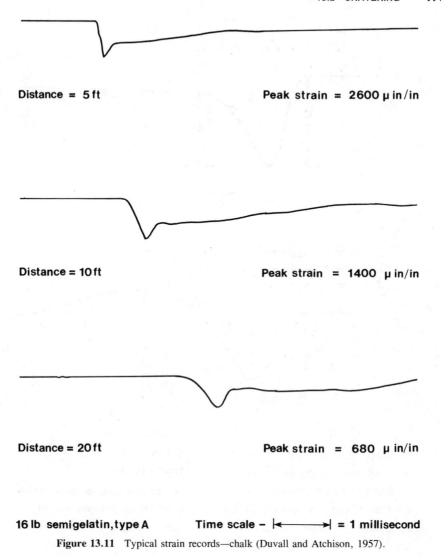

Distance = 5 ft          Peak strain = 2600 μ in/in

Distance = 10 ft          Peak strain = 1400 μ in/in

Distance = 20 ft          Peak strain = 680 μ in/in

16 lb semigelatin, type A      Time scale – |←——→| = 1 millisecond

**Figure 13.11** Typical strain records—chalk (Duvall and Atchison, 1957).

chalk; for sandstone the velocity was higher close to the charge (Figure 13.17), indicating an increase of velocity with stress. The average wave velocity in the laboratory was measured on core samples (Table 13.1).

High-speed motion pictures did not show the sequence of events until a few milliseconds after detonation or after most fragmentation is accomplished. The fly velocity of the broken rock is greater (Figure 13.18) for shallow charge depths. The initial slope of the curves gives the ejection (fly rock) velocity of the broken rock from the surface (Table 13.7).

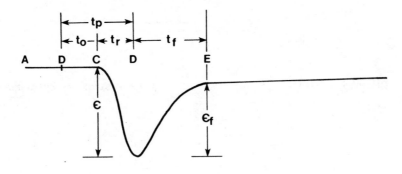

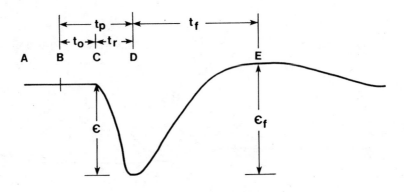

A = Start of trace

B = Detonation of charge

C = Start of strain pulse

D = Peak of compressive strain

E = End of fall strain

$t_0$ = Arrival time for start of pulse

$t_r$ = Rise time

$t_f$ = Fall time

$t_p$ = Arrival time for peak strain

$\epsilon$ = Peak compressive strain

$\epsilon_f$ = Fall strain

**Figure 13.12** Strain record measurements (Duvall and Atchison, 1957).

## 13.2.3 Analysis of Crater Data

The crater data (Tables 13.3–13.6; Figures 13.19–13.24) show that the size and shape of a crater are functions of the charge size, charge depth, and rock type, and that other variables such as loading density, explosive type, and charge length diameter ratio are of secondary importance within the range tested. Effects of variation in loading density and charge length diameter ratio were minimized. Some

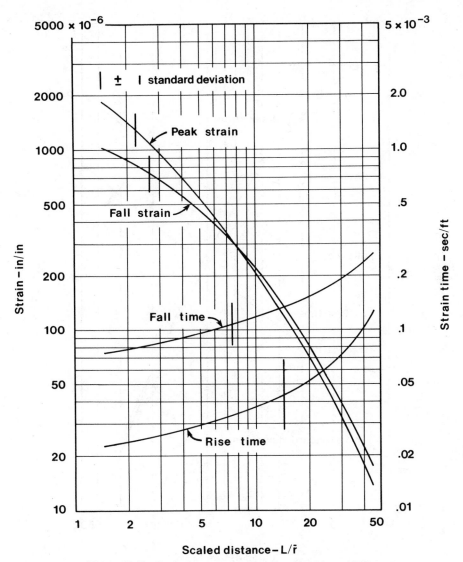

**Figure 13.13** Strain data—granite (Duvall and Atchison, 1957).

effects of explosive type were observed in the spread in data, which was also affected by differences in the local geologic structure and physical properties of the rock. The effect of charge size is normalized by scaling.

### 13.2.3.1 Crater Radius
The scaled crater radius curves for the four types of rock are similar in shape, but with some important differences. The maximum radius varies from about 3.0 for

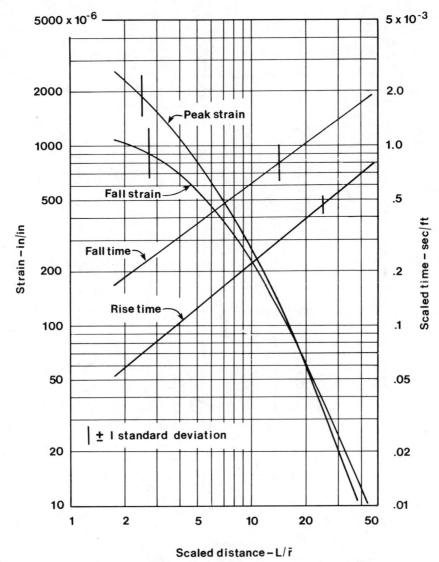

**Figure 13.14** Strain data—marlstone (Duvall and Atchison, 1957).

marlstone to 4.0 for sandstone, with the peaks occurring at scaled depths of 1.75 for granite to 2.75 for sandstone. These differences are relatively small, and the scaled depth of burial for optimum breakage varies only over a small range. In each case, the curves drop very rapidly after the peak is reached, indicating that the value for best blasting effectiveness is critical and should not be exceeded. The same arguments hold for crater depth and volume.

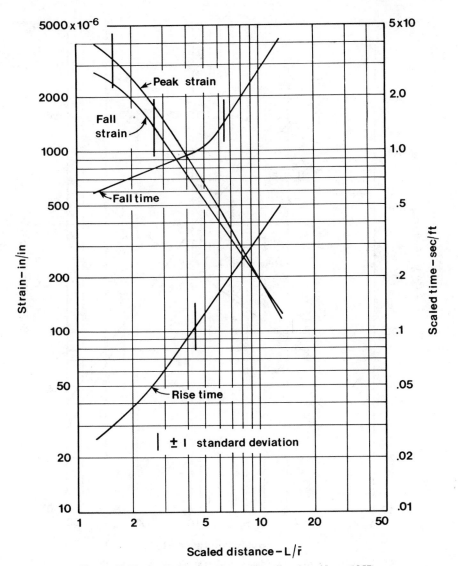

**Figure 13.15**  Strain data—sandstone (Duvall and Atchison, 1957).

### 13.2.3.2 *Crater Depth*

For all four test rocks the true scaled crater depth increases with charge depth to a maximum and then decreases. For marlstone, sandstone, and chalk, the crater depth is approximately equal to the charge depth to the peak of the curve. However, for granite, it is relatively constant and is less than the charge depth. The charge depth required for excavation of maximum depth varies with the rock type, and it

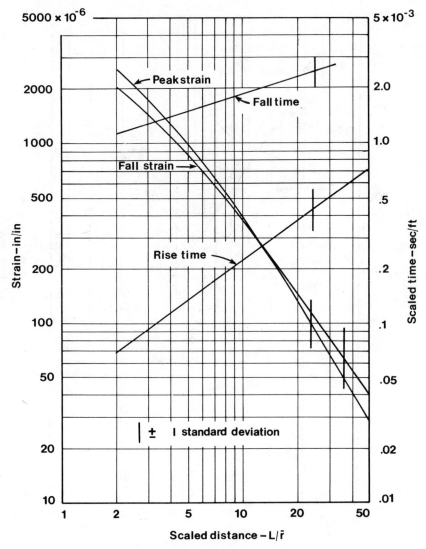

**Figure 13.16** Strain data—chalk (Duvall and Atchison, 1957).

also increases as the strength of the rock decreases. That is, the crater in granite, because of its strength, structure, and so on, is not broken to the charge depth, whereas in the other rocks they are even for small depths of burial.

For all large charge depths greater than critical, the crater depth is smaller than the charge depth; that is, the crater zone is above the explosive location, with solid rock remaining between it and the bottom of the crater. The remnants of the shot holes inhibited only crushing and cracking. The remaining solid rock betwen the

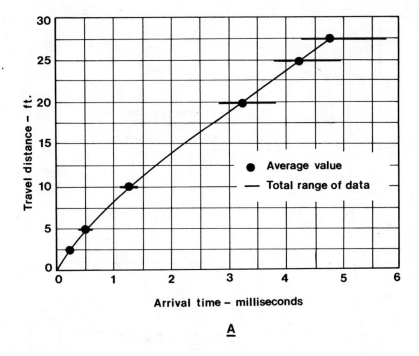

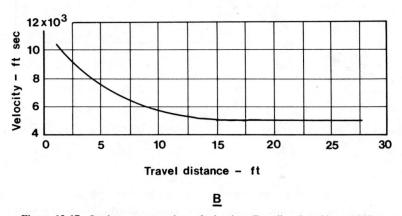

**Figure 13.17** Sandstone propagation velocity data (Duvall and Atchison, 1957).

charge and the crater transmitted a high level of explosive energy to the crater position without extensive fracturing.

To determine the effect of the late expansion of gases from an explosion in breaking rock, crater tests in chalk were made with horizontal holes drilled from a vertical face in a bench at depths greater than the critical burial distance below

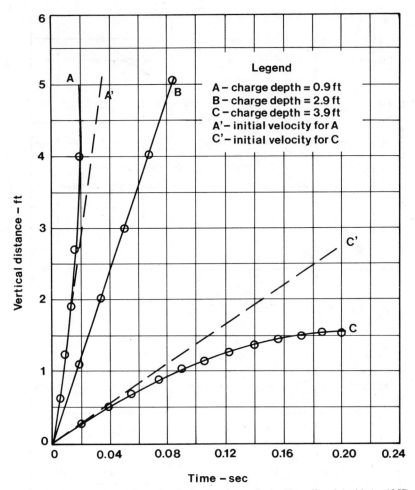

**Figure 13.18** Examples of fly rock velocity data in sandstone (Duvall and Atchison, 1957).

the horizontal surface. Thus, the scaled charge depths with respect to the horizontal surface were greater than the critical scaled charge depth. The holes were well stemmed, but explosive gases were not forced out of the fracture through the rock above the charge. After the blast, solid rock remained between the resulting craters and the explosive position (Figures 13.6, 13.7). Shallow craters were produced, but, without gas invading the crater zone, the stress wave in the rock transmitted the breakage energy from the explosive to the cratered rock. For large depths of burial, no crater was produced, but fracturing of the rock occurs. For example, no crater was produced in chalk (Figure 13.7), where the scaled charge depth was 7.0. However, surface fracturing occurred in the shape of a small crater. Some

**TABLE 13.7. Time of Cracking and Fly Rock Velocity Data—Sandstone**

| Charge Depth (ft) | Break Time (milliseconds) Horizontal Distance from Center of Charge Hole to Foil Strip | | | Fly Rock Velocity (ft/s) |
|---|---|---|---|---|
| | 1.0 ft | 2.5 ft | 5.0 ft | |
| Semi-gelatin, type A | | | | |
| 0.42 | 0.52 | 0.74 | 1.16 | |
| 0.42 | 0.35 | 1.05 | b | |
| 0.90 | 0.27 | 0.55 | 0.94 | 210 |
| 0.94 | 0.30 | 0.37 | 1.05 | |
| 1.8 | 0.35 | 0.50 | 0.95 | |
| 2.4 | 0.62 | 1.12 | 1.47 | 80 |
| 2.4 | 0.65 | 1.29 | 2.32 | 90 |
| 2.9 | 0.54 | 1.20 | 2.62 | 70 |
| 3.9 | 1.00 | 2.30 | 4.43 | 13 |
| 4.7 | 1.23 | 1.35 | 3.50 | 9 |
| Gelatin, type A | | | | |
| 1.8 | 0.40 | 0.65 | 1.65 | 120 |
| 2.5 | 0.55 | 2.33 | 5.46 | 70 |
| 2.7 | 0.45 | 0.90 | 1.23 | |
| 2.9 | 0.70 | 1.10 | 2.37 | 50 |
| 3.9 | 1.65 | 1.90 | 3.60 | 19 |
| Ammonia dynamite | | | | |
| 0.44 | 0.17 | 0.37 | 0.95 | 260 |
| 0.44 | 0.50 | 1.30 | | |
| 0.90 | 0.23 | 0.43 | 1.05 | |

*Source:* Duvall and Atchison (1957).

cracks occurred in the vertical face, but these did not propagate to the explosion cavity. Although these tests were conducted in chalk, the same type of slabbing and fracture without invasion of high-pressure gases into the crater zone may also occur in other types of rock. That is, slabbing and other breakage mechanisms accompanying reflection may play a vital part in such rock breakage.

### 13.2.3.3 Crater Volume

The general shape of these curves (Figures 13.23, 13.24) is similar to that for crater radius. The critical scaled charge depth, where the maximum occurs, increases as the strength of the rock decreases. Because the maximum volumes occur very close to the same scaled depth of burial as for the crater radius and depth,

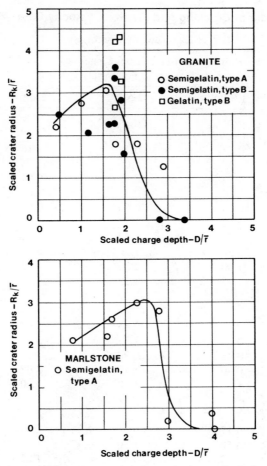

**Figure 13.19** Scaled crater radius versus scaled charge depth—granite and marlstone (Duvall and Atchison, 1957).

any of these three parameters may be used to predict the optimum scaled depth of burial of explosive for predicting optimum burden in bench blasting operations.

## 13.3 CRATERING AND THE REFLECTION THEORY OF ROCK BREAKAGE

The tests in the chalk bench show that (1) a crater is broken out at a free surface without gas penetrating the rock in the crater zone, and (2) solid rock remains between the crater zone and the blast point. Thus, the stress wave in the rock transmits energy from the explosive gases to the free surface, which is reflected and causes significant rock breakage.

The strain pulses (Figures 13.8–13.11) generated in the rock by an exploding charge were roughly triangular in shape. The initial part of the rock fracture may

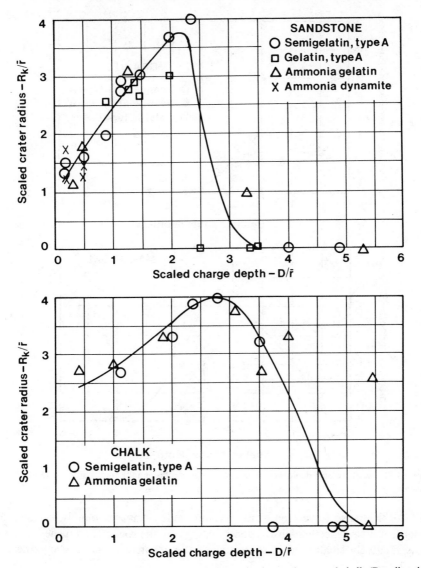

**Figure 13.20** Scaled crater radius versus scaled charge depth—sandstone and chalk (Duvall and Atchison, 1957).

be caused by the impingement of the pulse on the surface and more major breakage by its reflection. Thus, the detonating charge generates a compressive radial pulse that travels outward in the rock at a characteristic velocity. These examples illustrate a single compression with a steep front (or rapid rise time) and a lower decay (or fall time).

Very close to the detonating explosive, the magnitude of compressive stress is greater than the strength of the rock, and it is usually crushed and intensely frac-

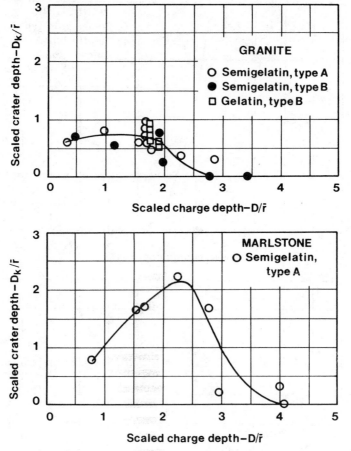

**Figure 13.21** Scaled crater depth versus scaled charge depth—granite and marlstone (Duvall and Atchison, 1957).

tured. As the pulse travels outward, its amplitude decreases rapidly because of geometrical divergence and absorption of energy by the rock. The crushed zone is usually relatively small, and in some rocks little apparent crushing occurs near a drill hole.

The high detonation pressure, $p_2$, exists only in the detonation wave, the important pressure in blasting being that exerted by the gases as they occupy the volume of the borehole at the borehole pressure, which is approximately one-half the detonation pressure. The gases exert a very high pressure against the stemming, generating a pressure pulse in it. With the loss of heat and even the small increase of hole volume resulting from compaction of stemming, crushing, and compression of the rock, the borehole gas pressure in the drill hole decreases rapidly. The pressure–volume curve for the gases at the high pressures and temperatures of the explosion state is such that a very small increase in volume results in a very large decrease in pressure.

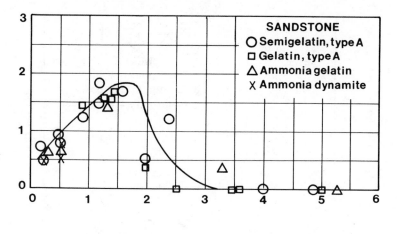

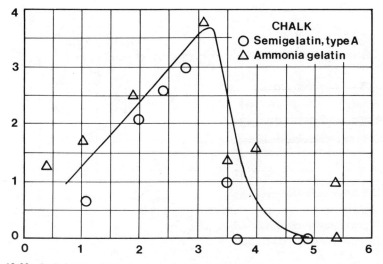

**Figure 13.22**   Scaled crater depth versus scaled charge depth—sandstone and chalk (Duvall and Atchison, 1957).

The close-in fall time curves (Figures 13.8–13.11) all have marked curvature except granite. Whereas Duvall and Atchison (1957) state that the scaled crater depth should be half of the scaled fall length,

$$\frac{D_k}{\bar{r}} = \frac{1}{2}\frac{ct_f}{\bar{r}}$$

(13.4)

it appears to be more appropriate to write this equation in terms of the fall time, $t_f$, to the point $t_b$ where the reflected tensile strain is equal to or greater than the breakage strain. That is,

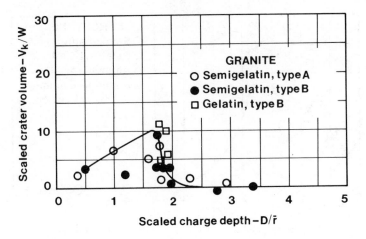

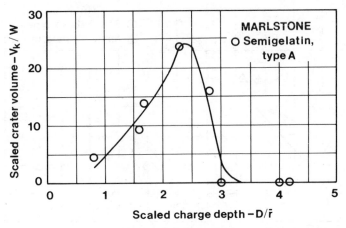

**Figure 13.23** Scaled crater volume versus scaled charge depth—granite and marlstone (Duvall and Atchison, 1957).

$$\frac{D_k}{\bar{r}} = \frac{ct_b}{\bar{r}} \tag{13.5}$$

and the first prediction equation holds only where $t_b = \frac{1}{2}t_f$.

The total distance $L$ traveled by the wave is from the charge to the surface and thence back to the point where slabbing occurs. Hence,

$$\frac{D_k}{\bar{r}} = \frac{L}{\bar{r}} - \frac{ct_b}{\bar{r}} \tag{13.6}$$

For $t_b = \frac{1}{2}t_f$, data from Figures 13.13–13.16, and the previous equations, the predicted crater depths were calculated for the craters (Figure 13.25). One proce-

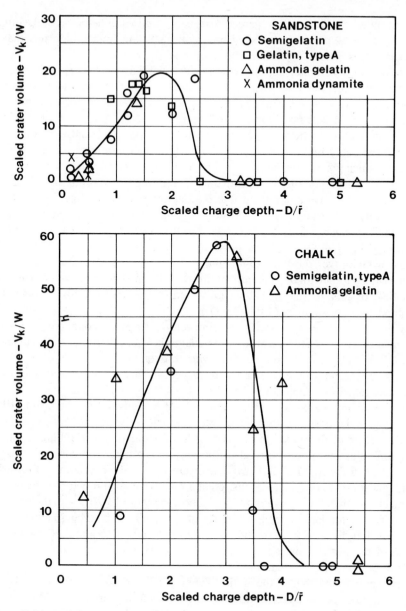

**Figure 13.24**  Scaled crater volume versus scaled charge depth—sandstone and chalk (Duvall and Atchison, 1957).

dure for calculating crater depths is to choose a scaled depth of charge, an approximate value of $t_f$ with the value of $c$ for a given rock (Table 13.1), and "iterate" by eye to determine a close value of $t_f/\bar{r}$ (Figures 13.13–13.16) for the approximate value of $L/\bar{r}$. With this $t_f$, calculate the crater depth from equation 13.6. The experimental crater curves were consistently lower than those calculated; however, those for granite and marlstone agreed reasonably well with the peaks of the curves. That the calculated curves for sandstone and chalk are higher indicates that the selected values for the fall times were too great.

The accuracy of crater depth predictions by slabbing from strain data thus depends on the correct choice of the longitudinal velocity measurement and of the fall time $t_f$. The close-in strain pulses exhibited a slow recovery to zero strain. Also, none of the fall time curves are linear; hence, the choice of value for fall time is somewhat arbitrary. Likewise, the wave velocity in sandstone was found to be considerably higher within 5 ft of the charges, where most of the breakage ocurred, than at greater distances. The time to first breakage by slabbing is a reliable measure for calculation of crater depth only if cratering is due primarily to slabbing. This mechanism plays a dominant role in cratering to a single free face.

An estimate of the dynamic tensile breaking strain of rock can be made as follows: The scaled charge depth is added to the crater depth (Table 13.8) to give the scaled travel distance of the pulse to the surface and back to the first slab surface, which gives the fall strain (Figures 13.13–13.16). The dynamic breaking strain is thus found to be one-half of the fall strain for these data. (The pulse velocities for granite and sandstone measured in the field are higher than those measured for core samples (Table 13.1).)

In some of the tests in sandstone, concentric strips of conducting foil were cemented to the rock surface around the charge hole at horizontal radial distances of 1.0, 2.5, and 5.0 ft. When the strips were broken by the fracture of the rock at the surface, the time was recorded with respect to the detonation and the related times and distances (Table 13.9).

The following analysis is in addition to that by Duvall and Atchison (1957): According to the postulate that slabbing occurs when the peak of reflection is one half the period of the pulse, the total time from detonation to slabbing is half the pulse time plus the time of arrival of the peak at the reflection surface. The time of breakage of the first foil at 1-ft radius distance from the borehole on the surface of the sandstone was found to be just greater than the arrival time of the peak strain and much less than the time required for slabbing (Figure 13.26), with the exception of two values for larger depth of burial of 3.9 and 4.7 ft. The fractures of the surface at 1-ft distance from the borehole takes place at 0.15 and 0.20 ms after the arrival of the peak of the strain wave, whereas the predicted slabbing times are about three times greater than this.

Also, it is noted that if the initial slabbing takes place at time $t_p + \frac{1}{2}t_f$, the additional time for the effect to this fracture to reach the surface would be at least another $\frac{1}{2}t_f$ if the fracture velocity is assumed to equal the wave velocity $c$. Hence, the minimum time for surface fracture resulting from slabbing would be $t_p + t_f$ (Figure 13.26).

Thus, larger times are required for slabbing than the initial surface fracture times. The small times required for initial surface fracture indicate that this breakage occurs about 0.10–0.20 ms after the arrival of the peak strain. That is, if the wave velocity in sandstone is assumed to be about 8,000 ft/s, the break at the surface takes place when the peak of the strain wave has been reflected about 0.80–

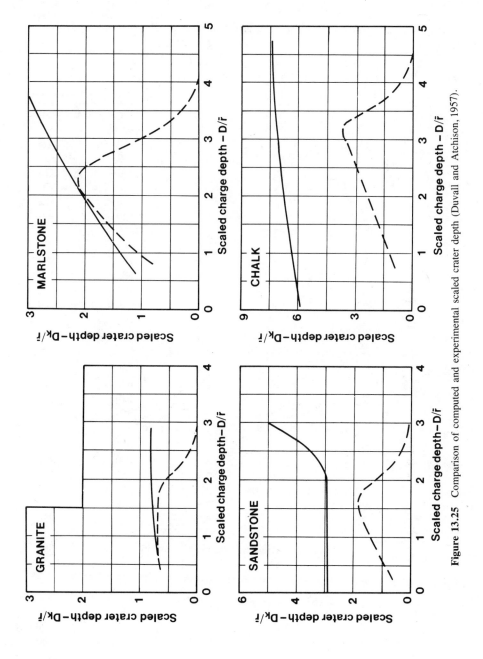

**Figure 13.25** Comparison of computed and experimental scaled crater depth (Duvall and Atchison, 1957).

**TABLE 13.8. Estimated Tensile Breaking Strain from Experimental Crater and Strain Data**

| Rock Type | Scaled Charge Depth | Scaled Crater Depth | Scaled Travel Distance | Fall Strain | One-Half[a] Fall-Strain |
|---|---|---|---|---|---|
| Granite | 1.8 | 0.70 | 2.5 | 760 | 380 |
| | 1.8 | .47 | 2.3 | 800 | 400 |
| | 1.8 | .60 | 2.4 | 790 | 400 |
| | 1.8 | .69 | 2.5 | 760 | 380 |
| | 1.8 | .75 | 2.6 | 740 | 370 |
| | 1.8 | .75 | 2.6 | 740 | 370 |
| | 1.8 | .76 | 2.6 | 740 | 370 |
| | 1.8 | .85 | 2.7 | 720 | 360 |
| | 1.9 | .56 | 2.4 | 790 | 400 |
| | 1.9 | .60 | 2.5 | 760 | 380 |
| | 1.9 | .63 | 2.5 | 760 | 380 |
| | 2.0 | .30 | 2.3 | 800 | 400 |
| | 2.3 | .33 | 2.6 | 740 | 370 |
| | 2.9 | .27 | 3.2 | 640 | 320 |
| | Estimated dynamic tensile breaking strain | | | | 380 |
| Marlstone | 2.3 | 2.1 | 4.4 | 630 | 320 |
| | 2.3 | 2.2 | 4.5 | 620 | 310 |
| | 2.8 | 1.7 | 4.5 | 620 | 310 |
| | Estimated dynamic tensile breaking strain | | | | 310 |
| Sandstone | 1.7 | 1.9 | 3.6 | 860 | 430 |
| | 2.0 | 0.4 | 2.4 | 1550 | 780 |
| | 2.0 | 0.5 | 2.5 | 1500 | 780 |
| | 2.4 | 1.2 | 3.6 | 860 | 430 |
| | 3.3 | 0.4 | 3.7 | 820 | 410 |
| | Estimated dynamic tensile breaking strain | | | | 560 |
| Chalk | 3.2 | 3.7 | 6.9 | 600 | 300 |
| | 3.5 | 0.7 | 4.2 | 1050 | 530 |
| | 3.5 | 1.4 | 4.9 | 880 | 440 |
| | 4.0 | 1.6 | 5.6 | 770 | 390 |
| | 5.4 | 1.0 | 6.4 | 650 | 330 |
| | Estimated dynamic tensile breaking strain | | | | 400 |

[a]Strains are in ($\mu$in./in.).
*Source:* Duvall and Atchinson (1957).

1.60 ft back into the rock. The distance would be even less if there is a finite time required for breaking of the tape.

The total displacement of the free surface for a pulse length equal to twice the depth of burial does not occur until the tail of the pulse has reached the surface of $t_p + t_f$. The ratio of $(t_b + t_p)/t_f$ varies from 0.05–0.28, with the exception of one data point, and this ratio divided by the travel distances varies from 0.033–0.058, with the exception of the first and the sixth data points. These ratios indicate that the initial surface fracture occurs when the peak of the strain wave has traveled approximately the same portion of the distance from the point of fracture on the surface back into the rock for each shot. Hence, it is concluded that the initial surface fracture in Navajo sandstone is not caused by the slabbing process but is due to the arrival of the peak strain at the surface. The first point where surface breakage time was measured was 1 ft from the borehole collar, and, inasmuch as the fracture velocity is approximately half that of the wave velocity, the fractures could have been initiated a fraction of a millisecond earlier at the borehole.

The cause of this type of fracture due to surface displacement is illustrated in the following: In the absence of a free face within a critical distance of a confined explosion—that is, within which the energy, stress, and strain are of a high enough level to cause breakage when a free face is present—the energy is otherwise dissipated by losses in the rock and dispersed owing to radial divergence. That is, for explosive charges buried beyond a critical depth, the only damage to the rock around the detonating charge is crushing and plastic deformation and some radial fracturing of the rock immediately around the hole.

If, however, a free face is present within a critical distance of the explosion, then the energy arriving at the free surface is reflected back into the rock. To review, when a longitudinal compressive pulse meets a free flat surface at an angle of incidence, $\alpha$, two pulses are reflected: (1) part of the compressive wave is reflected in tension and (2) part in shear, and the amount of energy and the directions of the waves are functions of $\alpha$. If the direction of the pulse is normal to the free surface, a compressive wave is completely reflected in tension. Also, as shown in Chapter 2, the partitioning of wave energy between kinetic and potential (strain) categories is not half-and-half when a reflected wave travels back through the incident wave. Thus, the amount and character of the energy in each type of reflected stress pulse constitute a strong function of the angle of incidence. The velocity of longitudinal pulses is greater than that of shear pulses, and, as the strength of rock in tension is much less than the strength of rock in compression (usually by a factor of 10–50 or more; see Table 13.1), the reflected pulse will break the rock in tension as it moves back away from the free face if the peak tensile strain exceeds the tensile breaking strain of the rock.

Consider the reflection of a plane triangular pulse at normal incidence to a plane free surface—that is (Figure 13.27), a triangular compressive strain pulse after equal successive units of time. The curves below the baseline represent the incident compressive pulse, and those above represent the reflected tensile pulse. The dotted lines represent the canceled portions of the pulses, and the solid line represents the resulting strain pulse. The maximum tension is developed at a distance from the

**TABLE 13.9. Predicted and Measured Time of Breaking in Sandstone**

| Charge Depth (ft) | Horizontal Distance (ft) | Travel Distance (ft) | Arrival Time for Front of Wave (ms) | Rise Time (ms) | Total Travel Distance (ft) | One-Half Fall-Time (ms) | Arrival Time for Peak (ms) | Arrival Time for Peak Plus One-Half Fall-Time (ms) | Measured Break Time (ms) |
|---|---|---|---|---|---|---|---|---|---|
| 1 | 2 | 3 | 4 | 5 | 6 | 7 | 8 | 9 | 10 |
| 0.4 | 1.0 | 1.1 | 0.10 | 0.04 | 5.7 | 0.82 | 0.14 | 0.96 | 0.39 |
|  | 2.5 | 2.5 | 0.24 | 0.05 | 7.9 | 0.94 | 0.29 | 1.23 | 0.87 |
|  | 5.0 | 5.0 | 0.54 | 0.09 | 10.8 | 1.12 | 0.63 | 1.75 | 1.50 |
| 0.9 | 1.0 | 1.4 | 0.13 | 0.04 | 6.2 | 0.85 | 0.17 | 1.02 | 0.27 |
|  | 2.5 | 2.7 | 0.27 | 0.06 | 8.1 | 0.94 | 0.33 | 1.27 | 0.46 |
|  | 5.0 | 5.1 | 0.56 | 0.10 | 10.9 | 1.14 | 0.66 | 1.80 | 1.05 |
| 1.8 | 1.0 | 2.1 | 0.20 | 0.05 | 7.3 | 0.90 | 0.25 | 1.15 | 0.38 |
|  | 2.5 | 3.1 | 0.30 | 0.06 | 8.6 | 0.97 | 0.36 | 1.33 | 0.58 |
|  | 5.0 | 5.3 | 0.58 | 0.10 | 11.1 | 1.15 | 0.68 | 1.83 | 1.30 |
| 2.5 | 1.0 | 2.7 | 0.27 | 0.05 | 9.0 | 0.99 | 0.32 | 1.31 | 0.57 |
|  | 2.5 | 3.5 | 0.35 | 0.07 | 9.1 | 0.99 | 0.42 | 1.41 | 1.40 |
|  | 5.0 | 5.6 | 0.63 | 0.11 | 11.4 | 1.20 | 0.74 | 1.94 | 2.60 |
| 2.9 | 1.0 | 3.1 | 0.30 | 0.06 | 8.6 | 0.97 | 0.36 | 1.33 | 0.62 |
|  | 2.5 | 3.8 | 0.38 | 0.07 | 9.5 | 1.01 | 0.45 | 1.46 | 1.15 |
|  | 5.0 | 5.8 | 0.65 | 0.12 | 11.6 | 1.23 | 0.77 | 1.93 | 2.50 |
| 3.9 | 1.0 | 4.0 | 0.41 | 0.07 | 9.7 | 1.02 | 0.48 | 1.50 | 1.33 |
|  | 2.5 | 4.6 | 0.48 | 0.08 | 10.4 | 1.08 | 0.56 | 1.64 | 2.10 |
|  | 5.0 | 6.3 | 0.71 | 0.13 | 12.1 | 1.30 | 0.84 | 2.14 | 4.02 |
| 4.7 | 1.0 | 4.8 | 0.52 | 0.09 | 10.6 | 1.11 | 0.61 | 1.72 | 1.23 |
|  | 2.5 | 6.3 | 0.58 | 0.10 | 11.7 | 1.24 | 0.68 | 1.92 | 1.35 |
|  | 5.0 | 6.9 | 0.80 | 0.15 | 12.8 | 1.42 | 0.95 | 2.37 | 3.50 |

*Source:* Duvall and Atchinson (1957).

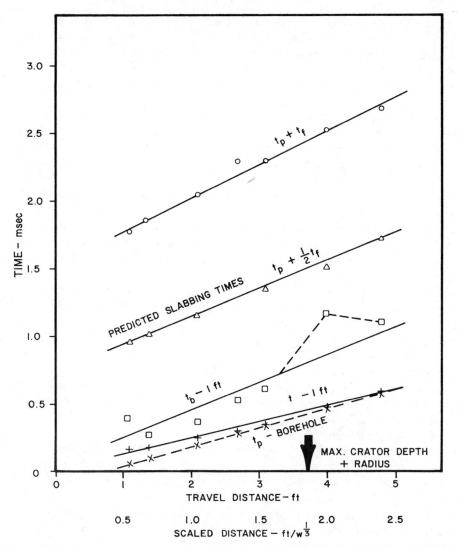

**Figure 13.26**  Relative times of peak strain arrival at surface and at distances for slabbing and initial tensile fracture at surface caused by reflected wave.

free surface equal to half of the fall length of the pulse; its magnitude depends on the shape of the decay. The shape of the rising portion of the pulse does not affect the magnitude or the location of the peak reflected tensile strain.

Thus, tensile fracture (Figure 13.28) results from the compressive strain reflected as a tensile pulse from a free surface. When the tensile strain equals the rock tensile breaking strain, a crack develops. The new surface then acts as a second free face from which the remainder of the pulse is reflected. The slab moves because of its kinetic energy, and, if the remaining tensile strain is great enough, another slab is formed. Both slabs move forward, the first with a high velocity since it has more kinetic energy. However, when the tensile strain is less than the breaking strain, the slabbing stops. The number of slabs equals the first whole number less than the ratio of the fall strain to the breaking strain. Such slabbing has been demonstrated experimentally for plan waves reflected from the ends of rock cores (Hino, 1959).

For charges to the size of those used in the crater tests described above, the wave front is approximately spherical. When a spherical compressive wave pulse is reflected from a plane surface, a three-dimensional set of tensions is developed: (1) stress normal to the reflected wave front just below the surface, (2) tension at the surface normal to radii of the shot hole, and (3) tangential tension at the surface normal to concentric circles around the shot hole. Thus, it was expected that pie-shaped fragments would result, and many of the craters formed in the Bureau of Mines' crater tests produced such pie-shaped fragments.

The calculation from the reflection theory that the depth of a crater should approximate half of the fall length implies that if half of the fall length of the strain pulse is greater than the charge depth, the reflected strain pulse will break rock only from the free surface back to the charge. When curved slabs develop in the rock near the crushed zone, the remaining incident pulse is reflected from the new free surface at larger angles of incidence, and therefore it develops smaller tensile strains. The crater data (Figures 13.21, 13.22) illustrate this where the crater depths are equal to the charge depths for the smaller depths of burial.

If slabbing takes place, when the fall strain is just greater than the breaking strain, a crater may be formed, and, if greater than twice the breaking strain, two slabs will form, the thickness of the second slab being equal to half of the fall length (Figure 13.27). For a fall strain just less than twice the breaking strain, only one slab is formed, with a thickness of less than half the fall length. This occurred in those craters produced at critical charge depths; that is, the thickness of the broken rock determined the crater depth, indicating the creation of one slab. Further, when the fall strain is 1–3 times the breaking strain, some spread in crater depths is caused by variations in the strength of the rock which influence the number of slabs, as do differences in the shape of the fall strain. Experimental data (Figures 13.21, 13.22) show considerable spread in crater depths where charge depths are greater than maximum crater depth.

The velocity with which the broken rock is ejected from the surface is approximately twice the average particle velocity of the wave pulse in it. From the data at hand, the peak particle velocity was estimated from the strain data. For a plane wave, particle velocity $v$ is related to strain $\epsilon$ and propagation velocity $c$ by

$$v = -\epsilon c \qquad (13.7)$$

The peak particle velocity in sandstone for various travel distances was computed from the strain and the wave propagation data. Twice these computed velocities (Figure 13.29) showed that for large charge depths they were larger than measured fly rock velocities.

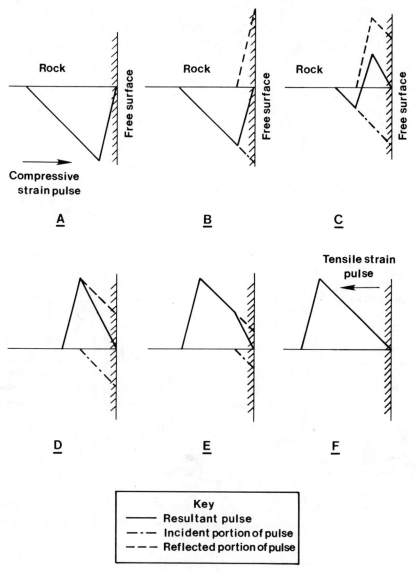

**Figure 13.27** Reflection of a triangular compressive strain pulse (Duvall and Atchison, 1957).

## 13.3.1 Stress Pulses and Energy

The relationship of the reflection slabbing mechanism to the wave energy partitioning—that is, between potential and kinetic energy—is of basic interest. For a *triangular pulse*, if slabbing occurs before the front of the reflected wave reaches the tail of the incident wave, most of the total pulse energy is trapped in the slab as kinetic energy. If slabbing occurs when the wave just overlaps itself, all of the

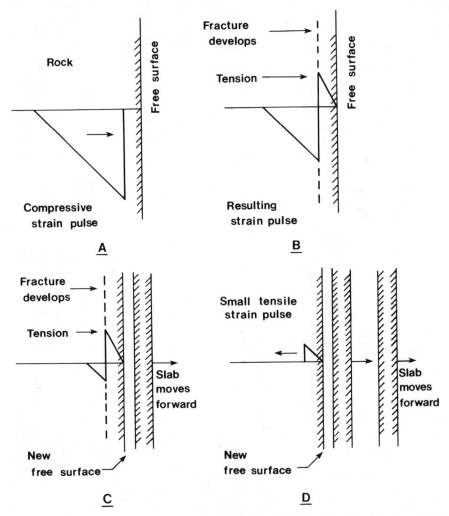

**Figure 13.28** Tensile fracture by reflection of a compressive strain pulse (Duvall and Atchison, 1957).

wave energy is trapped in the slab, all or most of it being kinetic. (See Chapter 2.)

An important factor is the history of the potential (strain) energy trapped in a slab after it is separated from the rock mass. The fragment is in tension in the direction of movement from the face, and the trapped wave will be reflected within the slab with rapid changes in kinetic and potential energy. Further fracture could be caused if weaknesses in the fragment rock are developed by the passage of the pulse, if existing weaknesses are exploited, or if the strain is intense enough to cause fracture.

Also, it is important to note that the basic mechanisms that are dominant in simple cratering are less observable in blasting with multiple charges with two or

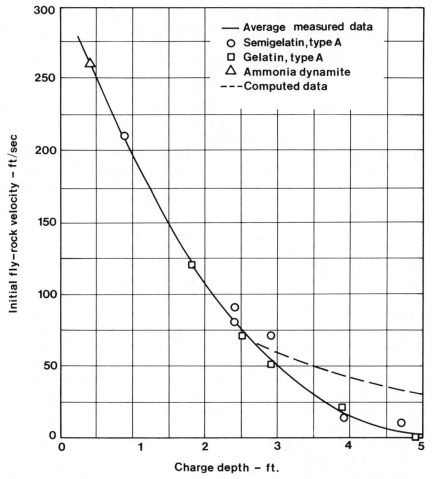

**Figure 13.29**   Measured and computed fly rock velocity in sandstone (Duvall and Atchison, 1957).

more free faces. The stress field of the reflected wave, which is superimposed upon that of the remainder of the outgoing wave, causes additional radial fracturing of rock within a cone angle which is determined by the rock and charge geometry and the wave parameters. The displacement relief provided by the outward movement of the rock at the surface due to the reflected wave is essential to the formation of additional fractures. That is, not only must the strain and kinetic energy of the rock be adequate to create the new surfaces of the fractures, but there must be enough energy to move the rock against confinement, to cause fracturing, and to overcome inertia. If the resistance of the confinement is too great, then the wave energy and the stress will not be great enough to cause effective breakage. Further, some fracturing may also be caused by the residual force of the high-pressure gases invading the fractures near the explosive cavity.

The throw (fragment velocity) of the broken material is due to the wave kinetic energy plus the (later) force of gases escaping through the fractures. Although it has not been possible to establish a mathematical relationship between the "free faces" and effective breakage, reliable empirical principles have been developed by blasting research practice. The effects of one free face are demonstrated by the cratering curves (Figures 13.29–13.34).

To illustrate further, an approximately spherically shaped boulder with a charge at the center at less than a critical breakage distance from the surface has a maximum of free faces for displacement relief. In a three-directional coordinate system a cubic boulder effectively has six free faces. Geometries of rock can be readily visualized that have five and four free faces, but these seldom occur in blasting. A cubic corner of a large mass of rock has three free faces, a bench in a quarry has two free faces, and the plane surface of a large mass or a tunnel face has one free surface.

The factors that all effective blasting geometries or processes have in common are the application of a high pressure on the walls of the explosive cavity, the generation of outward-traveling stress waves, the reflection of the stress waves from the free surface(s), and the accompanying fracturing of complex types.

Differences are caused by the shape of the explosion cavity, the loading density and consequent slope of the stress wave, the properties and structure of the rock, the depth of burden, the explosive properties, the overall geometry of the rock and explosive, the number and shape of free faces, and the number, timing, and sequence of firing of charges in multiple holes, plus other factors.

Research results in the laboratory and observations of the behaviorisms occurring in blasting processes in the field show that the details of explosive breakage processes are very complex, much of the complexity being due to the structure of rock. That is, most theories that have been developed and models that have been proposed are largely based on the assumption that the rock is homogeneous and free from gross imperfections. Rock masses other than small specimens, however, are characterized by imperfections such as fractures, joints, bedding planes, alterations, and similar structural features. Hence, the idealized models and theories can be used as guides only. For example, existing fractures will arrest growing cracks, cause reflections, serve as separation planes for tensile stresses, and often cause critical deviations from ideal behavior. In most rock blasting, enough explosive energy is used to override the effects of many imperfections that otherwise might reduce the effectiveness of the explosive fracture processes.

Much of the meaningful quantitative field experimentation that has been carried out has been based on spherical charge geometry—that is, cratering by (point) charges approximately spherical in shape with one free face present, for example, the U.S. Bureau of Mines' cratering experiments.

For a spherical wave impinging on a free surface, the analysis for a plane pulse can be applied only to the front of the wave, where it is traveling perpendicular to the surface. Thus, the reflected radial velocity and displacement along a line normal to the free face are, respectively, algebraically additive in the reflection process. Where a compressive wave reflects from the free surface at other than normal incidence, it divides into tensile and shear components. However, most of the reflected energy is in the tensile wave whose angle of reflection is equal to the angle of incidence. The front of the reflected pulse will intersect part of the outgoing pulse, and the resultant displacement and particle velocity will be the vector sum of the two quantities at any point (Figure 13.30).

For a single cratering charge that creates an approximately spherical wave in the surrounding rock, the limit of the crater breakage in a horizontal surface is determined by the direction, magnitude, and duration of the impinging and reflected pulse and the strength of the rock. The maximum stresses in the wave with normal incidence and its reflected counterpart should be approximately the same, although a decrease in magnitude with travel distance occurs because of spherical divergence. For the portion of the wave meeting the face at an angle other than 90°, the stress pattern is much more complex because of the division of the reflected wave into strong longitudinal and weaker shear components.

The breakage of an optimum crater is dependent upon the presence of a free face within a critical distance from the explosive charge. In addition to the breakage mechanism of reflection slabbing, a second is caused by the displacement relief accompanying the reflected wave, and a third is due to the initial and secondary radial fracturing radiating from the blasthole and the surface. In experimental measurements of strain waves generated by explosives in rock, the gages are usually placed in the rock outside of the breakage area in such a position that the gage senses only the outgoing strain wave. Laboratory tests have also been devised to measure the pulse generated by a confined explosive where there is no breakage except local crushing and limited radial fracturing around the charge location resulting from a wave trap placed around the test specimen to prevent reflection. In any case, however, the gages could not measure the total reflection strain history experienced by the rock that is broken out of a crater or a quarry face.

Reflection slabbing may occur not only if the explosive charge is close enough to the free face that the magnitude of the reflected tension is greater than the tensile strength of the rock at a point between the explosive charge and the face or behind the drill hole, but also at existing planes of weakness in the rock within a certain distance of the free face.

### 13.3.2 Radial Cracks from Surface

The peak strain of spherical and cylindrical waves decays with distance from the explosive cavity approximately as follows (Obert and Duvall, 1950; Kutter and Fairhurst, 1971):

Spherical:

1. Radial $1/r$
2. Tangential $1/(r^2)$

Cylindrical:

1. Radial $1/(r^{1/2})$
2. Tangential $1/(r^{1.5})$

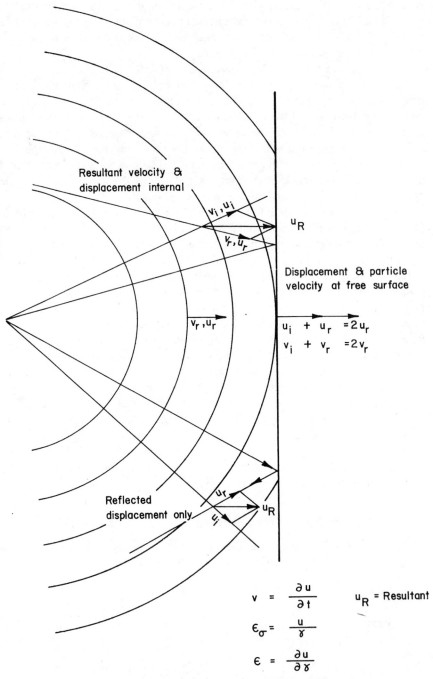

**Figure 13.30** Spherical wave parameters at a free surface.

That is, the spherical radial strain varies inversely as the distance traveled, but, as a wave is reflected at normal incidence, the radial strain at the surface is zero. As the reflected wave travels back through the incident wave, the radial strain and energy are initially small, most of the wave energy being in the form of kinetic energy.

The strain in a spherical wave (Figure 13.31) (Kutter and Fairhurst, 1971) is considerably less than that for a cylindrical wave. In each case, however, the tangential strain is equal to $u/r$ (where $u$ is the radial displacement). In addition to the action of an intense reflected stress wave in slabbing, it has been proposed that the radial component at the front of a spherical or cylindrical wave reflected in tension at angles other than normal incidence to a free surface (Figure 13.30) will reinforce the tangential stresses at the rear portion of the outgoing wave (Field and Ladegaarde-Pedersen, 1971).

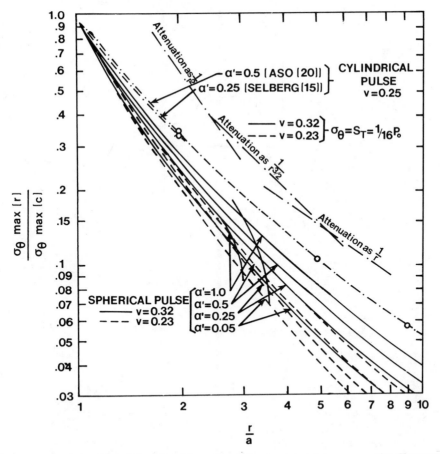

**Figure 13.31** Peak stress of the tangential tensile tail for an exponential pressure pulse (Kutter and Fairhurst, 1971).

### 13.3.3 Initial Surface Rock Fracture and Movement—Experimental Cratering

When a spherical radial wave reaches a surface, the radial displacement will attain a value, $2u$, which is twice the value that the displacement would be if no free face were present (Figure 13.30), and, as the fall portion of the wave is reflected, the particle velocity at the surface likewise increases to twice the value it would have if no free face were present. Thus, the final tangential strain at the surface is equal to $2u/r$, or to twice the value $\epsilon_\theta$ for a continuous medium.

Several theoretical and experimental studies have been made of the stress waves generated by explosive pressures exerted on the surface of spherical cavities in an infinite elastic medium, the field experiments having been performed utilizing these

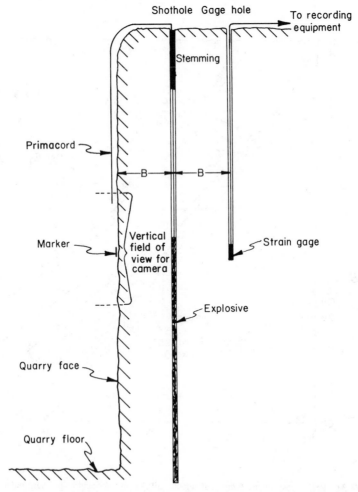

**Figure 13.32** Sectional view of instrumented quarry blast (Petkoff et al., 1961).

**TABLE 13.10. Table of Quarry Blast Parameters and Observed and Calculated Velocities**

| Shot No. | Shothole Diameter, (in.) | Explosive Loading Density, (lb/linear ft) | Rock Burden, (ft) | Measured Strain at Burden Distance, (10-in./in.) | Calculated Initial Fly Rock Velocity, (ft/sec) | Observed Fly Rock Velocity | | | | |
|---|---|---|---|---|---|---|---|---|---|---|
| | | | | | | No. of Observations | Initial, (ft/sec) Range | Initial Average | Final, (ft/sec) Range | Final Average |
| Granite, semi-gelatin explosive | | | | | | | | | | |
| L-1 | $4\frac{1}{2}$ | 7.5 | 12.0 | 1150 | 43 | 11 | 27–69 | 46 | 68–87 | 78 |
| L-2 | 3 | 2.9 | 8.2 | 1000 | 37 | 8 | 33–43 | 37 | 33–43 | 37 |
| L-3 | 3 | 3.0 | 8.5 | 750 | 28 | 6 | 20–35 | 29 | 38–71 | 53 |
| L-4 | $4\frac{1}{2}$ | 7.6 | 9.7 | 1350 | 50 | 4 | 31–83 | 52 | 71–148 | 111 |
| Marble, semi-gelatin explosive | | | | | | | | | | |
| T-1 | $3\frac{1}{2}$ | 3.5 | 14.0 | 270 | 11 | 12 | 10–27 | 17 | 13–42 | 35 |
| T-2A | 6 | 10.8 | 16.3 | 700 | 29 | 5 | 12–44 | 23 | 50–68 | 60 |
| T-2B | 6 | 10.8 | 21.5 | 510 | 22 | 2 | 16–17 | 17 | 45–50 | 48 |
| Limestone, ammonium nitrate prill and fuel oil | | | | | | | | | | |
| B-1 | $7\frac{3}{8}$ | 12.9 | 11.0 | 650 | 18 | 4 | 6–26 | 17 | 26–71 | 43 |

*Source:* Petkoff et al (1961).

studies as a basis for qualitative analyses. There are very few experimental measurements of stress, strain, or fracture times for either cylindrical or spherical waves reflected from a free surface. Plane wave geometry has been employed extensively to explain some of the phenomena of rock breakage in blasting, primarily to demonstrate the mechanism of reflection.

In their cratering studies, Duvall and Atchison (1957) utilized the data from wave strain, fly rock velocity, and fracture time measurements in sandstone to

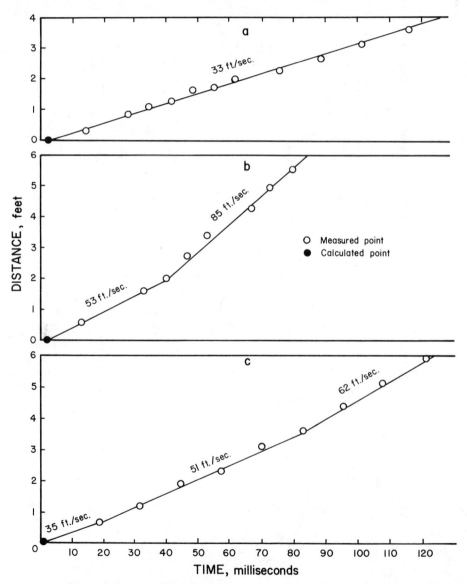

**Figure 13.33**   Typical distance–time plots for granite (Petkoff et al., 1961).

investigate the possible application of reflection slabbing. It was assumed that the theoretical depth of a crater equals approximately one half of the fall length (a distance corresponding to the fall time, $t_f$), provided half of the length is less than the charge depth and the fall strain is larger than the breaking strain of the rock. This assumption requires that the rise time $t_r$ of the pulse be relatively small. For granite and marlstone, the ratios of $t_f/t_r$ are about 3, whereas those for sandstone and marlstone are about 10 or larger. The predicted depths were close to the measured depths only for granite and marlstone (Figure 13.25).

For these, the short rise and fall times gave a first spall smaller than the charge depth, and the scaled depth was predicted quite accurately from the strain pulse data (Figure 13.25) to optimum crater depth. Thus, for rocks in which the strain pulse has a high magnitude and short rise and fall times, spallation may dominate crater breakage. Where the peak strain is lower and longer fall times occur, the

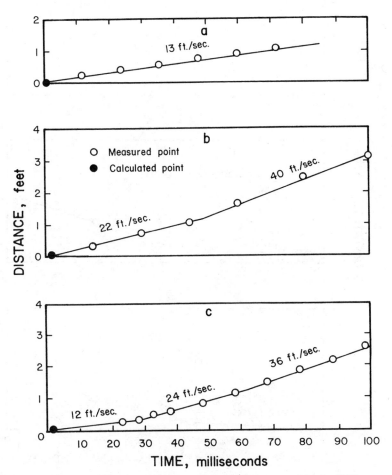

**Figure 13.34** Typical distance–time plots for marble (Petkoff et al., 1961).

first spall thickness is greater than the charge depth, and crater breakage may be influenced more by other mechanisms.

It has been found that the time between the detonation and the first observable movement by high-speed photography of the rock is 3–10 times that for the stress wave to travel from the explosive to the free face and back. However, his experiments were not designed to measure with millisecond accuracy.

### 13.3.4   Two Free Faces—Large Scale

Petkof et al. (1961) made a series of photographic and strain measurement studies of the movement of bench faces in quarries in granite, marble, and limestone,

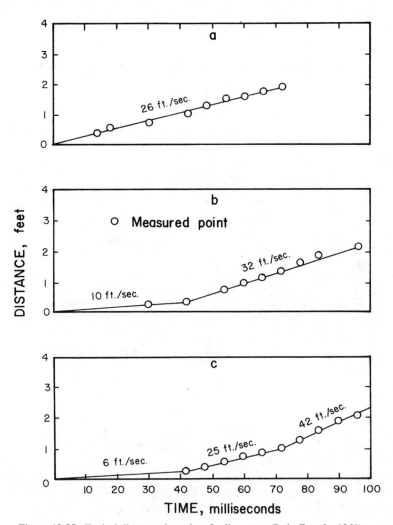

**Figure 13.35**  Typical distance–time plots for limestone (Petkoff et al., 1961).

whose physical properties were determined in the laboratory. High-speed movies were taken utilizing a marker on the quarry face, and strains were measured at a distance from the blasthole back into the rock equal to the burden (Figure 13.32). Initial maximum fly rock velocities were calculated from the strain wave values:

$$v_f = 2c\epsilon \qquad (13.8)$$

where $v_f$ = maximum fly rock velocity, $c$ = longitudinal wave velocity, and $\epsilon$ = peak strain of the incident pulse, compared with observed values (Table 13.10).

The measured initial fly rock velocities grouped around the calculated initial velocities, which was taken as evidence that the initial breakage was due to reflection slabbing. Most of the observed values were greater than the calculated values, and, in almost all of the observations, the fly rock velocity increased stepwise with distance of travel—that is, within the first foot or so of travel, or the first 100 ms after detonation (Figures 13.33–13.35). These increases were assumed to be accounted for by the in-flight impact of rock fragments. That is, the fragments from the points inside the face were traveling faster than those from the surface, and the impact from behind by the faster fragments increased the velocity of those in the front.

However, inasmuch as initial movement of the rock was measured by high-speed movie cameras, the first few microseconds when the rock began to move could not be accurately determined. Hence, the experiments yielded data that only partially define the mechanism of breakage.

## 13.4  PHOTOELASTIC STUDIES

In research by Barker et al. (1979), photoelastic plates were examined to define the mechanisms of explosive-induced fragmentation, particularly of the role of the stress waves in the fragmentation process. Studies were made using two-dimensional transparent polymeric models by means of high-speed photography and dynamic photoelasticity with full-field visualization of the stress caused by explosive loading, and the initiation and extension of cracks. The fragmentation mechanisms involve the initiation, extension, arrest, and branching of the cracks. Such fracture mechanics can accurately predict the behavior of cracks in this type of elastic material. In it the primary parameter that controls crack behavior is the stress intensity factor which is a measure of the stress intensity near the crack tip. A crack or flaw in a stressed material is stable and will not grow if $K$ is less than the fracture toughness of the material, $K_{Ic}$. However, if $K > K_{Ic}$, the crack will propagate with a velocity determined by $K_D$, the dynamic stress intensity factor. The successful branching or bifurcation of a crack will occur when $K_D$ becomes as large as $K_b$, the branching toughness, and the crack is arrested when $K_D$ falls below $K_a$, the arrest toughness.

The stress waves play a critical role in the explosive fragmentation process as found in photoelastic plates. The fragmentation pattern is created by the coalescence of cracks that are initiated at various locations and is quite complicated. It can best be understood by examining the various locations and processes by which the crack systems originate. The whole crack network results from a coalescence of the cracks initiated at both the borehole and remote sites.

## 13.4.1  Experimental Technique

High-speed photography and dynamic photoelasticity were used to visualize stress waves and the dynamic fracture process in plates of Homalite 100. Kobayashi and Dally (1977) determined the dynamic fracture characteristics of this material in terms of the instantaneous stress intensity factor as a function of crack velocity. The initiation toughness of Homalite 100 is less than that of Salem limestone, and the dynamic fracture characteristics of rock and Homalite 100 are quite similar, which at slow crack velocities does not exhibit stable crack growth. Cracks upon initiation jump to a very rapid velocity, greater than about 10,000 in./s, and arrest abruptly from velocities less than about 8000 in./s. Note that typically $K_a <$ $K_{Ic}$. However, slow, stable crack velocities have been found in rock, in which there is not distinct initiation or arrest velocity as there is in Homalite 100. The crack velocities expected in a typical rock fragmentation by blasting may be in the range where the fracture behavior in rock and Homalite 100 are similar. Thus, the details of the fragmentation process observed in Homalite 100 throw some light on the processes that occur in rock.

# 14

---

# FRAGMENTATION PREDICTION

In all of the methods of rock fragmentation and drilling, including those by mechanical means, explosives, thermal expansion, and water jets, as well as methods of comminution for fragmentation other than excavation, it would be advantageous in making feasibility studies and cost estimates for engineers to be able to predict from the results of relatively simple tests the fragmentation properties of rock and ore requiring excavation, its transport, and later processing where this is required.

The mechanical properties of rock have served as a basis for estimation of fragmentation resistance, particularly the unconfined compressive strength, but there are many cases where this has not by itself proved a reliable predictor, largely because factors other than rock strength, such as structure and inhomogeneity, may have dominating effects. Likewise, a static strength may not be a good indicator for a dynamic property, and virtually all fragmentation processes are dynamic in character. Further, the boundary or test conditions, such as shape of samples, test geometry, type of confinement, number of free faces, and so on, may not be easy to model in a simple test.

Special tests, such as that for the Protodiakonov Coefficient of Rock Strength (CRS), have proved to be good measures of the percussive drillability of certain brittle rocks but not for others and are not subject to a rational theory or analysis relating to fundamental mechanical processes and drillability. A combination of shear and compressive strength and other properties has been used in formulas to define tunnel borability, but this should also include factors such as cutter spacing. These tests also appear to be limited to rocks of specified ranges of properties only.

A knowledge of the fragment size distribution of the products of drilling, boring, blasting, and related processes is critical in the solution of the engineering problems involved in these operations (Figure 12.1). Broken material produced by conventional blasting should be of the best possible size obtainable by properly planned blasting procedures to meet the needs for the product, such as ore that requires further crushing, milling, or smelting; or fill for dams, roads, and railroads; or whether it is simply waste material that must be stored in spoil piles with proper regard for environmental considerations (Chapter 12).

The particle size produced by drilling and boring operations largely determines the energy required and is related to the machine performance. Hence, the available particle size data, their effects on excavation operations, and mathematical model design are important engineering design factors. With the data now available, it is possible to make useful predictions of the size limits and distributions to be expected for fragmentation processes in different types of rock and operations.

Rock fragmentation mechanisms are the result of some type of stress wave motion, usually of short duration, the stress pulses being accompanied by or producing different types of fracture phenomena. A compressive sawtoothed wave or similarly shaped pulse that is reflected from a free surface is a well-known example. This is reflected in tension, and, if the tensile strain in the pulse exceeds the breaking strain of the rock, then, ideally, the fracture occurs normal to the direction of wave propagation. Fracture is, of course, a propagating process with its own characteristic velocities and has been the subject of intense studies, particularly in metals that are relatively homogeneous. Rocks are usually far from ideal in composition and structure, and conventional fracture theories can be applied only as approximations.

However, for one to understand the complex processes of rock fragmentation, one must begin with a foundation of the basic science and engineering principles that have been utilized to describe the behavior of more idealized materials. Thus, the theory of elasticity serves as a most useful base for describing both static and dynamic behavior of material. Use of this theory must be accompanied by a working knowledge of the basic mathematics involved to provide, in turn, a means of modeling the stress fields, wave, and fracture behavior. Such analyses may require mathematics including differential equations, along with transform calculus, for the application of discontinuous boundary conditions.

## 14.1   PERCUSSIVE DRILLING

The USBM researchers have made extensive investigations of the applicability of physical properties to penetration rate prediction but found that other techniques were more applicable to the rocks tested by them.

### 14.1.1   Physical Properties

Physical properties of the rock were determined by USBM personnel in the laboratory on intact specimens for correlation with drill penetration rates. Data from earlier research (Paone et al., 1969; Bruce, 1968; Paone et al., 1966) with surface-set diamond bits had used the compressive strength of the rock as a correlation factor. The relationship between percussive drilling rates and compressive strength (Figure 14.1) has a trend similar to that for surface-set diamond bits (see Chapter 6 ''Diamond Drilling''). There is also a similar relationship demonstrated between penetration rate, tensile strengh, and Shore hardness (Figures 14.2, 14.3).

The scatter of data and the general decrease in the penetration rate for higher compressive strength do not provide a correlation consistent enough for the prediction of drillability of all rock types. For example, with a relatively weak dolomitic limestone, Mankato stone, incomplete removal of cuttings clogged the bit. This rock has the lowest compressive strength, yet it did not have the highest penetration rate. Also, Jasper quartzite was drilled at a rate higher than predicted from the compressive strength because quartzite is brittle and is more easily attacked by percussive drilling. A third exception, Dresser basalt, also is quite strong, lacks free quartz content and brittleness, and was drilled at a rate lower than predicted from its compressive strength.

Some of the low penetration rates may be due to resistance to fracture or ability to absorb impact energy. Young's modulus appears in part to take these phenomena into account (Figure 14.4), giving a somewhat better correlation with penetration rates. Of the other significant physical properties, tensile strength correlates well with compressive strength, and consequently its relationship to penetration rate has the same trend. Shore scleroscope hardness values also indicate that this property by itself is not a useful parameter for prediction of percussive drill penetration rates.

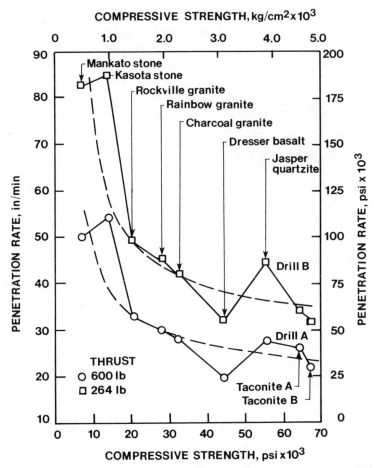

**Figure 14.1**  Penetration rate versus compressive strength: 100 psi (Paone et al., 1969).

Standard physical property tests do not measure all of the factors that influence fracturing and crushing under the drill bit. Rock characteristics such as grain size, grain bonding, mineral composition, and physical structure are important, and their influence on drilling rate varies from one rock type to another and for different samples of the same rock type. An ideal test for prediction of penetration rates would include all of the significant factors governing fragmentation at the bit–rock interface. This implies that the test should use fracture and crushing mechanisms approximating those of the impact by a percussive bit, for example.

### 14.1.2  Coefficient of Rock Strength

The Coefficient of Rock Strength (CRS) test, described by Protodiakonov (1962), has been used as a measure of the resistance of rock to fragmentation by impact.

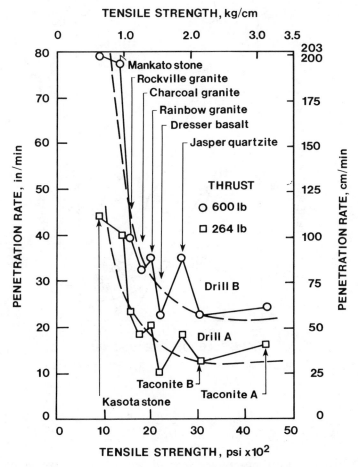

**Figure 14.2.**  Penetration rate versus tensile strength: 100 psi (Paone et al., 1969).

This test was modified by Paone et al. (1969) in conjunction with the laboratory drilling tests described above to evaluate the fracture resistance of the test rock fragments. For the modified test, a sample consisting of five irregularly shaped rock specimens was prepared, each with a volume of about 15 cm³. The total weight in grams of the five specimens was selected to be 75 times the specific gravity of the material, ±2 g. Each specimen was impacted individually in a hollow cylindrical drop tester with a 2.4-kg weight falling from a 0.6-m height.

The number of impacts required to crush each specimen varied from 3–40, depending on the strength of the rock. The broken material for each sample was combined and screened on a 0.5-mm (35-mesh) screen. The −0.5-mm material was weighed and divided by its specific gravity to determine its solid volume. The Coefficient of Rock Strength was then determined by dividing the volume of the −0.5-mm material into the number of impacts. That is

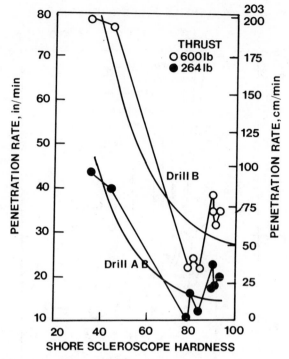

**Figure 14.3** Penetration rate versus Shore hardness: 100 psi (Paone et al., 1969).

$$f' = \frac{n \cdot \rho}{W_p} = \frac{n}{V_p} \tag{14.1}$$

where $n$ = number of impacts, $\rho$ = specific gravity, $W_p$ = weight of $-0.5$-mm ($-35$-mesh) material, $V_p$ = volume of $-0.5$-mm ($-35$-mesh) material, and $f'$ = Coefficient of Rock Strength (CRS). The above procedure was then repeated with larger and smaller numbers of drops to determine the minimum CRS. When the minimum number was ascertained, two additional tests were made for verification of this value.

A consistent relationship between penetration rates determined in the laboratory experiments with the CRS for the rocks tested (Figure 14.5) indicated that the CRS had potential as a useful indicator of penetration rate for percussion drilling. Some approximate similarities between the breakage mechanisms in the CRS test and the action of the percussive drill bit offer a partial explanation of the correlation. The fracture of small specimens by impact involves some of the rock properties that affect the percussive drilling process—grain size and composition, grain bond, existing fractures or microfractures, and brittleness of minerals, such as quartz.

The work rate, assuming that a maximum of the piston energy was transmitted

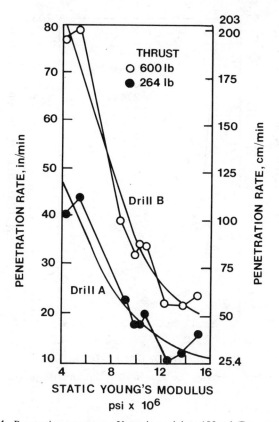

**Figure 14.4** Penetration rate versus Young's modulus: 100 psi (Paone et al., 1969).

to the rock, was calculated for operating pressures and thrust values used in earlier drillability experiments, and the energy required (specific drilling energy) to produce a unit volume of hole is

$$E_v = \frac{4E_r}{\pi D^2 PR} \tag{14.2}$$

where $E_v$ = energy per unit volume, in.-lb/in.$^3$ (specific energy); $E_r$ = work rate, in.-lb/min; $D$ = bit diameter, in.; and $PR$ = penetration rate, in./min.

The specific energy was calculated for all combinations of operating and feed pressures. For rainbow granite (Figure 14.6) the lowest specific energy was obtained for operating air pressures of 30–90 psi, indicating that this is the most efficient pressure range for this drill. However, higher pressures should result in

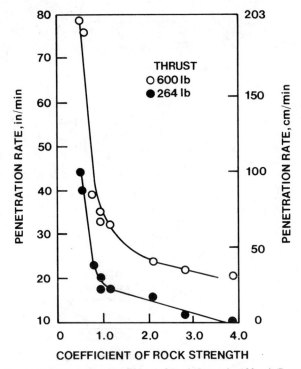

**Figure 14.5**   Penetration rate versus Coefficient of Rock Strength: 100 psi (Paone et al., 1969).

more rapid penetration; hence, the efficiency at lower pressure may be due to more effective removal of cuttings.

The correlation of specific energy for optimum drilling conditions with the CRS for each rock type (Figure 14.7) indicates that the specific energy is independent of the work rate. That is, the drill cuttings from drills A and B should have approximately the same particle size distribution. The proposed method of predicting penetration rate is to determine the CRS of a rock, and, from Figure 14.7, the specific energy for optimum conditions. The work rate can then be obtained either from the drill manufacturer or by calibration, and the penetration rate can be obtained from equation 4.19.

The previous procedure to determine the CRS was later modified by Tandanand and Unger (1975) as follows: Ten irregular-shaped representative fragments that average about 0.75 cm³ were sorted by hand or by screening to select the −1 in. (2.54 mm) plus $\frac{3}{4}$ in. (0.956 m). The volume of the samples was obtained by dividing their weight by the specific gravity.

The specimens were crushed two at a time in a drop tester where a weight of 5.3 lb (2.4 kg) was dropped 25 3/16 in. (0.60 m). As before, the number of drops

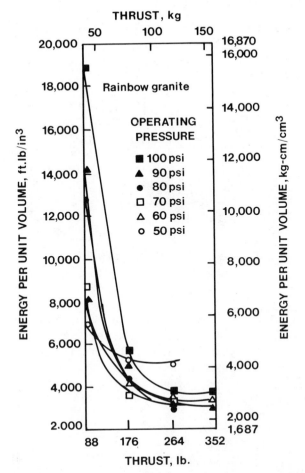

**Figure 14.6** Energy per unit volume versus thrust and operating air pressure, raindow granite, drill A (Paone et al., 1969).

varied from 3–40, depending on the crushing strength of the rock. Each of the five pairs of the samples was subjected to the same number of drops, and the broken material was combined and screened on a 35-mesh (0.5-mm) screen and shaken by hand for 40 s. The −35-mesh (0.5-mm) material was weighed and the solid volume calculated.

The CRS was then calculated by dividing the number of drops by the volume to obtain $f' = n/V_p$. This procedure was then repeated, varying the number of drops for each sample of five pairs and a curve of $n/V_p$ versus $n$ plotted (Figure 14.8). The minimum value of $n/v$ was taken as the CRS.

The CRS of the rocks tested does not correlate well with the compressive

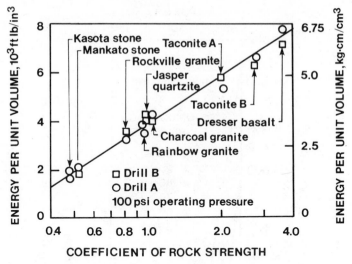

**Figure 14.7**   Specific energy versus the Coefficient of Rock Strength (Paone et al., 1969).

strength, but there is a better correlation with strain energy density equal to $\sigma_c^2/2E^4$, where $\sigma_c$ is the compressive strength. The toughness of rock is assumed to be represented by the area under a stress–strain curve for a uniaxial test, including plastic flow at failure. Toughness data were not available for the test rocks, so compressive strength was plotted against CRS (Figure 14.9). Assuming that rocks

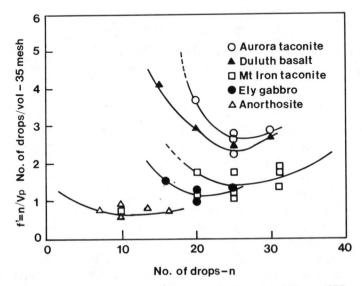

**Figure 14.8**   Typical curves for $n/v$ versus $n$ (Tandanand and Unger, 1975).

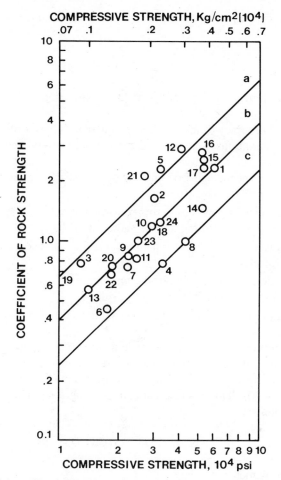

**Figure 14.9**  Coefficient of Rock Strength versus compressive strength (Tandanand and Unger, 1975).

that fail at the same magnitude of strain have the same breaking characteristics, rocks were assigned to a category of relative breaking characteristics. The three lines a, b, and c in Figure 14.9 indicate

| | |
|---|---|
| Tough rock | $f' = 0.65 \times 10^{-4}\ \sigma_c$ |
| Medium tough rock | $f' = 0.4\ \times 10^{-4}\ \sigma_c$ |
| Brittle rock | $f' = 0.23 \times 10^{-4}\ \sigma_c$ |

and the rocks may be classified as tough, medium tough, and brittle, accordingly.

To avoid the possible inaccuracies involved in assuming a value for the energy transfer coefficient $T_r$, Tandanand and Unger (1975) used the relationship between specific energy for laboratory drilling and the CRS (Figure 14.9) to obtain the equation

$$E_v = 3860 + 2244 \ln f' \text{ ft-lbits}/\text{in.}^3 \qquad (14.3)$$

which is based on drilling data for $1\frac{1}{2}$-in. (3.81-cm), diameter cross bit (Paone et al., 1969).

The proposed accuracy of application of equation 14.3 to drilling is founded on the following assumptions, some of which have not been proved:

1. Some of the mechanisms of breakage in the CRS test are similar to those that occur in percussive drilling.
2. The percentage energy loss in drop crushing at optimum specific energy and that for percussive drilling are the same for each type of rock.
3. The energy used for chip formation, crushing, and regrinding is the same for the two processes.
4. The particle size distribution of the product resulting from drop crushing for minimum $f'$ approximates that from drilling.
5. The overall energy partitioning is the same in the two processes.

Experimentation has shown that less specific energy is required for drilling larger holes. The method of calculating penetration rates thus requires a scaling factor to account for hole size effects. The relation of the scaling factor $K$ to its diameter is linear (Figure 14.10). This is close enough to being a straight line that it can be expressed as follows:

$$K = 1.583 - 0.389 \, D \qquad (14.4)$$

where $D$ = bit diameter-in.

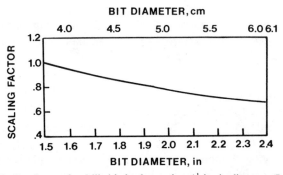

**Figure 14.10** Scaling factors for drilled holes larger than $1\frac{1}{2}$-in. in diameter (Paone et al., 1969).

The rate of penetration is expressed by

$$PR = \frac{E_p}{E_v A_H} \tag{14.5}$$

where $R$ = penetration rate, $E_p$ = power output, $A_H$ = area of drill hole, and $E_v$ = specific energy of drilling.

The specific energy so defined is the piston energy per unit volume of hole and does not represent the real specific energy of the drill cuttings. The correction for scaling is obtained by multiplying $E_v \times K$ (scaling factor), and the penetration rate equation may be written

$$\frac{E_p}{PR} = \frac{1}{AK\,(3860 + 2244\,f')} \quad \text{in.} / \text{ft-lb}$$

The drillability curves for five sizes of cross bits (Figure 14.11) demonstrate the effects of hole size on drilling rate and power. The maximum power output of the experimental drills A and B (Figure 14.12) may then be used to obtain the calculated rates of penetration based on the CRS. These calculated rates compare fa-

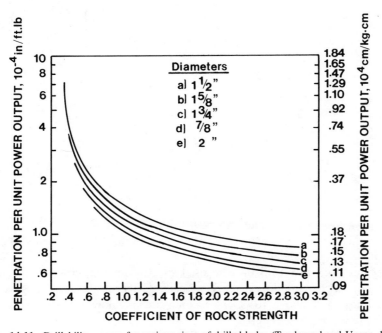

**Figure 14.11** Drillability curves for various sizes of drilled holes (Tandanand and Unger, 1975).

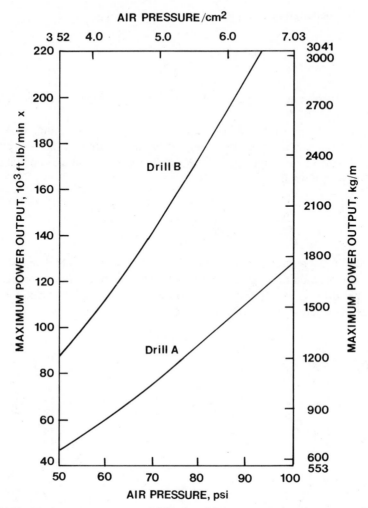

**Figure 14.12** Maximum power output of drills A and B versus operating air pressures (Tandanand and Unger, 1975).

vorably with the measured rates for both drills (Figures 14.13, 14.14) for $1\frac{1}{2}$-in. cross bits. There is more scatter in the data for 2-in. bits, especially at higher penetration rates (Figure 14.15).

Equations 14.3 and 14.6 may be used to estimate impact strength, and the percent free silica is used as another indicator. That is, a tough rock has about 40%, a brittle rock above 70%, and medium rocks below 40% silica, the latter including limestone and granite.

As described earlier, Paithankar and Misra (1980) performed percussive tests with microbits impacted with a drop weight and assumed that all of the impact energy was transmitted to the rock, which is only an approximation. The cuttings

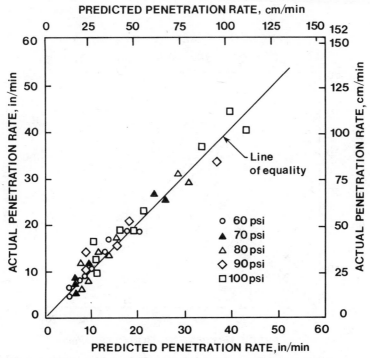

**Figure 14.13** Comparison of experimental and predicted penetration rate for drill A with $1\frac{1}{2}$-in. bits (Tandanand and Unger, 1975).

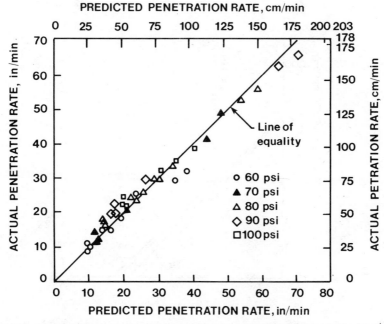

**Figure 14.14** Comparison of experimental and predicted penetration rate for drill B with $1\frac{1}{2}$-in. bits (Tandanand and Unger, 1975).

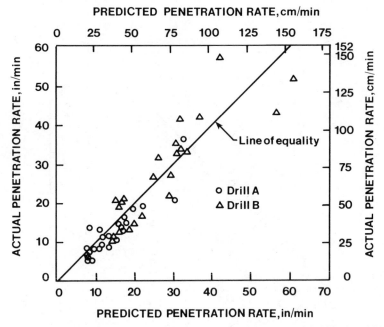

**Figure 14.15** Comparison of experimental and predicted penetration rates for drills A and B with 2-in. bits at 100-psi operating air pressure (Tandanand and Unger, 1975).

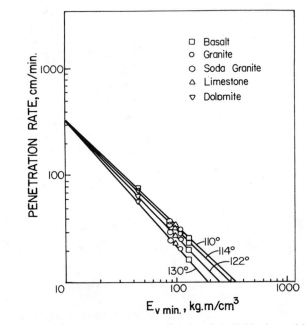

**Figure 14.16** $E_v$ min versus penetration rate (log–log plot) (Paithankar and Misra, 1980).

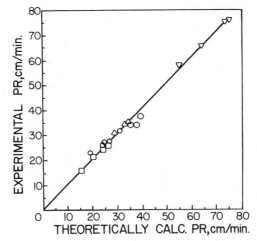

**Figure 14.17**  Experimental versus calculated penetration rate (Paithankar and Misra, 1980).

were taken out of the hole after each blow. The specific energy was found to be affected by the indexing. Also, the $PR$ versus $E_v$ curves were found to be straight lines on the log–log plot, whose equations are (Figure 14.16)

$$\log PR = m \log E_{v\min} + C \tag{14.7}$$

where $m = 0.0094\theta + 0.04$ and $C = 0.0089\theta + 2.544$, where $\theta$ is the angle of bit wear.

The experimental and calculated values of $PR$ agree well (Figure 14.17) for the rocks tested. It is stated that values so determined can be applied to other drills of known energy per blow $E_b$ and the hole cross-sectional area $A_H$.

## 14.2  FIELD TESTS (USBM)

Of the standard properties, only compressive strength and properties highly correlated with it, such as tensile strength and Young's modulus, showed good correlations with penetration rate (Figure 14.18) (the penetration rates are defined for maximum values at 100-psi operating pressure).

A least-squares best-fit curve for each drill is of the form

$$y = ax^{-b} \tag{14.8}$$

where $y$ = penetration rate and $x$ = compressive strength.

The compressive strength is less significant for rocks with strength in excess of 25,000 psi. A separate curve is required for each drill because values are not normalized for the drill power. These factors, together with the scatter in data, indicate that compressive strength is not the best measure of drillability.

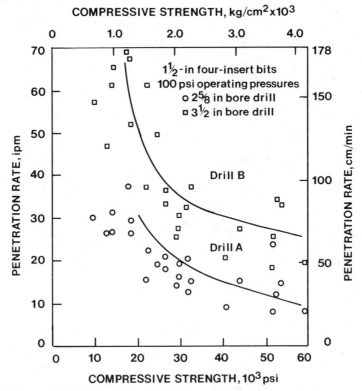

**Figure 14.18** Penetration rate versus compressive strength (Schmidt, 1972).

Similar tests were carried out by Hustrulid (1971) with three pneumatic drills for determining the drilling rate in quartzite. He found that the three drills, 3, 2 7/8, and 2 5/8 in. (7.62, 7.30, and 6.67 cm) diameter, drilled at about the same penetration rates in two different quartzites of 39,000 and 45,000-psi compressive strength, which was used to approximate the specific energy of drilling, together with an assumed value of 0.8 for the energy transfer coefficient $T_r$. The penetration rate as a function of free air consumption was found to be

$$PR = k \times (\text{cfm})^2 \qquad (14.9)$$

where $k$ varied from $6.25 \times 10^{-4}$ to $6.67 \times 10^{-4}$ for two quartzites (see Chapter 4 "Field Tests–Quartzite").

### 14.2.1 CRS

The CRS, in addition to the fact that the mechanics of breakage are by impact, has the advantage of being easy to determine and requires simple equipment.

Although better correlations were obtained (Figure 14.19) for penetration rate and CRS for both drills and there is less scatter than with compressive strength, there are two disadvantages in using the CRS by itself as a predictor: the penetration rates are insensitive to changes in values of CRS < 1.0, and the CRS does not account for differences in drill power. (See 14.4.)

## 14.2.2  Specific Energy

Better criteria to predict percussive drill rates are obtained using drill power and specific energy. This method also permits estimates for different bit sizes.

Specific energy, that required to drill a unit of volume of rock, is dependent, as described earlier, on parameters of the drill system such as bit geometry, indexing angle, type of breakage system (percussive, rotary, etc.), and rock properties.

For field investigation by Schmidt (1972), specific energy was treated as a constant for each rock. (This assumed that the particle size distribution of the drill cuttings is approximately the same for a given value of CRS.) To simplify the

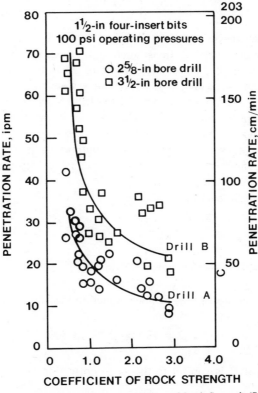

**Figure 14.19**   Penetration rate versus Coefficient of Rock Strength (Schmidt, 1972).

prediction of drillability, changes in specific energy due to drill size or bit geometry were also assumed to be negligible. This procedure was considered to be subject to improvements when more data became available.

The work rate is the blow energy multiplied by the blow rate, and when it is substituted into equation 3.15, one obtains

$$PR = (48) \, (E_w) \, (T_r)/(\pi) \, (E_r) \, (D^2) \qquad (14.10)$$

The work rate for a drill may be obtained from the manufacturer, or approximated with Ditson's equation:

$$E_w = (W_p \times 10^{-6}) \, (S \times f)^2/2.21 \qquad (14.11)$$

where $KE$ = kinetic energy per blow, $W_p$ = weight of piston and rifle nut (lb), and $S$ = working stroke of piston (in.).

Bruce and Paone (1969) found good agreement between experimentally determined values and those calculated from equation 14.11. Work rate is

$$E_w = (KE) \, (f) \qquad (14.12)$$

The two unknowns in equation 14.9 are $T_r$ and $E_r$. The transfer ratio is the percentage of the piston energy transmitted to the rock. Determination of this ratio is an involved procedure; therefore, a value of $T_r = 0.7$ was assumed. For the specific energy, the measured penetration rates were used to calculate apparent specific energies $E_{va}$, using equation 14.11 for drills A and B—that is, the maximum rate for each operating air pressure. The calculated values of $E_{va}$ were in good agreement, so it was assumed that the apparent specific energy is constant for each rock. The specific energy was approximated without drilling by use of the coefficient of work strength—that is, by plotting specific energy against its respective Coefficient of Rock Strength (Figure 14.20).

The best fit for the curve is given by

$$E_{va} = 29,700 \, f'^{\,0.554} \qquad (14.13)$$

where $E_{va}$ = apparent specific energy (in.-lb/cu in.) and $f'$ = Coefficient of Rock Strength.

For field calculations, a best-fit linear function may be used:

$$E_{va} = 13,900 \, f' + 15,500 \qquad (14.14)$$

Here, the value of $f'$ is subject to the same correction as in equation 14.3. The

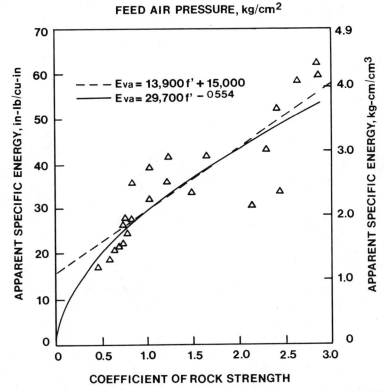

**FEED AIR PRESSURE, kg/cm²**

Within the plot:
- - - $E_{va} = 13{,}900\,f' + 15{,}000$
— $E_{va} = 29{,}700\,f'^{-0554}$

**Figure 14.20**   Apparent specific energy versus Coefficient of Rock Strength for rocks drilled in the field (Schmidt, 1972).

scatter in the data in Figure 4.20 may be largely due to the assumption that $T_r$ is the same for all rocks. The acoustic coupling of metal to rock varies for different rock types, and the mechanisms of breakage also are functions of rock properties. Hence, $T_r$ is logically a variable rather than a constant.

The procedure to predict percussive drillability using the foregoing is

1. Determine the CRS.
2. Use the CRS to determine the specific energy of the rock either from Figure 14.20 or from equation 14.14.
3. Obtain the $PR$ from equation 14.10 and the work rate $W$ either from the drill manufacturer or from equation 14.11.

An example follows: Assume drill A is to drill a dense basalt at 100-psi operating pressure with a $1\frac{1}{2}$-in. (3.81 cm) drill.

**1.** The CRS value is 3.94.

**2.** From equation 14.13, the specific energy is

$$E_{va} = 13,900\, f' + 15,500$$

$$= 13.900\, (3.94) + 15,500$$

$$= 70,266 \text{ in.-lb/in.}^3$$

**3.** The work rate of drill A is known to be 138,600 ft-lb/min. For an energy ratio of 0.7, the penetration rate is

$$PR = (48)(138,600)(0.7)/(\pi)(70,266)(1.5)^2$$

$$= 9.41 \text{ ipm} \tag{14.15}$$

The measured penetration rate of drill A into the basalt averaged 10.0 ipm.

This prediction procedure was used for several groups of unrelated drill data, that is, field and laboratory data not used in the development of the prediction equation.

Measured and calculated rates at six levels of operating pressure were compared for nine rocks that were drilled in the laboratory with $1\frac{1}{2}$-in. (3.81-cm), four-insert bits. Generally, the predictions correlate best for hard rocks, that is, those with strength coefficients greater than 1.0. The linear relationship in equation 14.14 (Figure 14.20) for simplifying the procedure gives calculated energy values for soft rocks that are usually too high.

Inasmuch as the prediction equation contains the bit area, it should give a good correlation for various bit sizes. Thus, penetration rates for 2-in. (5.04-cm), four-insert bits were calculated and compared with measured rates. The predictions are not as accurate as those for $1\frac{1}{2}$-in. (3.81-cm) bits, from which equation 14.15 was derived. The simplified procedure given, however, will serve as a guide as long as the user is aware of possible sources of error.

The penetration rates for the button bits were also calculated and compared with measured rates.

The correlation coefficients for the four series of predictions are

Drill A: $1\frac{1}{2}$-in. (3.81-cm), four-insert bits, laboratory rocks: 0.72
Drill B: $1\frac{1}{2}$-in. (3.81-cm), four-insert bits, laboratory rocks: 0.75
Drill A: $1\frac{1}{2}$-in. (3.81-cm) button bits, field rocks: 0.80
Both drills: 2-in. (5.08-cm), four-insert bits, field rocks: 0.62

## 14.3  INDEPENDENT ROTATION

For the drilling with independent rotation, as for standard percussive drilling tests, diamond drill core was used to make the following physical property measurements: compressive strength, tensile strength, Shore hardness, density, static Young's modulus, longitudinal velocity, bar velocity, torsional velocity, dynamic Young's modulus, Poisson's ratio, and shear modulus. The Coefficient of Rock Strength (CRS) was also determined for each rock type, but the $E_v$–$f$ relation was not linear for independent rotation drills.

Several properties affect the breaking of rock by percussive drilling, and the influence each property has on penetration rate varies with the kind of rock as in the earlier percussive drilling tests. The modified CRS test was used as before, but slightly different values were obtained for the same rock types.

The specific energy of percussive drilling can also be calculated by the equation

$$E_v = \frac{PT_r}{A(PR)} \qquad (14.16)$$

where $E_v$ = specific energy, ft-lb/in.$^3$; $P$ = power output of drilling system, ft-lb/min; $T_r$ = ratio between the energy transferred to the rock and the energy available for each blow; $A$ = hole cross-sectional area, in.$^2$; and $PR$ = penetration rate, in./min.

For independent rotation, the power of the drilling system, $P$, is taken as the sum of the outputs of the drill and air motor. For example, the total output of the system at 100-psi operating and rotation pressures and a rotational speed of 250 rpm is 227,000 ft-lb/min (Table 4.3) plus 11,000 ft-lb/min (Table 4.4). These values vary with air pressure and were obtained from the manufacturer. The transfer ratio $T_r$ was not determined in this experiment, but, since the energies were calculated to show only comparison between rocks, a transfer ratio of unity was used. Calculated energies are designated as apparent specific energies. The penetration rates used were the averages of those obtained at 100-psi operating and rotational pressures with a 2-in. bit.

The CRS test was again assumed to use energy to break the rock in a fashion similar to that of a percussive drill, the correlation between the rock strength number and energy per unit volume (Figure 14.21) being similar to that for standard percussion drills described earlier.

### 14.3.1  Particle Size

The particle size of drilling cuttings is related to total and specific energy and the efficiency of drilling. Cuttings were collected and analyzed from some of the drill holes. These were collected with a vacuum system and sized by standard procedures

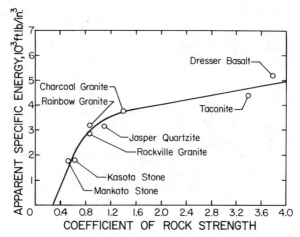

**Figure 14.21** Apparent specific energy versus Coefficient of Rock Strength: 100-psi operating and rotational pressures (Schmidt, 1972).

with a set of laboratory screens. Screen analysis results (Figure 14.22) from cuttings collecting from Mankato stone drilled at air motor operating pressures of 50 and 100 psi with other conditions kept constant showed that coarser cuttings were produced at the higher air motor operating pressures, which indicates more efficient drilling. The same result was obtained for cuttings collected during drilling of rainbow granite (Figure 14.23). Conclusions were mostly qualitative.

The penetration rates with air motor pressures of 100 psi were slightly higher than those with 50 psi in most of the rocks, the differences being more prominent in the softer rocks. More efficient drilling is indicated by the coarser cuttings produced with drilling with the higher air motor pressures.

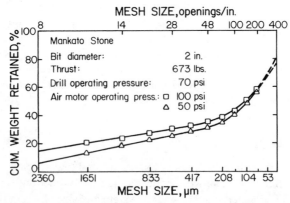

**Figure 14.22** Cumulative percentage retained versus mesh size: drill cuttings, Mankato stone (Schmidt, 1972).

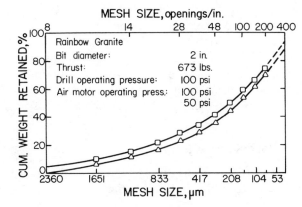

**Figure 14.23** Cumulative percent retained versus mesh size: drill cuttings, rainbow granite (Schmidt, 1972).

Thrust forces from 600–800 lb produced the maximum penetration rates for most of the rocks, the harder rocks showing better response. As with standard percussive drilling, although the physical properties of the rocks are an indication of their drillability characteristics, no one property correlates well with drillability. The CRS number correlates well with penetration rates and energy per unit volume for the independently rotated percussive drill, and the specific drilling energies for the various types of rock are lower for independent rotation.

## 14.4 EVALUATION OF CRS

As noted, experimental Coefficients of Rock Strength have been found to correlate well with the drillability of controlled drilling tests in the laboratory for the suite of brittle, hard rocks tested by the U.S. Bureau of Mines (just described) and, to a lesser extent, for the same or similar rocks drilled in the field. However, the CRS is not a number that can be determined in terms of physical units, and its real meaning has not been defined. Also, it has been found that there are many softer rocks that do not yield to the proper data for definition of a CRS number.

Serious questions revolve around the similarities and differences in the drop-crushing process and the mechanisms of breakage in drilling, the lack of similarity in the particle size distribution in drilling and crop weight crushing, the measurement of the cutoff size of −35 mesh, and other related factors, including the energy partitioning in each process. For example, in past reports and evaluations of CRS tests, it has always been assumed that all of the drop weight energy is consumed in crushing the rock specimens. However, in recent research (Selcuk, 1981), it was found that as much as 20% of the energy of a drop weight is absorbed by the equipment and that the percentage absorbed in crushing is a function of the number of drops, the particle size of the material at each stage of crushing, material compaction, rock properties, and other related factors.

Selcuk (1981) made a study of the CRS of Colorado red granite to measure the crushing energy and the particle size distribution resulting from crop crushing and drilling. His work established some of the important relationships between them and more quantitatively defined the meaning and limitations of the CRS and the basic physical parameters underlying its determination and consequent meaning.

## 14.4.1 Energy–Particle Size Relation

Important aspects of the rock-crushing process in the CRS tests may be analyzed by using energy–size relationships in comminution.

### 14.4.1.1 Comminution Laws

The most commonly used comminution laws are Kick's law, Rittinger's law, and Bond's third theory. All of these can be expressed mathematically by an empirical equation, which can be demonstrated experimentally but has no theoretical derivation (Charles, 1957):

$$dE = -C \frac{dx}{x^m} \qquad (14.17)$$

where $dE$ = infinitesimal energy change, $dx$ = infinitesimal size change, $x$ = size, $m$ = a constant, and $C$ = a constant.

The quantity $dE$ in equation 14.17 can also be defined as the infinitesimal energy requirement to reduce a solid particle of size $x$ to particles of size $(x - dx)$.

Rittinger's law states that energy of size reduction of a solid is proportional to the new surface area produced of a given quantity or weight and, hence, is inversely proportional to the particle size, stated mathematically as (Charles, 1957)

$$E_r = K(s_2 - s_1) \qquad (14.18)$$

where $E_r$ = energy input per unit volume, $K$ = a constant, $s_1$ = initial specific surface, and $s_2$ = final specific surface.

For the reduction of particles of size $x_1$ to size $x_2$, equation 14.18 becomes

$$E_r = K_i \left( \frac{1}{x_2} - \frac{1}{x_1} \right) \qquad (14.19)$$

where $K_i$ = a constant.

Equation 14.20 is derived from equation 14.17 by integration and assignment of a value of 2 to $m$:

$$\int_0^{E_r} dE = \int_{x_1}^{x_2} - C \frac{dx}{x^2} \tag{14.20}$$

$$E_r = K'\left(\frac{1}{x_2} - \frac{1}{x_1}\right) \tag{14.21}$$

where $K' = C$ in equation 14.20.

Experimental evidence in favor of Kick's law is minimal. On the other hand, in some mechanical processes, Kick's law seems to be fundamental to cutting, pressing, shaping, and rolling of metallic substances (Hukki, 1959), while Bond's theory is valid only in the fine grain operations of mineral dressing between certain particle size limits.

However, experimental evidence in favor of Rittinger's law has been presented by Piret et al. (1949), Schellinger (1952), and Gross and Zimmerley (1930). Gross and Zimmerley (1930) performed quartz sand crushing tests with a drop weight crusher similar to Protodiakonov's apparatus and measured the specific surface area of quartz particles by the method of dissolution. It was found that Rittinger's law is valid for this drop weight crushing process.

From the results of the CRS tests at Colorado School of Mines (described in the following section), it is believed proper to apply Rittinger's law to this crushing process.

### 14.4.1.2   *Energy and Particle Size Distribution in Comminution*

The size reduction energy can be determined if the particle size distribution is known, either graphically or analytically (Charles, 1957).

Size distribution of crushed material is described by the equation

$$y = f(x) \tag{14.22}$$

where $y$ = weight percent finer than size $x$, $x$ = size, and $f$ = a known function.

From equation 14.23, the energy required to reduce an element of weight of material $dy$ from size $x_m$ to size $x$ is

$$dE_1 = \int_{x_m}^{x} -C \frac{dx_1}{x_{m1}} dy \tag{14.23}$$

where $x_m$ = initial size.

The total energy required to reduce a weight of material of size $x_m$ to the distribution given by $y = f(x)$ is then the sum for all elements of weight of material $dy$:

$$E = \int_0^E \int_0^{100} dE_1 \, dy \qquad (14.24)$$

since

$$dy = f'(x) \, dx \qquad (14.25)$$

where $f'$ is the first derivative of $f(x)$, then

$$E = \int_{x_0}^{x_k} \int_{x_m}^{x} \left( -C \frac{dx_1}{x_1^m} \left( f'(x) \cdot dx \right) \right) \qquad (14.26)$$

Equation 14.26 can also be evaluated numerically using size distribution data. The first derivative of the cumulative size distribution function $dy$ is the weight percent retained on a screen of size $x$. Equation 14.26 can be rewritten as

$$E = \int_{x_0}^{x_k} K \left( \frac{1}{x^{m-1}} - \frac{1}{x_m^{m-1}} \right) P_i \, dx \qquad (14.27)$$

where $P_i = dy$, $x_k$ = largest size fraction, and $x_0$ = smallest size fraction, or in a summation form

$$E = K \sum_1 \left[ \frac{1}{x_i^{m-1}} - \frac{1}{x_m^{m-1}} \right] P_i \qquad (14.28)$$

or

$$E = K \left[ \sum_1 \frac{P_i}{x_m^{m-1}} - \frac{100}{x_m^{m-1}} \right] \qquad (14.29)$$

where $E$ = specific crushing energy per gram, $P_i$ = weight percent retained in $i$th screen, $x_i$ = average size of $i$th screen fraction, $x_m$ = initial size of the rock sample, $i$ = screen fraction designation, $K$ = a constant, and $m$ = a constant.

For Rittinger's law, equation 14.28 takes the following form

$$\text{Rittinger's law: } E_r = K_r \left[ \sum_1 \frac{P_i}{x_i} - \frac{100}{x_m} \right] \qquad (14.30)$$

### 14.4.2  Critical Review of CRS Analysis

In an attempt to develop a graphical method of determining the minimum $n/v$ ratio in the CRS tests, Tandanand and Unger (1975) made an analysis using the parameter involved. This analysis has at least three sources of error. The first is that Kick's law is assumed to apply to the drop-crushing process with no justification for the assumption. Rittinger's law has been shown to be more nearly applicable for granite, as is demonstrated below. Second, in using Charles's equation involving particle size and energy, in the middle of the derivation of equations, the terms for *individual particle sizes or volumes* are changed without explanation to have the meaning of *sample volume*, making the volume–energy incorrect as defined. Third, it is assumed that the crushing energy is proportional to the number of drops or that all or a constant portion of the drop energy is absorbed per drop in crushing. As is shown later, the percent crushing energy per drop is not constant. Also, the crushing energy of $-35$-mesh material cannot be determined unless the particle size distribution is known.

The method usually used assumes that all rocks display a minimum value of $n/v$ ratio. Brook (1977) criticizes this assumption, showing that $n/v$ versus $n$ curves for many rocks do not exhibit a minimum value, which causes difficulty in obtaining a CRS index (Figure 14.24).

#### 14.4.2.1  Experimental Analysis of CRS Tests

A series of CRS tests was made by Selcuk (1981) using an apparatus similar to Protokiakonov's original drop-weight crushing device and the same test procedure as described by Tandanand and Unger (1975):

**1.** Ten irregularly shaped rock specimens averaging 7.5 cm$^3$ were prepared.

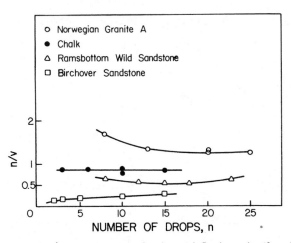

**Figure 14.24**  Typical $n/v$ versus $n$ curves showing no inflection point (from Brook, 1977).

Total weight of the ten specimens was 75 cm³ times the specific gravity of the Colorado red granite (approximately 202.5 g).

2. The sample was then divided into five groups of two specimens each.

3. Each group of specimens was crushed with a 2.4-kg weight falling from a height of 0.60 m.

4. Each group of specimens was subjected to the same number of drops until all five groups were broken. The broken material from the five groups was then combined on a 35-mesh screen and hand shaken.

5. The −35-mesh fraction screen was weighed (in grams), and the weight was divided by the specific gravity to determine the solid volume of the material ($v$).

6. The $n/v$ ratio was then determined by dividing number of drops used on each two-specimen group with the total volume of the −35-mesh (0.5-mm) material.

7. The above procedure was repeated for numbers of drops varying from 3–8, and an $n/v$ ratio for each test was obtained. For more complete results, each test was repeated three times. Then the average $n/v$ ratios of the three tests were plotted against corresponding $n$ values (Figure 14.25).

The $n/v$ versus $n$ curve displayed an inflection point where $n/v$ is minimum. This minimum $n/v$ value (0.60) was then taken as the CRS index for the Colorado red granite, which was selected because of its relatively uniform structure and availability.

### 14.4.2.2  Measurement of Net Crushing Energy

As noted earlier, all of the energy of each drop is not transmitted to the rock sample. This fact has been ignored in all previously reported investigations, but it

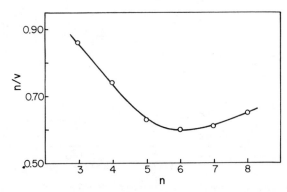

**Figure 14.25**   $n/v$ versus $n$ curve for Colorado red granite (Selcuk, 1981).

plays a very important role in defining the volume of fines versus number of drops and the specific energy–number of drops relationships.

Upon each impact, a certain amount of energy is delivered to the rock sample, but a significant percentage is absorbed by the apparatus. A condition in which the drop weight does not rebound after impact was established by placing relatively soft metal (lead) pellets between the mortar and a solid base. The apparatus was calibrated by measuring the deformation of the cylinders with no rock sample, which gives a measure of energy absorbed by the apparatus. When crushing is done, the deformation of the lead cylinders then gives a measure of the energy absorbed by the apparatus and that used in crushing. A calibration curve (Figure 14.26) was thus obtained for energy absorbed by the equipment as a function of pellet deformation. The standard drop weight (2.4 kg) was dropped successively from a distance of 60 cm, and, after each drop, deformation of the lead cylinders was measured and plotted against total energy in terms of number of drops.

For crushing tests, the deformation of the cylinders was measured, and the energy absorbed by the device was read from the calibration curve. The difference between the energy of the drop weight and the deformation energy was taken as the net energy used in crushing (Figure 14.27). These energy values divided by corresponding $n$ values gave the percent net crushing energy (Figure 14.28). The portion of input energy that is used in crushing increases to a maximum, then decreases and levels off with the increasing number of drops. The maximum point corresponds approximately the same to $n$ value where $n/v$ versus $n$ curve displayed an inflection point.

This trend of the net crushing energy may be one reason why $n/v$ versus $n$ curves have an inflection point where $n/v$ is minimum for certain rocks.

### 14.4.2.3 Screen Analysis
Crushed materials obtained from CRS tests for each number of drops were combined and mechanically screened using a complete set of USA standard testing

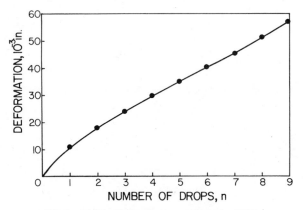

**Figure 14.26** Calibration curve (Selcuk, 1981).

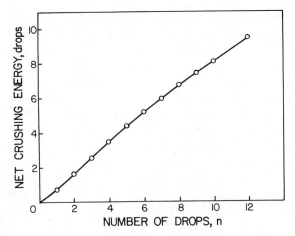

**Figure 14.27** Variation of net crushing energy with number of drops (Selcuk, 1981).

sieves. Material retained on each screen was weighed, the weight percents were calculated, and the cumulative weight percentage was determined ($-270$ mesh) to obtain the cumulative size distribution curves (Figure 14.29). These curves yielded straight parallel lines except for the upper portion of the curves. The particle size distribution for drilling is shown for comparison and is also almost an order of magnitude finer.

A statistical regression analysis showed the applicability of the Schuhman's size distribution equation to the drop crushing that was tested with satisfactory results. The slope of the regression lines ($\alpha$) showed some variation for the small number of drops, and the size modulus $k$ (the largest size fraction) decreased with increasing number of drops.

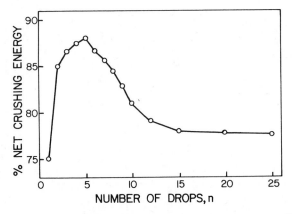

**Figure 14.28** Percent net crushing energy as a function of number of drops (Selcuk, 1981).

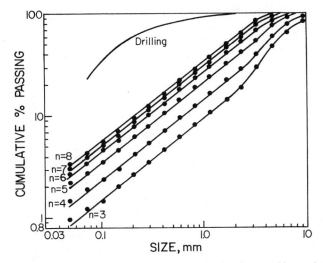

**Figure 14.29**  Cumulative size distribution curves for granite for drop crushing and drilling (Selcuk, 1981).

According to Charles (1957), the plot of log $k$ versus log $E$, the log of size reduction energy, should yield a straight line whose slope is $(1 - m)$. This hypothesis was tested for the data plot of log $K$ against log $E$, net crushing energy, yielding a straight line with a slope of $-0.81$ (Figure 14.30). Therefore, the value of $m$ in the CRS tests of Colorado red granite was 1.81.

Based on these experimental results, it is reasonable to state that the CRS comminution process nearly follows the Rittinger's law ($m = 0.81$ to $2.00$) for rocks similar to Colorado red granite.

### 14.4.2.4  Conclusions
The USBM method is not adequate to analyze the meaning and applicability of the Coefficient of Rock Strength, and it is shown by Brook (1977) that not all rock types display the inflection point on their $n/v$ versus $n$ curve. The need for a more definitive analysis with the CRS still remains.

A new expression of the relationship between volume of fines and number of drops in the CRS test has been derived based on experimental findings.

## 14.4.3  Specific Crushing Energy—CRS

To determine how much energy is used in crushing in a CRS test, it was necessary to determine the relationship between net specific crushing energy and input (drop weight) energy (Selcuk, 1981). If it is assumed that Rittinger's law is applicable to the CRS tests, the new specific surface of crushed rock can be used. Paithankar and Misra (1976–1980) used the procedure described below and concluded that

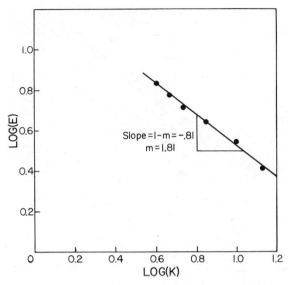

**Figure 14.30**   Log $K$ versus log $E$ (Selcuk, 1981).

specific surface energy of rock crushed in Protodiakonov tests was almost constant per drop for a give rock. Total surface area per unit mass was calculated using the following:

$$A_s = \frac{6}{\rho} \sum_i \frac{\lambda_i P_i}{x_i} \qquad (14.31)$$

where $A_s$ = total surface area per unit mass, $\lambda_i$ = shape factor, $P_i$ = ratio of mass of a screen fraction to the total mass of the sample, $x_i$ = average size of the particles in the $i$th fraction, and $\rho$ = density of rock.

Similarly, the initial specific surface area of the rock sample was calculated from

$$A_i = 6\lambda_o / \rho \cdot x_o \qquad (14.32)$$

where $A_i$ = initial specific surface per unit mass, $\lambda_o$ = shape factor, $x_o$ = size of initial rock sample, and $\rho$ = density.

New specific surface area is $A_s - A_i$. Total new surface area is then $(A_s - A_i)$ multiplied by the total mass of the sample. The shape factor ($\lambda$), defined as the ratio of actual surface per unit volume of sample to the surface area of spheres of the same size per unit volume, was experimentally determined by the air permeability method. The specific surface energy was then taken as the *total* impact energy divided by the total new surface area.

The plot of specific surface energy versus number of drops gave a flat trend except for very low values of $n$. It was concluded that the specific surface energy per drop was virtually constant for a given rock, and thus it was a better drillability index than the Coefficient of Rock Strength. This conclusion may be subject to considerable error.

This can be expressed with one equation. From equations 14.31 and 14.32

$$A_s - A_i = \frac{6}{\rho}\left[\sum_1 \frac{\lambda_i P_i}{x_i} - \frac{\lambda_o}{x_o}\right] \tag{14.33}$$

is obtained.

Total new surface area

$$A_t = w_t(A_s - A_i) = w_t \frac{6}{\rho}\left[\sum_1 \frac{\lambda_i P_i}{x_i} - \frac{\lambda_o}{x_o}\right] \tag{14.34}$$

where $w_t$ = total weight of sample.

Then, the total specific surface energies:

$$SE = \frac{E_T}{w_t \dfrac{6}{\rho}\left[\sum_1 \dfrac{\lambda_i P_i}{x_i} - \dfrac{\lambda_o}{x_o}\right]} \tag{14.35}$$

Paithankar and Misra (1980) erroneously assumed that all of the impact energy was used in crushing. Thus, equation 14.35 must be written as follows:

$$SE = \frac{E_n}{w_t \dfrac{6}{\rho}\left[\sum_1 \dfrac{\lambda_i P_i}{x_i} - \dfrac{\lambda_o}{x_o}\right]} \tag{14.36}$$

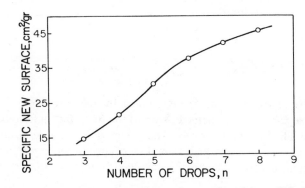

**Figure 14.31**   Plot of specific new surface against number of drops (Selcuk, 1981).

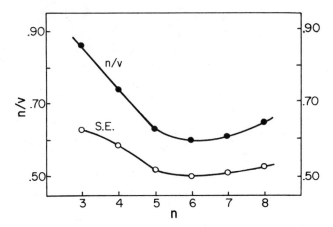

**Figure 14.32** Comparison of $n/v$ versus $n$ curve with specific energy versus $n$ curve for Colorado red granite (Selcuk, 1981).

where $E_n$ is the *net* crushing energy. This equation was used to calculate the specific energy of crushed Colorado red granite. A shape factor of $\lambda = 1$ was used, with the assumption that it is constant for all sizes. A plot of specific surface area against input energy (Figure 14.31) shows a trend similar to the $v$ versus $n$ and the net crushing energy versus number of drops curves (Figure 14.32). Most important of all, the specific energy versus input energy follows the same trend as for the $n/v$ versus $n$ curve.

The specific energy per unit size reduction of Colorado red granite is not constant as indicated by Paithankar and Misra (1980), but rather has an inflection point where the specific energy is a minimum. Thus, while the CRS as measured by standard tests does show a good correlation with drillability in brittle rocks, inasmuch as the Protodiakonov apparatus absorbs up to 20% of the drop weight energy, equation 14.3 relating $E_v$ with $f'$ and the interpretation of Figure 14.32 must be modified accordingly to improve its accuracy.

## 14.5   ROCK CRUSHING AND CHIP FORMATION

Efficient mechanical excavation is determined by the formation of large chips, which in turn depends on the characteristics of the rock, the cutter type, cutterhead disc array, and TBM operating levels. The indentation response or crushing of the rock also affects cutter wear, which should be evaluated separately from processes of chip formation, because the abrasion involved is different from that in the chipping process.

In the study by Nelson et al. (1984), the disc cutter–rock interaction was analyzed to evaluate the rock properties that affect chip formation and cutter wear. Project cutter replacement records were used to evaluate the effect of rock properties on the rate of cutter replacement and disc wear, and TBM muck samples were investigated to analyze the effects of machine operation, design, and rock properties on chip formation. A fracture mechanics approach to chip formation and TBM performance prediction offers some advantages over the use of standard physical properties such as shear strength.

As tool penetration begins, the rock immediately beneath the tool is crushed. As crushing continues, a zone of confined crushed material is formed (Figure 14.33)

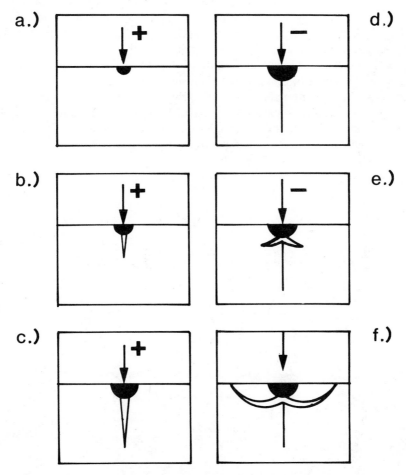

**Figure 14.33**  Fracture mechanics model of indentation and median vent and lateral crack formation (Lawn and Swain, 1975).

in which is created a state of stress that depends on rock and indenter characteristics. That is, the cutter edge angle and rock stiffness determine the stress concentration, which will cause a larger volume of lower-strength rock to fail than that in a higher-strength rock. However, if a rock is highly porous, it will be able to densify, and the level of stress in the crushed zone will be lower. For a dense rock, the failed material may behave dilatantly. The crushed zone continues to expand, until either (1) loading is terminated as the cutter moves or (2) the strength of the rock outside the crushed zone is exceeded and a vent crack develops, usually at the base of the crushed zone (Figure 14.33), where the tensile stresses are highest.

Possible mechanisms for the creation of chips include a limit equilibrium approach in the analysis and an assumed Mohr–Coulomb material failure criterion; others have assumed that chip formation occurs when the shear stresses on a failure plane exceed the shear strength. Korbin (1979) stated that chip formation in brittle rocks is not accurately described by the models in which failure is postulated by shear under high normal stress but that stresses created by the crushed zone material cause radial cracking and that chips are formed by tensile–shear fracture on planes originating at a center of pressure in the crushed zone.

Recent studies using fracture mechanics principles assume that the crushed zone grows until the median vent crack begins to form at a predictable applied load which is a function of the fracture toughness, $K_{Ic}$, of the rock. $K_{Ic}$ is a measure of the stress intensity required to initiate crack propagation. The median vent is a

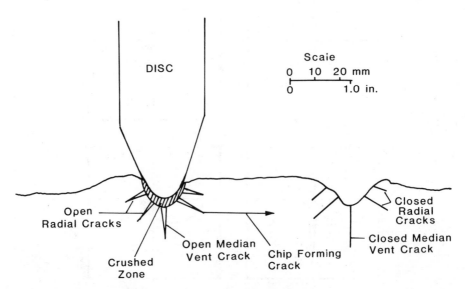

**Figure 14.34**  Cross-sectional view of disc cutter penetration and rock fracture between two disc cutter grooves (Nelson et al., 1984).

stable crack that will propagate straight downward from the load and is, therefore, not responsible for chip formation. Chips may form owing to unloading of the crushed zone as the crushed material starts to relax, and residual tensile stresses may be superimposed and later cracks initiated.

Conway and Kirchner (1980) considered various combinations of loads parallel and perpendicular to the surface treated as quasi-static loading conditions. They demonstrated that tensile stresses developed both in front of and perpendicular to the path of a moving indenter and caused chip formation by a rolling disc cutter.

With an idealized geometry (Figure 14.34) a chip may not form with every cutter pass but by multiple passes. Fractures are initiated by high tensile stresses, the direction of propagation is controlled by the tensile and shear stresses at the crack tip, and the crack propagates between cutter grooves.

The crushing is important in two ways:

1. Abrasive wear is caused by material in contact with the cutter, and, therefore, the mineralogy, grain size, and the state of stress in the crushed zone affect abrasion.
2. The energy required for excavation depends on the processes of chip formation and crushing.

Hypothetically, cutter wear may be reduced by keeping the contact between cutter and crushed zone to a minimum. For low-strength porous rock the crushed zone may become large before chips form, and, if the crushed material is abrasive, cutter wear can be severe. On the other hand, for brittle rock in which fractures can propagate between adjacent cutter grooves, the crushed zone may be of limited size. Crushing consumes more energy per unit than chip formation.

## 14.6  TBM PENETRATION PREDICTION WITH ROCK FRACTURE PARAMETERS

Nelson et al. (1984) selected three rock types likely to have formed chips (Figure 14.34): the Reynales limestone and the Romeo and Markgraf members of the Joliet Formation. Each is a relatively isotropic, crystalline limestone of dolostone of low porosity, beds are about 12 in. (305 mm) or greater in thickness, and the spacing of other discontinuities is many times the bedding thickness. The muck produced in the TBM excavation of these rocks contained many large chips, sometimes greater than 6 in. (152 mm) in maximum dimension.

Both fracture and chip formation consume energy in the creation of surface area. The fracture property used as a measure of such surface energy is $G_{I_c}$, defined as the critical energy release rate, or the critical crack driving force. Thus, it is

**TABLE 14.1. Modulus, Poisson's Ratio, Fracture Properties, and Machine Penetration Indices for Limestone and Dolostone**

| Rock Unit | Elastic Modulus $E_{tso}$ ksi (MPa) $\times 10^3$ | Poisson's Ratio $\nu$ | Fracture Toughness $K_{IC}$ ksi $\sqrt{in.}$ (MPa $\sqrt{m}$) | Critical Energy Release Rate $G_{IC}$ $\frac{lb}{in.}\left(\frac{N}{M}\right)$ $\frac{in.\,lb}{in.^2}$ or $\left(\frac{mN}{m^2}\right)$ | Average Cutter Thrust kips (kN) | Penetration Rate in./rev (mm/rev) | Field Penetration Index $R_f$ kips/in. (kN/mm) |
|---|---|---|---|---|---|---|---|
| Reynales limestone | 7.1 (49) | 0.23 | 1.88 (2.07) | 0.47 (83) | 31.8 (141) | 0.27 (6.8) | 119 (20.9) |
| Romeo dolostone | 13.1 (90) | 0.22 | 2.25 (2.47) | 0.37 (65) | 32.6 (145) | 0.32 (8.0) | 103 (18.0) |
| Markgraf dolostone | 8.8 (61) | 0.19 | 1.64 (1.80) | 0.29 (51) | 30.9 (137) | 0.37 (9.3) | 85 (14.8) |

*Source:* Nelson et al. (1984).

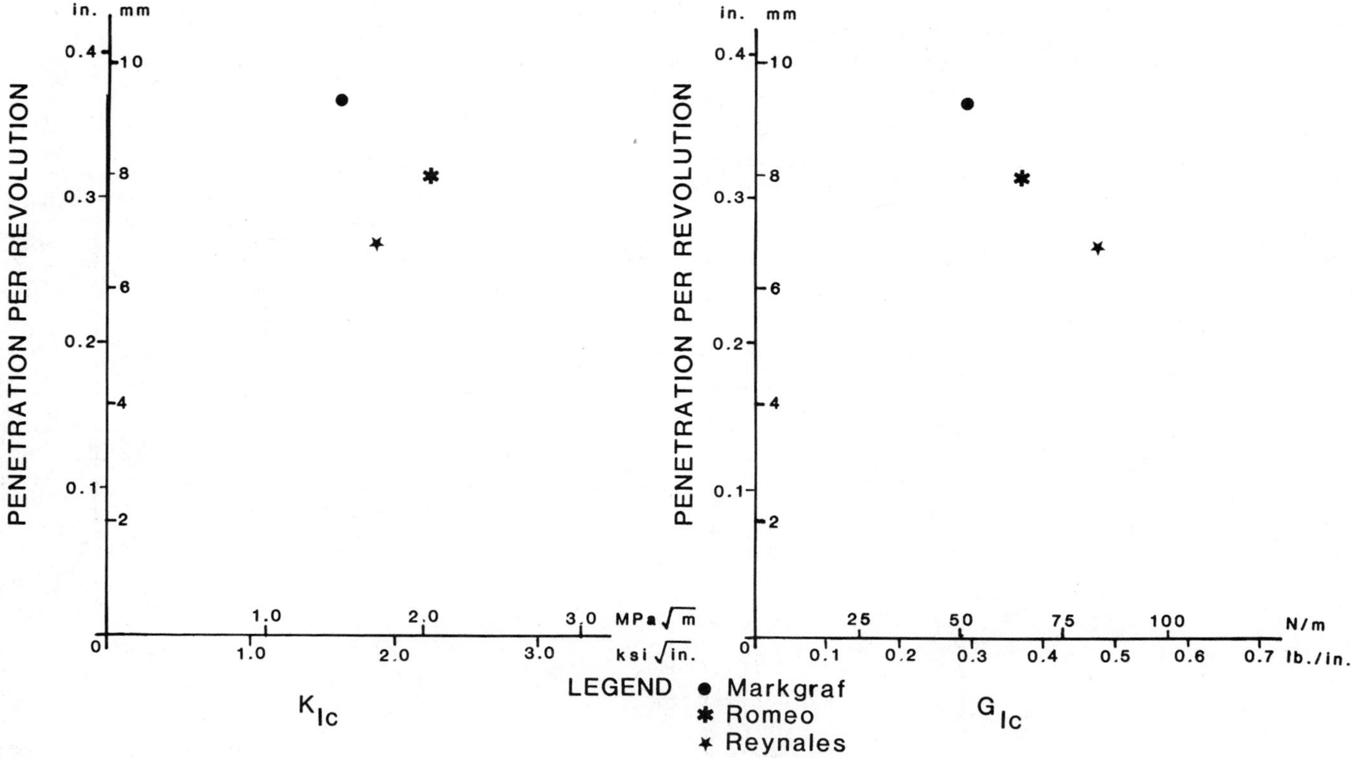

**Figure 14.35** Plot of penetration rate versus fracture toughness, $K_{Ic}$ (Nelson et al., 1984).

**Figure 14.36** Plot generation rate versus critical energy release rate, $G_{Ic}$ (Nelson et al., 1984).

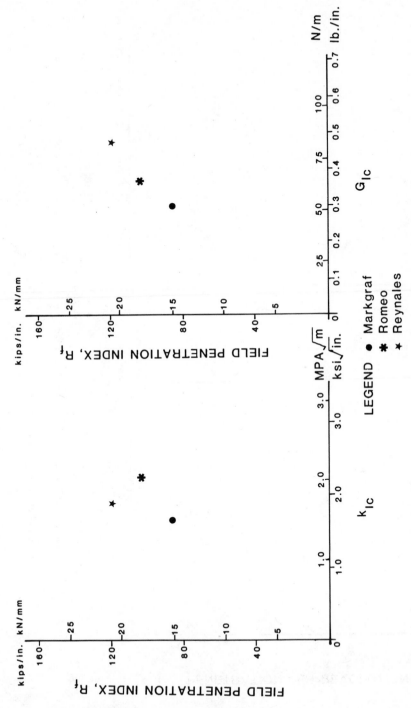

**Figure 14.37** Plot of field penetration index versus fracture toughness, $K_{Ic}$ (Nelson et al., 1984).

**Figure 14.38** Plot of field penetration index versus critical energy release rate, $G_{Ic}$ (Nelson et al., 1984).

assumed that, if chipping is a fracture process, then $G_{I_c}$ may be used for penetration rate prediction.

The value of $G_{I_c}$ is given by

$$G_{I_c} = \frac{K_{I_c}^2 (1 - \nu^2)}{E} \qquad (14.37)$$

where $\nu$ is Poisson's ratio and $E$ is the modulus of elasticity, both evaluated from uniaxial compressive strength testing results. The fracture toughness, $K_{I_c}$, is a measure of stress intensity required to initiate and propagate a fracture and has units of ksi $\sqrt{\text{in.}}$ ($MPa \sqrt{m}$) (Table 14.1).

These may be compared with cutter thrusts, penetration rates, and field penetration indices, $R_f$. The penetration rates and $K_{I_c}$ (Figure 14.35) show no simple mathematical relationship, whereas $G_{I_c}$ values (Figure 14.36) show a linear trend. Thus, the critical energy release rate, $G_{I_c}$, may be a useful predictor of penetration rate. The relation of $R_f$ values with $K_{I_c}$ (Figure 14.37) and $G_{I_c}$ (Figure 14.38) show no trend with the $K_{I_c}$, whereas a linear variation of $R_f$ with $G_{I_c}$ is apparent.

The average thrust per cutter for the three rock types was nearly constant (Table 14.1), and the torque capacities of the TBMs were similar. This indicates that the variation in penetration rate from 0.27–0.37 in./rev. (6.8–9.3 mm/rev.) must be due to changes in rock properties. The data for $G_{I_c}$ (Figures 14.36–14.38), which includes rock strength and stiffness factors, show that it may be related to the penetration rate for brittle, relatively high strength rock.

It is postulated that correlations between penetration rate and rock fracture indices are not likely to be demonstrated for rock types that are thinly bedded or highly jointed. In such rocks, some of the surface area required with chip formation may already exist in the rock discontinuities.

## 14.7 MUCK DISTRIBUTION AND TBM OPERATION

The relationships between muck size, machine operating factors, and rock properties give further quantitative information about the rock-cutting process for muck samples obtained at each of six tunnel sections (Nelson et al., 1984). Water content varied between 4 and 6% by weight. Each sample was wet-sieved by hand through No. 4 and No. 200 sieves (0.19 in. (4.76 mm) and 0.0029 in. (0.074 mm)). The muck retained on the No. 4 sieve was dried and sized with a series of sieves from the 1.5 in. (38.1 mm) to the No. 4 sieve. The material on the No. 200 sieve was also analyzed. The amount of particles finer than 200 mesh was determined by the weight loss during the initial wet sieving. The size of the largest muck chip in each sample was measured, and the number of chips on the 1.5-in. (38.1-mm) and 1.0-in. (25.4-mm) sieves was counted.

**TABLE 14.2. Summary of Results of Muck Gradation Analyses**

| Location | Tunnel | Station | Coarseness Index | Percent of Sample Passing #4 (476 mm) Sieve | $d_{50}$ in. (mm) | Largest Chip Dimension in. (mm) | Number of Chips Retained on 1.0 in. (25.4 mm) Sieve per lb |
|---|---|---|---|---|---|---|---|
| Buffalo | C11 Outbound | 61+75 | 172 | 40 | 0.33 (8.4) | 3.0 (76) | 1.7 (0.39) |
| | | 67+45 | 186 | 28 | 0.37 (9.5) | 3.8 (97) | 0.6 (0.14) |
| | | 91+60 | 159 | 38 | 0.30 (7.7) | 4.8 (122) | 1.0 (0.23) |
| | | 98+50 | 296 | 10 | 0.79 (20.0) | 4.0 (102) | 3.8 (0.86) |
| | C11 Inbound | 29+00 | 171 | 34 | 0.41 (10.5) | 3.3 (84) | 0.7 (0.15) |
| | | 74+00 | 316 | 7 | 0.83 (21.0) | 4.4 (112) | 3.2 (0.71) |
| | | 90+10 | 179 | 39 | 0.32 (8.0) | 5.5 (140) | 0.91 (0.20) |
| | | 100+69 | 285 | 8 | 0.72 (18.2) | 4.0 (102) | 2.9 (0.66) |
| | C31 Outbound | 71+30 | 141 | 45 | 0.24 (6.0) | 2.8 (71) | 0.7 (0.17) |
| | C31 Inbound | 39+07 | 156 | 45 | 0.25 (6.4) | 3.3 (84) | 1.4 (0.31) |
| | | 42+61 | 171 | 40 | 0.30 (7.5) | 3.5 (89) | 1.7 (0.39) |
| | | 50+50 | 122 | 55 | 0.16 (4.0) | 4.3 (109) | 0.6 (0.14) |
| | | 63+15 | 143 | 49 | 0.22 (5.6) | 3.8 (97) | 0.7 (0.16) |
| | | 66+80 | 173 | 38 | 0.33 (8.4) | 6.0 (152) | 1.3 (0.29) |
| | | 70+70 | 166 | 39 | 0.33 (8.3) | 3.8 (97) | 1.1 (0.25) |
| Rochester | Culver–Goodman | 14+81 | 129 | 61 | 0.04 (1.0) | 5.5 (140) | 1.1 (0.25) |
| | | 62+00 | 181 | 43 | 0.32 (8.0) | 4.0 (102) | 1.8 (0.40) |
| Chicago | TARP | Muck pile | 187 | 41 | 0.24 (6.0) | 5.6 (142) | 2.1 (0.47) |

Source: Nelson et al. (1984).

**TABLE 14.3. Summary of TBM Operating Levels, Rock Support, and Rock Unit at the Heading During Muck Sampling**

| Location | Tunnel | Station | Average Cutter Thrust Face kips (kS) | Number of Cutterhead Motors | Operating Average | Rock Support | Rock Unit[a] |
|---|---|---|---|---|---|---|---|
| Buffalo | C11 Outbound | 61+75 | 27.8 (124) | 6 | — | Bolts | 40% Oatka |
| | | 67+45 | 11.4 ( 51) | 6 | Low speed[b] | Steel sets | 40% Oatka |
| | | 91+60 | 24.0 (107) | 6 | — | Bolts/straps | 80% Oatka |
| | | 98+50 | 25.9 (115) | 6 | — | Bolts/straps | 60% Oatka |
| | C11 Inbound | 29+00 | 22.8 (101) | 6 | — | Bolts | 100% Oatka |
| | | 74+00 | 25.3 (113) | 6 | — | Bolts/straps | 40% Oatka |
| | | 90+10 | 17.7 ( 79) | 6 | — | Steel sets | 80% Oatka |
| | | 100+69 | 26.5 (118) | 6 | — | Bolts/straps | 50% Oatka |
| | C11 Outbound | 71+30 | 21.8 ( 93) | 6 | 150 | Bolts | 60% Oatka |
| | C31 Inbound | 39+07 | 18.6 ( 83) | 4 | 200 | Steel sets | 50% Oatka |
| | | 42+61 | 17.2 ( 77) | 5 | 200 | Bolts | 50% Oatka |
| | | 50+50 | 18.6 ( 83) | 5 | 200 | Bolts | 75% Oatka |
| | | 63+15 | 16.5 ( 73) | 4 | 175 | Bolts | 60% Oatka |
| | | 66+80 | 15.8 ( 70) | 4 | 175 | Bolts | 60% Oatka |
| | | 70+70 | 18.6 ( 83) | 4 | 170 | Bolts | 60% Oatka |
| Rochester | Culver–Goodman | 14+81 | 26.3 (117) | 4 | 175 | Bolts/straps | Lower Grimsby |
| | | 62+00 | 23.3 (184) | 5 | 230 | Bolts/straps | Upper Grimsby |
| Chicago | TARP | Muck pile | — | | — | — | Romeo and Markgraf |

[a]For the Buffalo tunnels, only the percent of the tunnel face in the Oatka Member of the Bertie Formation is listed; the remainder of the rock face was composed of the Falkirk Member of the Bertie Formation.

[b]The lower of the two cutterhead rotation rates was used when mining in this area.

*Source:* Nelson et al. (1984).

A log–log plot of percent material retained versus the sieve opening size shows the cumulative distribution and the median particle diameter of the sample (Tables 14.2, 14.3) with the penetration rate, average thrust per cutter, number of operating motors, amperage, initial rock support, and the rock type in the heading. A coarseness index was determined by summing the cumulative percentage weight retained on the sieves from the 1.5-in. (38.1-mm) sieve to the No. 4 sieve (4.76 mm). Since there were five sieves in this range used in the study by Nelson et al. (1984), the maximum coarseness index, CI, could be 500, indicating that the cumulative percentage weight on each sieve was 100%, or all retained on the largest sieve. A CI of zero indicates that all of the sample passed through the No. 4 (4.76-mm) sieve.

The particle size distributions of 14 of the 18 samples (Table 14.2) are very similar, the coarseness indices varying only from about 120–190. The coarseness index and other size distribution factors are largely independent of the penetration rate, cutter loads, operating torque, and cutterhead rotation speed. The muck from stations 74+00 and 100+69 in the C11 inbound tunnel and at station 98+50 in the C11 outbound tunnel contained relatively coarse particles probably owing to the jointing of the rock, whereas that from station 14+81 in the Culver–Goodman tunnel is characterized by finer particles, which may be due to both the worn condition of the cutters and the poor cementation of the rock.

The cumulative weight percentages for 14 samples with similar characteristics fall within a fairly limited zone which only approximates the Schuhmann distribution equation. The average cutter thrust for these samples varied from 11.4 kips (51 kN) to more than 27.8 kips (124 kN), and the average penetration rate varied from 0.20–0.42 in./rev. (5.1–10.7 mm/rev.). Thus, the data do not show that significant changes in particle size occur as a function of either thrust or penetration. Hence, the muck sample evaluations from these rock formations do not show increased particle size with increased thrust and penetration as proposed by some investigators.

The size distribution of the sample from station 74+00 in the C11 inbound tunnel was similar to those obtained at station 100+69 in the same tunnel and at station 98+50 in the C11 outbound tunnel. The rock types at these three locations were similar to those at other places in the tunnels, but, at these locations, the average bedding thickness of the Falkirk Member was 4–6 in. (102–152 mm), and the beds of both the Falkirk and Oatka Members were densely cut by short, discontinuous joints. For this rock condition, the rock fragments bounded by the discontinuities could be mined without extensive chipping, and the muck consequently had a low content of fines.

Of the two muck samples from the Grimsby Sandstone, the particle size distribution of the sample taken at station 62+00 occurs within the shaded zone (Figure 12.16). The curve for the gradation of the sample taken at station 14+81 is above the others because it contained a high percentage of fine material. The differences between these two distributions result from the change in rock type. The lower

Grimsby is a poorly cemented, coarse-grained sandstone which was easily crushed under the cutters. On the other hand, the upper Grimsby is a well-cemented, medium- to fine-grained sandstone which chipped more readily. Further, the cutter wear in each rock varied markedly and may have affected the particle size. That is, at station $14+81$, the cutters were well worn, and most of them were replaced during the next shift, whereas the cutters at station $62+00$ had just been replaced.

# 15

# GRAVITY EFFECTS ON FRAGMENTED ROCK MATERIALS

When rock is fragmented, either by natural or man-induced processes, there are many situations, such as those found in spoil piles, broken ore masses in mining caving systems, rock slides, broken rock masses, explosive crater slopes, and similar, related cases, where the stability and flow of the rock are determined largely by the force of gravity. Where fragmented rock, including that which is jointed and fractured, forms part of underground structures, either the loads on such structures are directly gravity-induced, or the stress field is due to the gravity force on the overlying rock.

For experimentation with gravity-loaded rock structures, loads applied by standard testing machines do not give a satisfactory simulation, and hence, where possible, geotechnical centrifuges have been used to determine the behavior of rock and soil where the principal load is due to gravity. Though research on soils has been carried out in several countries, typical work on spoil piles in the Soviet Union has been summarized by Malushitsky (1981), other research in soil mechanics in other countries has been reported at several conferences (e.g., Craig, 1984). Significant research has also been done in rock mechanics (Clark, 1981), some of which is concerned with the stability of broken and jointed rock masses. Another stability condition that is of importance in gravity loading is the arching of geologic materials. All of the reported research in this area has dealt only with the arching of sands under applied loads. The true response factors of broken rock masses will only be determined experimentally where gravity loading is simulated. As is emphasized later in the development of the theory of centrifugal loading, there are several reasons why applied loading does not simulate body force or gravity loading.

Centrifugal loading has been applied in many areas of science and engineering, particularly in the high-$g$ loading of space vehicle instruments and components. It has also been applied successfully to static and dynamic problems in soil and rock mechanics, as well as bubble and wave behavior occurring in underwater explosions.

## 15.1  PRINCIPLES OF CENTRIFUGAL TESTING

The principles involved in the solution of problems of modeling the behavior of earth and rock models in their response to gravity are well known to those who have done research in this area of engineering. The research on rock beams by Bucky and Panek and on rock masses by Hoek (Clark, 1981) successfully demonstrated the applicability of centrifugal testing to rock mechanics, and this type of research in soil mechanics has become well established in science and engineering. Only limited research has been done in the centrifugal testing of coarsely fragmented rock in the various conditions in which it is found naturally or created by excavation operations.

One of the main advantages of centrifugal testing is found in the fact that, if the body force on the model is increased in inverse proportion to the scale ratio, the same material may be used in the model as is found in the prototype, provided the grain size is not too large. Although complete modeling of all of the detailed behaviorisms of earth materials cannot be simulated in centrifugal testing, many of the critical factors related to engineering behavior can be found by such experimentation. In the event that it is desired, and if the capabilities of a given centrifuge permit, it is possible to carry out tests at a multiple of $g$'s other than the $n$ factor prescribed by the properties of the material. Then another material may be used, which involves the "modeling of models." That is, if the natural materials are not to be used in modeling, then the size of the model, the number of $g$'s, and the properties of the materials can theoretically be varied to obtain the desired results. However, the adjustment of the properties of artificial modeling materials is a difficult problem which requires considerable painstaking research and experimentation. Also, distorted modeling may be employed. This latter practice is more common in engineering design than is true modeling—that is, in ship and aircraft design, for example.

## 15.2  ELASTIC SIMILITUDE LAWS*

Dimensional analysis may be used to obtain the general (ideal) form of the equations determining the relationship between a model and its prototype. The structural geometry of the model and prototype must be similar in three dimensions, and the other parameters are the material elastic properties and the nature of the applied loads, whether uniform, concentrated, or body loads. Temperature effects are usually neglected.

The behavior of a point in an earth mass of coordinates $x$, $y$, and $z$ depends on the geometry of the structure—that is, its shape, the shapes of its constituent members, and so forth, which, in turn, are defined by $L$ (a typical dimension) and $_rL$ (a set of dimensionless ratios relating other dimensions, such as span, depth, width or possible grain size). For ideal analysis the properties of the material are $\rho$ (density), $E$ (Young's modulus), and $\nu$ (Poisson's ratio).

If the structure is made up of more than one material or the properties, or state of stress, depend on direction or position in the structure, then these properties are related to those at a given point by dimensionless ratios $_r\rho$, $_rE$, and $_r\nu$. Mechanical stresses in the model are those due to externally applied loads, and those internally applied and stress are due to gravitational loads or body force (Hoek, 1965):

$Q$ = an externally applied load,

$P$ = an externally applied stress,

$\sigma$ = an internal stress,

$g$ = the acceleration of the entire body which results in gravitational body forces,

$u_o$ = a displacement of a point in the structure,

$\gamma$ = an acceleration imposed upon and unit mass of the structure, and

$\epsilon$ = strain at a point, dimensionless.

*15.2 Elastic Similitude Laws from Hoek, (1965).

The applied loads, stresses, strains, displacements and accelerations are related to those at a point by the corresponding dimensionless ratios.

Thus, the behavior of a point $x$, $y$, or $z$ at a time $t$ is defined by $\sigma$, the stress, and $u$, the displacement. The magnitudes of $\sigma$ and $u$ are functions of the parameters $x$, $y$, $a$, $t$, $L$, $\rho$, $E$, $Q$, $P$, $\sigma_o$, $g$, $u_o$, and $\gamma$ and the dimensionless ratios. These quantities are functions of the three fundamental units of length, $L$, mass, $M$, and time, $t$. Poisson's ratio is dimensionless and is considered with the other dimensionless ratios such as dimensionless strain. The dimensions of the 15 parameters (Table 15.1) give a matrix of rank 3, from which 12 dimensionless independent products are obtained:

$$\frac{u}{L}, \frac{\sigma L^2}{Q}, \frac{x}{L}, \frac{y}{L}, \frac{z}{L}, \frac{t\gamma^{1/2}}{L^{1/2}}, \frac{EL^2}{Q}, \frac{\rho\gamma L^3}{Q}, \frac{g}{\gamma}, \frac{PL^2}{Q}, \frac{u_o}{L}, \frac{\sigma_o L^2}{Q}$$

Other sets are possible, and the maximum control over the dimensionless products is obtained if the original variables can be regulated so that they occur in only one product. For example, $E$ occurs in one group that can be regulated by changing $E$, and none of the others are affected. It is not possible to include each variable in only one quantity, and one must choose the variables and their grouping.

Buckingham's second theorem states that a dimensionally homogeneous equation relates the displacement $u$ to the other parameters:

$$u = f\left(x, y, z, t, L, \rho, E, Q, P, \sigma_o, g, r, \nu, {}_rL, \cdots , {}_r\nu\right) \qquad (15.1)$$

which may be reduced to

$$\frac{u}{L} = F\left(\frac{x}{L}, \frac{y}{L}, \frac{z}{L}, \frac{t\gamma^{1/2}}{L^{1/2}}, \frac{EL^2}{Q}, \frac{\rho\gamma L^3}{Q}, \frac{g}{\gamma}, \frac{PL^2}{Q}, \frac{u_o}{L}, \frac{\sigma_o L^2}{Q},\right.$$
$$\left. \nu, {}_rL, {}_rE, {}_r\nu, {}_rQ, {}_rP, {}_ru_o, {}_r\rho, {}_r\sigma \right) \qquad (15.2)$$

Also, the stress may be related to the other parameters by

$$\frac{\sigma L^2}{Q} = F^1\left(\frac{x}{L}, \cdots , {}_r\gamma\right) \qquad (15.3)$$

where $F$ and $F^1$ are undetermined functions.

For a model and prototype to be physically similar, the arguments of the functions $F$ and $F^1$ must be the same. The equality of $x/L$, $y/L$, and $z/L$ represents the same point in the model and prototype, and the equality of ${}_rL$ implies geometrical similarity. The equality of the ratios ${}_rE$, ${}_r\nu$, and ${}_r\sigma$ stipulates the same

**TABLE 15.1. Rank Matrix**

|   | $u$ | $\sigma$ | $x$ | $y$ | $z$ | $t$ | $E$ | $\rho$ | $g$ | $P$ | $u_o$ | $\sigma_o$ | $L$ | $Q$ | $\gamma$ |
|---|---|---|---|---|---|---|---|---|---|---|---|---|---|---|---|
| $L$ | 1 | $-1$ | 1 | 1 | 1 | 0 | $-1$ | $-3$ | 1 | $-1$ | 1 | $-1$ | 1 | 1 | 1 |
| $M$ | 0 | 1 | 0 | 0 | 0 | 0 | 1 | 1 | 0 | 1 | 0 | 1 | 0 | 1 | 0 |
| $t$ | 0 | $-2$ | 0 | 0 | 0 | 1 | $-2$ | 0 | $-2$ | $-2$ | 0 | $-2$ | 0 | $-2$ | $-2$ |

distribution of material properties, and the applied loads, stresses, accelerations, and displacements have the same distribution in both systems.

The dimensionless Poisson's ratio cannot be included in the dimensionless groups, and differences in Poisson's ratio in the model and prototype cannot be compensated for by changes in the other parameters. For example, changes in $E$ can be balanced by changes in $Q$ or $L$ to maintain the equality of the group $EL^2/Q$. Hence, Poisson's ratio in model and prototype must be the same, which can contribute to difficult problems in elastic model analysis and construction. Model materials that conform to other requirements seldom have the same value of Poisson's ratio, but the effect on stress distribution is generally small and can be neglected.

The remaining dimensionless groups or combinations permit some latitude in choice of scales, materials, and loading conditions. Provided that the equality of the functions $F$ and $F^1$ is maintained, the parameters can be chosen within practical ranges. Again, it must be emphasized that, if loading conditions permit the use of the same material in model and prototype, then similarity is automatically obtained, with few exceptions.

For an elastic body in thermal equilibrium, three material properties will establish similitude—namely, $E$, $\rho$ and $\nu$. Poisson's ratio of model and prototype must be the same:

$$\nu_m = \nu_p \qquad (15.4)$$

where subscripts $m$ and $p$ refer to model and prototype, respectively. The other properties $E$ and $\rho$ can be related by

$$\frac{\rho\gamma L^3}{Q} \cdot \frac{g}{\gamma} \cdot \frac{Q}{EL^2} = \frac{\rho g L}{E} \qquad (15.5)$$

which results in

$$\frac{\rho_m g_m L_m}{E_m} = \frac{\rho_p g_p L_p}{E_p} \qquad (15.6)$$

or

$$\frac{L_p}{L_m} = \frac{E_p}{E_m} \cdot \frac{\rho_m}{\rho_p} \cdot \frac{g_m}{g_p} \qquad (15.7)$$

This holds for the condition that the stress $\sigma$ and the strain $\epsilon$ be identical at similar points in model and prototype.

In experimental model studies, it is not possible to obtain an exact reproduction of the prototype stresses and displacements in the model. To satisfy the sensitivity requirements of measuring techniques, it may be necessary to load the model so that the model stresses and displacements are a fraction or multiple of the prototype values. If this scale factor is $\alpha$, the following conditions can be imposed:

$$\sigma_p = \alpha\sigma_m \qquad (15.8)$$

and

$$u_p = \alpha u_m \qquad (15.9)$$

To maintain equality of the arguments of the functions $F$ and $F^1$, the model and prototype scales become

$$\frac{\alpha\rho_m g_m L_m}{E_m} = \frac{\rho_p g_p L_p}{E_p} \qquad (15.10)$$

or

$$\frac{L_p}{L_m} = \frac{\alpha E_p}{E_m} \cdot \frac{\rho_m}{\rho_p} \cdot \frac{g_m}{g_p} \qquad (15.11)$$

This important concept is embodied in equation 15.11. If the gravitational field $g_p$ of the prototype is altered to a value $g_m$ of the model, the model scale may be chosen accordingly. Or, if the prototype material is used in this model, then the model scale is inversely proportional to $g_m$.

Applied or induced stresses are defined by the parameters $Q$, $P$, $\sigma_o$, $u_o$, and $\gamma$. These appear in the dimensionless groups

$$\frac{t^2\gamma}{L}, \frac{EL^2}{Q}, \frac{L^3\gamma\rho}{Q}, \frac{g}{\gamma}, \frac{PL^2}{Q}, \frac{u_o}{L}, \text{ and } \frac{\sigma_o L^2}{Q} \qquad (15.12)$$

This leads to the equation relating model and prototype. From

$$\frac{P}{E} = \frac{PL^2}{Q} + \frac{Q}{EL^2} \qquad (15.13)$$

one obtains the relationships between externally applied loads $Q$ and stresses $P$ in model and prototype:

$$\frac{Q_m}{Q_p} = \frac{E_m L_m^2}{E_p L_p^2} \qquad (15.14)$$

and

$$\frac{P_m}{P_p} = \frac{E_m}{E_p} \tag{15.15}$$

From

$$\frac{\sigma_o}{E} = \frac{\sigma_o L^2}{Q} \cdot \frac{Q}{EL^2} \tag{15.16}$$

the stresses $\sigma_o$ and displacements $u_o$ in model and prototype are related by

$$\frac{\sigma_{om}}{\sigma_{op}} = \frac{L_m}{L_p} \tag{15.17}$$

and

$$\frac{u_{om}}{u_{op}} = \frac{L_m}{L_p} \tag{15.18}$$

Inertia and gravity forces in model and prototypes are characterized by the dimensionless groups

$$\frac{\gamma \rho L^3}{Q} \text{ and } \frac{\gamma}{g} \tag{15.19}$$

which were used in deriving equation 15.5, and which governs gravity forces in the model.

The time scale can be derived from

$$\frac{\rho L^2}{Et^2} = \frac{\rho \gamma L^3}{Q} \cdot \frac{Q}{EL^2} \cdot \frac{L}{\gamma L^2} \tag{15.20}$$

resulting in

$$\frac{t_m}{t_p} = \left[ \frac{\rho_m E_p}{\rho_p E_m} \cdot \frac{L_m}{L_p} \right] \tag{15.21}$$

In the studies of stresses in earth structures it is generally assumed that they result from gravitational body forces. When an underground excavation, for example, is small in relation to its depth and proximity to other excavations, a reasonably accurate solution of the stress field, before failure, can be obtained by replacing the body forces by externally applied forces. The stress gradient due to increasing depth is ignored, and the condition is governed by equation 15.19.

However, when the excavation extends over an appreciable vertical distance, compared with its depth, or when it is near other excavations, the above simplification is no longer applicable, and body forces must be simulated by body force loading of the models in a centrifuge.

In a photoelastic model, the number of isochromatic fringes is proportional to maximum shear stress $\tau_{max}$, and equation 15.8 may be used to related model and prototype stresses. Thus, equation 15.9 becomes

$$\tau_{maxp} = \alpha \tau_{maxm} \tag{15.22}$$

The centrifugal acceleration to be applied is obtained from equation 15.23:

$$\frac{g_m}{g_p} = \frac{1}{\alpha} \cdot \frac{L_p}{L_m} \cdot \frac{\rho_p}{\rho_m} \cdot \frac{E_m}{E_p} \tag{15.23}$$

As noted previously, for correct similitude Poisson's ratio of model and prototype must be identical. Poisson's ratio for rock usually lies between 0.1 and 0.2, and, since photoelastic plastics for stress freezing tests have a Poisson's ratio approaching 0.5, this condition cannot be satisfied.

However, it is possible to reach some important conclusions from an examination of the stress distributions that can be anticipated in earth structures. An important effect of Poisson's ratio is in the later stresses induced in a body the lateral deformation of which is restrained. The relation between the vertical and the lateral stress is given by

$$\sigma_L = \frac{\nu}{1 - \nu} \cdot \sigma_\nu \tag{15.24}$$

where $\sigma_L$ = the lateral stress, $\sigma$ = the vertical stress, and $\nu$ = Poisson's ratio.

Large differences in the lateral stress may occur with a variation in Poisson's ratio. Correction of the stress field for $\nu$ in the model is time-consuming. One experimental method is to reduce lateral restraint by supporting a model on a liquid-filled rubber bag. The model is then free to deform laterally, and the stress state is uniaxial. For lateral loads, rubber bags can be filled with a liquid of the correct density and placed around the model to induce the required stress and stress gradient.

The stresses around an excavation in a known triaxial stress field are, as a first approximation, independent of the elastic constants of the material. This problem is similar to the problems investigated photoelastically by Clutterbuck (1958) and Fessler and Lewin (1960). It was concluded by them that errors due to Poisson's ratio are smaller than the photoelastic experimental errors, which are about 5%.

## 15.3   BODY FORCE LOADING

Most conventional strength and failure analyses of stresses in engineering structures are concerned with applied forces, and, even though the stresses are really induced by gravity loading, the gravity loads are assumed to act as static applied loads. This usually serves as a useful approximation. Known examples are the loads on structures such as elements of bridges and buildings. Where the stresses in a structure result from some type of dynamic force such as wind, earthquakes, or explosions, the simulation of such stresses by applied static forces is now used for engineering design. This is not only true of synthetic structures but equally true of rock and geologic engineering structures that are subjected to these three types of forces.

In some cases a structure may be subject to more than one of these types of forces simultaneously, and, if the effects of each are linear, the resulting stresses, strains, displacements, and similar quantities may be superimposed. The primary concern herein is with broken or fragmented rock masses. In these types of masses of earth materials the effects of the restraint of the confining mass are also an important factor.

## 15.4   STABILITY OF EARTH WASTE EMBANKMENTS

The stability and behavior of natural and artificial earth slopes, such as relatively intact and broken natural rock slopes, embankments of blasted rock from mining operations, and crater slopes in soil and rock, are determined by the characteristics of the material and its resulting behavior in the earth's gravity field. The only satisfactory method of experimentation with such slopes is by means of a geotechnical centrifuge. The results of such experimentation with soil for civil engineering purposes are too extensive to review here.

However, in the Soviet Union the experimentation with the stability of soil embankments created by surface mining of coal has been intensively investigated. The behavior of such embankments directly involves the strength and failure mechanisms of soil and rock slopes, and the techniques of centrifugal have been successfully applied to each. Malushitsky (1981) presented the results of about 30 yr of Russian research on the stability of man-made embankments, giving methods of testing, the results of research, the solution of the problems of centrifugal testing, and the engineering application of the test results. The centrifuge used had a capacity of 320 $g$'s, an effective radius of 2.5 m, and a model bulk weight capacity of 2.0 tons.

As a result of this Russian experimentation it was concluded that centrifugal experimentation permitted a good initial approximation in the determination of the influence of moisture content, initial density, and rate and method of heap formation in the characterization of mass stability:

1. The stability of a waste heap depends on five basic factors: composition, moisture content, initial density, and rate and method of deposition.
2. The centrifugal method provides an experimental means for determination of the critical height of a waste dump as a function of the five parameters listed above.
3. The test of influence of the five factors gave data for basic reliable engineering approximations.
4. Qualitative differences in experimental results were determined for the failure of clayey and sandy materials versus critical heights, and engineering approximations were established for the field reduction of critical heights of clayey dumps.
5. Test results also established the influence of initial conditions and drainage on soft, moist soils.
6. Further, field conditions were established for the safe deposition of waste dumps on weak (semi-liquid) bases, making it possible to use dumping areas of old hydraulic waste heaps.
7. Centrifugal experimentation established the local conditions and influence of properties of particular clayey materials and sand.
8. It also determined conditions for the stability of clearable waste heaps for the reduction of volume of material required in their reexcavation.

In this research with centrifugal loading, the principles of similitude were followed wherever possible and utilized the force loading of the centrifuge to create simulated field conditions. The magnitude of loading was increased, for example, until slope failure occurred. This method provides a body force testing process which is not possible with conventional testing machines. It was further concluded that ''from a comparison of the natural state with the models the correspondence for the conditions of similarity is confirmed and its degree of accuracy can be considered established by the results of the comparative investigations of models of different scales of the order of 7–8%.'' Also, ''the factorial experiment method enables us to find the minimum critical height of a waste-heap with guaranteed accuracy.''

## 15.5 ARCHING OF GEOLOGIC MATERIALS

Because practically all solid, intact rock is relatively weak in tension, and fractured, fragmented, or particulate rock masses cannot support tensile stresses, a structure of such earth mass either is weak in tension or will not support tensile stresses. Thus, the Mohr–Coulomb criterion is usually applied to materials where a compressive stress exists, which, in turn, assumes that shear exists across a plane at an angle defined by the internal friction. This basic mechanism is critical in the analysis of rock structures subject to either applied loads, body forces, or dynamic loads.

For example, the lower half of a simple intact rock beam loaded by vertical, lateral forces, and by its own weight is usually in tension in the lower part of the beam and can support only nominal tensile stresses. Vertical fractures in the beam result in a different stress distribution, and the loads, then, may be supported by arching.

Much of the stress analysis and investigation of stability of rock masses is made on the assumption that rock behaves in an ideal manner, either elastically, plastically, or in another simplified manner. The response of a geologic structure is due to the weight of the overlying rock, including, in some cases, tectonic forces. Because rocks are not ideally elastic, solutions to problems based on that assumption may give only approximations. In the case of underground openings, the material may be fractured by excavation operations or cut by joints and faults, or may be composed of fragmented or particulate-size breccia, similar to gravels, sands, or soils. The weight of the intradosal material may play an important part in local structure stability and the support required in underground openings, particularly where arching of the material is involved.

One definition of arching is the action taking place in intact or broken geological materials when an opening is made therein and the process of failure of the roof takes place to a point where a condition of stability or equilibrium is reached and failure ceases. In certain types of mining in appropriate geologic structures, this process was studied under "dome theories" (Fayol, 1885; Irving, 1946; Dinsdale, 1940; Mohr, 1956; Isaacson, 1958), where much of the load provided by the material in the dome was considered to be transferred to the abutment of the opening.

The dome theory and equations usually express an induced abutment load only in terms of the weight of the rock included in the dome and, hence, do not include the effects of pressure of all of the overlying material, which is usually considered a vital parameter in studies or arching. This is true particularly in materials such as fine sands, the materials used in most quantitative experiments reported to date.

A necessary condition for geologic materials around underground openings to possess structural stability is that the materials be capable of supporting compressive, shear, and often tensile stress in critical positions in the structure. Gradual failure may be due to flow or creep.

Geologic materials *in situ* vary from massive, very strong rocks that have high strengths to saturated clays that are almost hydraulic in nature with virtually no tensile or shear strength. Most materials over underground openings that possess strength will fail under critical load until they form a stronger arch over an opening within the mass of the material.

Arching has been observed underground in both tunnels and mines (Terzaghi, in Proctor and White, 1977; Isaacson, 1958), experimentally in sands (Terzaghi,

1936; McNulty, 1965), in ore in caving methods of mining (Kvapil, 1973), in stratified rock models loaded in a centrifuge (Sutherland, 1982), and in models of solid rock stressed in a centrifuge (Bucky, 1931; Stephansson, 1971). Research into the behavior of particulate materials in chutes and bins and idealized experimentation and analysis of cylindrical particles have added to the understanding of the behavior of loose materials in arching.

Arching in geological materials may be considered in the following order: (1) dome theory and arching; (2) arches, masonry; (3) arching of fine sand; (4) rock and ore—large fragments; (5) fragmented geologic structures; (6) arching in intact materials—model studies (competent rocks); (7) idealized particulate arching.

### 15.5.1  Dome Theory

Basically, the dome theory assumed that the load of the weight of the rock of a hemispherical shape above the opening is transferred to the abutments. This means that the induced load must be equal to the difference between the column over the opening and the hemisphere, or $\rho l(h - (\pi l)/4)$. When $h = (\pi l)/4$, the load from the intradosal rock should be zero, and the maximum load for constant $h$ is where $l = (2h)/\pi$. Some of the assumptions on which the theory is based, plus the fact that the load becomes negative for large values of $l$, do not agree with observations.

### 15.5.2  Masonry and Boulder Arches

Some of the factors that govern the stability of masonry arches ideally illustrate factors that determine the arching of irregular blocks of ore in chutes and bins, caving masses of ore and rock, and assorted rubble created by explosive processes, as well as the factors of structural stability affecting movement or support during and after excavation.

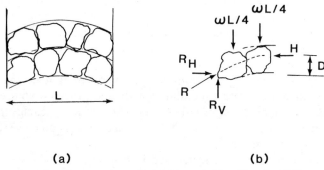

**(a)**                                         **(b)**

**Figure 15.1**   Analysis of a boulder arch (Kvapil et al., 1971).

Simple analysis of a boulder arch (Figure 15.1) shows that it fails in three ways: (1) by blocks falling out owing to tensile stress; (2) by crushing at areas of contact; and (3) by slippage due to low frictional resistance between boulders at the abutment. For stability, the line of action of the thrust between the boulders must create a moment that will hold them in place, much like a keystone in a masonry arch. Similar analyses have been made considering the rock structure as a Voussoir beam by Haycocks (1962), utilizing models tested in a centrifuge. Arch analyses are done largely in terms of empirical rules.

### 15.5.3  Arching of Sand

Most of the research on quantitative arching has been done with relatively firm sand, and the results have been used to predict the response of tunnel excavations in loose materials near the earth's surface. Illustrative experimental methods of arching with sand are the following:

**1.** Use of uniform sand, loaded by its own weight, approximately 40 mesh, in a bin with a trap door in the bottom. Parameters measured were downward movement of the door and the total natural vertical pressure on the door for com-

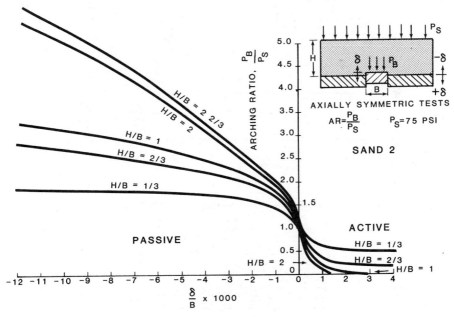

**Figure 15.2**  Comparison of several active and passive arching curves (McNulty, 1965).

pacted and loose sand (Terzaghi, 1936). For small movements of the door of 1 mm, the pressure was reduced by one-half.

**2.** Use of coarse quartz sand 0.8–2.0 mm with surface pressure on the sand and measurement of effects on a buried structure instead of a trap door. Parameters were elastic properties of medium geometry—including depth of sand, shape of buried structure, and elastic properties of structure. It was found that there were increased arching effects with depth of sand, increase of load, higher density and rigidity of sand, and changes in arching with abutment conditions.

**3.** Use of buried structure in 20–30 Ottawa sand, density 96–104 lb/ft$^3$ with surface pressure on a buried disc structure on a vertical cylinder. There was a stress increase on the horizontal disc, called passive arching. This increased with height of cylinder, increased with depth of sand, and decreased with sand bulk density, and active arching of the sand increased with bulk density (Triandafilidis et al., 1964).

**4.** Model studies were made of sand approximately 20–100 mesh in a pressure vessel with cylindrical trap door moving up and down. Arching ratios defined a pressure $P_B$ on door divided by surface pressure $P_S$ measured as a function of deflection of door. Active arching was obtained with release of door, and passive, with raising of door (Figure 15.2) (McNulty, 1965).

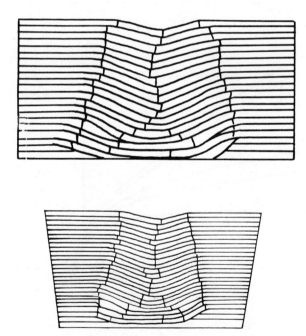

**Figure 15.3** Failure of strata due to body force loading on a centrifuge (Sutherland, 1982).

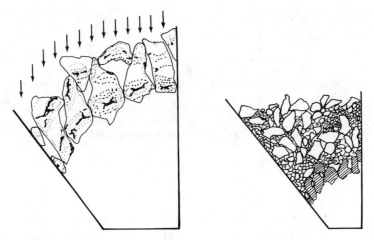

Figure 15.4  Irregular fragment arching in bins (Kvapil et al., 1966).

### 15.5.4  Arching—Bedded and Jointed Structures

Similar arching occurred in the models of stratified rock loaded in a centrifuge (Figure 15.3) (Sutherland, 1982). In processes related to the dome theory described previously, Terzaghi (Proctor and White, 1977) described several types of arching that occur over tunnels. As a result of the experimental findings of arching in sand, it was proposed that loose materials would arch to a height equal to the diameter of a circular tunnel and stratified material in an inverted V shape. In most tunnels in fractured rock, a half-dome is formed over the longitudinal section at the end of the tunnel, and this must be taken into account in providing support.

### 15.5.5  Arching of Blasted Ore Fragments

Kvapil (1966) made extensive qualitative studies of the mechanism of flow and arching of fragments of blasted ore and rock up to 40 in. in diameter as it occurs

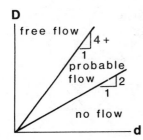

Figure 15.5  Effect of particle size, $d$, in relation to diameter of opening, $D$, on flow conditions of sand in small-diameter pipes (Ayataman, 1960).

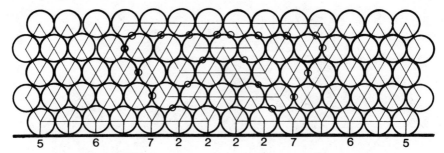

**Figure 15.6** Arching over a hole in the base (Jenkins, 1931).

in sublevel caving and movement of broken material through ore passageways and chutes (Figure 15.4). Parameters in his analysis include the angle of inclination of the sides of a chute, angle of friction, and diameter of fragments. He found that (1) arching restricts or stops gravity flow, (2) the arch is formed by the lowest layer of blocks, (3) the arch is parabolic in form, (4) a low arch is more stable, (5) arches form more easily with irregular particles, (6) arches form more easily in less mobile material, and (7) small openings are more conducive to archng.

Smaller-scale simulated flow in vertical pipes observed by Ayataman (1960) showed that where the diameter $D$ of the pipe was 4.21 times the diameter $d$ of the grains, vertical flow under gravity took place, probable flow took place only between ratios of 2.0 and 4.21, and no flow occurred below 2.0 (Figure 15.5).

### 15.5.6 Arching—Cylindrical Disc Grains

Jenkins (1931) recognized that dilatancy affects the behavior of granular materials and that the pressure system in a granular mass depends on its weight, boundary geometry, the geometry of points of contact, and the coefficient of friction. Changes in volume due to pressure and rearrangement of the grains, or dilatancy, is a critical

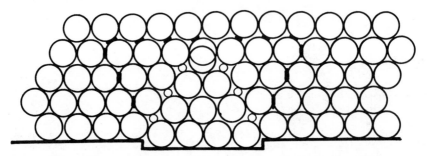

**Figure 15.7** Arching over a hole; solution without friction (Jenkins, 1931).

parameter. It is vital to note that most of the weight of the column of granular material enclosed by vertical walls is borne by the walls and not by the base of the container.

Jenkins (1931) used cylindrical discs of steel to obtain different packaging arrangements to investigate the pressure distributions in two dimensions. For hexagonal packing and where the bottom pressure was released by a hole in the place, similar to the Terzaghi sand experiments, the discs assumed a new pattern, with gaps between a line of discs near the hole (Figure 15.6). The system of forces is determinate (Figure 15.7), and the lines of action form an arch over the hole. The discs forming the arch do not descend, and the disturbance is not propagated upward. Similarly, if the bottom in the middle of the base in a grain silo is moved slightly downward, a single arch formed by gaps appears at the sides.

## 15.6 MATHEMATICAL FLOW MODELS

Although there has been no comprehensive theoretical method established for computing the parameters of muck gravity flow, the development of reliable models and their verification will yield valuable information for pertinent engineering design. This includes computation of the effect of all the important muck flow conditions, studies of the stability and movement of natural slopes, and analysis of the behavior of fragmented masses of rock. Pertinent state-of-the-art research results have been summarized by Yenge (1980, 1981), and recent laboratory research has been done by Peters (1984).

The usual parameters of flow of muck that are subjected to theoretical and experimental study in mining operations deal with (1) the movement of fragments and their flow paths in the possible dilution of ore by waste rock and (2) the clogging and bridging of material in ore chutes. The behavioral mechanics involve the angle of internal friction and the friction between the walls of the passageway and the muck as well as the relative sizes of the fragments and the effective diameter of the chutes. When a chute becomes clogged, it is important in mining and similar operations to devise a means of upsetting the stability of the clogging structure and reestablishing the muck flow. Impact tools or explosives may be used for this purpose.

The behaviorisms of muck movement in commercial mining are similar to those in most industrial excavation processes. Further, most of the experimentation on flow and arching of geologic materials has been done in the earth's gravity field. In model studies, arching effects of horizontal ground pressures were accomplished with applied loads. However, these effects are physically and mechanically different from those of body forces. Also, the effects of scaling of such factors as geometries, fragment sizes, densities, strengths, and friction on muck flow and its simulation have not been addressed in required detail. Thus, the most effective theoretical and experimental approach to the solution of these problems will be by means of large geotechnical centrifuges.

The process of scaling model results to full-size prototype (Peters, 1984) was concerned only with the size of openings, spacing of draw points, and fragment size as related to $1\frac{1}{2}$-in. material.

### 15.6.1   Physical Modeling and Similitude: Centrifugal Testing

As with the design of soil and rock models in multiple-$g$ fields, in fragmented rock the pressures must be the same at corresponding points in the model and prototype. Also, if the grain size does not affect the model behavior, the same material may be used in the model as exists in the prototype. In studies of muck flow it may be necessary to scale fragment size as well as other geometric factors.

Most Swedish and U.S. laboratory experimentation in centrifugal testing has utilized materials of relatively uniform size. Russian research (Malushitsky, 1981) was done on mode spoil piles with natural clays and sands from prototype spoil piles. One-$g$ research, however, could not be used for simulation and scaling of results. Physical models to date, using fragments up to $1\frac{1}{2}$ in. in size, were designed primarily to verify simple theories of single factors. The largest known model to date, at the Colorado School of Mines, consists of a steel frame with transparent plastic front $3 \times 20 \times 15$ ft, with up to four draw points. Batches of uniform crushed igneous rock are loaded by hand with coded fragments of known location, whose path and history of travel are determined as material is drawn. No attempt was made to measure pressure distribution in the material or on the walls or floor of the bin (Peters, 1984).

### 15.6.2   Mathematical Analysis and Modeling

Most of the mathematical analyses to 1984 (Yenge, 1980, 1981; Peters, 1984) have been concerned with the angle of draw, draw-column width with respect to draw-point width, angle of repose, angle of friction of the fragmented mass, and the relation of these parameters to flow characteristics such as the envelope of flow, spacing and size of draw points, size of fragments, flow pattern, and special types of flow such as rat-holing and chimney flow. In some arching tests of sand, it was possible to measure critical pressures, but with larger fragments, new techniques are needed to make meaningful peak and average pressure measurements. Some efforts have also been directed at estimating the ratio of vertical to horizontal pressures, dead zones, conditions for free flow, effect of cohesion, bridging or arching, obstruction, effective yield strength, movement on inclined planes, kinematics of chimney movement, and other critical factors.

Simple mathematical expressions have been developed to describe some of the specific factors affecting muck behavior (Yenge, 1980):

Vertical pressure in bins
Cross-sectional loads on silo walls

Distribution of load on flat-bottomed silos

Friction between material and wall

Failure state of material and load on wall

Hoop stresses induced in silo walls

Ratio of horizontal to vertical pressures

Cohesion of mass in silo as a function of height

Mass flow pattern in silos, in bins, and *in situ*

Conditions of continuous flow

Material yield characteristics

Wall yield locus

Obstruction to flow in hoppers

Analysis of yield strength

Classification of bulk solids

Flow factors

Time effects on consolidation

Rat-holing and chimney flow

Flow in inclined openings

Draw ellipse in massive solids

Systems for free flow

Drawing patterns for caved rock

Gravity flow in sublevel caving

Based on a detailed review of existing experimental and mathematical studies, Chen (1984) utilized the ellipsoidal theory of bulk flow, the Bergmark–Roos equations of gravity flow, and the Monte-Carlo method of predicting granular flow to establish a computer model. In this study a stochastic theory was employed as a predictive tool. In the latter the downward particle flow is represented by an equivalent upward-biased random flight or void, in which a derived equation is formally identical to the diffusion equation. A finite difference computer method was used to code the stochastic model, and the agreement with the physical model was within 10%. Mathematical models have been designed in an attempt to describe some of these phenomena and are summarized by Chen (1984). He developed them further but utilized only part of the factors affecting muck flow, along with previously developed codes to produce a more complete mathematical model.

### 15.6.3  Scaling of Muck Flow Parameters

To be of real engineering value, the results of model tests where the earth's gravity is a dominant influence, or where applied pressures are used to create interfragment pressure, or where a centrifuge is used for simulation of muck flow under the influence of gravity, and the pertinent similitude principles must be known and properly applied.

An idealized two-dimensional analysis of forces and pressures of uniform cylindrical grains (Jenkins, 1931) with square and hexagonal packing was based on the principle of dilatancy and present useful concepts. However, three-dimensional arrangements were too difficult to analyze.

In full-scale muck flow situations, the action of the particulate mass is due to the effect of gravity on the whole mass, including the action causing the pressures at depth, the friction between the particles and the retaining walls and the weight of the particles themselves, and the weight of the particles forming a free boundary or arch.

# COMPUTER CODE FOR EXPLOSIVES

The mathematical logic for this computer code was devised by Dr. Cook and colleagues for the calculation of the ideal detonation and explosion state parameters of near-oxygen-balanced explosives. It does not take into account such factors as grain size, critical diameter, or diameter effects of nonideal explosives. It has been used extensively for research and teaching of the methods of calculation of near-oxygen-balanced explosives, ANFO, slurries and other blasting agents, as well as explosives with small amounts of metallic elements. The major calculation of the material balance is by iteration, which requires a library code for the solution of four simultaneous, nonlinear equations and thermo-chemical data (Tables A-1 to A-5). The matrix for usual compositions of oxygen-balanced explosives is quite stable. There is also a small subroutine for the program for the iteration of the values of $\beta$ and $v_2$.

*Symbols—explosive calculations*

$D$ = detonation velocity

$W$ = average velocity of gaseous products in direction of the detonation front

$v_1$ = specific volume of the explosive (L/kg) or cm$^3$/g

$v_2$ = specific volume of the gaseous products in the detonation head

$v_3$ = specific volume of the explosion = $v_1$

$v_4$ = specific volume of the gases in the final state (1 atm)

$p_1$ = initial pressure of explosive (atm)

$p_2$ = detonation pressure (atm)

$p_3$ = explosion pressure (atm)

$p_4$ = final pressure = 1 atm

$\rho_1$ = density of explosive (kg/L) or g/cm$^3$

$\rho_2$ = density of gases in detonation front

$\alpha$ = co-volume (L/kg) of gases

$E$ = total energy content

$E_a$ = activation energy

$\overline{C}_v$ = average heat capacity per kilogram at constant volume between $T_1$ and $T$

$C_v$ = heat capacity per kilogram at constant volume

$T_1, T_2, T_3, T_4$ = absolute temperatures corresponding to $v_1, v_2, v_3, v_4$

$t$ = centigrade temperature

$Q$ = heat of explosion per kilogram at constant volume

$s$ = subscript for adiabatic conditions

$n$ = number of moles of gaseous products per kilogram of explosives (0.00198 when A is in kcal/kg)

$R$ = gas constant (0.08207 L atm $\cdot$ mole$^{-1}$ $\cdot$ K$^{-1}$ for pressure calculation 8.315 J $\cdot$ mole$^{-1}$ $\cdot$ K$^{-1}$ for detonation velocity calculation)

$K$ = degrees absolute, or Kelvins

$\beta$ = $(nR + C_v)/C_v - d\alpha/dv_2$ where $d\alpha/dv_2$ is the slope of the $\alpha, v$ curve at the point corresponding to the density considered

```
C       MIN.ENG. 507-DRILLING AND BLASTING - DR. CLARK
C       PART ONE READ INPUT DATA
        DIMENSION DELH(20),AK(24,4),FG(2,4),CV(20,3),CVB(20,3)
        DIMENSION AV(3,2),CAY(24),GAMMA(15),Y(15),A(4,4),B(4)
        DIMENSION WKAREA(28),ALPHA(4)
        DATA DELH/0.0000000,0.0000000,94.05000,26.42000,
     :    57.80000,11.04000,17.89000,48.08000,
     :    86.67000,0.0000000,-10.06000,-21.60000,
     :   -52.09000,-85.09000,-59.16000,70.50000,
     :    100.0000, 155.00000,0.0000000,0.0000000/
        DATA AK/  0.0000000,0.0000000, 2.471053,0.0000000,
     :    3.495409, 3.499324,0.0000000,0.0000000,
     :    10.73168,0.0000000,0.0000000,-16.28025,
     :    7.059063, 6.896320, 4.969711,0.0000000,
     :   0.0000000,0.0000000,0.0000000,0.0000000,
     :    9.182171,-4.161277,0.0000000, 8.564920,
     :   0.0000000,0.0000000,-22.73629,0.0000000,
     :   -4.001861,-39.74432,0.0000000,0.0000000,
     :   -58.53599,0.0000000,0.0000000, 17.42642,
     :   -32.08833,-40.14065,0.7614930,0.0000000,
     :   0.0000000,0.0000000,0.0000000,0.0000000,
     :    3.181536,-10.87662,0.0000000,-55.07017,
     :   0.0000000,0.0000000,-6.220603,0.0000000,
     :   0.2038000E-02,-6.211124,0.0000000,0.0000000,
     :   -12.42984,0.0000000,0.0000000, 16.21497,
     :   -6.213579,-6.233845,-8.133611,0.0000000,
     :   0.0000000,0.0000000,0.0000000,0.0000000,
     :   -16.05504, 8.341794,0.0000000,-12.45950,
     :   0.0000000,0.0000000, 3.551435,0.0000000,
     :  -0.1249000E-02,3.536575,0.0000000,0.00000,
     :    7.086523,0.0000000,0.0000000,-6.962426,
     :    3.538873, 3.557053, 3.500181,0.0000000,
     :   0.0000000,0.0000000,0.0000000,0.0000000,
     :    6.851758,-3.623585,0.0000000, 7.108882/
        DATA FG/  -2.394740, 4.772180, 7.984487,-9.310743,
     :   -4.058194, 11.33966, 2.250761,-2.401737/
        DATA CV/   8.487000, 7.478000, 14.61600,
     :    7.478000, 12.22100, 18.11300,
     :    24.08900, 29.02500, 24.03400,
     :    9.235000, 8.642000, 7.759000,
     :    2.981000, 6.476000, 4.029000,
     :   0.0000000,0.0000000,0.0000000,
     :   0.0000000,0.0000000,-5.962000,
     :   -2.060000,-5.692000,-1.994000,
     :   -2.758000,-2.341000,-2.434000,
     :   -2.212000,-1.898000,-5.995000,
     :   -7.084000, -3.157000, 0.0000000,
     :  -18.84000, -5.326000,  0.0000000,
     :   0.0000000, 0.0000000, 0.0000000,
     :   0.0000000,-0.4582000,-0.3408000,
     :  -0.5344000,-0.4157000, 1.225000,
     :    2.793000 ,3.782000 , 4.892000,
     :    3.089000,-0.5193000,-0.7045000,
     :  -0.7906000, 0.0000000, -1.288000,
     :   -1.256000, 0.0000000, 0.0000000,
     :   0.0000000, 0.0000000, 0.0000000/
        DATA CVB/  7.638000, 7.193000, 13.85100,
     :    7.204000, 11.88700, 17.85400,
     :    23.83000, 28.79400, 23.82900,
     :    8.470000, 7.679000, 7.321000,
```

```
     :    2.981000,   3.744000,   3.279000,
     :    0.0000000,  0.0000000,  0.0000000,
     :    0.0000000,  0.0000000, -7.429000,
     :   -3.634000,  -8.569000,  -3.379000,
     :   -10.75100,  -15.56400, -21.67100,
     :   -24.00300,  -14.94800,  -7.664000,
     :   -7.274000,  -3.437000,  0.0000000,
     :   -4.210000,  -1.498000,  0.0000000,
     :    0.0000000,  0.0000000,  0.0000000,
     :    0.0000000, -0.8281000, -0.5823000,
     :   -0.6243000, -0.5623000, -0.6200000,
     :   -0.5278000, -0.4431000, -0.4282000,
     :   -0.4356000, -0.8948000, -0.8506000,
     :   -0.6433000,  0.0000000, -1.498000,
     :   -1.372000,   0.0000000,  0.0000000,
     :    0.0000000,  0.0000000,  0.0000000/
      DATA ALPHA/ 1.250604,-0.8541108, 0.3417540, -0.5957424E-01/
      DATA AV/ -0.1000000E+00,-0.1000000E-01,
     :    0.1100000,  0.5000000,
     :    0.4000000,  0.3000000/
      DATA EX,EV,ET/.0005,0.0005,10.00/
      DATA X1,X2,X3,X4/.1,1.,3.,.1/
      DATA IN,IOUT/4,4/
      WRITE(IN,1001)
      READ(IN,100)CARB
      WRITE(IN,1002)
      READ(IN,100)HYDRO
      WRITE(IN,1003)
      READ(IN,100)GEN
      WRITE(IN,1004)
      READ(IN,100)OXYG
      WRITE(IN,1005)
      READ(IN,100)QR
      WRITE(IN,1006)
      READ(IN,100)RHO1
      WRITE(IN,1007)
      READ(IN,100)T1
      WRITE(IN,1008)
      READ(IN,100)T2
      WRITE(IN,1009)
      READ(IN,100)T3
      WRITE(IOUT,101)
      IDIOT=1
      V1=1./RHO1
      V2=V1*(.72+.1*(RHO1-.90))
C        PART TWO COMPUTE K(I)
   34 T2P=1000./T2
      DO 3 I=3,24
      IF(AK(I,1)-0.)2,1,2
    1 CAY(I)=0.0
      GO TO 3
    2 CY=AK(I,2)+T2P*(AK(I,3)+AK(I,4)*T2P)
      CY=AK(I,1)+T2P*CY
      CAY(I)=EXP(CY)
    3 CONTINUE
C        PART THREE COMPUTE FUGACITY
      RHO2=1./V2
   30 N=2
      IF(RHO2-1.1)4,10,5
    4 IF(RHO2-0.5)22,22,6
```

```
    5 IF(RHO2-2.5)10,10,23
    6 N=1
   10 F=FG(N,2)+RHO2*(FG(N,3)+FG(N,4)*RHO2)
      F=FG(N,1)+RHO2*F
      F=EXP(F)
      WRITE(IOUT,100)F
      FSQR=F**.5
      FISQR=1./FSQR
      FSQRR=FSQR**.5
C     PART FOUR COMPUTE GAMMA(I)
      GAMMA(1)=FISQR
      GAMMA(2)=1.
      GAMMA(3)=GAMMA(2)
      GAMMA(4)=GAMMA(1)
      GAMMA(5)=CAY(5)
      GAMMA(6)=FSQRR/CAY(15)
      GAMMA(7)=CAY(12)
      GAMMA(8)=FSQR/CAY(21)
      GAMMA(9)=FSQR*CAY(22)
      GAMMA(10)=CAY(9)
      GAMMA(11)=CAY(13)/FSQRR
      GAMMA(12)=CAY(14)
      GAMMA(13)=CAY(3)/(FSQRR*FSQR)
      GAMMA(14)=CAY(6)*GAMMA(1)
      GAMMA(15)=CAY(24)*GAMMA(1)
C     PART FIVE COMPUTE Y(I)
   16 Y(1)=GAMMA(1)*X1
      Y(2)=X2*X2
      Y(3)=X3*X4
      Y(4)=GAMMA(4)*X4
      Y(5)=GAMMA(5)*X1*X3
      Y(6)=GAMMA(6)*X2*X1**(3./2.)
      Y(7)=GAMMA(7)*X1*X1*X4/X3
      Y(8)=GAMMA(8)*X1*X1*X4
      Y(9)=GAMMA(9)*X1*X3*X4
      Y(10)=GAMMA(10)*X3*X3
      Y(11)=GAMMA(11)*X3*X1**.5
      Y(12)=GAMMA(12)*X2*X3
      Y(13)=GAMMA(13)*X1**.5
      Y(14)=GAMMA(14)*X2
      Y(15)=GAMMA(15)*X3
C     PART SIX COMPUTE A(I,J)
      A(1,1)=2.*(Y(1)+Y(5)+Y(9))+.5*(Y(11)+Y(13))+8.*(Y(7)+Y(8))+4.5*Y(6
     2)
      A(1,2)=3.*Y(6)
      A(1,3)=2.*(Y(5)+Y(9))-4.*Y(7)+Y(11)
      A(1,4)=4.*(Y(7)+Y(8))+2.*Y(9)
      B(1)=HYDRO-(2.*(Y(1)+Y(5)+Y(9))+3.*Y(6)+4.*(Y(7)+Y(8))
     1    + Y(11) + Y(13) )
      A(2,1)=.5*A(1,2)
      A(2,2)=4.*Y(2)+Y(6)+Y(12)+Y(14)
      A(2,3)=Y(12)
      A(2,4)=0.0
      B(2)=GEN-(2.*Y(2)+Y(6)+Y(12)+Y(14))
      A(3,1)=Y(5)+2.*(Y(8)+Y(9))+.5*Y(11)
      A(3,2)=A(2,3)
      A(3,3)=2.*(Y(3)+Y(9))+4.*Y(10)+Y(5)+Y(11)+Y(12)+Y(15)
      A(3,4)=2.*(Y(3)+Y(9))+Y(4)+Y(8)
      B(3)=OXYG-(2.*(Y(3)+Y(9)+Y(10))+Y(4)+Y(5)+Y(8)+Y(11)+Y(12)+Y(15))
      A(4,1)=.5*A(1,4)
```

```
            A(4,2)=0.0
            A(4,3)=Y(3)+Y(9)-Y(7)
            A(4,4)=Y(3)+Y(4)+Y(7)+Y(8)+Y(9)
            B(4)=CARB-(Y(3)+Y(4)+Y(7)+Y(8)+Y(9))
             CALL LEQT1F(A,1,4,4,B,0,WKAREA,IER)
              IF(IER.GT.128)GO TO 998
              GO TO 999
      998     WRITE (IOUT,303)
              STOP
        999 CONTINUE
            IF(ABS(B(1))-EX)11,14,14
         11 IF(ABS(B(2))-EX)12,14,14
         12 IF(ABS(B(3))-EX)13,14,14
         13 IF(ABS(B(4))-EX)15,14,14
      C     PART SEVEN COMPUTE   X=X+X*DELX
         14 IF(B(1))141,143,143
        141 IF(B(1)+1.)142,142,143
        142 B(1)=B(1)/2.
            GO TO 141
        143 IF(B(2))144,146,146
        144 IF(B(2)+1.)145,145,146
        145 B(2)=B(2)/2.
            GO TO 144
        146 IF(B(3))147,149,149
        147 IF(B(3)+1.)148,148,149
        148 B(3)=B(3)/2.
            GO TO 147
        149 IF(B(4))151,153,153
        151 IF(B(4)+1.)152,152,153
        152 B(4)=B(4)/2.
            GO TO 151
        153 X1=X1+X1*B(1)
            X2=X2+X2*B(2)
            X3=X3+X3*B(3)
            X4=X4+X4*B(4)
            WRITE(IOUT,200)X1,X2,X3,X4,IER
            GO TO 16
         15 CONTINUE
      C     PART EIGHT COMPUTE   Y(J)*CV(J)
            EY=0.0
            DO 18 J=1,15
         18 EY=Y(J)+EY
            EY=10.*EY
            GO TO (1718,31),IDIOT
       1719 EYR=0.00198*EY
            EYC=0.0
            DO 17 J=1,15
         17 EYC=Y(J)*(CV(J,1)+CV(J,2)*T2P*(1.+CV(J,3)*T2P))+EYC
            EYC=EYC*0.01
            EYCI=1./EYC
      C     PART NINE ITERATE ON V2 AND BETA
            SUM=0.0
         29 N=2
            IF(RHO2-1.2)19,26,21
         19 IF(RHO2-0.9)20,26,26
         20 IF(RHO2-0.5)22,22,24
         21 IF(RHO2-2.2)25,25,23
         24 N=1
            GO TO 26
         25 N=3
```

```
   26 DELAV=AV(N,1)+RHO2*AV(N,2)
      BETA=EYCI*EYR+1.-DELAV
      ALFA=ALPHA(1)+RHO2*(ALPHA(2)+RHO2*(ALPHA(3)+RHO2*ALPHA(4)))
   27 DELV2=-((V1-V2)*BETA-V2+ALFA)/(DELAV-BETA-1.)
      V2=V2+DELV2
      RHO2=1./V2
      SUM=DELV2+SUM
      IF(ABS(DELV2)-EV)28,29,29
C     PART TEN COMPUTE EYCB AND EYDELH
   28 IF(ABS(SUM)-EV)31,30,30
   31 EYDELH=0.0
      DO 32 J=1,15
   32 EYDELH=Y(J)*DELH(J)+EYDELH
      EYDELH=10.*EYDELH
      EYCB=0.0
      DO 33 J=1,15
   33 EYCB=Y(J)*(CVB(J,1)+(CVB(J,2)*T2P)*(1.+CVB(J,3)*T2P))+EYCB
      EYCB=EYCB*0.01
C     PART ELEVEN COMPUTE Q2 AND T2
      QP=EYDELH
      Q2=QP-QR
      GO TO (3337,41),IDIOT
 3337 TP=(BETA*(Q2+T1*EYCB))/(BETA*EYCB-0.5*EYR)
      IF(TP-1500.)38,36,37
   37 IF(TP-6000.)36,36,39
   36 CONTINUE
      DELT2=TP-T2
      T2=TP
      IF(ABS(DELT2)-ET)35,34,34
   35 CONTINUE
C     PART TWELVE COMPUTE P2 AND D2
      P2=.08207*EY*T2/(V2-ALFA)
      D2=V1/(V2-ALFA)*(EY*8.315*T2*BETA)**.5
   46 WRITE(IOUT,104)
      WRITE(IOUT,105)(Y(I),I=1,5)
      WRITE(IOUT,106)(Y(I),I=6,10)
      WRITE(IOUT,107)(Y(I),I=11,15)
      WRITE(IOUT,108)X1,X2,X3,X4
      WRITE(IOUT,109)T2
      WRITE(IOUT,201)P2
      WRITE(IOUT,202)Q2
      GO TO (47,99),IDIOT
   47 WRITE(IOUT,203) D2
      IDIOT=IDIOT+1
      WRITE(IOUT,204)
      V3=V1
      V2=V3
   40 T2=T3
      GO TO 34
   41 Q3=Q2
      TP=Q3/EYCB+T1
      IF(TP-1500)38,43,42
   42 IF(TP-6000.)43,43,39
   43 DELT3=TP-T3
      T3=TP
      IF(ABS(DELT3)-ET)45,40,40
   45 ALFA=ALPHA(1)+RHO2*(ALPHA(2)+RHO2*(ALPHA(3)+RHO2*ALPHA(4)))
      P3=.08207*EY*T3/(V2-ALFA)
      P2=P3
      GO TO 46
```

```
     99 CONTINUE
        STOP
     22 WRITE(IOUT,102)
        STOP
     23 WRITE(IOUT,103)
        STOP
     38 WRITE(IOUT,301)
        STOP
     39 WRITE(IOUT,302)
        STOP
    100 FORMAT(G)
    101 FORMAT(1X,'DET STATE',///)
    102 FORMAT (1X,50HRHO2 IS LESS THAN OR EQUAL TO 0.5-ERROR-CALL EXIT )
    103 FORMAT (1X,41HRHO2 IS GREATER THAN 2.2-ERROR-CALL EXIT )
    104 FORMAT (//1X,31HGAS COMPOSITION (MOLES/100 GM.) ,//)
    105 FORMAT(7X,8HHYDROGEN,7X,8HNITROGEN,7X,8HCARBDICK,8X,7HCARBMNX,10X,
        25HWATER,/5E15.8/)
    106 FORMAT(8X,7HAMMONIA,8X,7HMETHANE,7X,8HMETHANOL,5X,10HFORMICACID,
        29X,6HOXYGEN,/5E15.8/)
    107 FORMAT(7X,8HHYDROXYL,9X,6HNITROX,7X,8HATOMIC H,7X,8HATOMIC N,
        27X,8HATOMIC O,/5E15.8/)
    108 FORMAT(13X,2HX1,13X,2HX2,13X,2HX3,13X,2HX4,/4E15.8/)
    109 FORMAT (2X,14HTEMPERATURE = ,F9.2,16H DEGREES KELVIN ,/)
    200 FORMAT(5G)
    201 FORMAT (5X,11HPRESSURE = ,F10.2,6H ATM. ,/)
    202 FORMAT (1X,15HQ (RELEASED) = ,F10.3,15H KCAL/1000 GM. ,/)
    203 FORMAT (5X,11HVELOCITY = ,F10.2,13H METERS/SEC. )
    204 FORMAT(1X,///,1X15HEXPLOSION STATE,//)
    301 FORMAT(1X,39HERROR-T2 IS LESS THAN 1500 DEG. KELVIN )
    302 FORMAT(1X,42HERROR-T2 IS GREATER THAN 6000 DEG. KELVIN )
    303 FORMAT(1X,'ERROR-NOT A SINGULAR SOLUTION')
   1001 FORMAT('      ENTER THE CARBON. (GRAM ATOM/100GM)',20X,$)
   1002 FORMAT('      ENTER THE HYDROGEN. (GRAM ATOM/100GM)',18X,$)
   1003 FORMAT('      ENTER THE NITROGEN. (GRAM ATOM/100GM)',18X,$)
   1004 FORMAT('      ENTER THE OXYGEN.  (GRAM ATOM/100GM)',19X,$)
   1005 FORMAT('      ENTER THE HEAT OF FORMATION OF EXPLOSIVES.
      : (KCAL/KGM)',2X,$)
   1006 FORMAT('      ENTER THE DENSITY OF THE EXPLOSIVES.
      :  (GM/CC)',10X,$)
   1007 FORMAT('      ENTER THE AMBIENT TEMP. (298K)',25X,$)
   1008 FORMAT('      ENTER THE APPROXIMATE DENOTATION TEMP.
      :(DEG.K)',9X,$)
   1009 FORMAT('      ENTER THE APPROXIMATE EXPLOSION TEMP. (DEG.K)',10X,$)
        STOP
        END
```

## TABLE A.1. Covolume—Specific Volume Relationships

| Density | Density Factor (1/a) | Specific Volume (v) | Co-volume ($\alpha$) | $a = \dfrac{1}{v - \alpha}$ | $\left(\dfrac{d\alpha}{dv}\right)_s$ | $\displaystyle\int_{v_4}^{v} \dfrac{d\alpha}{v - \alpha}$ |
|---|---|---|---|---|---|---|
| 0.5000 | 0.9182 | 2.000 | 0.911 | 1.0890 | 0.150 | |
| 0.5500 | 1.062 | 1.818 | 0.876 | 0.9420 | 0.175 | |
| 0.6000 | 1.221 | 1.667 | 0.848 | 0.8190 | 0.220 | |
| 0.6500 | 1.389 | 1.538 | 0.818 | 0.7200 | 0.225 | |
| 0.7000 | 1.577 | 1.429 | 0.795 | 0.6340 | 0.250 | 0.100 |
| 0.7500 | 1.783 | 1.333 | 0.772 | 0.5610 | 0.275 | 0.150 |
| 0.8000 | 2.000 | 1.250 | 0.750 | 0.5000 | 0.300 | 0.200 |
| 0.8500 | 2.247 | 1.177 | 0.732 | 0.4450 | 0.325 | 0.250 |
| 0.9000 | 2.525 | 1.111 | 0.715 | 0.3960 | 0.350 | 0.300 |
| 0.9500 | 2.809 | 1.053 | 0.697 | 0.3560 | 0.370 | 0.350 |
| 1.000 | 3.125 | 1.000 | 0.680 | 0.3200 | 0.390 | 0.400 |
| 1.050 | 3.467 | 0.9524 | 0.664 | 0.2884 | 0.410 | 0.450 |
| 1.100 | 3.845 | 0.9091 | 0.649 | 0.2601 | 0.430 | 0.500 |
| 1.150 | 4.219 | 0.8700 | 0.633 | 0.2370 | 0.450 | 0.550 |
| 1.200 | 4.623 | 0.8333 | 0.617 | 0.2163 | 0.470 | 0.600 |
| 1.250 | 5.076 | 0.8000 | 0.603 | 0.1970 | 0.485 | 0.675 |
| 1.300 | 5.580 | 0.7692 | 0.590 | 0.1792 | 0.500 | 0.750 |
| 1.350 | 6.072 | 0.7407 | 0.576 | 0.1647 | 0.515 | 0.825 |
| 1.400 | 6.523 | 0.7143 | 0.561 | 0.1533 | 0.530 | 0.900 |
| 1.450 | 7.158 | 0.6897 | 0.550 | 0.1397 | 0.545 | 0.975 |
| 1.500 | 7.710 | 0.6667 | 0.537 | 0.1297 | 0.560 | 1.050 |
| 1.550 | 8.319 | 0.6452 | 0.525 | 0.1202 | 0.575 | 1.150 |
| 1.600 | 8.929 | 0.6250 | 0.513 | 0.1120 | 0.590 | 1.250 |
| 1.650 | 9.699 | 0.6061 | 0.503 | 0.1031 | 0.605 | 1.375 |
| 1.700 | 10.50 | 0.5882 | 0.493 | 0.0952 | 0.620 | 1.500 |
| 1.800 | 11.82 | 0.5556 | 0.471 | 0.0846 | 0.650 | 1.750 |
| 1.900 | 13.40 | 0.5263 | 0.450 | 0.0763 | 0.670 | 1.950 |
| 2.000 | 14.29 | 0.5000 | 0.430 | 0.0700 | 0.700 | 2.150 |
| 2.100 | 15.58 | 0.4762 | 0.412 | 0.0642 | 0.720 | 2.400 |
| 2.200 | 17.09 | 0.4545 | 0.396 | 0.0585 | 0.740 | 2.700 |

## TABLE A.2. Constants for Fugacity Calculations of Equilibrium Concentrations

| T(°K) | $K_3$ | $K_5$ | $K_6$ | $K_8$ | $K_9$ | $K_{12}$ | $K_{13}$ | $K_{14}$ |
|---|---|---|---|---|---|---|---|---|
| 1000 | 7.813 (−11) | 6.031 (−1) | 8.965 (−18) | 2.216 (−3) | 4.229 (−24) | 3.295 (+4) | 6.659 (−13) | 1.791 (−16) |
| 1200 | 6.669 (−9) | 1.175 | 1.303 (−14) | 5.513 (−2) | 2.719 (−19) | 2.182 (+2) | 2.702 (−10) | 2.784 (−13) |
| 1400 | 1.579 (−7) | 1.892 | 2.337 (−12) | 5.346 (−1) | 7.220 (−16) | 6.353 | 1.948 (−8) | 5.238 (−11) |
| 1600 | 1.679 (−6) | 2.704 | 1.135 (−10) | 2.885 | 2.623 (−13) | 4.642 (−1) | 4.776 (−7) | 2.638 (−9) |
| 1800 | 1.049 (−5) | 3.570 | 2.310 (−9) | 1.056 (+1) | 2.535 (−11) | 6.238 (−2) | 5.712 (−6) | 5.520 (−8) |
| 2000 | 4.516 (−5) | 4.458 | 2.560 (−8) | 2.948 (+1) | 9.715 (−10) | 1.281 (−2) | 4.136 (−5) | 6.254 (−7) |
| 2200 | 1.484 (−4) | 5.348 | 1.823 (−7) | 6.767 (+1) | 1.901 (−8) | 3.570 (−3) | 2.080 (−4) | 4.536 (−6) |
| 2400 | 3.987 (−4) | 6.223 | 9.324 (−7) | 1.342 (+2) | 2.249 (−7) | 1.250 (−3) | 7.960 (−4) | 2.356 (−5) |
| 2600 | 9.167 (−4) | 7.074 | 3.698 (−6) | 2.381 (+2) | 1.807 (−6) | 5.212 (−4) | 2.470 (−3) | 9.467 (−5) |
| 2800 | 1.866 (−3) | 7.896 | 1.202 (−5) | 3.870 (+2) | 1.073 (−5) | 2.489 (−4) | 6.503 (−3) | 3.110 (−4) |
| 3000 | 3.448 (−3) | 8.685 | 3.328 (−5) | 5.868 (+2) | 4.997 (−5) | 1.325 (−4) | 1.501 (−2) | 8.696 (−4) |
| 3200 | 5.887 (−3) | 9.440 | 8.099 (−5) | 8.410 (+2) | 1.913 (−4) | 7.690 (−5) | 3.114 (−2) | 2.134 (−3) |
| 3400 | 9.421 (−3) | 1.016 (+1) | 1.772 (−4) | 1.151 (+3) | 6.228 (−4) | 4.795 (−5) | 5.919 (−2) | 4.704 (−3) |
| 3600 | 1.429 (−2) | 1.085 (+1) | 3.548 (−4) | 1.517 (+3) | 1.773 (−3) | 3.172 (−5) | 1.046 (−1) | 9.480 (−3) |
| 3800 | 2.070 (−2) | 1.150 (+1) | 6.593 (−4) | 1.936 (+3) | 4.507 (−3) | 2.204 (−5) | 1.738 (−1) | 1.772 (−2) |
| 4000 | 2.887 (−2) | 1.212 (+1) | 1.150 (−3) | 2.405 (+3) | 1.041 (−2) | 1.597 (−5) | 2.741 (−1) | 3.108 (−2) |
| 4200 | 3.897 (−2) | 1.271 (+1) | 1.900 (−3) | 2.919 (+3) | 2.215 (−2) | 1.198 (−5) | 4.134 (−1) | 5.160 (−2) |
| 4400 | 5.112 (−2) | 1.328 (+1) | 2.997 (−3) | 3.474 (+3) | 4.391 (−2) | 9.273 (−6) | 6.001 (−1) | 8.173 (−2) |
| 4600 | 6.543 (−2) | 1.381 (+1) | 4.538 (−3) | 4.064 (+3) | 8.185 (−2) | 7.366 (−6) | 8.426 (−1) | 1.243 (−1) |
| 4800 | 8.196 (−2) | 1.432 (+1) | 6.632 (−3) | 4.694 (+3) | 1.446 (−1) | 7.987 (−6) | 1.149 | 1.823 (−1) |
| 5000 | 1.008 (−1) | 1.481 (+1) | 9.395 (−3) | 5.329 (+3) | 2.437 (−1) | 4.963 (−6) | 1.527 | 2.491 (−1) |
| 5200 | 1.218 (−1) | 1.527 (+1) | 1.295 (−2) | 5.993 (+3) | 3.939 (−1) | 4.187 (−6) | 1.984 | 3.581 (−1) |
| 5400 | 1.451 (−1) | 1.571 (+1) | 1.741 (−2) | 6.673 (+3) | 6.135 (−1) | 3.588 (−6) | 2.526 | 4.830 (−1) |
| 5600 | 1.706 (−1) | 1.613 (+1) | 2.291 (−2) | 7.363 (+3) | 9.246 (−1) | 3.116 (−6) | 3.160 | 6.371 (−1) |
| 5800 | 1.983 (−1) | 1.653 (+1) | 2.956 (−2) | 8.059 (+3) | 1.353 | 2.740 (−6) | 3.889 | 8.241 (−1) |
| 6000 | 2.280 (−1) | 1.692 (+1) | 3.747 (−2) | 8.759 (+3) | 1.928 | 2.436 (−6) | 4.718 | 1.047 |

Note: 4.229 (−24) stands for $4.229 \times 10^{-24}$, etc.

| $T(°K)$ | $K_{15}$ | $K_{16}$ | $K_{21}$ | $K_{22}$ | $K_{23}$ | $K_{24}$ | $K_{25}$ | $K_{26}$ | $K_{27}$ |
|---|---|---|---|---|---|---|---|---|---|
| 1000 | 2.992 | 2.570 (−5) | 2.351 (+1) | 3.302 (−5) | 1.526 (−3) | 1.534 (−23) | 1.445 (−3) | 5.893 | 1.232 (−2) |
| 1200 | 7.587 | 3.235 (−4) | 1.134 (+2) | 6.914 (−5) | 1.438 (−3) | 5.547 (−19) | 4.120 (−3) | 1.092 | 1.529 (−3) |
| 1400 | 1.440 (+1) | 1.975 (−3) | 3.326 (+2) | 1.200 (−4) | 1.412 (−3) | 9.762 (−16) | 8.915 (−3) | 3.438 (−1) | 3.702 (−4) |
| 1600 | 2.287 (+1) | 7.672 (−3) | 7.192 (+2) | 1.849 (−4) | 1.417 (−3) | 2.605 (−13) | 1.620 (−2) | 1.498 (−1) | 1.348 (−4) |
| 1800 | 3.232 (+1) | 2.204 (−2) | 1.274 (+3) | 2.623 (−4) | 1.442 (−3) | 1.981 (−11) | 2.613 (−2) | 8.068 (−2) | 6.407 (−5) |
| 2000 | 4.215 (+1) | 5.127 (−2) | 1.968 (+3) | 3.508 (−4) | 1.478 (−3) | 6.266 (−10) | 3.873 (−2) | 5.029 (−2) | 3.653 (−5) |
| 2200 | 5.191 (+1) | 1.023 (−1) | 2.761 (+3) | 4.492 (−4) | 1.522 (−3) | 1.048 (−8) | 5.394 (−2) | 3.479 (−2) | 2.371 (−5) |
| 2400 | 6.127 (+1) | 1.819 (−1) | 3.603 (+3) | 5.561 (−4) | 1.572 (−3) | 1.088 (−7) | 7.162 (−2) | 2.598 (−2) | 1.692 (−5) |
| 2600 | 7.006 (+1) | 2.961 (−1) | 4.457 (+3) | 6.705 (−4) | 1.625 (−3) | 7.827 (−7) | 9.163 (−2) | 2.056 (−2) | 1.296 (−5) |
| 2800 | 7.815 (+1) | 4.495 (−1) | 5.290 (+3) | 7.914 (−4) | 1.682 (−3) | 4.225 (−6) | 1.138 (−1) | 1.701 (−2) | 1.049 (−5) |
| 3000 | 8.551 (+1) | 6.455 (−1) | 6.078 (+3) | 9.182 (−4) | 1.741 (−3) | 1.813 (−5) | 1.380 (−1) | 1.457 (−2) | 8.858 (−6) |
| 3200 | 9.213 (+1) | 8.859 (−1) | 6.806 (+3) | 1.050 (−3) | 1.802 (−3) | 6.456 (−5) | 1.640 (−1) | 1.283 (−2) | 7.735 (−6) |
| 3400 | 9.803 (+1) | 1.171 | 7.465 (+3) | 1.186 (−3) | 1.864 (−3) | 1.973 (−4) | 1.917 (−1) | 1.155 (−2) | 6.939 (−6) |
| 3600 | 1.033 (+2) | 1.502 | 8.052 (+3) | 1.326 (−3) | 1.928 (−3) | 5.308 (−4) | 2.209 (−1) | 1.059 (−2) | 6.363 (−6) |
| 3800 | 1.079 (+2) | 1.875 | 8.566 (+3) | 1.470 (−3) | 1.992 (−3) | 1.283 (−3) | 2.516 (−1) | 9.859 (−3) | 5.940 (−6) |
| 4000 | 1.119 (+2) | 2.290 | 9.009 (+3) | 1.617 (−3) | 2.057 (−3) | 2.832 (−3) | 2.835 (−1) | 9.292 (−3) | 5.628 (−6) |
| 4200 | 1.154 (+2) | 2.744 | 9.385 (+3) | 1.767 (−3) | 2.123 (−3) | 5.783 (−3) | 3.167 (−1) | 8.849 (−3) | 5.398 (−6) |
| 4400 | 1.184 (+2) | 3.235 | 9.698 (+3) | 1.919 (−3) | 2.190 (−3) | 1.104 (−2) | 3.509 (−1) | 8.501 (−3) | 5.230 (−6) |
| 4600 | 1.209 (+2) | 3.759 | 9.953 (+3) | 2.074 (−3) | 2.257 (−3) | 1.989 (−2) | 3.862 (−1) | 8.228 (−3) | 5.113 (−6) |
| 4800 | 1.231 (+2) | 4.314 | 1.016 (+4) | 2.230 (−3) | 2.324 (−3) | 3.406 (−2) | 4.223 (−1) | 8.015 (−3) | 5.034 (−6) |
| 5000 | 1.250 (+2) | 4.896 | 1.031 (+4) | 2.389 (−3) | 2.392 (−3) | 5.577 (−2) | 4.594 (−1) | 7.849 (−3) | 4.988 (−6) |
| 5200 | 1.265 (+2) | 5.504 | 1.043 (+4) | 2.549 (−3) | 2.460 (−3) | 8.778 (−2) | 4.972 (−1) | 7.723 (−3) | 4.969 (−6) |
| 5400 | 1.277 (+2) | 6.133 | 1.050 (+4) | 2.711 (−3) | 2.528 (−3) | 1.334 (−1) | 5.358 (−1) | 7.630 (−3) | 4.972 (−6) |
| 5600 | 1.287 (+2) | 6.781 | 1.055 (+4) | 2.874 (−3) | 2.597 (−3) | 1.965 (−1) | 5.750 (−1) | 7.565 (−3) | 4.995 (−6) |
| 5800 | 1.295 (+2) | 7.446 | 1.056 (+4) | 3.038 (−3) | 2.666 (−3) | 2.815 (−1) | 6.149 (−1) | 7.523 (−3) | 5.035 (−6) |
| 6000 | 1.301 (+2) | 8.126 | 1.055 (+4) | 3.204 (−3) | 2.735 (−3) | 3.933 (−1) | 6.554 (−1) | 7.501 (−3) | 5.091 (−6) |

Note: 2.570 (−5) stands for $2.570 \times 10^{-5}$, etc.

## TABLE A.3. Fugacity Factors[a]

| $\rho$ | a | z | $e^z$ | $\dfrac{e^z}{a} = F$ |
|---|---|---|---|---|
| 0.500 | 1.0890 | 0.950 | 2.586 | 2.375 |
| 0.650 | 0.7200 | 1.370 | 3.935 | 5.465 |
| 0.700 | 0.6340 | 1.520 | 4.572 | 7.211 |
| 0.750 | 0.5610 | 1.685 | 5.392 | 9.611 |
| 0.800 | 0.5000 | 1.850 | 6.360 | 1.272 (+1) |
| 0.850 | 0.4450 | 2.035 | 7.652 | 1.720 (+1) |
| 0.900 | 0.3960 | 2.220 | 9.207 | 2.325 (+1) |
| 0.950 | 0.3560 | 2.445 | 1.153 (+1) | 3.239 (+1) |
| 1.000 | 0.3200 | 2.670 | 1.434 (+1) | 4.385 (+2) |
| 1.050 | 0.2884 | 2.900 | 1.817 (+1) | 6.170 (+1) |
| 1.100 | 0.2610 | 3.130 | 2.287 (+1) | 8.792 (+1) |
| 1.150 | 0.2370 | 3.390 | 2.967 (+1) | 1.252 (+2) |
| 1.200 | 0.2163 | 3.650 | 3.848 (+1) | 1.779 (+2) |
| 1.250 | 0.1970 | 3.925 | 5.065 (+1) | 2.571 (+2) |
| 1.300 | 0.1792 | 4.200 | 6.669 (+1) | 3.722 (+2) |
| 1.350 | 0.1647 | 4.495 | 8.957 (+1) | 5.438 (+2) |
| 1.400 | 0.1533 | 4.790 | 1.203 (+2) | 7.848 (+2) |
| 1.450 | 0.1397 | 5.090 | 1.624 (+2) | 1.162 (+3) |
| 1.500 | 0.1297 | 5.390 | 2.192 (+2) | 1.690 (+3) |
| 1.550 | 0.1202 | 5.720 | 3.049 (+2) | 2.537 (+3) |
| 1.600 | 0.1120 | 6.050 | 4.241 (+2) | 3.787 (+3) |
| 1.650 | 0.1031 | 6.415 | 6.109 (+2) | 5.925 (+3) |
| 1.700 | 0.0952 | 6.780 | 8.801 (+2) | 9.245 (+3) |
| 1.800 | 0.0846 | 7.440 | 1.703 (+3) | 2.013 (+4) |
| 1.900 | 0.0763 | 8.010 | 3.011 (+3) | 3.946 (+4) |
| 2.000 | 0.0700 | 8.600 | 5.432 (+3) | 7.760 (+4) |
| 2.100 | 0.0642 | 9.150 | 9.414 (+3) | 1.466 (+5) |
| 2.200 | 0.0585 | 9.690 | 1.616 (+4) | 2.762 (+5) |

[a] +1 indicates $10^1$ and so on.

**TABLE A.4. Ideal Molal Heat Capacities ($C_v(300° \leq T \leq 6000°)$)(cal · mole$^{-1}$ · K$^{-1}$)**

| T(°K) | H$_2$[1] | N$_2$[1] | CO$_2$[1] | CO[1] | H$_2$O[1] | NH$_3$[2] | CH$_4$[2] |
|---|---|---|---|---|---|---|---|
| 400 | 4.987 | 5.004 | 7.884 | 5.026 | 6.198 | 7.84 | 7.71 |
| 600 | 5.021 | 5.210 | 9.324 | 5.289 | 6.690 | 9.39 | 10.52 |
| 800 | 5.091 | 5.525 | 10.313 | 5.637 | 7.267 | 10.68 | 13.07 |
| 1000 | 5.230 | 5.829 | 11.008 | 5.945 | 7.874 | 11.82 | 15.17 |
| 1200 | 5.417 | 6.076 | 11.503 | 6.181 | 8.426 | 12.82 | 16.85 |
| 1400 | 5.623 | 6.266 | 11.857 | 6.362 | 8.922 | 13.67 | 18.17 |
| 1600 | 5.827 | 6.412 | 12.029 | 6.494 | 9.346 | 14.34 | 19.17 |
| 1800 | 6.017 | 6.525 | 12.344 | 6.598 | 9.721 | 14.89 | 19.96 |
| 2000 | 6.188 | 6.615 | 12.515 | 6.678 | 10.021 | 15.33 | 20.58 |
| 2200 | 6.341 | 6.687 | 12.656 | 6.743 | 10.269 | 15.70 | 21.08 |
| 2400 | 6.477 | 6.746 | 12.776 | 6.797 | 10.476 | 15.99 | 21.45 |
| 2600 | 6.595 | 6.794 | 12.879 | 6.839 | 10.649 | 16.22 | 21.76 |
| 2800 | 6.705 | 6.837 | 12.973 | 6.877 | 10.796 | 16.42 | 22.03 |
| 3000 | 6.804 | 6.874 | 13.056 | 6.911 | 10.926 | 16.60 | 22.26 |
| 3200 | 6.890 | 6.905 | 13.130 | 6.939 | 11.028 | 16.74 | 22.42 |
| 3400 | 6.968 | 6.933 | 13.197 | 6.964 | 11.118 | 16.85 | 22.57 |
| 3600 | 7.039 | 6.959 | 13.258 | 6.987 | 11.195 | 16.95 | 22.70 |
| 3800 | 7.104 | 6.981 | 13.315 | 7.008 | 11.262 | 17.04 | 22.81 |
| 4000 | 7.164 | 7.002 | 13.368 | 7.028 | 11.321 | 17.12 | 22.91 |
| 4200 | 7.219 | 7.021 | 13.418 | 7.046 | 11.371 | 17.18 | 22.99 |
| 4400 | 7.270 | 7.039 | 13.465 | 7.064 | 11.415 | 17.24 | 23.06 |
| 4600 | 7.317 | 7.056 | 13.510 | 7.080 | 11.457 | 17.29 | 23.13 |
| 4800 | 7.361 | 7.073 | 13.555 | 7.095 | 11.496 | 17.34 | 23.18 |
| 5000 | 7.402 | 7.089 | 13.599 | 7.109 | 11.534 | 17.38 | 23.24 |
| 5200 | 7.440 | 7.104 | 13.633 | 7.123 | 11.565 | 17.42 | 23.28 |
| 5400 | (7.475) | (7.119) | (13.665) | (7.137) | (11.594) | 17.45 | 23.31 |
| 5600 | (7.510) | (7.133) | (13.696) | (7.150) | (11.620) | 17.48 | 23.35 |
| 5800 | (7.543) | (7.144) | (13.725) | (7.163) | (11.645) | 17.51 | 23.39 |
| 6000 | (7.577) | (7.154) | (13.752) | (7.175) | (11.667) | 17.53 | 23.42 |

[1]Adapted from National Bureau of Standards, "Selected Values of Chemical Thermodynamic Properties Series III"
[2]Calculated from Statistical Mechanics

**TABLE A.4.** (Continued)

| T (°K) | CH$_3$OH$^{(2)}$ | CH$_2$O$_2^{(2)}$ | O$_2^{(1)}$ | OH$^{(1)}$ | NO$^{(1)}$ | H$^{(1)}$ |
|---|---|---|---|---|---|---|
| 400 | 9.94 | 11.62 | 5.207 | 5.087 | 5.175 | 2.981 |
| 600 | 13.77 | 14.57 | 5.683 | 5.066 | 5.481 | 2.981 |
| 800 | 16.89 | 16.62 | 6.077 | 5.163 | 5.846 | 2.981 |
| 1000 | 19.33 | 18.12 | 6.348 | 5.346 | 6.139 | 2.981 |
| 1200 | 21.21 | 19.29 | 6.543 | 5.564 | 6.355 | 2.981 |
| 1400 | 22.65 | 20.07 | 6.689 | 5.785 | 6.511 | 2.981 |
| 1600 | 23.76 | 20.73 | 6.814 | 5.986 | 6.627 | 2.981 |
| 1800 | 24.58 | 21.24 | 6.930 | 6.165 | 6.715 | 2.981 |
| 2000 | 25.25 | 21.64 | 7.043 | 6.321 | 6.784 | 2.981 |
| 2200 | 25.78 | 21.96 | 7.153 | 6.456 | 6.841 | 2.981 |
| 2400 | 26.20 | 22.22 | 7.262 | 6.574 | 6.887 | 2.981 |
| 2600 | 26.54 | 22.36 | 7.367 | 6.675 | 6.926 | 2.981 |
| 2800 | 26.83 | 22.60 | 7.468 | 6.767 | 6.962 | 2.981 |
| 3000 | 27.07 | 22.75 | 7.565 | 6.851 | 6.994 | 2.981 |
| 3200 | 27.25 | 22.86 | 7.654 | 6.925 | 7.023 | 2.981 |
| 3400 | 27.43 | 22.96 | 7.736 | 6.994 | 7.049 | 2.981 |
| 3600 | 27.57 | 23.05 | 7.812 | 7.059 | 7.074 | 2.981 |
| 3800 | 27.69 | 23.13 | 7.882 | 7.119 | 7.098 | 2.981 |
| 4000 | 27.79 | 23.20 | 7.946 | 7.175 | 7.120 | 2.981 |
| 4200 | 27.89 | 23.26 | 8.002 | 7.228 | 7.141 | 2.981 |
| 4400 | 27.96 | 23.31 | 8.052 | 7.278 | 7.161 | 2.981 |
| 4600 | 28.03 | 23.35 | 8.097 | 7.327 | 7.181 | 2.981 |
| 4800 | 28.09 | 23.39 | 8.136 | 7.374 | 7.201 | 2.981 |
| 5000 | 28.15 | 23.42 | 8.170 | 7.419 | 7.221 | 2.981 |
| 5200 | 28.20 | 23.45 | 8.202 | 7.461 | 7.240 | 2.981 |
| 5400 | 28.24 | 23.48 | (8.233) | (7.502) | (7.259) | 2.981 |
| 5600 | 28.28 | 23.50 | (8.262) | (7.542) | (7.279) | 2.981 |
| 5800 | 28.32 | 23.52 | (8.291) | (7.580) | (7.298) | 2.981 |
| 6000 | 28.34 | 23.54 | (8.318) | (7.616) | (7.317) | 2.981 |

| T (°K) | N[1] | O[1] | HCN[2] | CH$_2$O[2] | C$_2$H$_4$[2] | C$_2$H$_6$[2] |
|---|---|---|---|---|---|---|
| 400 | 2.981 | 3.147 | 7.40 | 7.35 | 10.99 | 12.61 |
| 600 | 2.981 | 3.062 | 8.48 | 9.48 | 15.18 | 18.63 |
| 800 | 2.981 | 3.028 | 9.31 | 11.36 | 18.27 | 23.46 |
| 1000 | 2.981 | 3.012 | 9.99 | 12.79 | 20.61 | 27.16 |
| 1200 | 2.981 | 3.002 | 10.52 | 13.87 | 22.42 | 29.99 |
| 1400 | 2.981 | 2.997 | 10.96 | 14.67 | 23.78 | 32.08 |
| 1600 | 2.981 | 2.993 | 11.30 | 15.27 | 24.90 | 33.70 |
| 1800 | 2.981 | 2.991 | 11.56 | 15.73 | 25.70 | 34.87 |
| 2000 | 2.982 | 2.991 | 11.77 | 16.08 | 26.35 | 35.82 |
| 2200 | 2.984 | 2.991 | 11.94 | 16.36 | 26.87 | 36.56 |
| 2400 | 2.988 | 2.994 | 12.08 | 16.58 | 27.27 | 37.16 |
| 2600 | 2.996 | 3.000 | 12.18 | 16.75 | 27.60 | 37.62 |
| 2800 | 3.007 | 3.007 | 12.27 | 16.90 | 27.88 | 38.02 |
| 3000 | 3.024 | 3.017 | 12.35 | 17.02 | 28.11 | 38.35 |
| 3200 | 3.050 | 3.031 | 12.41 | 17.12 | 28.29 | 38.61 |
| 3400 | 3.082 | 3.046 | 12.46 | 17.20 | 28.45 | 38.84 |
| 3600 | 3.123 | 3.064 | 12.51 | 17.27 | 28.59 | 39.04 |
| 3800 | 3.171 | 3.083 | 12.55 | 17.33 | 28.72 | 39.21 |
| 4000 | 3.227 | 3.104 | 12.58 | 17.38 | 28.82 | 39.35 |
| 4200 | 3.294 | 3.127 | 12.61 | 17.43 | 28.90 | 39.47 |
| 4400 | 3.368 | 3.151 | 12.64 | 17.46 | 28.97 | 30.58 |
| 4600 | 3.448 | 3.175 | 12.66 | 17.50 | 29.04 | 39.67 |
| 4800 | 3.534 | 3.199 | 12.68 | 17.53 | 29.10 | 39.76 |
| 5000 | 3.624 | 3.223 | 12.70 | 17.56 | 29.16 | 39.83 |
| 5200 | 3.720 | 3.247 | 12.72 | 17.58 | 29.21 | 39.89 |
| 5400 | (3.819) | (3.271) | 12.73 | 17.60 | 29.25 | 39.95 |
| 5600 | (3.921) | (3.295) | 12.74 | 17.62 | 29.29 | 40.01 |
| 5800 | (4.025) | (3.318) | 12.75 | 17.63 | 29.32 | 40.06 |
| 6000 | (4.131) | (3.340) | 12.76 | 17.65 | 29.35 | 40.10 |

**TABLE A.4. (Continued)**

| T(°K) | CH₂O⁽²⁾ | C₂H₅OH⁽²⁾ | C(s) | Al₂O₃(s) | AlO(g') | Al₂O (g) |
|---|---|---|---|---|---|---|
| 400 | 7.35 | 16.69 | 2.851 | | | |
| 600 | 9.48 | 23.31 | 4.02 | | | |
| 800 | 11.36 | 28.18 | 4.74 | | | |
| 1000 | 12.79 | 31.84 | 5.13 | 27.89 | 6.658 | 10.868 |
| 1200 | 13.87 | 34.62 | 5.41 | 30.70 | 6.743 | 11.150 |
| 1400 | 14.67 | 36.75 | 5.66 | 31.34 | 6.800 | 11.339 |
| 1600 | 15.27 | 38.36 | 5.82 | 31.83 | 6.832 | 11.463 |
| 1800 | 15.73 | 39.60 | 5.94 | 32.28 | 6.857 | 11.553 |
| 2000 | 16.08 | 40.57 | 6.04 | 32.66 | 6.875 | 11.620 |
| 2200 | 16.36 | 41.34 | 6.13 | 32.92 | 6.889 | 11.670 |
| 2400 | 16.58 | 41.96 | 6.21 | 33.15 | 6.899 | 11.709 |
| 2600 | 16.75 | 42.45 | 6.28 | 32.29 | 6.907 | 11.736 |
| 2800 | 16.90 | 42.87 | 6.35 | 32.44 | 6.913 | 11.760 |
| 3000 | 17.02 | 43.20 | 6.41 | 33.55 | 6.919 | 11.783 |
| 3200 | 17.12 | 43.48 | 6.47 | 33.64 | 6.924 | 11.799 |
| 3400 | 17.20 | 43.71 | 6.53 | 32.71 | 6.927 | 11.813 |
| 3600 | 17.27 | 43.91 | 6.59 | 33.76 | 6.929 | 11.825 |
| 3800 | 17.33 | 44.09 | 6.65 | 33.79 | 9.932 | 11.834 |
| 4000 | 17.38 | 44.24 | 6.71 | 33.83 | 6.934 | 11.843 |
| 4200 | 17.43 | 44.37 | 6.77 | 33.85 | 6.936 | 11.850 |
| 4400 | 17.46 | 44.48 | (6.83) | 33.87 | 6.938 | 11.857 |
| 4600 | 17.50 | 44.58 | (6.89) | 33.87 | 6.940 | 11.863 |
| 4800 | 17.53 | 44.67 | (6.95) | 33.89 | 6.941 | 11.868 |
| 5000 | 17.56 | 44.75 | (7.01) | 33.90 | 6.942 | 11.873 |
| 5200 | 17.58 | 44.81 | (7.07) | 33.91 | 6.943 | 11.877 |
| 5400 | 17.60 | 44.88 | (7.13) | 33.92 | 6.944 | 11.881 |
| 5600 | 17.62 | 44.94 | (7.19) | 33.92 | 6.944 | 11.884 |
| 5800 | 17.63 | 44.99 | (7.24) | 33.93 | 6.946 | 11.887 |
| 6000 | 17.65 | 45.04 | (7.30) | 33.94 | 6.946 | 11.889 |

**TABLE A.5.** $C_v$ Average Ideal Molal Heat Capacities (from $300°-TK$) (cal $\cdot$ mole$^{-1}$ $\cdot$ K$^{-1}$)

| T(°K) | H$_2$ | N$_2$ | CO$_2$ | CO | H$_2$O | NH$_3$ | CH$_4$ | O | HCN | CH$_2$O | C$_2$H$_4$ | C$_2$H$_6$ | C$_2$H$_5$OH |
|---|---|---|---|---|---|---|---|---|---|---|---|---|---|
| 1000 | 5.054 | 5.323 | 9.395 | 5.400 | 6.867 | 9.62 | 11.04 | 3.073 | 8.59 | 9.81 | 15.53 | 19.36 | 23.92 |
| 1200 | 5.114 | 5.464 | 9.812 | 5.548 | 7.153 | 10.22 | 12.15 | 3.058 | 8.96 | 10.60 | 16.87 | 21.42 | 26.00 |
| 1400 | 5.187 | 5.593 | 10.153 | 5.681 | 7.430 | 10.77 | 13.14 | 3.047 | 9.29 | 11.27 | 18.01 | 23.18 | 27.77 |
| 1600 | 5.270 | 5.708 | 10.437 | 5.796 | 7.694 | 11.27 | 13.99 | 3.039 | 9.57 | 11.85 | 18.99 | 24.67 | 29.28 |
| 1800 | 5.357 | 5.810 | 10.678 | 5.896 | 7.941 | 11.72 | 14.74 | 3.033 | 9.82 | 12.33 | 19.83 | 25.96 | 30.58 |
| 2000 | 5.445 | 5.900 | 10.884 | 5.984 | 8.168 | 12.12 | 15.39 | 3.028 | 10.04 | 12.75 | 20.56 | 27.06 | 31.70 |
| 2200 | 5.532 | 5.979 | 11.063 | 6.060 | 8.377 | 12.48 | 15.96 | 3.024 | 10.23 | 13.12 | 21.20 | 28.02 | 32.68 |
| 2400 | 5.615 | 6.049 | 11.221 | 6.128 | 8.567 | 12.80 | 16.47 | 3.021 | 10.40 | 13.44 | 21.75 | 28.87 | 33.53 |
| 2600 | 5.695 | 6.111 | 11.358 | 6.187 | 8.738 | 13.08 | 16.91 | 3.019 | 10.55 | 13.72 | 22.24 | 29.59 | 34.27 |
| 2800 | 5.771 | 6.168 | 11.485 | 6.241 | 8.899 | 13.34 | 17.31 | 3.018 | 10.68 | 13.98 | 22.69 | 30.26 | 34.95 |
| 3000 | 5.845 | 6.219 | 11.600 | 6.290 | 9.045 | 13.58 | 17.67 | 3.017 | 10.80 | 14.20 | 23.09 | 30.86 | 35.56 |
| 3200 | 5.914 | 6.265 | 11.702 | 6.333 | 9.177 | 13.79 | 17.99 | 3.018 | 10.91 | 14.39 | 23.46 | 31.37 | 36.09 |
| 3400 | 5.979 | 6.308 | 11.795 | 6.373 | 9.299 | 13.98 | 18.28 | 3.019 | 11.01 | 14.57 | 23.77 | 31.85 | 36.57 |
| 3600 | 6.041 | 6.346 | 11.862 | 6.410 | 9.413 | 14.16 | 18.54 | 3.021 | 11.10 | 14.74 | 24.05 | 32.28 | 37.01 |
| 3800 | 6.100 | 6.382 | 11.963 | 6.444 | 9.517 | 14.33 | 18.79 | 3.024 | 11.18 | 14.88 | 24.31 | 32.67 | 37.41 |
| 4000 | 6.157 | 6.415 | 12.039 | 6.475 | 9.614 | 14.48 | 19.01 | 3.028 | 11.26 | 15.02 | 24.55 | 33.04 | 37.78 |
| 4200 | 6.210 | 6.445 | 12.108 | 6.504 | 9.702 | 14.61 | 19.21 | 3.033 | 11.32 | 15.14 | 24.77 | 33.36 | 38.12 |
| 4400 | 6.260 | 6.474 | 12.172 | 6.531 | 9.804 | 14.74 | 19.39 | 3.038 | 11.39 | 15.25 | 24.97 | 33.66 | 38.42 |
| 4600 | 6.308 | 6.501 | 12.234 | 6.556 | 9.861 | 14.86 | 19.56 | 3.043 | 11.45 | 15.36 | 25.16 | 33.94 | 38.71 |
| 4800 | 6.354 | 6.526 | 12.292 | 6.580 | 9.933 | 14.97 | 19.73 | 3.050 | 11.50 | 15.45 | 25.34 | 34.19 | 38.97 |
| 5000 | 6.398 | 6.550 | 12.347 | 6.602 | 10.001 | 15.07 | 19.88 | 3.057 | 11.55 | 15.54 | 25.50 | 34.43 | 39.22 |
| 5200 | 6.440 | 6.572 | 12.398 | 6.623 | 10.064 | 15.17 | 20.02 | 3.064 | 11.60 | 15.62 | 25.65 | 34.65 | 39.45 |
| 5400 | 6.479 | 6.593 | 12.448 | 6.642 | 10.124 | 15.25 | 20.14 | 3.072 | 11.64 | 15.69 | 25.79 | 34.87 | 39.67 |
| 5600 | 6.518 | 6.608 | 12.495 | 6.661 | 10.181 | 15.33 | 20.26 | 3.080 | 11.68 | 15.77 | 25.92 | 35.06 | 39.86 |
| 5800 | 6.555 | 6.633 | 12.541 | 6.679 | 10.235 | 15.42 | 20.38 | 2.088 | 11.73 | 15.84 | 26.04 | 35.23 | 40.04 |
| 6000 | 6.590 | 6.651 | 12.586 | 6.697 | 10.286 | 15.49 | 20.48 | 3.096 | 11.76 | 15.90 | 26.16 | 35.41 | 40.22 |

**TABLE A.5.** (Continued)

| T(°K) | CH₃OH | CH₂O₂ | O₂ | OH | NO | H | N | C(s) | Al₂O₃(s) | AlO(g) | Al₂O(g) | SO₂ | CaO | Al₂O₃ |
|---|---|---|---|---|---|---|---|---|---|---|---|---|---|---|
| 1000 | 14.27 | 14.74 | 5.746 | 5.137 | 5.587 | 2.981 | 2.981 | 3.31 | 26.67 | 6.22 | 9.69 | | | |
| 1200 | 15.61 | 15.62 | 5.903 | 5.207 | 5.735 | 2.981 | 2.981 | 4.31 | 27.48 | 6.33 | 9.98 | | | |
| 1400 | 16.76 | 16.36 | 6.033 | 5.292 | 5.862 | 2.981 | 2.981 | 4.53 | 28.13 | 6.41 | 10.21 | | | |
| 1600 | 17.76 | 16.98 | 6.143 | 5.384 | 5.971 | 2.981 | 2.981 | 4.72 | 28.66 | 6.47 | 10.40 | | | |
| 1800 | 18.61 | 17.52 | 6.241 | 5.476 | 6.065 | 2.981 | 2.981 | 4.88 | 28.11 | 6.52 | 10.54 | | | |
| 2000 | 19.35 | 17.98 | 6.328 | 5.567 | 6.146 | 2.981 | 2.981 | 5.01 | 29.51 | 6.56 | 10.67 | 10.4 | | |
| 2200 | 20.00 | 18.38 | 6.409 | 5.654 | 6.216 | 2.981 | 2.981 | 5.13 | 29.86 | 6.60 | 10.77 | | | |
| 2400 | 20.58 | 18.73 | 6.485 | 5.736 | 6.278 | 2.981 | 2.982 | 5.23 | 30.16 | 6.63 | 10.86 | | | |
| 2600 | 21.07 | 19.04 | 6.557 | 5.812 | 6.332 | 2.981 | 2.983 | 5.31 | 30.42 | 6.65 | 10.94 | | | |
| 2800 | 21.52 | 19.32 | 6.627 | 5.885 | 6.381 | 2.981 | 2.984 | 5.40 | 30.64 | 6.67 | 11.00 | | | |
| 3000 | 21.93 | 19.57 | 6.693 | 5.954 | 6.426 | 2.981 | 2.986 | 5.47 | 30.85 | 6.69 | 11.06 | 11.1 | 15.2 | 27.5 |
| 3200 | 22.29 | 19.79 | 6.755 | 6.018 | 6.465 | 2.981 | 2.990 | 5.53 | 31.03 | 6.71 | 11.11 | | | |
| 3400 | 22.61 | 19.98 | 6.816 | 6.079 | 6.502 | 2.981 | 2.995 | 5.60 | 31.21 | 6.72 | 11.15 | | | |
| 3600 | 22.91 | 20.17 | 6.874 | 6.137 | 6.536 | 2.981 | 3.001 | 5.66 | 31.36 | 6.73 | 11.19 | | | |
| 3800 | 23.18 | 20.34 | 6.930 | 6.191 | 6.567 | 2.981 | 3.009 | 5.71 | 31.50 | 6.74 | 11.23 | | | |
| 4000 | 23.43 | 20.50 | 6.984 | 6.243 | 6.597 | 2.981 | 3.019 | 5.77 | 31.63 | 6.75 | 11.26 | | | |
| 4200 | 23.65 | 20.64 | 7.034 | 6.292 | 6.624 | 2.981 | 3.032 | 5.82 | 31.74 | 6.76 | 11.29 | 11.3 | | 28.0 |
| 4400 | 23.86 | 20.78 | 7.082 | 6.339 | 6.650 | 2.981 | 3.046 | 5.87 | 31.84 | 6.76 | 11.32 | | | |
| 4600 | 24.05 | 20.90 | 7.128 | 6.384 | 6.675 | 2.981 | 3.063 | 5.91 | 31.94 | 6.77 | 11.35 | | | |
| 4800 | 24.23 | 21.00 | 7.173 | 6.426 | 6.697 | 2.981 | 3.082 | 5.96 | 32.04 | 6.78 | 11.37 | | | |
| 5000 | 24.40 | 21.10 | 7.215 | 6.468 | 6.719 | 2.981 | 3.103 | 6.00 | 32.11 | 6.79 | 11.39 | | | 28.3 |
| 5200 | 24.55 | 21.20 | 7.254 | 6.508 | 6.740 | 2.981 | 3.126 | 6.05 | 32.19 | 6.80 | 11.41 | | | |
| 5400 | 24.69 | 21.29 | 7.292 | 6.546 | 6.760 | 2.981 | 3.152 | 6.09 | 32.26 | 6.80 | 11.43 | | | |
| 5600 | 24.83 | 21.37 | 7.328 | 6.583 | 6.779 | 2.981 | 3.179 | 6.13 | 32.32 | 6.81 | 11.45 | | | |
| 5800 | 24.96 | 21.45 | 7.362 | 6.619 | 6.798 | 2.981 | 3.208 | 6.17 | 32.38 | 6.82 | 11.47 | | | |
| 6000 | 25.08 | 21.52 | 7.396 | 6.653 | 6.816 | 2.981 | 3.238 | 6.21 | 32.43 | 6.82 | 11.48 | | | |

# REFERENCES

Adamson, M. G., R. Cooper, and B. Perel, "The Wearing of Rockdrill Metals: How to Ease the Problem," *S. Afr. Mining Eng. J.* **89,** 62–70 (1980).

Aeberli, H. U., and H. Wanner, "On the Influence of Discontinuities on the Application of Tunneling Machines," in *Proc. 3rd Int. Congr. Int. Assoc. Eng. Geol.*, Sec. III, Vol. 2 (Madrid), 1978, pp. 7–14.

Appl, F. C., and D. S. Rowley, "Drilling on Drag Bit Cutting Edges." University of Minnesota, School of Mines and Metallurgy, 5th Symposium on Rock Mechanics, Minneapolis, MN, 1963, pp. 119–136.

Ayataman, V., "Causes of Hanging in Ore Chutes." *Can. Mining J.* **81,** 77–81 (1960).

Azo, K., "Phenomenon Involved in Presplitting by Blasting." Unpublished doctoral disseration, Stanford University, Palo Alto, CA, 1966.

Banks, D. C., "Selected Methods for Analyzing the Stability of Crater Slopes." U.S.A. Waterways Experiment Station, Miscellaneous Paper S-68-8, 1968.

Banks, D. C., and B. N. McIver, "Variation in Angle of Internal Friction with Confining Pressure." U.S.A. Waterways Experiment Station, Miscellaneous Paper S-69-12, 1969.

Barendsen, P., and R. G. Cadden, "Machine-Bored Small-Sized Tunnels in Rock with Some Case Studies." *Proc. Tunnelling '76,* 423–433 (1976).

Barker, D. B., W. L. Fourney, and D. C. Holloway, Photoelastic Investigation of Flaw Initiated Cracks and Their Contribution to the Mechanisms of Fragmentation. 21st Symposium on Rock Mechanics Austin, TX, June 1979.

Batten, C., D. G. A. Thomas, and T. Wainwright, "The Measurement of Rotary Drill Bit Wear." *Colliery Eng.* **31** (359), 21–25 (1954).

Becker, K. R., C. M. Mason, and R. W. Watson, "Bureau of Mines Instrumented Impact Tester, Preliminary Studies." U.S. Bureau of Mines Report of Investigations No. 7670, 1972.

Benjumea, R., and D. L. Sikarskie, "A Note on the Penetration of a Rigid Wedge into a Nonisotropic Grittle Material." *Int. J. Rock Mechanics Mining Sci.* **6** (4), 343–352 (1965).

Bickel, J. O., and T. R. Kuesel, Eds., *Tunnel Engineering Handbook.* Van Nostrand Reinhold, New York, 1982.

Birkhoff, G., and E. H. Sarantello, *Jets, Wakes and Cavities.* Academic Press, New York, 1957.

Blake, R., "Cutting Action of Diamond Drill Bit," unpublished master's dissertation, University of Minnesota, Minneapolis, MN, 1951.

Blindheim, O. T., "Preinvestigations, Resistance to Blasting and Drillability Predictions in Hard Rock Tunneling," Mechanical Boring or Drill and Blast Tunnelling, First U.S.–Swedish Underground Workshop, Stockholm, 1976, pp. 81–97.

Blindheim, O. T., "Drillability Predictions in Hard Rock Tunnelling," *Proc. Tunnelling '79,* London, 1979, pp. 284–289.

Bloemsma, J. H., R. Ramsay, and O. Deane, "Some Experiments with Tungsten Carbide Tipped Drill Steel." *J. Chem. Metallurgical Mining S. S. Afr.* **47,** 243–283 (1947).

Blomberry, R. I., C. M. Perrot, and P. M. Robinson, "Abrasive Wear of Tungsten Carbide–Cobalt Composites. I. Wear Mechanisms." *Metals Sci. Eng.* **13,** 93–100 (1974).

Bowden, F. P., and D. Tabor, *The Friction and Lubrication of Solids.* Part II. Oxford University Press, Oxford, U.K., 1950.

Bowden, F. P., and A. D. Yoffe, *Initiation and Growth of Explosions in Liquid and Solids*. Cambridge University Press, Cambridge, U.K., 1952.

Brinkley, S. R. Jr., and R. W. Smith Jr., *Proceedings of Seminar on Scientific Computation*. International Business Machines Corporation, New York, 1949, p. 58.

Brook, N., "The Use of Irregular Specimens for Rock Strength Tests." *Int. J. Rock Mechanics Mining Sci.* **14**, 193–202 (1977).

Brook, N., and D. A. Summers, "The Penetration of Rock by High-Speed Water Jets." *Int. J. Rock Mechanics Mining Sci.* **6**, 249–258 (1969).

Brown, F. W., "Theoretical Calculations for Explosives." U.S. Bureau of Mines Technical Publication No. 632, 1941.

Bruce, W. E., "Bureau of Mines Conducts Diamond Drilling Experiments." *Mines Magazine* **58** (4), 17–21 (1968).

Bruce, W. E. and R. J. Morrell, "Principles of Rock Cutting Applied to Mechanical Boring Machines." American Institute of Mining and Metallurgical Engineers, Second Symposium on Rapid Excavation, 1969a, pp. 314–332.

Bruce, W. E. and R. J. Morrell, "Principles of Rock Cutting Applied to Mechanical Boring Machines," *Proceedings 2nd Symposium on Rapid Excavation*. Sacramento State College, Sacramento, CA, Oct. 1969b, pp. 3-1 to 3-43.

Bruce, W. E. and J. Paone, "Energetics of Percussive Drills." U.S. Bureau of Mines Report of Investigations No. 7235, 1969.

Brune J., "Drillability Studies in Oil Shale." Unpublished master's dissertation, Colorado School of Mines, Golden, CO, 1983.

Bucky, P. B., "The Use of Models for the Study of Mining Problems." American Institute of Mining and Metallurgical Engineers, Technical Publication No. 425, 1931.

Bullock, R. L., "Industry-wide Trends Toward All-Hydraulically Powered Rock Drills." *Mining Congr. J.* **60** (10), 54–65 (1974).

Bullock, R. L., "Technological Review of All-Hydraulic Rock Drills." Transactions of the American Institute of Mining, Metallurgical, and Petroleum Engineers–Society of Mining Engineers, preprint No. 75-Au-42, 1975.

Bullock, R. L., "An Update of Hydraulic Drilling Performance." American Institute of Mining, Metallurgical, and Petroleum Engineers, Rapid Excavation and Tunneling Conference, Las Vegas, 1976, pp. 627–648.

Calaman, J. J., and H. C. Rolseth, "Jet Piercing." In *Surface Mining*. American Institute of Mining, Metallurgical, and Petroleum Engineers, New York, 1968, pp. 325–337.

Carboni, E., "Un Contributo Alla di un Metodo Otto a Valutare la Resista Lorgio per Abrasione dei Carbuni Sinterizatti." *Metallurgia Ital.* **61**, 593–601 (1969).

Charles, R. J., "Energy–Size Reduction Relationship in Comminution." *Trans. Am. Inst. Mining Metallurgical Petroleum Eng.* **208**, 80–88 (1957).

Cheetham, W. R., and E. W. Inett, "Factors Affecting the Performance of Percussive Drills." *Trans. Inst. Mining Metallurgy* **63**, 45–74 (1953–55).

Chen, G., "Computer Modeling Studies on Gravity Flow of Coarse Granular Material." Unpublished master's thesis, Colorado School of Mines, Golden, CO, 1984.

Chollette, H., G. B. Clark and T. Lehnhoff, "Fracture Stresses Induced by Rock Splitters." *Int. J. Rock Mechanics Mining Sci.* **13**, 281–287 (1976).

Clark, G. B., "Mathematics of Explosives Calculations." Fourth Symposium on Mining Research. University of Missouri School of Mines and Metallurgy Bulletin Technical Series No. 97.32-80, 1959.

Clark, G. B., "Principles of Rock Drilling." *Colo. School Mines Q.* **74** (2) (1979).

Clark, G. B., "Geotechnical Centrifuges for Model Studies and Physical Property Testing of Rock Structures." *Colo. School Mines Q.* **76** (4) (1981).

Clark, G. B., R. F. Bruzewski, J. J. Yancik, J. E. Lyons, and R. Hopler, "Particle Characteristics of Ammonium Nitrate and Blasting Agent Performance." *Colo. School Mines Q.* **56,** 183–198 (1961).

Clark, G. B., R. F. Bruzewski, J. G. Stites, J. E. Lyons, and J. J. Yancik, "Performance Parameters of Densified Micro-Prilled Ammonium Nitrate–Fuel Oil Blasting Agents." *Univ. MO. School Mines Metallurgy Int. Symp. Mining Res.* **1,** 29–45, 1962.

Clark, G. B., and H. Maleki, "Basic Operational Parameters of an Automated Plug and Feather Rock Splitter." Sponsored by National Science Foundation, NSF Ap73-07486-A02, Colorado School of Mines, 1978.

Clutterbuck, M., "The Dependence of Stress Distribution on Elastic Constants." *Br. J. Appl. Phys.* **9,** 323–329 (1958).

Conway, J. C., Jr., and H. P. Kirchner, "The Mechanics of Crack Initiation and Propagation Beneath A Moving Sharp Indentor." Jl. of Materials Science Vol. 15, No. 11, Nov. 1980, pp. 2879–2883.

Cook, M. A., Class Lecture Notes, unpublished, University of Utah, 1954.

Cook, M. A., *The Science of High Explosives.* Reinhold, New York, 1958.

Cook, M. A., *The Science of Industrial Explosives.* Graphics Service, Salt Lake City, Utah, 1974.

Cook, N. G. W., "Analysis of Hard Rock Cuttability of Machines." Conference on Tunnel and Shaft Excavation, University of Minnesota, 1968.

Cook, N. G. W., N. C. Toughlin, and G. A. Weibols, "Rock Cutting and Its Potentialities as a New Method of Mining." *J. S. Afr. Inst. Mining Metallurgy* 435–454 (1968).

Craig, W. H. (Ed.), *The Application of Centrifuge Modelling to Geotechnical Design.* University of Manchester, Manchester, U.K., 1984.

Critchfield, J., Personal communication in Nelson, et al. 1984.

Crow, S. C., "A Theory of Hydraulic Rock Cutting." *Int. J. Rock Mechanics Mining Sci.* **10,** 567–584 (1973).

Crow, S. C., and G. H. Hurlburt, "The Mechanics of Hydraulic Rock Cutting." *2nd International Symposium on Jet Cutting Technology.* Paper E1, Cambridge, U.K., 1974.

Cunningham, C., "The Kuz-Ram Model for Prediction of Fragmentation from Blasting." First International Symposium on Rock Fragmentation by Blasting, Lulea, Sweden, Vol. 2, pp. 439–452, 1983.

Dahlin, C., "Factors Influencing the Life of Drill Steel Equipment." *Univ. MO. School Mines Metallurgy Int. Symp. Mining Res.* **1,** 351–370 (1962).

Dallavalle, J. M., *Micromeritics.* Pitman, New York, 1948.

Daneshy, H. A., "Numerical Inversion of the LaPlace Transformation and the Solution of Viscoelastic Wave Equations." Unpublished doctoral dissertation, University of Missouri, Rolla, MO, 1969.

Dawihl, W., and B. Frisch, "Wear Properties of Tungsten Carbide and Aluminum Oxide Sintered Materials." *Wear* **12,** 17–25 (1968).

Dawihl, W., and G. Altmeyer. *Z. Metallkunde* **55,** 231 (1964).

Descoeudres, F., and G. Rechsteiner, "Etude de Correlacions entre la Geologie, les Properties Mechaniques et la Forabilite des Roches de Crespera-Gemmo." *Schweizerische Banzeitouq* **48** (3), 9–16 (1973).

Dick, R. A. and L. R. Gletcher, "A Study of Fragmentation from Bench Blasting in Limestone at a Reduced Scale." U.S. Bureau of Mines Report of Investigations No. 7704, 1973.

Dinsdale, J. R., "Ground Failure Around Excavations." *Trans. Inst. Mining Metallurgy* **L,** (1940).

Ditson, J. D., "Determining Blow Energy of Rock Drills." *Compressed Air Magazine* **53** (1), 15–16 (1948).

Dobson, P. S. and H. Wilman, "Friction and Wear and Their Interrelationship in Abrasion of a Single Crystal of Brittle Nature." *Br. J. Appl. Phys.* **14,** 132 (1963).

Doeg, H. H., "Cement Hard Metals—Their Basis with Particular Reference to the Tungsten Carbide Cobalt System." *J. S. Afr. Inst. Mining Metallurgy* **60,** 663 (1959-1960).

Dollinger, G. L., Personnal communication, 1982 (cited in Nelson et al., 1984).

Dutta, P. K., "The Determination of Stress Wave Forms Produced by Percussive Drill Pistons of Various Geometrics Design." *Int. J. Rock Mechanics Mining Sci.* **5,** 501-508 (1968).

Duvall, W. I., "Strain Wave Shapes in Rock Near Explosions." *Geophysics* **18,** 310-323 (1953).

Duvall, W. I., and T. C. Atchison, "Rock Breakage by Explosives." U.S. Bureau of Mines Report of Investigations No. 5356, 1957.

Exner, H. E., "The Influence of Sample Preparation on Palmqvist's Method of Toughness Testing of Cemented Carbides." *Trans. Am. Inst. Mining Metallurgical Petroleum Eng.* **245,** 677 (1969).

Exner, H. E., and J. Gurland, "A Review of Parameters Influencing Some Mechanical Properties of Tungsten Carbide–Cobalt Alloys." *Powder Metallurgy* **13,** 13 (1970).

Eyring, H., R. E. Powell, G. H. Duffey, and R. B. Parlin, "Stability of Detonation." *Chem. Rev.* **45,** 69 (1949).

Faddeenkov, N. N., "Application of the Rozin–Rammler Law to the Analysis of the Grain Size Composition of Blasted Rock." Institute of Mining, Novosibirsk Fizio-Technisheskie Problemy Razabotke Poleznhk Iskopacymkh, No. 6, Nov.–Dec. 1974, No. 1, Jan.–Feb. 1975.

Fairhurst, C., "Wave Mechanics in Percussive Drilling." *Mining Quarry Eng.* **27,** (1961).

Fairhurst, C. and Lacabanne, W. D., 1957, *Hard Rock Drilling Techniques, Mining, and Quarry Engineering*, Vol. 23, pp. 157, 174.

Fayol, M., *Revue de l'Industrie Minerale*, Vol. XIV, 1885.

Fessler, H., and H. Lewin, "A Study of Large Strains and the Effect of Different Values of Poisson's Ratio." *Br. J. Appl. Phys.* **11,** (1960).

Field, J. E., and A. Ladegaarde-Pedersen, "The Importance of the Reflected Wave in Blasting." *Int. J. Rock Mechanics Mining Sci.* **8,** 213-226 (1971).

Fischmeister, H., and H. E. Exner, "Gefuegebhaenigkeit der eigenschaften von Wolfram-Carbid-Kobalt-Hartlegierungen." *Arch. Eisenwhuettenwesen.* **37,** 499-510 (1966).

Fish, B. G., "A Comparison of Percussive, Rotary, and Percussive–Rotary Drilling." *Trans. Inst. Mining and Metallurgy* **116,** 775-789 (1956-57).

Fish, B. G., "Studies in Percussive-Rotary Drilling." *Colliery Eng.* **34,** 101-104 (1957a).

Fish, B. G., "Studies in Percussive-Rotary Drilling." *Colliery Eng.* **34,** 141-146 (1957b).

Fish, B. G., "Studies with Water and Air as Flushing Media in Rock Drilling." *Mine Quarry Eng.* **23,** 306-310, (1957c).

Fish, B. G., "Measuring Abrasive Bit Wear." *Mine Quarry Eng.* **24,** 264-267 (1958).

Fish, B. G., and J. S. Barker, "The Design of Rotary Drilling Tools." *Colliery Eng.* **34,** 513-518 (1957).

Fish, B. G., G. A. Suppy, and J. T. Ruben, "Abrasive Wear Effects in Rotary Rock Drilling." *Trans. Inst. Mining Metallurgy* **68,** 357-383 (1959).

Fogelson, D. E., W. I. Duvall, and T. C. Atchison, "Strain Energy in Explosive-Generated Strain Pulses." U.S. Bureau of Mines Report of Investigations No. 5514, 1959.

Foreman, S. E., and G. A. Secor, "The Mechanics of Rock Failure Due to Water Jet Impingement." Proceedings of the 6th Conference on Drilling and Rock Mechanics, Society of Petroleum Engineers, Austin, TX, 1973.

Fowkes, R. S., and J. S. Wallace, "Hydraulic Coal Mining Research." U.S. Bureau of Mines Report of Investigations No. 7090, 1968.

Gaye, F., "Efficient Excavation." *Tunnels Tunneling* **4,** Pt. I, No. 1, 34-48 (1972).

Getzler, F., A. Komornek, A. Mazwicks, "Model Study on Arching Above Buried Structures," *J. Soil Mech. Found.*, ASCE, Sept. 1968.

Glasstone, S., *Textbook of Physical Chemistry.* Nostrand, Princeton, NJ, 1946.

Goldberg-Soino Associates of New York, P.C., "Buffalo Light Rail Rapid Transit Project, Buffalo, New York, Geotechnical Interpretive Report," Aug. 1978, 58 p.

Gross, J., and S. R. Zimmerly, "Crushing and Grinding. III. Relation of Work Input to Surface Produced in Crushing Quartz." *Trans. Am. Inst. Mining Metallurgical Petroleum Eng.* **86,** 35–50 (1930).

Gstalder, S., and J. Raynal, "Measurement of Mechanical Properties of Rocks and Their Relationship to Rock Drillability." *J. Petroleum Technol.* **18,** 991–996 (1966).

Gurland, J., and P. Bardzil, "The Relation of Strength, Composition, and Grain Size of Sintered WC–Co Alloys." *Trans. Am. Inst. of Mining Metallurgical Petroleum Eng.* **7,** 31 (1955).

Haimson, B., "High Velocity, Low Velocity, and Static Bit Penetration Characteristics in Tennessee Marble." Unpublished master's thesis, University of Minnesota, 1965.

Haimson, B., "Theoretical and Experimental Study of Percussive Drilling of Rock." Unpublished doctoral dissertation, University of Minnesota, 1968.

Haimson, B. C., and C. Fairhurst, "Some Bit Penetration Characteristics in Pink Tennessee Marble." Proceedings 12th Symposium of Rock Mechanics, University of Missouri School of Mines, 1970, pp. 547–559.

Haller, H. F., H. C. Pattison, and O. C. Baldonado, "Interrelationship of In-Situ Rock Properties, Excavation Method, and Muck Characteristics." Holmes and Narver, Technical Report HN-8121.2, 1973.

Hamilton, W. H., and G. L. Dollinger, "Optimizing Tunnel Boring Machine and Cutter Design for Greater Boreability." *Proc. Excavation Tunneling Conf.* **1,** 280–296 (1971).

Harris, C. C., "The Application of Size Distribution Equations to Multi-event Comminution Processes." *Trans. Am. Inst. Mining Metallurgical Petroleum Eng.* **241,** 343–358 (1968).

Hartman, H. L., "Basic Studies of Percussive Drilling." *Trans. Am. Inst. Mining and Metallurgical Petroleum Eng.* **215,** 68–75 (1959).

Hartman, H. L., "The Effectiveness of Indexing in Percussion and Rotary Drilling." *Int. J. Rock Mechanics Mining Sci.* **3,** 265–278 (1966).

Hartman, H. L., "Principles of Drilling." In Pfleider E.P., *Surface Mining*, American Institute of Mining, Metallurgical and Petroleum Engineers, N.Y, N.Y., 1968.

Hassialis, M. D., and H. A. Behre, "Sampling and Testing." In A. F. Taggart, *Handbook of Mineral Dressing Ores and Industrial Minerals.* Wiley, New York, 1945, pp. 19-01 to 19-208.

Haley and Aldrich, Inc., "Wastewater Facilities Plan—Combined Sewer Overflow Abatement Program," Vol. V, Geotechnical Report, Rochester, NY, Dec. 1976, p. 55.

Haley and Aldrich, Inc., "Final Report on Geotechnical Investigations for the Proposed Culver-Goodman Tunnel," Rochester, NY, Sept. 1978, p. 37.

Haycocks, C., "Mechanics of a Voussior Arch." Unpublished master's thesis, University of Missouri School of Mines, 1962.

Hino, K., *Theory and Practice of Blasting*, Nippon Kayaku Co., Ltd., 1959.

Hoek, E., "The Design of a Centrifuge for Simulation of Gravitational Force Fields in Mine Models." *J. S. Afr. Inst. Mining Metallurgy* **65** (9), 1965.

Holliday, R. P. M., "Changes in Tungsten Carbide–Tipped Drill Steel and Drilling Practice Likely to Result from the Introduction of Anfex." *J. S. Afr. Inst. Mining Metallurgy* **68,** 582–598 (1968).

Hornsey, E. E., and G. B. Clark, "Comparison of Spherical Elastic Voigt and Observed Wave Forms for Large Underground Explosions." Tenth Symposium on Rock Mechanics, Austin, Texas, 1968.

Howarth, D. F., "Groove Deepening with Disc Cutters." Proceedings of the Rapid Excavation and Tunneling Conference, San Francisco, 1981, Vol. 2, pp. 1352–1369.

Hukki, R. T., "Proposal for a Solomonic Settlement Between the Theories of Von Rittinger, Kick, and Bond." *Trans. Am. Inst. Mining Metallurgical Petroleum Eng.* **220,** 403–408 (1959).

Hustrulid, W. A., "A Study of Energy Transfer to Rock and Prediction of Drilling Rates in Percussive Drilling." Unpublished master's thesis, University of Minnesota, MN, 1965.

Hustrulid, W. A., "Theoretical and Experimental Study of Percussive Drilling of Rock." Unpublished doctoral dissertation, University of Minnesota, MN, 1968.

Hustrulid, W. A., "Development of a Tunnel Boreability Index." U.S. Bureau of Mines Contract No. 14-06-D-6848, 1970.

Hustrulid, W. A., "The Percussive Drilling of Quartzite." *J. S. Afr. Inst. Mining Metallurgy* **71,** (12), 245–270 (1971a).

Hustrulid, W. A., "A Theoretical and Experimental Study of the Percussive Drilling of Rock." *Int. J. Rock Mechanics Mining Sci.* **8,** 311–333, (1971b).

Innett, E. W., "Survey of Rotary Percussive Drilling." *Mine Quarry Eng.* **27,** 2, 62, 106 (1957).

Institute of Makers of Explosives, "Suggested Code of Regulations for the Manufacture, Transportations, Storage and Use of Explosive Materials." Publication No. 3, New York, 1970, p. 38.

Irving, C. J., "Some Aspects of Ground Movement." *J. Chem. Metallurgical Mining Soc. S. Afr.* **XLVI** pp. 11–12 (1946).

Isaacson, E. de St. Q., *Rock Pressure in Mines.* Mining Publications Ltd., London, 1958.

Keifer and Associates, Inc., "Preliminary Design Report for the Calumet System of the Tunnel and Reservoir Plan," Chicago, IL, Dec. 1976a, 108 p.

Keifer and Associates, Inc., Geotechnical Design Report for the Calumet System of the Tunnel and Reservoir Plan," Chicago, IL, Dec. 1976b.

Jackson, I. F., and H. L. Hartman, "An Investigation of Hard Metal Inserts for Cutting Slates." *Trans. Am. Inst. Mining Metallurgical Petroleum Eng.* **223,** 255–266 (1962).

Jenkins, C. F., "The Pressure Exerted by Granular Materials." *Proc. R. Soc. (Lond.) Ser. A* **131** (1931).

Jenni, J. P. and M. Balissat, "Tock Testing Methods Performed to Predict the Utilization Possibilities of a Tunnel Boring Machine." *Proc. 4th Congr. Int. Soc. Rock Mechanics* **2,** 267–274 (1979).

Jost, W., *Explosion and Combustion Processes in Gases.* McGraw-Hill, New York, 1946.

Just, G. D., "Rock Fragmentation in Blasting." *Can. Inst. Mining Bull.* Vol. 72, 143–147 (1979).

Keifer and Associates, Inc., "Preliminary Design Report for the Calumet System of the Tunnel and Reservoir Plan," Chicago, IL, Dec. 1976a, 108 p.

Keifer and Associates, Inc., Geotechnical Design Report for the Calumet System of the Tunnel and Reservoir Plan," Chicago, IL, Dec. 1976b.

Korbin, G. E., "Factors Influencing the Performance of Full-Face Hard Rock Tunnel Boring Machines." Report No. UMTA-CA-06-0211079-1, Urban Mass Transportation Administration, Washington, 1979.

Kobayashi, T. and J. W. Dalliy, "The Relation Between Crack Velocity and Stress-Intensity Factor in Birefringent Polymers," ASTM STP 627 Symp. on Fast Fracture and Crack Arrust, 1977.

Kreimer, G. S., *Strength of Hard Alloys.* Consultant's Bureau, New York, 1968.

Kurschov, M. M. et al., *Abrasive Wear.* Trudy Intituta Gornogo Dela Akademici Nauk, Moscow, USSR, 1970.

Kutter, H. K. and C. Fairhurst, "On the Fracture Process in Blasting." *Int. J. Rock Mechanics Mining Sci.* **8,** 181–202 (1971).

Kuzmich, I. A., "Some Relationships in the Coal Penetration by High Pressure Water Jets." First International Symposium on Jet Cutting Technology. Paper E1, Coventry, 1972.

Kuznetsov, V. M., "The Mean Diameter of the Fragments Formed by Blasting Rock." *Soviet Mining Sci.* **9,** 144–148 (1973).

Kvapil, R., "Gravity Flow of Granular Materials in Hoppers and Bins in Mines." *Int. J. Rock Mechanics Mining Sci.* **2,** 227 (1973).

Lagerquist, M., "A Study of the Thermal Fatigue Crack Propagation of WC–Co Cemented Carbide." *Powder Metallurgy* **18,** 71–88 (1975).

Langefors, U., and B. Kihlstrom, *Rock Blasting,* 3rd Ed. Wiley, New York, 1978.

Lardener, E. W., "Isostatically Hot Presses Cemented Carbide and Its Utilization in Mining Tools." *Mining Magazine* **130,** 35 (1974).

Larsen-Basse, J., "Wear of Hard Metals in Rock Drilling: A Survey of the Literature." *Powder Metallurgy* **16** (31), 1–32 (1973).

Larsen-Basse, J., et al., "Abrasive Wear of Tungsten-Carbide–Cobalt Composites in Rotary Drilling Tests." *Metals Sci. Eng.* **13,** 83–91 (1974).

Latin, A., "Properties of Tungsten Carbide–Cobalt Alloys Used for Mineral Cutting Tools." *Metallurgia* **64** (385), 211–216, and (386), 267–273 (1961).

Lean, D. J., and G. G. Paine, "Preliminary Blasting for a Bucket Wheel Excavator Operation at CQCA Coonyella Mine, Central Queensland." Australian Mineral Foundation Workshop Course 152/81, March 1981.

Lee, H. B., and R. L. Akre, *Blasting Process* (patent). U.S. Patent 2,703,528.

Litchfield, E. L., M. H. Hay, and J. S. Monroe, "Electrification of Ammonium Nitrate Pneumatic Loading." U.S. Bureau of Mines Report of Investigations No. 7139, 1968.

Lundquist, R. G., "Rock Drilling Characteristics of Hemispherical Insert Bits." Unpublished master's thesis, University of Minnesota, MN, 1968.

Lounds, C. M., "Computer Modelling of Fragmentation from an Array of Shotholes." First International Symposium on Rock Fragmentation by Blasting." Sweden, 1986, Vol. 2, pp. 455–468.

Lundquist, R. G., and C. F. Anderson, "Energetics of Percussive Drills—Longitudinal Strain Energy." U.S. Bureau of Mines Report of Investigation No. 7329, 1969.

Maiokang, Z., and O. Johnson, "On Dynamic Failure of Rock in Percussive Drilling." Technical Report 1983, 55T, Lulea University, Sweden, 1983.

Malushitsky, Y. N., *The Centrifugal Model Testing of Waste Heap Embankments.* Cambridge University Press, Cambridge, U.K., 1981.

Manufacturing Chemists Association, "Fertilizer-Grade Ammonium Nitrate—Properties and Recommended Methods for Packaging, Handling, Transportation, Storage, and Uses." Sheet A-10, Supplement 1, 1962.

Mason, C. M., and E. G. Aiken, "Methods of Evaluating Explosives and Hazardous Materials." U.S. Bureau of Mines Information Circular No. 8541, 1972.

Mason, C. M., J. Rabovish, and R. W. Van Dolah, "Studies on the Bullet Sensitivity of Ammonium Nitrate–Fuel Oil Mixtures," U.S. Bureau of Mines Report of Investigations No. 6203, 1963.

Maurer, W. C., *Advanced Drilling Techniques.* Maurer Engineering Co., Houston, Texas, 1979.

McFeat-Smith, I., and P. J. Tarkoy. "Assessment of Tunnel Boring Machine Performance." *Tunnels Tunnelling* **11** (10), 33–37 (1979).

McNulty, J. W., "An Experimental Study of Arching in Sand." U.S.A. Waterways Experiment Station Technical Report No. 1-674, 1965.

McWilliams, J. R., "Diamond Drilling with Impregnated Bits." Seventh Annual Drilling Symposium, University of Minnesota, 1957.

Miller, R. J., "Laboratory Analysis of Small and Large Scale Cutting of Rock for Improvement of Tunnel Boreability Prediction and Machine Design," Unpublished doctoral dissertation, Colorado School of Mines, Golden, CO, 1974.

Miller, R. J., and F. D. Wang, "Rock Cutter Boreability Parameters." Report to NSF, No. APR 73-07776-A03, Colorado School of Mines, Golden, CO, 1975.

Mohr, F., "Influence of Mining on Strata." *Mine Quarry Eng.* **22** (4), 140–152, 178–189, 225–233 (1956).

Montgomery, R. S., "The Mechanism of Percussive Wear of Tungsten Carbide Composites." *Wear* **12,** 309–329 (1968).

Montgomery, R. S., "Percussive Wear Properties of Cemented Carbides." *Trans. Am. Inst. Mining Metallurgical Petroleum Eng.* **244,** 153–156 (1969).

Montgomery, R. S., A. Hara, and T. Ikeda, "Influence of Carbide Grain on Percussive Wear of Cemented Tungsten Carbide Rock Inserts" 1970.

Morgan, J. M., Barret, D. A., and J. A. Hudson, "Tunnel Boring Machine Performance and Ground Properties." Supplementary Report No. 469, Transport and Road Research Laboratory. Crowthorne, U.K., 1979.

Morrell, R. J., W. E. Bruce, and D. A. Larson, "Disk Cutter Experiments in Sedimentary and Metamorphic Rocks—Tunnel Boring Technology." U.S. Bureau of Mines Report of Investigations No. 7410, 1970.

Morrell, R. J., and D. A. Larsen, "Disk Cutter Experiments in Metamorphic and Igneous Rocks—Tunnel Boring Technology." U.S. Bureau of Mines Report of Investigations No. 7691, 1974.

Muirhead, I. R., and L. G. Glossop, "Hard Rock Tunneling Machines." *Trans. Inst. Mining Metallurgy* **77,** A1–A21 (1957).

Nathan, G. D., and W. J. D. Jones, "Influence of Hardness of Abrasive on Abrasive Wear of Metals." Int. Mechanical Eng. Proc. 5th Conv. Lubrication Wear **181,** 215–221 (1966–67).

National Fire Protection Association, "Blasting Agents." In *Code for the Manufacture, Transportation, Storage, and Use of Explosives and Blasting Agents*, No. 495, 1962, pp. 25–29.

Nelson, P. N., T. D. O'Rourke, R. F. Flanagan, F. H. Kulhawy, and A. R. Ingraffa, "Tunnel Boring Machine Performance Study." DOT-TSC-UMTA-83-54, U.S. Department of Transportation, 1984.

Nevill, H. F., and J. G. D. Cron, "Wear of Rotary Bits in Granite Rock." *Trans. Inst. Mining Metallurgy* **121,** 249–262 (1962).

Nikonov, G. P., and Y. A. Goldin, "Coal and Rock Penetration by Fine, Continuous High Pressure Water Jets." First International Symposium on Jet Cutting Technology, BHRA, Coventry, U.K., Paper E2, 1972.

Obert, L., and W. I. Duvall, "Generation and Propagation of Strain Waves in Rock," Pt. I. U.S. Bureau of Mines Report of Investigations No. 4683, 1950.

Osborn, H. J., "Wear of Rock Cutting Tools." *Powder Metallurgy* **12,** 471–502 (1969).

Ozdemir, L., "A Laboratory and Field Investigation of Tunnel Boreability." Unpublished master's thesis, Colorado School of Mines, Golden, CO, 1975.

Ozdemir, L., "Development of Theoretical Equations for Predicting Tunnel Boreability." Unpublished doctoral dissertation, Colorado School of Mines, Golden, CO, 1977.

Ozdemir, L., R. J. Miller, and F. D. Wang, "Rock Cutter Boreability Parameters." Second Annual Report to NSF, grant No. APC-73-0776-A03, Colorado School of Mines, Golden, CO, 1976a.

Ozdemir, L., R. J. Miller, and F. D. Wang, "Mechanical Tunnel Boring Prediction and Machine Design." Report to NSF, contract No. APR 73-0776-A03, Colorado School of Mines, Golden, CO, 1976b.

Ozdemir, L., R. J. Evans, and R. J. Miller, "CSM–USBM Cutting Machine: Research Capabilities and Trials for Improving the Performance of Tunnel, Raise and Shaft Boring Machines." *Rapid Excavation Tunneling Conf.* **1,** 106–130 (1983).

Paithankar, A. G., and G. B. Misra, "A Critical Appraisal of the Protodiaknov Index." *Int. J. Rock Mechanics Mining Sci.* **13,** 249–251 (1976).

Paithankar, A. G., and G. B. Misra, "Drillability of Rocks in Percussive Drilling from Energy per Unit Volume as Determined by a Microbit." *Mining Eng.* **21** (9), 1407–1410 (1980).

Palmqvist, S., *Arch. Eisenhuettenwesen* **33** (1962).

Palovitch, E. R., and W. T. Malenka, "Hydraulic Mining Research—A Progress Report." *Mining Cong. J*, **50**, 66–73 (1964).

Paone, J., and W. E. Bruce, "Drillability Studies—Diamond Drilling." U.S. Bureau of Mines Report of Investigations No. 6324, 1963.

Paone, J., and D. Madson, "Drillability Studies: Impregnated Diamond Bits." U.S. Bureau of Mines Report of Investigations No. 6776, 1965.

Paone, J., and S. Tandanand, "Inelastic Deformation of Rock Under a Hemispherical Drill Bit." U.S. Bureau of Mines Report of Investigations No. 6838, 1966.

Paone, J., W. E. Bruce, and P. R. Virciglio, "Drillability Studies, Statistical Regression Analysis of Diamond Drilling." U.S. Bureau of Mines Report of Investigations No. 6800, 1966.

Paone, J., D. Madson, and W. E. Bruce, "Drillability Studies, Laboratory Percussive Drilling." U.S. Bureau of Mines Report of Investigations No. 7300, 1969.

Peele, R., and J. H. Church, *Mining Engineers Handbook,* Sec. 10. Wiley, New York, 1941.

Peters, D. C., "Physical Modeling of Draw Behavior of Broken Rock in Caving." *Colo. School Mines Q.* **79** (1), 1984.

Peterson, C. R., *Roller Cutting Forces*, preprint, paper member SPE 2393, Fourth Conference on Drilling and Rock Mechanics, University of Texas, Austin, 1969, p. 115.

Petkoff, B. T., T. C. Atchison, and W. I. Duvall, "Photographic Observation of Quarry Blasting." U.S. Bureau of Mines Report of Investigations No. 5849, 1961.

Pfleider, E. P., and W. D. Lacabanne, "High Air Pressure for Bottom Hole Percussion Drills." *Colo. School Mines Q.* **56** (1), 1961, p. 97–114.

Piret, E. L., et al., "Energy—New Surface Relationship in the Crushing of Solids." *Chem. Eng. Prog.* **45**, 508, 655, 708 (1949).

Pons, L., et al., "Sur la Fragilisation Superficielle au Cours de Frottement de Carbures de Tungstene Frittes." Academie des Sciences, Paris, Vol. 225, 1962, p. 2100.

Protodiakonov, M. M., "Mechanical Properties and Drillability of Rock." Proceedings Fifth Symposium on Rock Mechanics, University of Minnesota, 1962, pp. 103–118.

Rad, P. F., and R. C. Olson, "Tunneling Machine Research, Interaction Between Disc-Cutter Grooves in Rock." U.S. Bureau of Mines Report of Investigations No. 7881, 1974a.

Rad, P. F., and R. C. Olson, "Tunneling Machine Research: Size Distribution of Rock Fragments Produced by Rolling Disc Cutters." U.S. Bureau of Mines Report of Investigations No. 7882, 1974b.

Rae, D., "Safety in Mines Establishment," Research Report No. 236, So. Africa, 1966.

Rehbinder, G., "Slot Cutting in Rock with a High-Speed Water Jet." *Int. J. Rock Mechanics Mining Sci.* **14**, 229–234 (1977).

Rehbinder, G., "Erosion Resistance of Rock." Fourth International Symposium on Jet Cutting Technology, BHRA, Canterbury, U.K., 1978.

Rosin, P., and E. Rammler, "Laws Governing the Fineness of Powdered Coal." *J. Inst. Coal* **7**, 29–36 (1933).

Ross, N. A., "Theoretical and Experimental Analysis of Tunnel Boreability." Unpublished doctoral dissertation, Colorado School of Mines, 1970.

Roxborough, F. F., and A. Rispin, "Rock Excavation by Disc Cutter." Report to Transport and Road Research Laboratory of the Department of Environment, University of Newcastle-upon-Tyne, U.K., 1972.

Roxborough, F. F., and H. R. Phillips, "Rock Excavation by Disc Cutter." *Int. J. Rock Mechanics Mining Sci.* **12**, 361–366 (1975).

Roxborough, F. F., "Rock Cutting Research for the Design and Operation of Tunnelling Machines." *Tunnels Tunnelling* **1** (3), 125–128 (1969).

Rozeneau, L., "Fatigue Wear as a Rate Process." *Wear* **6,** 337 (1963).

Ryd, E., and J. Holdo, *Manual on Rock Blasting*. Sec. 12, "Percussive Rock Drills." Atlas Deisel and Saudvikens, Stockholm, Sweden, 1953.

Sasaki et al., "Studies on the Cutting Resistance of Rock." Translators Y. Kojima and W. Hustrulid, from Japanese. Interim Final Report, Twin Cities Mining Research Center, 1972.

Schellinger, K., "Solid Surface Energy and Calorimetric Determination of Surface–Energy Relationships for Some Common Minearals." *Trans. Am. Inst. Mining Metallurgical Petroleum Eng.* **193,** 369–374 (1952).

Schmidt, R. L., "Drillability Studies, Percussive Drilling in the Field." U.S. Bureau of Mines Report of Investigations No. 7648, 1972.

Selberg, J. L., "Transient Compression Waves from Spherical and Cylindrical Cavities." *Ark. Tsik.* **5,** 97–108 (1951).

Selcuk, S., "Analysis of Coefficient of Rock Strength, and Measurement of Percussive Penetration Rates and Bit Wear." Unpublished master's thesis, Colorado School of Mines, Golden, CO, 1981.

Selim, A. A., and W. E. Bruce, "Prediction of Penetration Rate for Percussive Drilling." U.S. Bureau of Mines Report of Investigations No. 7396, 1970.

Sharpe, J. A., "The Production of Elastic Waves by Explosion Pressure. I. Theory and Empirical Field Observation." *Geophysics* **17** (3), 144–155 (1942).

Simon, R., "Digital Machine Computations of the Stress Waves Produced by Striker Impact in Percussive Drilling Machines." In *Fifth Symposium on Rock Mechanics*. Pergamon, Oxford, U.K., 1963.

Simon, R., "Theory of Rock Drilling." Sixth Annual Drilling and Blasting Symposium, University of Minnesota, 1956.

Snowden, R. A., M. D. Ryley, and J. Temporal, "A Study of Disc Cutting in Selected British Rocks." *Int. J. Rock Mechanics Mining Sci.* **19** (3), 107–121 (1982).

Somerton, W. H., F. Esfandari, and A. Singhal, "Further Studies of the Relation of Physical Properties of Rock to Drillability." Fourth Conference on Drilling and Rock Mechanics, University of Texas, Austin, 1969.

Stack, B., *Handbook of Mining and Tunneling Machinery*. Wiley, New York, 1982.

Stephansson, O., "Stability of Single Openings in Horizontally Bedded Rock." *Eng. Geol.* **5,** 5–71 (1971).

Stelloh, R. T., "Explosive Performance Parameters." Unpublished master's thesis, University of Missouri School of Mines, Roloa, 1961.

Stites, J. G., M. D. Barnes, and R. F. McFarlin, "A Survey of the Physical and Chemical Characteristics of Fertilizer Grade Ammonium Nitrate." Fifth Annual Symposium on Mining Research, Bulletin 98, University School of Mines, 1959.

Stjernberg, K. G., U. Fischer, and N. I. Hogoson, "Wear Mechanisms Due to Different Rock Drilling Conditions." *Powder Metallurgy* **18,** 89–106 (1975).

Sutherland, H. J., "Centrifuge Simulation of Subsidence of Coal Mines." High Gravity Simulation Workshop, Colorado School of Mines, Golden, CO, 1982.

Tandanand, S., "Principles of Drilling." In *SME Mining Engineering Handbook*, A. B. Cummins and I. A. Given, Eds. American Institute of Mining, Metallurgical and Petroleum Engineers, New York, 1973.

Tandanand, S., and H. F. Unger, "Drillability Determination, a Drillability Index of Percussion Drills." U.S. Bureau of Mines Report of Investigations No. 8073, 1975.

Tarkoy, P. J. J., "Predicting TBM Penetration Rates in Selected Rock Types." In *Proceedings*, 9th *Canadian Rock Mechanics Symposium*, Montreal, 1973, pp. 263–274,

Tarkoy, P. J., "A Study of Rock Properties and Tunnel Boring Machine Advance Rates in Two Mica Schist Formations." In *Proceedings, 15th Symposium on Rock Mechanics*, Custer State Park, SD, 1975, pp. 415–447.

Tarkoy, P. J., "Predicting Raise and Tunneling Boring Machine Performance: State of the Art." In *Proceedings, Rapid Excavation and Tunneling Conference*, Atlanta, 1979, Vol. 1, pp. 333–352.

Tarkoy, P. J., and A. J. Hendron, "Rock Hardness Index Properties and Geotechnical Parameters for Predicting Boring Machine Parameters." Univ of Illinois, Urbana, IL, RANN Report, 1975.

Taylor, J., *Detonation in Condensed Explosives*. Oxford University Press, Oxford, U.K., 1952.

Taylor, J. W. Jr., "Safe Handling of Blasting Agents." *Nat. Fire Protection Assoc. Q.* **56,** 261–266 (1963).

Teale, R., "The Mechanical Excavation of Rock—Experiments with Roller Cutters." *Int. J. Rock Mechanics Mining Sci.* **1** (10), 63–78 (1964).

Teale, R., "The Concept of Specific Energy in Rock Drilling." *Int. J. Rock Mechanics Mining Sci.* **2** (1), 57–74 (1965).

Terzaghi, K. V., "Load Supports in Running Ground." In R. V. Proctor and T. L. White (Eds.), *Rock Tunneling with Steel Supports*. C.S. Co., Hanover, Germany, 1977.

Terzaghi, R. V., "Stress Distribution in Dry and in Saturated Sand above a Yielding Trapdoor." *Proc. 1st Int. Conf. Soil Mechanics Foundations Eng.* **1,** 306 (1936).

Te Water, L. H., and A. J. Mihulka, "The Possibilities of Increasing Productivity by Using Ammonium Nitrate–Fuel Oil Mixtures and Narrow Diameter Holes." Unpublished, cited in Hustrulid, 1971.

Timoshenko, S., *Theory of Elasticity*. McGraw-Hill, New York, 1934.

Tournay, W. E., E. J. Murphy, G. H. Damon, and R. W. Van Dolah, "Some Studies in Ammonium Nitrate–Fuel Oil Compositions." In *Fourth Annu. Symp. Mining Res.*, University of Missouri School of Mines Bulletin No. 97, 1959, pp. 164–174.

Tretyakov, V. I., *Sintered Hard Alloys*. Metallurgia Italiana, 1962.

Triandafilidis, G. E., E. Hampton, and N. Spanovich, "An Experimental Evaluation of Soil Arching." Proceedings Symposium on Soil–Structure Interaction, University of Arizona, 1964.

Unger, H. F., and R. R. Fumanti, "Percussive Drilling with Independent Rotation." U.S. Bureau of Mines Report of Investigations No. 7692, 1972.

U.S. Bureau of Mines, "Safety Recommendations for Sensitized Ammonium Nitrate Blasting Agents." U.S. Bureau of Mines Information Circular No. 8179, 1963.

Van Dolah, R. W., and J. S. Malesky, "Fire and Explosion in a Blasting Agent Mix Building, Norton, Virginia." U.S. Bureau of Mines Report of Investigations No. 6015, 1962.

Van Dolah, R. W., N. E. Hanna, E. J. Murphy, and G. H. Damon, "Further Studies of ANFO Compositions." In *Fifth Annu. Symp. Mining Res.*, University of Missouri School of Mines Bulletin No. 98, 1959, pp. 90–101.

Van Dolah, R. W., F. C. Gibson, and J. N. Murphy, "Further Studies on Sympathetic Detonations." U.S. Bureau of Mines Report of Investigations No. 6903, 1966a.

Van Dolah, R. W., F. C. Gibson, and J. N. Murphy, "Sympathetic Detonation of Ammonium Nitrate and Ammonium Nitrate–Fuel Oil." U.S. Bureau of Mines Report of Investigations No. 6746, 1966b.

Van Dolah, R. W., C. M. Mason, F. J. P. Perzak, and D. R. Forshey, "Explosion Hazards of Ammonium Nitrate under Fire Hazard." U.S. Bureau of Mines Report of Investigations No. 6773, 1966c.

Vijay, M. M., and W. H. Brierly, "Drilling of Rocks with Rotating High Pressure Water Jets: An Assessment of Nozzles." In *5th Int. Symp. Jet Cutting Technol.*, Hanover, Germany, 1980.

Vijay, M. M., Brierly, W. H., and P. E. Gratton-Bellow, "Drilling of Rocks with Rotating High Pressure Water Jets: Influence of Rock Properties." In *6th Int. Symp. Jet Cutting Technol.*, Guilford, England 1982.

Wahl, H., and G. Kantenwein, "Verschleiss Beim Gesteinbonram." *Wear* **4** (3), 234–245 (1961).

Wang, F. D., and R. J. Miller, "A Theoretical and Experimenal Study of Tunnel Boring by Machine with an Emphasis on Boreability Prediction and Machine Design." U.S. Bureau of Mines Contract No. H0210043, Colorado School of Mines, 1974.

Wanner, H., "On the Influence of Geological Conditions on the Application of Tunnel Boring Machines." *Bull. Int. Assoc. Eng. Geol.* **12,** 21–28 (1975).

Wanner, H., and U. Aeberli, "Tunneling Machine Performance in Jointed Rock." In *4th Congr. Int. Soc. Rock Mechanics* Montreux, 1979, Vol. 1, pp. 573–580.

Watson, R. W., "Card-Gap and Projectile Impact Sensitivity Measurements, A Compilation." U.S. Bureau of Mines Information Circular No. 8605, 1973.

Wells, E.S., "Penetration Speed References for the Drillability of Rock." In *Proc. Australasian Institute Mining Metallurgy* 158–159, 1950.

Wells, E. S., "Penetration Speed of Percussion Drill Bits." *Chem. Eng. Mineral Rev.* **41** (10), 362–364 (1949).

Whitbread, J. E., "Bit Temperatures in Rotary Drilling." *Colliery Eng.* **37,** 25–59 (1960a).

Whitbread, J., *Sheffield Mining Magazine*. Sheffield University, London, 1960b.

Wheby, F. T. and E. M. Cikanck, "A Computer Program for Estimating Costs of Tunneling." *Proc. Rapid Excavation and Tunneling Conference* Vol. 1 p 185, June 1974

Winzer, R. R., and A. P. Ritter, "Effect of Delays in Fragmentation in Large Limestone Blocks." Martin Marietta Laboratories Report No. MML TR 80-25, 1980.

Yancik, J. J., R. F. Bruzewski, and G. B. Clark, "Some Detonation Properties of Ammonium Nitrate." University of Missouri School of Mines Bulletin No. 98, 1959, pp. 67–89.

Yenge, L. I., "Analysis of Bulk Flow of Materials under Gravity." Pt. I. "Sublevel Caving in Relation to Flow in Bins and Bunkers." *CSM Q.* **75** (4) (1980). Pt. II. "Theoretical and Physical Modeling of Gravity Flow of Broken Rock." *CSM Q.* **76** (3) (1981).

# INDEX